Design of Power Management Integrated Circuits

Design of Power Management Integrated Circuits

Bernhard Wicht
Leibniz University Hannover, Germany

The right of Bernhard Wicht to be identified as the author of the editorial material in this work has been asserted in accordance with law.

Registered Offices
John Wiley & Sons, Inc., 111 River Street, Hoboken, NJ 07030, USA
John Wiley & Sons Ltd, The Atrium, Southern Gate, Chichester, West Sussex, PO19 8SQ, UK

For details of our global editorial offices, customer services, and more information about Wiley products visit us at www.wiley.com.

Wiley also publishes its books in a variety of electronic formats and by print-on-demand. Some content that appears in standard print versions of this book may not be available in other formats.

Library of Congress Cataloging-in-Publication Data applied for:

Hardback ISBN: 9781119123064

Cover Design: Wiley
Cover Image: © fotograzia/Getty Images

Set in 9.5/12.5pt STIXTwoText by Straive, Chennai, India
Printed and bound by CPI Group (UK) Ltd, Croydon, CR0 4YY

C9781119123064_130824

To my beloved family,
Sabine, Luise, Friederike, Konstantin,
and to my parents
Siegrid and Eberhard.

Contents

Preface

In the March 2020 issue of the IEEE Solid-State Circuits Magazine, IEEE fellow Marcel Pelgrom writes in one of his excellent and inspiring Associate Editor's View columns entitled *Standing on Shoulders* about the timeless need for textbooks: "…where is the next disruptive view of our field, something that is desperately needed after this dazzling journey from 10-micron devices to nanometer electronics for quantum computing?" This book is my take on the field of power management.

The book delves into the fascinating world of power management IC design. This field has seen rapid growth in recent years, with the increasing demand for energy-efficient electronics, in particular, portable battery-operated devices. Power management integrated circuits are used for highly efficient power supplies and controlling power switches. It is incredible how these technologies have gained tremendous importance in making electronic solutions for global growth areas such as renewable energies, transportation, and communications more compact, energy-efficient, and reliable. Future machine learning and AI applications will only be possible with intelligent power management to supply complex processors and sensors.

I got into power management when I joined the industry in the early 2000s and went along a steep learning curve on all kinds of power management systems and design aspects. When I became a professor in 2010, I created a new course dedicated to power management IC design. However, no comprehensive textbook was available, and I had to rely on scientific papers and application notes.

A few years later, the idea of this book came up. Since then, I have been fascinated and challenged by the fast pace of progress and innovation in power management. The book provides a complete resource for those interested in power management IC design, covering basic concepts, advanced topics, and recent innovations in this rapidly evolving field. It is intended for students, educators, professors, and new and experienced engineers who want to learn about power management IC design, providing valuable insight and practical guidance for designing power management circuits and systems. Each chapter is organized to make it easy to find specific sub-topics, with numerous real-world examples illustrating key design concepts and techniques.

When I teach an entire course on power management IC design at a Master's or advanced Bachelor's level, I reduce the content and follow this outline: (1) Introduction (applications, challenges, physical implementation); (2) Linear Voltage Regulators; (3) Charge Pumps and Capacitive DC–DC Converters; (4) Power Transistors; (5) Gate Drivers; (6) Protection and Sensing; (7) Inductive DC–DC Converters; (8) Hybrid Converters. I also offer a design lab based on Spice simulation accompanying the lecture. Starting the class with the linear regulator right after the introduction allows the lab to begin early in the semester. I turned a lot of lab assignments and exercises into the many examples in this book (to my respected future students: I hope I'm not giving too much away.).

In his column, Marcel Pelgrom also writes about the burden of writing textbooks. And indeed, this book is the result of an investment of uncountable hours over several years. Writing a book also means that there will be missing content, on purpose but also by mistake. My fellow readers: Despite careful review, there will be mistakes, and I apologize in advance. Any feedback is highly appreciated. Please get in touch.

This book would not be in your hands without careful review, invaluable feedback, and encouragement by many people. I want to thank my former and recent Ph.D. students, in particular, Peter Renz (who read through the entire draft) and Tobias Funk, Saurabh Kale, Maik Kaufmann, Tim Kuhlmann, Jens Otten, Christoph Rindfleisch, and Jürgen Wittmann. Thanks to Markus Henriksen for his feedback on switched-capacitor (SC) and hybrid converters. Hartmut Grabinski ensured that Maxwell's equations were correctly applied to interconnections and printed circuit board (PCB) layout. Detlev Habicht and Niklas Deneke supported in capturing photographs for the book. I am grateful to the team at Wiley, in particular, to Sandra Grayson, Juliet Booker, Kavipriya Ramachandran, and Jeevaghan Devapal for their excellent support and patience. I wish to thank many more people.

Writing such a book is impossible without the support of my family. I want to thank my parents, Siegrid and Eberhard, who have supported my fascination with microelectronic circuits since I was nine. Thanks go to my children, who became real fans of my book project, even though they often had to take a back seat, especially when finalizing the manuscript over the last 1–2 years. I am indebted to my wife, Sabine. This book would not have been possible without her love and understanding.

Enjoy the exciting journey of exploring the design of power management integrated circuits!

August 2023, Gehrden   *Bernhard Wicht*

1

Introduction

Power management integrated circuits (PMICs) are essential in today's electronic devices. They manage power delivery and consumption, provide efficient power supplies, and drive power switches that control actuators and motors, as illustrated in Fig. 1.1. PMICs can be integrated into complex integrated circuits (ICs) or implemented as dedicated ICs. In this book, the term PMIC will refer to any type of power integrated circuit.

The importance of PMICs has grown significantly in recent years, driving innovation and progress in various industries, from consumer electronics to automotive and industrial applications. With the progress of machine learning and artificial intelligence (AI), intelligent power management is critical to supplying complex processors and sensors.

PMICs have enabled the development of smaller, more energy-efficient, and reliable electronic solutions. They also play an essential role in environmental aspects and sustainability. By regulating the power supply of electronic devices, PMICs can reduce energy consumption and carbon emissions. Moreover, PMICs are crucial for the development of renewable energies, such as solar and wind power, by enabling efficient power conversion and management.

1.1 What Is a Power Management IC and What Are the Key Requirements?

A PMIC is an electronic component that delivers one or more supply voltages to other circuit blocks at a sufficient power level out of an electrical energy source, as shown in Fig. 1.1. The power conversion can happen in a linear way (usually the more straightforward method) or a switched-mode fashion, delivering energy portions at a specific frequency (usually the more energy-efficient approach).

The PMIC aims to utilize the energy source at maximum efficiency, while the input and output voltage may vary during operation. It also reacts to varying load currents from a few microamperes (standby) to several amperes (full-power operation).

The voltage conversion ratio is the relation V_{out}/V_{in} between the output and input voltage. The input voltage V_{in} can be greater or lower than the output voltage V_{out}, defining a step-down converter (buck converter) or a step-up converter (boost converter). Buck-boost converters allow V_{in} to vary over a wide range below and above V_{out}.

Design of Power Management Integrated Circuits, First Edition. Bernhard Wicht.
© 2024 John Wiley & Sons Ltd. Published 2024 by John Wiley & Sons Ltd.

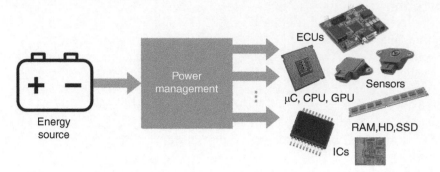

Figure 1.1 The role of power management: placed between the energy source and the electronics, it provides one or multiple supply voltages at the correct power level required by the application. Source: Brunbjorn/Adobe Stock; daniiD/Adobe Stock; Ruslan Kudrin/Adobe Stock; estionx/Adobe Stock.

The power conversion efficiency η (sometimes also called *PCE*) is defined as the ratio between the output power P_{out} delivered to the load and the input power P_{in} dissipated from the energy source,

$$\eta = \frac{P_{out}}{P_{in}} = \frac{P_{out}}{P_{out} + P_{loss}}, \tag{1.1}$$

where P_{loss} accounts for the power dissipated within the power management circuit. It needs to be delivered from the input but does not contribute to the output power. We want to keep P_{loss} as low as possible. For $P_{loss} = 0$, the efficiency reaches its maximum, $\eta = 1$. It is common to express the efficiency in percent. In that case, we multiply Eqn. (1.1) by 100%.

PMICs typically include various features like voltage regulators, battery chargers, and power management control algorithms. They may also include monitoring and protection against over-current, overheating, and other failure cases. In some applications such as automotive, PMICs are alternatively called smart power ICs, emphasizing the combination of power devices with smart control and monitoring features, all integrated on a single chip.

One major trend is the increasing integration of PMICs. As more functions are combined onto a single chip, the resulting system becomes smaller, more efficient, reliable, and less expensive.

To summarize, the key requirements of PMICs are

- *Size, volume, footprint, and weight*: The PMIC, including external passive components, must often fit into a confined space like in smartphones or wearables. In portable devices, also the weight is critical. The lower weight is also crucial in automotive as it reduces gas and energy consumption.
- *Power conversion efficiency*: High efficiency means low losses. The lower the power losses, the longer the battery time. It also causes reduced heat and lower cooling effort, which, in turn, reduces the size and weight of the power management solution.
- *Reliability, no disturbances, and low noise*: PMICs are noise sources that may impact other sensitive electronic parts due to their switching nature. Handling high voltages and currents causes stress and reliability issues at the component, package, and assembly levels.
- *Cost*: Like most microelectronic products, there is always some pressure to reduce the cost of the IC and the overall bill of materials at the system level. Power management is not always considered a key differentiator. At the same time, physics cannot be cheated, and PMICs are a fastly growing market with good margins.

1.2 The Smartphone as a Typical Example

Looking at Fig. 1.2a), it is impressive to see how far mobile phones have come since the early 1990s. Back then, phones could only make voice calls and had a standby time of about a day or less. The picture is not to scale, but it was bulky and about 500 g in weight. It is incredible to think about all the features and functions that modern smartphones have today, illustrated in Fig. 1.2b). It is a remarkable example of the outstanding advancements in modern microelectronics. Today's smartphones are much smaller, lighter (typically 150 g), and more powerful. They have considerable computing power, 4K video capture, high-end gaming, virtual reality functions, and higher display resolution. This achievement in performance is thanks to ultra-low-power microelectronics and dedicated power management. Additionally, it is noteworthy that making a phone call is no longer the primary use case for these advanced devices.

Now we do what we usually do not want to; we drop our precious smartphone and look at the electronics inside. Figure 1.2c) shows a printed circuit board of the iPhone 13. The entire electronics is implemented on a layered motherboard sandwich of which Fig. 1.2c) shows a major part. The white frame boxes indicate some of the many PMICs inside the phone. There are more PMICs on the reverse side and other printed circuit board (PCB) parts, including ICs for the audio amplifier and wireless charging. PMICs are a considerable part of the smartphone. Connected to the Li-ion battery with a typical cell voltage of 3.7 V, multi-phase DC–DC converters supply the application processor that comprises multicore CPU and GPU blocks. The voltage levels are dynamically scaled in the range of typically 0.25–1.5 V at load currents of more than 10 A (see dynamic voltage and frequency scaling in Section 1.7). The typical power consumption is in the range of a few watts. In comparison, desktop PC processors dissipate more than 100 W. Running at high switching frequencies of tens of MHz, the voltage converters achieve small size, ultralow profile, and near-load integration at high conversion efficiency. No active cooling is required.

Looking closely, we identify hundreds of tiny passive components surrounding the ICs, mainly capacitors and inductors. As they are energy-storing components, their size can be reduced by decreasing the storing times, in other words, by increasing the switching frequency of the power conversion. It defines one of the leading research goals of today's power management solutions – achieving faster switching while keeping the conversion efficiency high. We will continuously address this topic throughout this book.

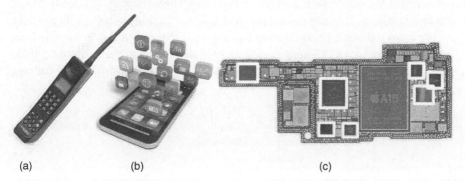

(a) (b) (c)

Figure 1.2 a) The mobile phone in the early 1990s, b) the smartphone today, and c) the electronics of the iPhone 13 with PMICs marked by white boxes. Source: a,b) aquatarkus/Adobe Stock; c) ifixit.

1.3 Fundamental Concepts

There are different ways to implement DC–DC converters that convert an input DC voltage to another voltage level. To keep it more practical, we consider a scenario of how to convert 12 to 2 V.

1.3.1 Using a Resistor – The Linear Regulator

We can use a simple resistor to convert 12 to 2 V, as shown in Fig. 1.3. For a load current of 1 A, a resistor of 10 Ω results in $V_{out} = 2$ V. In reality, the resistor is replaced by a controlled transistor such that its conductance is adjusted depending on the operating conditions like input voltage and load current. This approach works very well. However, the voltage drop between input and output is converted into heat. That is why there is significant power dissipation in the resistor, $10\,\text{V} \cdot 1\,\text{A} = 10\,\text{W}$ in this example. The power loss is even larger than the output power $P_{out} = 2$ W. In terms of energy efficiency, this concept has a significant drawback.

Nevertheless, it is the fundamental principle of a linear voltage regulator and, by far, the most used power management circuit today. On the positive side, besides its simplicity, it gives a "clean" output voltage with a fast transient response.

Without the excessive losses, there would be no need for alternative power conversion concepts, as discussed in Sections 1.3.2–1.3.4 below. The lower the voltage drop across the resistor (the controlled transistor), the lower the power loss. For this reason, linear regulators are often called low-dropout regulators with the widely used short-term LDO. Chapter 7 is dedicated to linear regulators.

1.3.2 Using Switches and an Inductor – The Inductive DC–DC Converter

To overcome the limited efficiency of the linear regulator, we again ask the question, how can we convert 12 to 2 V? We now use switches as shown in Fig. 1.4a). The switches are combined with an inductor, forming an inductive DC–DC converter as a typical switched-mode power supply (SMPS) implementation. As there is no resistive element in the power path, this concept has the potential to achieve much higher power conversion efficiency compared to a linear regulator.

The operation is as follows: The two switches turn on periodically in a complementary way. They are connected to the so-called switching node. The voltage V_{sw} at that node sees a square wave with an amplitude equal to V_{in} (12 V in this case), as shown in Fig. 1.4b). The switching node feeds into an L-C low-pass with two functions: filtering and energy storing. The filtering characteristic provides the average V_{sw} at the converter's output. The average corresponds to the area under the switching node transient curve. Hence, V_{out} is a DC voltage; see Fig. 1.4b). By changing the on-time t_{on} of S1, the area under the square wave, and, consequently, the level of V_{out} can be varied. This concept

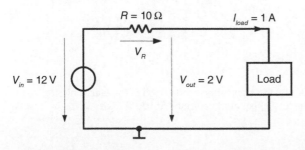

Figure 1.3 Conversion of 12 to 2 V by a simple resistor. This is the concept of a linear voltage regulator.

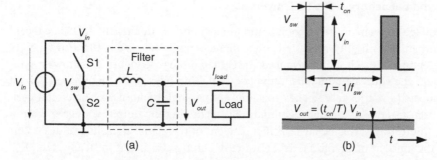

Figure 1.4 Voltage conversion using switches and an inductor L achieving high conversion efficiency: a) the fundamental step-down converter; b) waveforms of the switching node voltage V_{sw} and the output voltage V_{out}.

is called pulse-width modulation (PWM), the most popular control method in DC–DC converters. The duty cycle D defines the ratio between the on-time and the period time,

$$D = \frac{t_{on}}{T}. \tag{1.2}$$

For the step-down converter in Fig. 1.4, the duty cycle determines the voltage conversion ratio:

$$\frac{V_{out}}{V_{in}} = D = \frac{t_{on}}{T} \tag{1.3}$$

There are other topologies of DC–DC converters that have different conversion ratios.

The energy-storing characteristic of the L-C network is required in two ways. If S1 is turned on (S2 is off), energy is brought into the system. The capacitor C buffers V_{out} in case of varying load currents (load transients). C is called a bypass capacitor because it bypasses the actual regulator during instantaneous load steps before the control loop can respond. Alternatively, C is referred to as the output buffer capacitor. The inductor L delivers the load current if S1 is active and in the second switching phase when S2 turns on (S1 is off). Due to the switching nature of the DC–DC converter, there will always be some finite output voltage ripple. It is a significant disadvantage compared to linear regulators (Section 1.3.1). The ripple can be reduced by enlarging L and C at the expense of larger size and reduced power density. Another way of reducing the ripple and, at the same time, increasing the output power is to use multiple parallel DC–DC converters. Such multi-phase converters operate in a time-interleave scheme, delivering multiple energy packages during each cycle.

When discussing energy efficiency, it is essential to note that in a steady state, there should be no power loss P_{loss} at a switch. If the voltage across the switch is V and the current through the switch is I, the loss is $P_{loss} = V \cdot I$:

$$\text{Switch turned on: } V = 0, I = I_{load} \rightarrow P_{loss} = V \cdot I = 0 \tag{1.4}$$

$$\text{Switch turned off: } V = V_{max}, I = 0 \rightarrow P_{loss} = V \cdot I = 0 \tag{1.5}$$

V_{max} is the (maximum) blocking voltage of the switch, which is equal to V_{in} in Fig. 1.4a). The relationship in Eqns. (1.4) and (1.5) is the fundamental reason, why switched-mode operation is widely used in power electronics. In actual designs, there will be various loss contributions, such as the finite on-resistance of the switches. There will also be switching losses and losses in the passive components. However, these losses are usually much lower compared to a linear regulator introduced in Section 1.3.1. Conversion efficiencies of more than 90% can be achieved. Chapter 10 covers inductive DC–DC converters comprehensively.

1.3.3   Switches and Capacitors – The SC Converter

Another way of voltage conversion is the combination of switches with capacitors. This concept has become very attractive for highly integrated power management designs due to the availability of high-density integrated capacitors in advanced CMOS technologies. Figure 1.5 shows a typical circuit along with the equivalent circuits in the two switching phases. During phase φ_1, both capacitors are connected in series to V_{in}. The capacitors are parallel in phase φ_2. The circuit periodically changes from a series to a parallel configuration, reducing the output voltage to half of the input voltage. Interestingly, this behavior is independent of the actual capacitor values. Their values determine the amount of charge shared between switching cycles, but in steady state, V_{out} will be exactly half of V_{in}. Note that ideally, no losses have occurred so far.

How can we convert 12 to 2 V? We can use two conversion stages. The first results in 6 V, the second one gives 3 V. How do we get to the target value of $V_{out} = 2$ V? We take advantage of the fact that C_1 can deliver only a limited charge. In other words, we let the load current discharge the output capacitor C until V_{out} reaches exactly 2 V. Most easily, this can be achieved by adjusting the clock frequency. Unfortunately, this is when the SC converter introduces power loss due to charge redistribution ($\sim CV^2$). The output voltage drop from 3 to 2 V can be seen as a voltage drop across an equivalent output resistance. Significant research has been dedicated to finding improved SC converter topologies and control mechanisms that minimize these losses. SC converters are further explored in Chapter 9.

1.3.4   Switches and Capacitors and Inductors – The Hybrid Converter

This approach takes the SC converter of Fig. 1.5 and adds an inductor, as illustrated in Fig. 1.6. The combination of L and C leads to the name hybrid converter. It utilizes the benefits of the inductive and capacitive conversion concepts presented in Sections 1.3.2 and 1.3.3. Two mechanisms help improve conversion efficiency. The inductor ensures soft charging of the capacitor, which

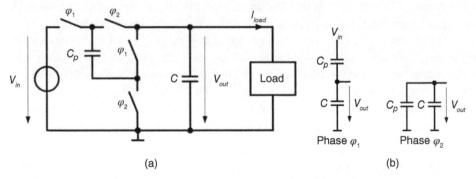

(a)   (b)

Figure 1.5   Voltage conversion using capacitors: a) the fundamental switched-capacitor voltage converter; b) the equivalent circuits in phases φ_1 and φ_2, alternating between series and parallel configurations.

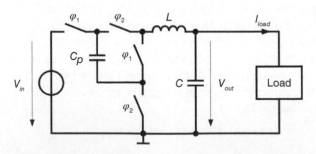

Figure 1.6   A hybrid converter formed by adding an inductor L to an SC converter resulting in higher conversion efficiency.

eliminates charge redistribution losses and minimizes the equivalent output resistance. L and C also form a resonant tank, which can achieve zero-current switching of the power switches. This way, the switching losses can be significantly reduced. On the downside, this concept has higher complexity than an inductive or capacitive converter. However, handling complexity is one of the great benefits of advanced microelectronics. Chapter 11 covers hybrid DC–DC converters comprehensively.

1.4 Power Management Systems

Electronic systems usually do not require only one supply voltage. Instead, various functional blocks have different supply voltage and power requirements. Figure 1.7 shows two examples of power management systems. Dedicated converters are assigned to supply each block at the point of load (PoL). State-of-the-art power management systems comprise multiple PoL regulators and DC–DC converters to supply microcontroller units (MCU), processors (CPU, GPU, and DSP), as well as analog and mixed-signal circuits.

In space-constraint applications such as smartphones, multiple voltage regulators are implemented in a single IC to minimize their footprint. Known as the multirail power supply (MRPS), it is the leading PMIC type with a market share of 20% (see Section 1.9).

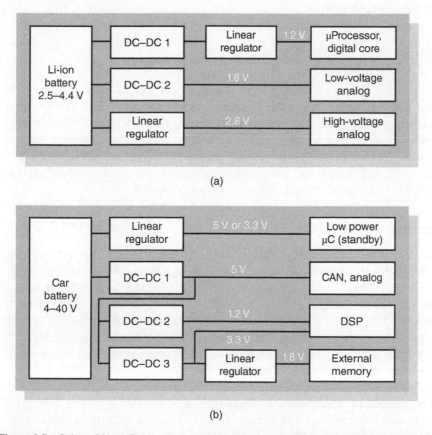

(a)

(b)

Figure 1.7 Point-of-load (PoL) power management systems: a) handheld devices operating from a Li-ion battery have lower step-down ratios as compared to b) automotive, industrial, and data center applications that operate from a higher supply voltage like a car battery.

For large step-down ratios, linear regulators are not efficient. However, switched-mode DC–DC converters may not fulfill strict supply ripple and noise requirements for sensitive analog circuits such as sensor front ends. Hence, various conversion stages are combined, as illustrated in Fig. 1.7.

Combining a DC–DC converter as a first conversion stage with a subsequent linear regulator (LDO) is beneficial. The DC–DC converter guarantees high efficiency, while the LDO ensures a "clean" output voltage with a fast transient response. If the intermediate voltage V_{mid}, provided by the DC–DC converter, is close to the target output voltage, the linear voltage regulator will also show an acceptable efficiency. We can calculate the overall efficiency by multiplying the efficiency values of each stage:

$$\eta = \frac{P_{out}}{P_{in}} = \frac{P_{out}}{P_{mid}} \cdot \frac{P_{mid}}{P_{in}} = \eta_1 \cdot \eta_2 \tag{1.6}$$

For a large step-down ratio, i.e., if V_{in} is much greater than V_{out}, achieving high efficiency will be more challenging.

As illustrated in Fig. 1.7a), a Li-ion battery typically supplies portable devices such as smartphones and wearables with a voltage range of 2.5–4.4 V. There is a moderate step-down ratio. The automotive application shown in Fig. 1.7b) operates from a 12 V lead-acid battery. The board net voltage can vary a lot, for instance, from 4 to 40 V. Hence, the step-down ratio is much larger than in devices supplied by a Li-ion battery. A step-up converter (boost converter) is typically inserted that kicks in if the board net drops toward 4 V and stabilizes the input voltage of the subsequent stages to typically 10 V (not shown in Fig. 1.7b) for simplicity). This way, the power management system can still provide 5 V at the output.

1.5 Applications

The application defines the energy source on the left of Fig. 1.1 with several examples shown in Fig. 1.8. Consequently, the applications also determine the system voltages, which are the power management input voltages. There is a trend toward higher voltages, as discussed in the following subsections, motivated by higher energy efficiency. For the same reason, the IC-level voltages of the electronics on the right of Fig. 1.1 scale into the opposite direction and reach levels of 1 V and lower. More details on IC-level supply voltages follow in Section 1.6. We discuss various applications below, starting with the lowest voltages at the bottom left of Fig. 1.8.

1.5.1 IoT Nodes and Energy Harvesting

Internet-of-Things (IoT) wireless nodes, installed in smart homes and office spaces and used in an industrial environment, are often designed with 10-year battery lifetime targets [1]. To prolong the life of a 3 V-CR2032 coin cell, the average current needs to be kept below 2.5 μA (7.5 μW). However, this can be challenging since the wireless node draws around 5 mA when it is in active mode (transmit). Several blocks contribute to this average power dissipation, including the CPU, the transceiver, sensors, and power management. Fortunately, continuous operation is not required, and duty cycling reduces the average current consumption to typically 6 μA. The sleep current in the order of 1 μA consists of leakage current, memory retention, analog circuits, such as a power-on reset, and a low-frequency sleep clock.

Energy harvesting can be used to expand the battery time. The ultimate goal is to remove the battery at all. In addition to the positive environmental impact, it can significantly reduce maintenance effort and cost. Besides the IC, the battery is the most expensive part of the wireless sensor

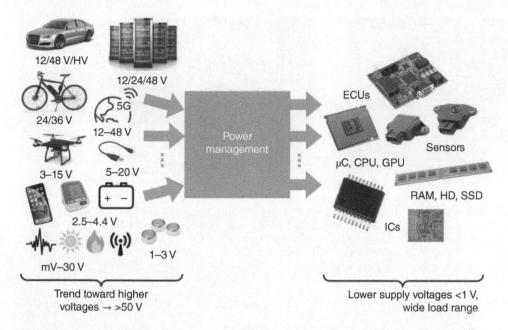

Figure 1.8 The system voltages in various applications define the input voltage for the power management circuits. On the load side, the supply voltages follow the trend of decreasing supply voltages toward 1 V and below with a wide load current range. Source: dezign56/Adobe Stock; ifeelstock/Adobe Stock; Pawinee/Adobe Stock; Scanrail/Adobe Stock, kornkun/Adobe Stock.

node (about 20% [1]). With the decreasing cost of the IC, the battery cost is expected to contribute a growing percentage of the total node cost over time.

Energy harvesting converts mechanical energy (kinetic energy and vibration), light (via solar cells), thermal energy, and the energy of radio frequencies (RFs) into usable voltages. RF energy harvesting is the lowest cost option, but the available power levels are the lowest. Hence, a promising approach is multisource energy harvesting. Nevertheless, the typical output power is in the order of 1 mW and below.

Power management circuits in energy harvesting applications operate from very low voltages in the millivolt range and even below. Charge pumps or similar techniques bring these low input levels to an intermediate voltage of ~0.5–1 V, just above the transistor threshold voltage. At this point, another step-up DC–DC converter (inductive or capacitive) kicks in. It boosts this voltage to 1.8 V, for instance, suitable for supplying functional electronic blocks. Once the voltage reaches the power-on reset level, the IoT node transmits a data packet and shuts off until enough power is available again.

1.5.2 Portable Devices, Smartphones, and Wearables

Many designs run from low-voltage batteries (button cells) in the range of 1–3 V. Li-ion batteries with a cell voltage of 2.5–4.4 V have become the primary battery technology for portable devices like laptops, mobile phones, and wearables. Laptops use two or three battery cells in series to supply input voltages of 7.4 or 11 V. Smartphones are covered as the introductory example in Section 1.2. The growing field of wearables includes applications like smartwatches, fitness trackers, smart headphones, glasses, medical monitoring devices, and implants. The power consumption of wearables is typically a few hundred milliwatts, an order of magnitude lower than a smartphone. It

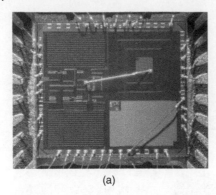
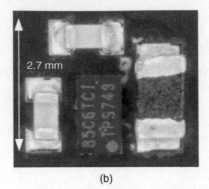

(a)   (b)

Figure 1.9 Miniaturized DC–DC converters for wearables: a) a fully integrated PMIC with on-chip integrated passives. Source: Peter Renz et al. [2]/from IEEE; b) A DC–DC converter in a 1.6 mm × 0.9 mm, 8-ball wafer-level chip-scale package (WCSP) with discrete passives (TPS627431, Texas Instruments). Both photographs are depicted to scale. The size of the objects in the photographs is represented in relation to one another.

benefits from a smaller display with fewer pixels and a scaled-down CPU that runs at a lower frequency.

PMICs for portable devices must provide high power conversion efficiency to ensure long battery times. As the primary requirement, the PMIC designs need to be ultracompact and lightweight in addition to providing high power conversion efficiency. These advantages are achieved by fast-switching DC–DC converters with miniaturized passive components (see Section 1.8.3 and more details on integrated passives in Chapter 4).

Figure 1.9 shows two PMIC examples for wearables. Figure 1.9a) is a hybrid DC–DC converter with fully integrated passives [2]. The output buffer capacitor has a capacitance of 10 nF (lower right corner of the die photo), and the inductor is a square spiral coil of 9 nH (upper right, see Fig. 4.10 for an enlarged picture). Due to these small values, the converter operates at switching frequencies of up to 47.5 MHz. The power management circuits support low quiescent currents during sleep modes (power levels <25 mW) to achieve high efficiency and long battery times. The DC–DC converter in Fig. 1.9b) fits in a wafer-level chip-scale package (WCSP). The device is optimized to operate with a 2.2 µH inductor and 10 µF output capacitor, operating at 1.2 MHz. It offers ultralow quiescent current of typically 360 nA. Despite the compact design, the DC–DC converters in Fig. 1.9 reach peak efficiencies of more than 85% and 95%, respectively.

Example 1.1 *We want to estimate the battery time of a smartwatch supplied by a Li-ion battery of 270 mAh. The battery voltage of $V_{bat} = 3.6$ V is converted to $V_{out} = 1.8$ V to supply the internal electronics at an output power of $P_{out} = 100$ mW. We distinguish two types of voltage conversion: a) an LDO (linear regulator) with an efficiency of 50%; b) a hybrid converter with an efficiency of 85%.*

With an efficiency of less than 100%, the input power P_{in} will be higher than P_{out}. P_{in} determines the current drawn out to the battery. From Eqn. (1.1), we obtain for case a)

$$P_{in} = \frac{P_{out}}{\eta} = \frac{100 \text{ mW}}{0.5} = 200 \text{ mW}, \tag{1.7}$$

$$I_{in} = \frac{P_{in}}{V_{in}} = \frac{200 \text{ mW}}{3.6 \text{ V}} = 56 \text{ mA}. \tag{1.8}$$

Now that we know the current, we can determine the battery time for case a) with the LDO:

$$t_{bat} = \frac{270\,\text{mAh}}{I_{in}} = \frac{270\,\text{mAh}}{56\,\text{mA}} = 4\,\text{h}\,48\,\text{min} \tag{1.9}$$

For case b), we repeat the calculations and obtain P_{in} = 118 mW, I_{in} = 33 mA, which results in t_{bat} = 8 h 11 min.

The example shows the importance of conversion efficiency. The hybrid converter's battery time is almost twice as long as using a linear regulator.

1.5.3 Universal Serial Bus (USB)

The USB interface (Universal Serial Bus) connects a computer, a laptop, or a smartphone with peripheral devices, such as chargers, speakers, and headsets, using only a single cable. This connection transfers data and power. Many devices use the reversible USB-C interface, which supports much higher data rates and power levels. Also, the cable orientation and the current-carrying capability are detected. The early USB 2 standard supports data rates of 480 MB/s and a maximum power of 500 mW in a 5 V scheme. The USB 4 standard supports 40 GB/s and 8K video resolution using a USB-C cable.

Regarding power delivery, one of the main tasks is connecting a charger. The USB charger recognizes the connected device and ensures that the right amount of power is delivered to charge the battery as quickly as possible.

Within the USB-C ecosystem, the USB power delivery standard (USB-C PD) supports up to 100 W of power transfer for the growing number of applications with increasing complexity. For low-power levels below 15 W (chargers for smartwatch and wireless headphones), the bus voltage is kept at 5 V. At 15 W, the load current reaches 3 A. If the output power further increases, the output current is held at 3 A while the voltage goes up, from 5 V to 9 V, 15 V and, finally, 20 V. This way, USB-C can deliver power levels of about 30 W for fast-charging smartphones and tablet PCs. At 3 A · 20 V = 60 W, the output current linearly increases from 3 to 5 A reaching the maximum power of 100 W to supply desktop PCs. Figure 11.27 in Chapter 11 shows a PMIC design for bidirectional power delivery in a USB cable. The PMIC and all passives are integrated into the Type-C connector.

1.5.4 Drones

Drones, miniature pilotless aircraft, are a fast-growing market with many commercial, government (defense), and consumer applications. Main business areas include infrastructure, agriculture, transportation, media and entertainment, telecommunication, and mining. The distance that drones can fly is limited by the payload and its weight, including the size of the energy source. The electronics consist of a microprocessor, a CMOS camera, Wi-Fi, USB interfaces, and various sensors like ultrasonic altimeters and micro-electromechanical systems (MEMS) gyroscopes. Because of their excellent power-to-weight ratio, drones are often supplied with lithium polymer batteries. Typical power levels of 15–25 W are achieved by stacking several battery cells of ~3 V each resulting in a supply voltage range of 3–15 V with a trend up to 36 V.

1.5.5 Telecommunication Infrastructure

Telecommunication infrastructure must handle significant data rates and large capacity at very high reliability. The 5G standard approaches transmission bandwidths up to 71 GHz utilizing a

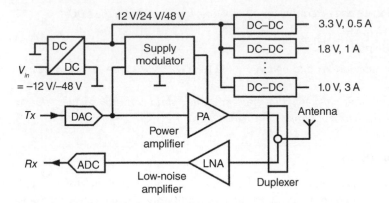

Figure 1.10 A 5G base station with the transmit and receive path (*Tx*, *Rx*). The power management part at the top consists of a galvanically isolated DC–DC converter that provides 12 or 48 V, a supply modulator for the power amplifier, and multiple point-of-load DC–DC regulators.

millimeter wave spectrum. It opens up completely new applications besides making mobile phones faster. It will be the underlying technology for cloud computing, autonomous driving, and the IoT. However, increasing frequencies lead to higher operating expenses, a big part related to power consumption. Within a 5G base station, the baseband unit (BBU) consumes 300 W. Each BBU is connected to multiple separated remote radio heads (RRH). With more frequency bands, the total power consumption is expected to reach 10–20 kW per base station site.

Figure 1.10 shows the power management architecture of a 5G base station. The power supply comes from a DC telecom voltage of −24 or −48 V. This voltage is negative to prevent the equipment from lightning strokes because a cloud in the air is negatively charged at the bottom, requiring a positive potential to discharge. A galvanically isolated DC–DC converter generates a system voltage of typically 12, 24, or 48 V, which is positive with respect to ground. This voltage is further stepped down in subsequent DC–DC conversion stages of many subrails ranging from 3.3 V to less than 1 V to supply the ICs used for baseband processing. The main electronic blocks in a 5G base station include the main power amplifier (PA), various data converters (DAC, ADC), low-noise amplifiers (LNA), and digital signal processing components like CPUs and Field-Programmable Gate Arrays (FPGAs).

In the transmit path (*Tx*), as shown in Fig. 1.10, the supply of the PA provides an output power in the range between 100 mW and 90 W. The PA is modulated based on envelope tracking, a power management technique that allows for more efficient use of power by dynamically adjusting the voltage supplied to the PA. This technique minimizes the voltage difference between the transmit signal and the modulated PA supply, corresponding to losses dissipated as heat. Due to the high-speed, high-power, and high-voltage requirements, wide-bandgap GaN transistors (see details in Section 3.1.4) have become an essential technology for PAs in 5G base stations.

Unlike earlier standards, 5G allows the ability to analyze the traffic load and go into sleep mode for durations as short as a few milliseconds to save power. The power supply unit (PSU) has to be able to switch between quiescent and normal power states repeatedly.

1.5.6 E-Bikes

E-bikes are growing in popularity, offering environmentally friendly mobility. The e-bike power architecture consists of a battery, an electric motor, a power control block with several power transistors used to drive the motor, and various sensor electronics required for vehicle operation. Inductive DC–DC buck converters or charge pumps (capacitive DC–DC converters) provide the supply voltage for the gate drivers of the motor control block and the microcontroller. For safety reasons, diagnostic functions are an integral part of the electronics, such as the detection of overvoltage, undervoltage, and overcurrent, as well as temperature monitoring in the battery and the power stage of the motor driver.

Most e-bikes use batteries of 24, 36, or 48 V that deliver a few hundred watts of motor power. E-bikes with 250 W motors typically come with 36 V batteries. E-bikes usually have an efficient pedal-assisted brushless direct current (BLDC) motor mounted either in the wheel hub or between the pedals at the bike's bottom bracket. Three half-bridges with six power switches are required for three-phase operation. Current sensing of the entire power stages or per branch supports different motor control schemes. Efficiency is one of the key requirements of the power control system, as it ensures long battery times, greater mileage, and lower operating costs. Attention is paid to the solution size to ensure it fits into the given box volume. The electronics need to ensure quiescent currents in the order of 100 µA and below to reduce the power consumption during idle time, for example, at traffic lights. New generations of e-bikes also include regenerative braking to recharge the battery, extending battery time and mileage.

1.5.7 Automotive

The board net in a car is supplied conventionally by a 12 V lead-acid battery, which delivers a peak power in the order of 3–5 kW. The car battery voltage varies between 9 and 18 V. During cold crank, the board net may reach voltages as low as 4–6 V. This effect is due to the voltage drop across the battery's source resistance and the interconnection to the starter generator. Assume that battery has a source resistance of 20 mΩ and that the starter generator requires a start current of 300 A. Neglecting the resistance of the interconnections, the voltage drop across the source resistance will be 20 mΩ · 300 A = 6 V. Toward the upper end, the voltage is determined by the effect called load dump. It relates to a failure case where a load disconnects. The board net represents a distributed R-L-C network. The finite energy (current) in the cable harnish causes the voltage to rise steeply during load dump. Transient suppressing diodes clamp the voltage to levels of 36–60 V. Voltages below the limit of 60V are considered safety extra-low voltage (SELV), which eases the handling requirements. The load dump voltage transient takes a few hundred milliseconds, long enough to consider this a DC parameter. The typical device specification of a PMIC connected to the automotive board net defines a voltage range from as low as V up to 60 V. Also, the ambient temperature specifications are challenging, covering a range of −40 to 125 °C. Due to power losses, the junction temperature in automotive PMICs may reach more than 150 °C.

With the trend of adding electrical support to mechanical components or replacing them with electrical counterparts, for example, related to braking, steering, clutch, and oil pump, the power demand has grown significantly. Power is also required to increase the safety functions of advanced driver assistance systems (ADAS). Enormous computing power will be needed to enable fully autonomous vehicles. More and more connected infotainment systems further contribute to increasing power requirements.

To overcome the power limitations, a new board net with a 48 V battery has been introduced, delivering power levels of 8–15 kW. Using a higher voltage allows for the power distribution at much lower currents, decreasing the cable's resistive losses and reducing the cable cross section by as much as 75%, resulting in lighter vehicles and emission reduction. Recent cars are equipped with the 48 V battery in addition to the 12 V board net, but a permanent shift to a single 48 V system may occur in the future.

The dual-architecture 12/48 V automotive board net is depicted in Fig. 1.11. Applications with lower power demand are supplied from the 12 V battery utilizing existing components designed for

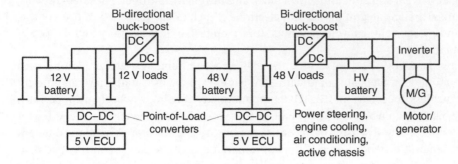

Figure 1.11 The automotive 12 V/48 V/HV boardnet.

that voltage domain. High-power applications, including power steering, engine cooling, and air conditioning, are moved to the 48 V board net. There is a central bidirectional DC–DC converter between both domains for safety reasons and to ensure an emergency operation. Similar to the 48 V domain, the 48 V board net varies a lot during operation, typically over a range of 24–54 V. Using the Li-ion technology for both voltage domains, both batteries and the DC–DC converter can fit into the size of the conventional 12 V lead-acid battery.

Figure 1.11 includes the drive train in electric cars, which is supplied from a high-voltage Li-ion battery of about 400 V with a trend to much higher voltage levels. A bidirectional DC–DC converter connects the high-voltage and the 48 V domains. An inverter generates the AC supply to drive the motor, which can also act as a generator to recharge the battery.

Today's cars contain more than 100 electronic control units (ECUs) that are supplied from both the 12 V and the 48 V board net. Each ECU has various point-of-load (PoL) DC–DC converters (see Section 1.4) that generate the required supply voltages for microcontrollers, sensors, motor and solenoid drivers, interface controllers (LIN, CAN, and other), radar transceivers, displays, radio receivers, and many more. The standard system voltage in the ECU is still 5 V. In that regard, automotive stands out from the general trend of decreasing supply voltages. There are several reasons why the 5 V persist. It provides sufficient noise margin, and the physical layer of the Controller Area Network (CAN) standard refers to 5 V. Out of the 5 V, subregulators generate lower voltages as illustrated in Fig. 1.7b).

One fast-growing automotive application is advanced driver assistance systems (ADAS). Functions include automatic emergency parking, lane departure warning and assistance, and vehicle and pedestrian detection. The increasing number of sensors, cameras, real-time data processing units, and high-speed communications lead to excessive power dissipation. Recent ADAS ICs consume 20–30 W. Compact front-end (12/48 V) DC–DC converters are required that need to be able to deliver currents in the range of 10 A. Therefore, ADAS is an exciting and challenging PMIC design field.

A particular concern in automotive is electromagnetic interference (EMI). See Section 12.3 for an introduction on EMI. Standards strictly define the maximum amount of conducted and radiated emissions on IC and module levels. For this reason, automotive DC–DC converters are preferred to operate at constant switching frequency. Techniques like spread-spectrum frequency modulation helps minimize EMI by continuously varying the switching frequency according to a predefined scheme.

Automotive ICs must comply with the Automotive Safety Integrity Level (ASIL) standards, ensuring tighter protections and accuracy, redundant references, fail-safe on open and shorted pins, and several more diagnostics. Automotive ICs are designed for an operating life of at least 15 years.

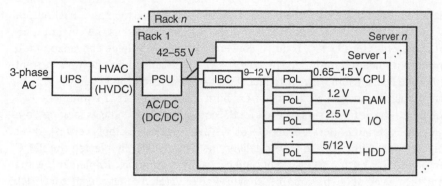

Figure 1.12 Data center 48 V power distribution architecture.

1.5.8 Data Centers

Data centers provide a secure, scalable, and reliable infrastructure for storing, processing, and analyzing large amounts of data ranging from local edge enterprise facilities to global hyper-scale cloud data centers. They are the backbone for streaming services, 5G, the IoT, social media, and many more applications. By hosting high-powered processors (CPUs), high-speed memory, and specialized hardware such as graphics processor units (GPUs), data centers can accelerate complex calculations in machine learning algorithms and support big data and AI applications. These workloads are data- and compute-intensive and require a massive power consumption of 10 MW and more per facility.

Figure 1.12 shows the power system in a data center. An uninterpretable power supply (UPS) is between the main power outlet and the IT equipment. The UPS protects against fluctuations of the AC mains and malfunctions like surges, sags, and brownouts.

The power is distributed via a high-voltage AC bus or, more recently, a DC bus toward 9-inch server racks. One thousand racks or more are installed per data center. The power per rack ranges from below 10 kW to over 60 kW in high-performance computing clusters that support AI applications. It results in supply currents up to 2.5 kA distributed in a standard 9-inch rack. High power efficiency is challenging due to copper and distribution losses in the backplane.

Within the rack, there are three power conversion stages: (1) a power supply unit (PSU, also called power entry module (PEM)), (2) an intermediate bus converter (IBC), and (3) a set of point-of-load DC–DC converters (PoL). All three stages provide incredibly high performances and peak efficiencies well above 90%.

Each rack contains up to 10 chassis, which hold the server blades. Older architectures have one PSU per chassis. Newer racks have one central PSU. It is much more efficient to transmit power using a high voltage and a low current. For this reason, the rack-level distribution voltage has been increased from 12 V to nominally 48 V. This approach is similar to the one used in the automotive industry, as discussed in Section 1.5.7. In March 2016, Google leveraged the Open Compute Project to release the specification for a high-efficiency direct conversion 48 V board for server racks.

The increase to 48 V leads to a four times lower current. The power distribution losses (I^2R) reduce by a factor of 16. However, achieving high efficiency for step-down ratios from 48 V to ~1 V is difficult and defines one of the challenges in the design of PMICs (see Sections 1.7 and 12.8). Nevertheless, converters in the 48 V domain see significant efficiency and power density benefits from using wide-bandgap GaN power transistors over silicon counterparts.

With the intermediate bus architecture shown in Fig. 1.12, the entire system can benefit from a significant reduction in energy consumption, contributing to cost savings and a more sustainable operation of the data center. At the server level, the IBC converts the semi-regulated PSU output voltage of 42–54 V into 9–12 V. Being a fixed-ratio converter, the IBC is used as the current multiplier. As its input voltage of typically 48 V is down-converted by a fixed voltage conversion ratio $M = 1/4$ or $1/6$, the output current is multiplied by $1/M = 4$–6. At the same time, the input current is only a fraction M of the load current, further reducing the power distribution losses.

Each server has a motherboard with CPUs, RAM, I/Os, hard disks, and other IT equipment. PoL voltage regulators generate the different supply levels for the equipment with voltages ranging from 12 V down to computation core supply levels of about 0.65 V. Advanced designs apply vertical power delivery to minimize the impedance of the power delivery network (PDN) by placing the PMIC and passives under the processor using 3-D assembly and packaging techniques. Implementing the voltage regulator and the processor on the same die, known as an integrated voltage regulator (IVR), further improves efficiency and dynamic response. See Sections 1.7 and 12.8 for more details on vertical power delivery and IVR. Single- or multi-phase buck converters are used, which can handle the relatively wide input range while providing excellent load transient response. They follow the concept of Fig. 1.4 with details provided in Chapter 10.

The power usage effectiveness (PUE) describes how well a data center delivers energy to IT equipment by calculating the ratio between the total facility energy to the IT equipment energy, distributed via the high-voltage bus. In the ideal case, the PUE would be unity, meaning that all power goes to the IT load. However, a significant amount of energy is needed for the UPS and other electrical equipment as well as for cooling, ventilation, and air conditioning, resulting in typical PUE values between 1.3 and 2. Cooling is usually the most significant contributor. A PUE of 2 means that half the energy would be consumed by the overhead of running the data center and supporting the power conversion. Optimizing the PUE shifts the focus of power loss optimization toward the efficiency of the PSU and each PoL converter in the IT equipment on the rack level.

1.6   IC Supply Voltages

In the application overview of Fig. 1.8, the integrated circuits in the electronic blocks on the right are supplied by the power management. The development of supply voltages for ICs has come a

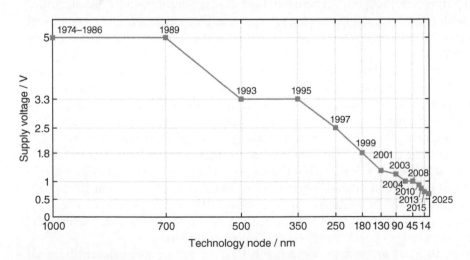

Figure 1.13   IC supply voltages versus technology node and year of introduction.

long way since the 1970s, as shown in Fig. 1.13. The x-axis shows the corresponding technology dimension, which, in recent years, refers to the digital core supply voltage.

Until the end of the 1980s, voltages were typically in the range of 5 V (or more) with process dimensions in the order of 1 μm. However, as technology progressed, the need for lower voltages became apparent. It led to the development of 3.3 and 2.5 V supply voltages in the 1990s.

As the demand for even lower power consumption grew, supply voltages continued to decrease. By the early 2000s, 1.8 V became the standard for many integrated circuits. 1.8 V and also 1.2 V are still widely used in many applications. However, in the advanced Complementary Metal-Oxide-Semiconductor (CMOS) technologies like 22, 14, 7 nm core supply voltages are below 1 V, typically 0.7 V. Beyond 2025, process nodes of 1–2 nm are expected to operate at a nominal core supply voltage of 0.65 V.

As the main benefit, lower supply voltages reduce power and energy consumption, which is critical not only in applications where battery life is a concern. More details are provided in Section 1.7. Lower voltages also reduce heat dissipation, increasing the reliability and lifetime of integrated circuits.

1.7 Power Delivery

As computing technology advances, power management systems for microprocessors have become increasingly important. Power delivery is not limited to PCs, servers, and data centers. Microcontrollers and processors are essential in various applications (see the overview in Section 1.5). It is becoming increasingly challenging to manage power delivery for high-performance microprocessors as power levels, and the number of power rails continue to grow.

The major challenges in designing PMICs for compute power delivery are

- high supply currents in the order of 10–1000 A,
- high-speed response to line and load variations, and
- power conversion efficiency (heat dissipation and cooling).

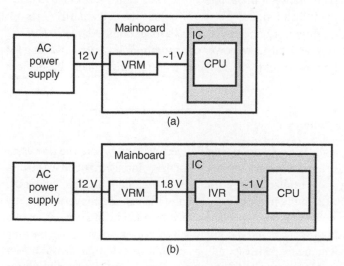

(a)

(b)

Figure 1.14   Power delivery in a desktop PC:
a) for a conventional CPU and b) using an integrated voltage regulator (IVR).

Figure 1.14a) shows the power delivery in a desktop PC. An AC power supply converts the mains voltage of 120VAC or 230VAC into a DC voltage of 12 V that supplies the mainboard. A Li-ion battery may deliver the input voltage in smartphones and other portable devices. An inductive DC–DC converter (see Fig. 1.4) forms a voltage regulator module (VRM) that provides the CPU core voltage of approximately 1 V. The VRM is a typical PMIC that acts as a PoL power supply (see Section 1.4).

In advanced CMOS technologies, the reduction in the size of transistors does not lead to an increase in speed anymore. As a result, the industry has shifted its focus toward increasing the core count in processors instead of the operating frequency of a single core [3]. However, this has brought new challenges in managing workloads between cores and ensuring optimal power consumption and thermal behavior of the processor. High-end multicore processors that perform complex computations and data analysis at ultrahigh speed require sophisticated power management schemes with tens of different power rails. The current flowing into the processor package increased dramatically with the rising power density and falling supply voltage. Processors may draw over 100 A from the voltage regulator.

Modern processors are highly dynamic, adapting their behavior depending on the workload. Sudden spikes in the current draw can cause a voltage droop and brownout failures. Voltage converters need to be able to respond extremely fast. However, they are usually slower than the load current transients caused by switching activity, which can be as fast as 5 A/ns [4]. For processor core voltages as low as 0.65 V, the droop voltage margin becomes extremely tight (typically <10% of the supply level).

For this reason, bypass or decoupling capacitors (decaps) are placed at as many steps of the power delivery network (PDN, also called power distribution network), particularly as close as possible to the processor. These capacitors quickly provide power to stabilize the voltage while the regulator is just beginning to respond. On the IC level, this is addressed by filling up all white space with capacitors of various types, including Metal-Oxide-Semiconductor capacitors (MOS capacitors) (high density, but limited voltage ratings) and trench capacitors (very high density, but expensive and not always available). Details related to physical implementation can be found in Chapter 12. On-chip capacitors are covered in Chapter 4.

Nevertheless, there are still differences between the applications. Data centers, for instance, are optimized for efficiency and high CPU utilization. Processors for notebooks and smartphones are typically optimized to deliver maximum performance over only short periods. About 90% of the time, the mobile processor is idle with practically zero power consumption (e.g., while waiting for user input). The power delivery design must meet the form-factor requirements of portable devices while keeping costs low.

1.7.1 Lateral and Vertical Power Delivery

Most of today's voltage regulator designs use a lateral power delivery system, where the processor and the regulator(s) are placed next to each other on the top side of the printed circuit board (PCB). In that case, the interconnect resistance is limited to a few hundred $\mu\Omega$, which makes it challenging to deliver currents in the order of 100 A to the processor. By placing the voltage converter directly under the processor die, the power delivery path can be shortened by a factor of 10, significantly reducing resistance and related losses [5]. This vertical approach to power delivery has become increasingly crucial in meeting the demand for fast-responding and more efficient processor power supplies. Chapter 12 covers the physical implementation of power management systems with more details on lateral and vertical power delivery (see Section 12.8.1).

Example 1.2 *Calculate the system power conversion efficiency for the architecture in Fig. 1.14a).*
The VRM has an input voltage of 12 V. It delivers a current of $I_{load} = 100$ A to the processor at $V_{out} =$
1 V and an efficiency of $\eta_{VRM} = 90\%$. The routing resistance (mainboard and package) between the
VRM and the processor is $R_{PDN} = 1.5$ mΩ.

The VRM needs to deliver the processor power of $P_{out} = V_{out} \cdot I_{load} = 100$ W and the routing
losses of

$$P_{PDN} = I_{load}^2 R_{PDN} = (100 \text{ A})^2 \cdot 1.5 \times 10^{-3} \text{V/A} = 15 \text{ W}. \tag{1.10}$$

Hence, the output power of the VRM is

$$P_{VRM} = P_{out} + P_{PDN} = 100 \text{ W} + 15 \text{ W} = 115 \text{ W}. \tag{1.11}$$

Based on Eqn. (1.1), we determine the input power of the VRM,

$$P_{in} = \frac{P_{VRM}}{\eta_{VRM}} = \frac{115 \text{ W}}{0.9} = 128 \text{ W} \tag{1.12}$$

and finally, the total power efficiency of the system,

$$\eta_{tot} = \frac{P_{out}}{P_{in}} = \frac{100 \text{ W}}{128 \text{ W}} = 78.3\%. \tag{1.13}$$

Even though the VRM has a large efficiency of 90%, the system efficiency is almost 12% lower. We see
that the PDN routing resistance significantly impacts the overall system efficiency.

1.7.2 Integrated Voltage Regulator (IVR)

IVRs expand the idea of vertical power delivery by placing one or more inductive or hybrid DC–DC
converters on the processor die [3, 4, 6, 7]. Example 1.2 shows the significance of maintaining
low routing losses in the PDN. IVRs deliver the processor power at a higher voltage, reducing the
current flow through the PDN and minimizing routing losses. Implemented locally beside the processor, IVRs enable granular power management with a swift dynamic response. This behavior is
essential to leverage the benefits of dynamic voltage and frequency scaling (DVFS), as explained
furtherlater.

The power delivery architecture with IVR is shown in Fig. 1.14b). As a first stage, the voltage
regulator module VRM (PMIC) provides an intermediate voltage of typically 1.8 V. This voltage
is distributed across the mainboard to the processor die with the IVR. The IVR is responsible for
the last power conversion stage and converts the voltage to the processor core level of around 1 V
and below. The IVR's power stage and control loop circuits are part of the processor IC, while the
power inductor is typically implemented as an air-core device using package or PCB traces. As part
of a vertical power delivery setup, IVRs minimize the impedance of the PDN. Hence, they provide a
very fast response and achieve high efficiency while operating at frequencies in the 100 MHz range.

Section 12.8.2 further examines the benefits of IVR regarding the physical implementation of
the PDN.

Example 1.3 *Repeat Example 1.2 and calculate the system power conversion efficiency for the*
architecture in Fig. 1.14b) with an IVR. The IVR has an efficiency of $\eta_{IVR} = 90\%$. Its input voltage
is $V_{IVR} = 1.8$ V. The IVR supplies the processor at $V_{out} = 1$ V and delivers $I_{load} = 100$ A. Due to the
higher intermediate voltage, the VRM can be implemented more efficiently. Because the IVR provides
fine regulation of the processor voltage, the VRM is implemented as a fixed-ratio converter (e.g., an SC
or hybrid converter) that achieves an efficiency of $\eta_{VRM} = 97\%$. The VRM input voltage remains 12 V.

The input power of the IVR is

$$P_{in} = \frac{P_{out}}{\eta_{IVR}} = \frac{100\,\text{W}}{0.9} = 111\,\text{W}. \tag{1.14}$$

Its input current I_{IVR} flows through the PDN. We can determine it from Eqn. (1.1),

$$\eta_{IVR} = \frac{P_{out}}{P_{in}} = \frac{V_{out} \cdot I_{load}}{V_{IVR} \cdot I_{IVR}}, \tag{1.15}$$

which results in

$$I_{IVR} = \frac{V_{out} \cdot I_{load}}{V_{IVR} \cdot \eta_{IVR}} = \frac{1\text{V} \cdot 100\,\text{A}}{1.8\,\text{V} \cdot 0.9} = 61.7\,\text{A}. \tag{1.16}$$

Knowing I_{IVR}, we can determine the PDN routing losses:

$$P_{PDN} = I_{IVR}^2 R_{PDN} = (61.7\,\text{A})^2 \cdot 1.5 \times 10^{-3}\text{V/A} = 5.7\,\text{W}. \tag{1.17}$$

Note that this value is nearly three times lower than the value in Example 1.2 (see Eqn. (1.10)) because the IVR operates from 1.8 V.

The VRM delivers the input power of the IVR and the routing losses:

$$P_{VRM} = P_{IVR} + P_{PDN} = 111\,\text{W} + 5.7\,\text{W} = 117\,\text{W} \tag{1.18}$$

Interestingly, this result is about the same as in Eqn. (1.11) in Example 1.2 for the single-VRM architecture. However, a significant portion of the power dissipation (11 W) is shifted from the PDN to the IVR.

With an efficiency of $\eta_{VRM} = 97\%$ its input power becomes

$$P_{in} = \frac{P_{VRM}}{\eta_{VRM}} = \frac{117\,\text{W}}{0.97} = 120\,\text{W}, \tag{1.19}$$

which translates to a total system power efficiency of

$$\eta_{tot} = \frac{P_{out}}{P_{in}} = \frac{100\,\text{W}}{120\,\text{W}} = 83\%. \tag{1.20}$$

Compared to the architecture without IVR, the overall efficiency is about 5% higher.

Example 1.3 illustrates the various mechanisms that contribute to improving the overall power efficiency by using an IVR. In addition, the IVR provides a fast dynamic response (see, for instance, multi-phase DC–DC converters in Section 10.14) that enables further power and energy savings using DVFS, as explained next.

1.7.3 Dynamic Voltage and Frequency Scaling (DVFS)

Changing the processor frequency f_{clk} and voltage V_{DD} depending on the workload or the operating conditions, known as DVFS, is a major technique to minimize the dynamic power consumption in processors as defined by

$$P = CV_{DD}^2 f_{clk}, \tag{1.21}$$

where C is the total switching capacitance [8].

Compute-intensive tasks such as video processing require short latency, translating into high clock frequency f_{clk}. On the other hand, low-speed tasks are not timing critical. By lowering the clock frequency, their execution can be expanded over a longer time frame t. Assuming that the task takes N clock cycles:

$$t = \frac{N}{f_{clk}} \tag{1.22}$$

In the conceptual plot of Fig. 1.15, the overall power (dynamic and static) reduces to 20% by lowering the frequency by 10.

From Eqns. (1.21) and (1.22), we can determine the energy consumption:

$$E = P \cdot t = NCV_{DD}^2 \tag{1.23}$$

According to Eqn. (1.21), the power dissipation P shows a linear dependency on the clock frequency f_{clk}. However, the energy consumption E remains the same because the task takes longer. This relationship is confirmed by Eqn. (1.23), which is independent of f_{clk}. Reducing only the frequency does not improve the battery time, which is critical in portable devices. However, energy can be saved by lowering the processor supply voltage V_{DD}. Due to the quadratic relationship shown in Eqn. (1.23), this can significantly reduce energy consumption.

To achieve maximum power and energy savings, f_{clk} and V_{DD} are dynamically varied depending on the instantaneous workload. As illustrated by the lower curve in Fig. 1.15, controlling power utilization has become a crucial aspect of hardware management. Hardware power managers have become fundamental in controlling power utilization by implementing various strategies to minimize energy waste. Additionally, these strategies aim to maintain a safe thermal environment for optimal performance. Operating systems communicate with different hardware power managers via an advanced configuration and power interface (ACPI) [9]. The ACPI standard defines P-states as DVFS operating points to reduce the active power for compute-intensive tasks. Illustrated in Fig. 1.15, higher P-states save more power. Likewise, ACPI defines C-states when the CPU is in idle sleep mode. P-state transients can occur at speeds of 50 mV/10 ns. The maximum processor clock frequency may be in to order of 3 MHz. DVFS aims to span a power reduction by a factor of 100 when scaling down the clock frequency by 10 times as shown in Fig. 1.15.

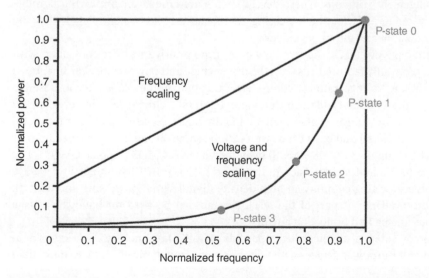

Figure 1.15 Dynamic voltage and frequency scaling: processor power dissipation versus clock frequency. Reducing only the frequency reduces the dynamic power only. Scaling the processor supply voltage, in addition, reduces both the power and energy consumption significantly. P-states indicate DVFS operating points.

1.7.4 Near-Threshold Computing

The technique of near-threshold computing scales the supply voltage down to levels around the transistor's threshold voltage [10]. Due to the quadratic relationship (see Eqn. (1.23)), this technique yields an energy reduction of 10 times. As a trade-off, the delay increases by approximately the same factor. It is crucial not to go into subthreshold because the substantial rise in leakage will dominate over the reduction in switching energy. In addition, reducing V_{DD} causes the gate delay to increase exponentially.

Example 1.4 *To what level do we need to decrease the nominal supply voltage of $V_{DD} = 1.8$ V to achieve a reduction of the switching energy by a factor of 10?*

Denoting the supply voltage before and after lowering by V_{DD1} and V_{DD2}, respectively, and, likewise, the energy by E_1 and E_2, Eqn. (1.23) yields

$$\frac{E_2}{E_1} = \frac{1}{10} = \left(\frac{V_{DD2}}{V_{DD1}}\right)^2 \quad \rightarrow \quad V_{DD2} = \frac{V_{DD1}}{\sqrt{10}} = \frac{1.8\,\text{V}}{3.16} = 569\,\text{mV}. \tag{1.24}$$

1.8 Technology, Components, and Co-integration

Power management designs benefit from suitable semiconductor technologies with advanced integrated and discrete components, compact passives suitable for high-frequency operation, and advanced packaging technologies enabling heterogeneous co-integration. An overview is given in this section, while dedicated chapters of this book provide the details.

1.8.1 Semiconductor Technology

Significant semiconductor and packaging technology advancements have allowed for more complex power management solutions. These developments are crucial for improving portable electronics' efficiency and form factor and increasing the robustness and reliability of challenging applications in the industrial and automotive fields.

There exist different types of PMICs. Voltage regulators or motor drivers can be controllers with an off-chip power stage or fully featured ICs with the driver stage and power transistor integrated on-die. Depending on the voltage ratings, controllers can often be developed using baseline nanometer- Complementary Metal-Oxide-Semiconductor (CMOS) technologies. However, highly integrated PMICs suitable for voltages higher than 5 V require dedicated semiconductor technologies capable of integrating signal and power transistors on the same die. Such a specialized power IC process is called BDC (bipolar-CMOS-DMOS) or high-voltage BiCMOS (bipolar-CMOS). The development of these technologies dates back to the 1980s [11, 12]. BCD combines low-voltage CMOS devices for precise analog circuits and high-density digital signal processing and control. Bipolar transistors are available as part of the mask set and can be used for bandgap voltage references and various analog and protection circuits.

Figure 1.16 gives an overview of a typical device portfolio of a BCD technology. A variety of analog, digital, and high-voltage/power devices with voltage ratings from typically 1.2 V to more than 100 V are available [13–15].

The high-voltage and power part comprises drain-extended and Double-Diffused Metal-Oxide-Semiconductor (DMOS) transistors (see Chapter 3). Some technologies offer lateral insulated-gate bipolar transistors (IGBT).

High voltage and power	Capacitors
DMOS, Drain-extended MOS 5 V, 12 V,..., 80 V	MOS 1.8 V, 3 V 2–5 fF/μm^2 MOM 80 V 0.5 fF/μm^2 Poly-Poly, MIM 1 fF/μm^2
Low voltage transistors	Resistors
Digital CMOS 1.2, 1.8 V, 3 V Analog CMOS 1.8 V, 3 V, 5 V Bipolar vertical, lateral 5 V, 20 V	Poly low-/very-high sheet 5–1000 Ω/□ Metal 70 mΩ/□ Diffusion 100 Ω/□
Diodes	Memory, isolation, ESD
p+/nwell, n+/pwell 5 V, 12 V Zener 6 V, 12 V Schottky, power diodes	EEPROM, Flash, OTP memory Buried-layer, Trench isolation, SOI ESD protection 1.2–80 V

Figure 1.16 Component set of a typical BCD technology.

The low-voltage transistors include digital core and I/O devices and analog CMOS transistors. There are also bipolar junction transistors (BJT), especially isolated vertical bipolar NPN and lateral PNP transistors. Besides regular PN-diodes, Schottky and Zener diodes are available (see Section 3.5). The component set comprises various types of resistors. Very-high sheet resistance is essential to address increasingly strict low-power requirements. The capacitor types support different voltage classes with varying densities in the order of 1 fF/μm^2.

Nonvolatile memory, EEPROM, and Flash are available to store trimming values (e.g., for voltage and current references, oscillator frequency), production IDs, and program code [16]. One-time programmable memory (OTP) operates like a fuse and stores a low number of bits.

Buried layer and trench isolation prevent failures due to parasitic bipolar effects (see Section 3.3.1). Each IC pin must be protected against electrostatic discharge (ESD). The preferred method for protection against over-voltage spikes is through the use of silicon-controlled rectifiers (SCR), while diodes are utilized to protect against voltages that are below ground. As technology advances and transistor dimensions shrink, developing ESD protection measures against ESD is becoming more challenging, primarily due to thinner gate oxides.

Customization of the process for supporting specific application requirements is the driving factor behind BCD technologies [15], rather than the reduction of the lithography node (see also Fig. 1.13). Nevertheless, also BCD technologies see the need to go from state-of-the-art 0.13–0.18 μm BCD technologies to sub-100 nm nodes. This trend is driven by the growing demand for high-density and high-speed digital circuits and interfaces [17].

Some power management designs require a complete set of more than 30 mask levels. Cost can be reduced by leaving out specific masks. Saving two out of 30 masks gives an average cost reduction of 6.7%. An example is the mask for implanting the p-doped base of a vertical NPN bipolar junction transistor. The vertical BJT is still available without that mask, but its current gain reduces from more than 50 to < 10.

Chapters 3 and 4 cover integrated power transistors, diodes, capacitors, and inductors.

1.8.2 Discrete Power Transistors

For applications requiring higher power levels, discrete power transistors are used. They are controlled by PMICs, which can be stand-alone gate drivers or sophisticated designs with complex

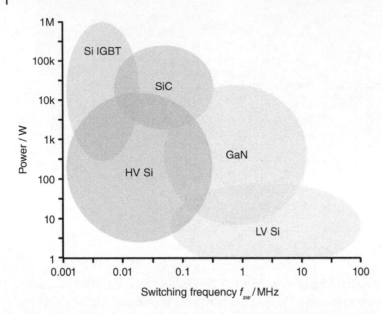

Figure 1.17 Power transistor technologies: output power versus switching frequency.

control. Figure 1.17 maps different technologies for power transistors concerning their power capabilities and achievable switching frequency f_{sw}. Generally, the higher the power level, the lower the switching frequency. This relationship is mainly due to the switching characteristics of the devices. The amount of energy consumed for switching determines the switching losses and limits the switching frequency.

Silicon transistors are most commonly used, covering various voltage, load current, and on-resistance specifications. High-voltage types support power levels up to the 10 kW range and reach switching frequencies of a few hundred kHz. Low-voltage silicon devices ≤ 5 V, such as integrated transistors described in Section 1.8.1, can operate at switching frequencies of more than 100 MHz. They can handle a wide power range from extremely low power levels in the nanowatt domain toward the 100 W domain. A typical scenario is single-digit watt levels at $f_{sw} = 1$ MHz.

Above 350 V, the insulated-gate bipolar transistor (IGBT) combines low on-state voltage drop (low losses) with high blocking voltages. The IGBT performs at high power levels while operating in the 1–10 kHz range.

Wide-bandgap devices, such as gallium nitride (GaN) and silicon carbide (SiC) transistors, offer several advantages over traditional silicon power switches regarding conversion efficiency and solution size. Their high switching frequencies reduce the size of passive components (see Section 1.8.3) and, consequently, the system cost. They also have higher thermal conductivity and can operate at higher temperatures reducing the cooling effort. For these reasons, wide-bandgap devices are becoming increasingly popular in applications such as electric vehicles, renewable energy systems, and data centers, despite their higher component cost and more complex manufacturing processes.

For a deeper understanding of different types of discrete power transistors, refer to Chapter 3.

1.8.3 Passive Components

If we take a look at a typical inductive DC–DC converter like the one shown in Fig. 1.18a), we immediately recognize that the passive components L and C in the power path dominate the size.

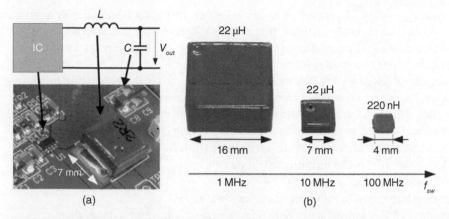

Figure 1.18 Size impact of passive components in DC–DC converters: a) the inductor and capacitor dominate the size of a step-down converter; b) scaling of the coil versus switching frequency.

In Fig. 1.18a), $L = 2.2\,\mu H$ and $C = 22\,\mu F$. Depending on the application, the inductor may be larger, for instance, $22\,\mu H$. In the example of Fig. 1.18a), the IC contains the power stage and the control part of the DC–DC converter. It can handle load currents up to 3A and still fits into a tiny package of approximately $1.5\,mm \times 2.2\,mm$.

The size of the passives depends on various parameters, including their current capabilities and voltage ratings. On the system side, various parameters like load current, input, output voltage, the inductor current ripple, and output voltage ripple determine the inductance and capacitance, respectively. However, the main influence is the PWM switching frequency f_{sw}. Both the inductor and the capacitor are energy-storing components. By decreasing the storing times, the inductance and the capacitance can be reduced. Shorter storing times correspond to higher switching frequencies f_{sw} where $L, C \propto 1/f_{sw}$.

State-of-the-art DC–DC converters operate at frequencies of 1 MHz or single-digit megahertz. Figure 1.18b) illustrates how the size of the inductor scales over frequency starting from $22\,\mu H$ at $f_{sw} = 1$ MHz at a size of 16 mm. Ten times the frequency allows to go for $L = 2.2\,\mu H$, which reduces the size to 7 mm. At $f_{sw} = 100$ MHz, we can use an inductor of 220 nH, which requires a side length of only 4 mm. We see that high switching frequencies are the key to miniaturization.

Why are not all DC–DC converters operating at 100 MHz? There are several significant challenges:

- *Losses*: The fundamental principle of switched-mode conversion in Section 1.3.2 emphasizes the benefit of low losses. However, there are various switching losses, which all scale with frequency. For instance, during every switching transition, there is a short time when the drain current and the voltage drop across the power transistor are nonzero (known as hard-switching). Also, charging of any parasitic capacitance causes charging losses, which also scale proportional to f_{sw}. Techniques like soft-switching and dead time control can eliminate some of these losses enabling faster switching.
- *Speed*: Section 1.8.1 explains that technologies for PMICs are primarily optimized to support various application requirements. Power transistors have excellent area-specific on-resistance and voltage ratings of 100 V and more. However, as a trade-off, these devices are relatively slow. Consider the buck converter of Fig. 1.4 with a voltage conversion ratio of 1/10. At $f_{sw} = 30$ MHz Eqn. (1.3) results in an on-time as low as ~ 3 ns. This timing is challenging given the fact that the switching node sees a full swing between the ground and the input voltage, which can be

large, for instance, $V_{in} = 48$ V. The transition needs to happen within less than 1n s for a pulse duration of $t_{on} \sim 3$ ns. This example confirms that switching frequencies of more than 30 MHz are difficult to achieve, especially at higher input voltages. In any case, it requires fast and robust circuits such as gate drivers and level shifters, covered in Chapter 5.

- *Substrate coupling, noise, EMI*: Steep switching transitions, as outlined in the previous item, cause much disturbance. Capacitive charging currents from spacious components and isolation wells can be in the order of 100 mA, causing substrate debiasing of the PMIC during each switching transition. This effect can lead to malfunctions of not only analog circuits (noise) but also logic gates (bit flips). It also causes electromagnetic interference (EMI) at IC, PCB, and system levels (see Section 12.3). These effects can be mitigated on various levels:
  - assembly and packaging (see Chapter 12),
  - isolation options (trench and junction isolation, see Section 3.3),
  - slew rate control of the power stage reducing the transition slope (slope shaping, see Section 2.5) and resonant operation (sine wave instead of rectangular switching node voltage, see Section 2.9.1).

1.8.4 Co-integration

Advances in 3-D-integration technologies offer minimized power management designs by combining one or multiple ICs with the passive components in a single package [18]. Figure 1.19 shows a DC–DC converter with a package-integrated inductor called system-in-package (SiP). Only the input and output bypass capacitors and two feedback resistors are external. The total solution size is approximately 35 mm².

Co-integration also allows combining a control and gate driver IC with a discrete power transistor in the same package [19]. Advanced 3-D integration can minimize interconnection and integration parasitics, allowing for an optimum balance between cost and performance. This benefit is significant for modern devices that demand smaller form factors and higher power density. Multiple ICs from different technologies can be stacked and connected through copper-filled silicon vias to support low resistance and low inductance interconnect. More details on SiP and heterogenous integration can be found in Section 12.2.5, which covers the physical implementation of PMIC designs.

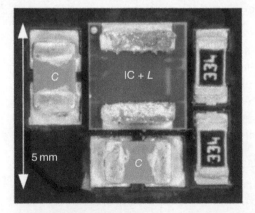

Figure 1.19 A step-down converter consisting of an IC with a package-integrated inductor. Together with the input and output capacitor and two resistors of the feedback divider, the entire converter fits on 35 mm² (TPS82150 Texas Instruments Incorporated).

1.9   A Look at the Market

The PMICs market has seen a massive growth surge in recent years. It has been driven by the trend toward electric-powered transportation and the increasing demand for battery-operated devices, including smartphones, laptops, and wearables. PMICs are essential in major growth areas such as automotive, industrial, mobile communication, IoT, and health care. Power management accounts for approximately 25% of the market volume of analog integrated circuits (IC Insights, 2022), representing a total revenue of more than 100 Billion USD by the mid-2020s. Market research institutions like Yolé Development estimate the total PMIC market to increase from 25–30 Billion USD to more than 56 Billion USD by 2031.

Figure 1.20a) shows the PMICs market by segments. Smartphones and consumer electronics account for the largest market share (45%). This is primarily due to the growing adoption of smartphones and other portable devices that require efficient power management solutions. Industrial applications like factory and building automation account for 18% of the market. Among all segments, automotive has the highest growth rate of about 9% [20] driven by automotive electrification and autonomous driving. Since transportation is the sector that causes the most pollution, adopting electrification will substantially impact the reduction of CO_2 emission. Further applications include ADAS and in-vehicle infotainment. Indicated in Fig. 1.20a), the recent market share of automotive power ICs is 15%. Computing makes up 11% of the business, while telecommunication accounts for 8%. Medical is a minor but still significant contributor, making up 3%.

Among the different types of PMICs, nearly half of the products provide voltage conversion according to the fundamental concepts introduced in Section 1.3, as shown in Fig. 1.20b). They split into linear regulators (LDO) (12% of the PMIC market), DC–DC converters (18%), mainly conventional inductive step-down converters, and multirail power supplies (MRPS, 20%). An MRPS includes several linear regulators and DC–DC converters on the same die or within the same package. Therefore, they are mainly used in space-constraint applications, such as smartphones and ADAS, to supply several independent loads with different voltage and current requirements as described in Section 1.4. Battery management ICs (BMS) take 17% market share. They are driven by the electrification of transport and the trend to more battery-operated power and medical tools.

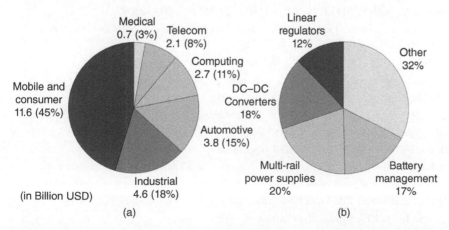

Figure 1.20   Market for PMICs by a) segment and b) by type [21]. The market is expected to grow from about 25 Billion USD (2025) to more than 56 Billion USD in 2031.

In addition to the above market numbers, it is worth noting that many intricate integrated circuits have on-chip power management integrated with all their functional blocks. This level of integration can lead to more efficient power usage and improved overall performance.

References

1 Griffith, D. (2022) Toward Zero: Power consumption trends in low data rate wireless connectivity. *IEEE Solid-State Circuits Magazine*, 14 (4), 51–60, doi: 10.1109/MSSC.2022.3195122.

2 Renz, P., Kaufmann, M., Lueders, M., and Wicht, B. (2019) 8.6 A fully integrated 85%-peak-efficiency hybrid multi ratio resonant DC-DC converter with 3.0-to-4.5 V input and 500 µA-to-120 mA load range, in *2019 IEEE International Solid- State Circuits Conference (ISSCC)*, IEEE, San Francisco, CA, USA, pp. 156–158, doi: 10.1109/ISSCC.2019.8662491.

3 Radhakrishnan, K., Swaminathan, M., and Bhattacharyya, B.K. (2021) Power delivery for high-performance microprocessors - challenges, solutions, and future trends. *IEEE Transactions on Components, Packaging and Manufacturing Technology*, 11 (4), 655–671, doi: 10.1109/TCPMT.2021.3065690.

4 Sturcken, N., Petracca, M., Warren, S., Mantovani, P., Carloni, L.P., Peterchev, A.V., and Shepard, K.L. (2012) A switched-inductor integrated voltage regulator with nonlinear feedback and network-on-chip load in 45 nm SOI. *IEEE Journal of Solid-State Circuits*, 47 (8), 1935–1945, doi: 10.1109/JSSC.2012.2196316.

5 Hayes, C. (2018) Package prepares for AI computing. *Electronic Specifier*. Vicor.

6 Burton, E.A., Schrom, G., Paillet, F., Douglas, J., Lambert, W.J., Radhakrishnan, K., and Hill, M.J. (2014) FIVR - Fully integrated voltage regulators on 4th generation Intel(R) Core(TM) SoCs, in *2014 IEEE Applied Power Electronics Conference and Exposition - APEC 2014*, pp. 432–439, doi: 10.1109/APEC.2014.6803344.

7 Kar, M., Carlo, S., Krishnamurthy, H., and Mukhopadhyay, S. (2014) Impact of process variation in inductive Integrated Voltage Regulator on delay and power of digital circuits, in *2014 IEEE/ACM International Symposium on Low Power Electronics and Design (ISLPED)*, pp. 227–232, doi: 10.1145/2627369.2627637.

8 Ma, D. and Bondade, R. (2010) Enabling power-efficient DVFS operations on silicon. *IEEE Circuits and Systems Magazine*, 10 (1), 14–30, doi: 10.1109/MCAS.2009.935693.

9 Hogbin, E.J. (2015) *ACPI: Advanced Configuration and Power Interface*, CreateSpace Independent Publishing Platform, North Charleston, SC, USA.

10 Dreslinski, R.G., Wieckowski, M., Blaauw, D., Sylvester, D., and Mudge, T. (2010) Near-threshold computing: Reclaiming moore's law through energy efficient integrated circuits. *Proceedings of the IEEE*, 98 (2), 253–266, doi: 10.1109/JPROC.2009.2034764.

11 Andreini, A., Contiero, C., and Galbiati, P. (1986) A new integrated silicon gate technology combining bipolar linear, CMOS logic, and DMOS power parts. *IEEE Transactions on Electron Devices*, 33 (12), 2025–2030, doi: 10.1109/T-ED.1986.22862.

12 Murari, B., Contiero, C., Gariboldi, R., Sueri, S., and Russo, A. (2000) Smart power technologies evolution, in *Conference Record of the 2000 IEEE Industry Applications Conference. 35th IAS Annual Meeting and World Conference on Industrial Applications of Electrical Energy (Cat. No.00CH37129)*, vol. 1, pp. P10–P19, doi: 10.1109/IAS.2000.880960.

13 Chil, M.N., Yang, O.S., Ke, D., Mo, K.J., Kun, L., Tiong, M., Purakh, R.V., and Nair, R. (2016) Advanced 300 mm 130 nm BCD technology from 5 V to 85 V with Deep-Trench Isolation, in

2016 28th International Symposium on Power Semiconductor Devices and ICs (ISPSD), pp. 403–406, doi: 10.1109/ISPSD.2016.7520863.

14 van der Pol, J., Ludikhuize, A., Huizing, H., van Velzen, B., Hueting, R., Mom, J., van Lijnschoten, G., Hessels, G., Hooghoudt, E., van Huizen, R., Swanenberg, M., Egbers, J., van den Elshout, F., Koning, J., Schligtenhorst, H., and Soeteman, J. (2000) A-BCD: An economic 100 V RESURF silicon-on-insulator BCD technology for consumer and automotive applications, in *12th International Symposium on Power Semiconductor Devices & ICs. Proceedings (Cat. No.00CH37094)*, pp. 327–330, doi: 10.1109/ISPSD.2000.856836.

15 Murari, B., Bertotti, F., and Vignola, G. (2002) *Smart Power ICs*, Springer-Verlag, Berlin Heidelberg, New York, 2nd edn.

16 Ma, X., Thant, Z.T., Jiang, H., Mun, N., Chong, K.F., Yeoh, E.E., Nguyen, B.Y., Dong, K., Liao, H.C., Zhou, J., Yang, Z., Ong, S.Y., Koo, J.M., Siah, S.Y., Cuevas, L., Tadayoni, M., Norman, J., Tkachev, Y., Lemke, S., Hsueh, S.H., and Nguyen, H. (2019) Integration of split-gate flash memory in 130 nm BCD technology for automotive applications, in *2019 Electron Devices Technology and Manufacturing Conference (EDTM)*, pp. 318–320, doi: 10.1109/EDTM.2019.8731154.

17 Rose, M. and Bergveld, H.J. (2016) Integration trends in monolithic power ICs: Application and technology challenges. *IEEE Journal of Solid-State Circuits*, 51 (9), 1965–1974, doi: 10.1109/JSSC.2016.2566612.

18 Pendharkar, S. (2016) Smart power technologies enabling power SOC and SIP, in *2016 IEEE Symposium on VLSI Technology*, pp. 1–2, doi: 10.1109/VLSIT.2016.7573394.

19 Pendharkar, S. (2020) Mixed signal and power semiconductor technology for industrial and automotive electronics, in *2020 32nd International Symposium on Power Semiconductor Devices and ICs (ISPSD)*, pp. 10–13, doi: 10.1109/ISPSD46842.2020.9170153.

20 Villamor, A. and Ly, A. (2021) Status of Power IC: Technology, Industry and Trends 2021. *Yolé Development Market & Technology Report*.

21 Villamor, A. and Ly, A. (2022) Power ICs: A $21B market evident in all applications. *Semiconductor Digest*.

2

The Power Stage

Power stages comprise the power transistor, its gate driver, and control circuits, such as the level shifter. Figure 2.1 shows the fundamental configurations with the gate driver indicated by the triangular symbol. We will explore the power transistor, its switching behavior, associated losses, and parasitic effects in this chapter [1–3]. The actual semiconductor devices are covered in Chapter 3 followed by a dedicated chapter on gate drivers, including level shifters and supporting circuits (Chapter 5).

2.1 Introduction

Power transistors control the flow of energy in many different applications. They differ in power and voltage level as well as in switching characteristics. While the transistor is used as a switch in most power electronics circuits, there is also a need for linear operation (transistor in saturation region) like in a linear voltage regulator. The associated current through the power transistor can be large as there is always a certain power level. Hence, its on-resistance is a crucial parameter. The lower the transistor's on-resistance, the lower the voltage drop across and the power loss in the transistor. Due to the higher carrier mobility, n-channel devices are preferred over p-channel transistors. Nevertheless, for lower power levels, p-channel transistors are well-suitable.

Depending on the location of the power switch, a low-side switch (Fig. 2.1a) and a high-side switch (Fig. 2.1b) can be distinguished. There are two versions of a high-side switch: an n-type and a p-type device.

The power stage's output connects the transistor and the load. It is referred to as a switching node with the associated voltage V_{sw}, referred to ground. If the switch is turned on, current flows through the load, and the load gets activated. In the case of the low-side switch, the switching node V_{sw} is pulled to ground (zero) if the switch is turned on. If the switch is off, the load pulls V_{sw} to the supply, usually a battery V_{bat}. For the high-side switch in Fig. 2.1b), V_{sw} changes the opposite way.

In the case of a n-channel high-side switch, a particular challenge arises if the transistor turns on. As the goal is to reach $V_{sw} = V_{bat}$ (assuming negligible low on-resistance with zero voltage drop), the gate potential of the switch needs to be higher than $V_{sw} = V_{bat}$ by at least the threshold voltage of the power transistor. Consequently, a gate supply greater than V_{bat} is required. A charge pump, a boost converter, or a bootstrap circuit can generate this supply. Despite this effort, the n-type power transistor is preferred for integrated power stages for higher power levels (e.g., currents higher than 100 mA) as it gives lower on-resistance per chip area (cost). Nearly all external discrete power transistors are n-type devices for the same reason.

Design of Power Management Integrated Circuits, First Edition. Bernhard Wicht.
© 2024 John Wiley & Sons Ltd. Published 2024 by John Wiley & Sons Ltd.

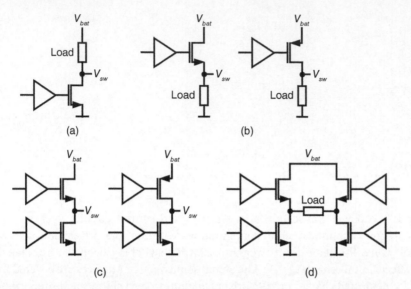

Figure 2.1 Power stage configurations: (a) low-side switch, (b) high-side switch (n- and p-type version), c) half-bridge (n- and p-type version), and d) full-bridge (H-bridge) (n-type version).

The implementation of the high-side switch will take more work. Nevertheless, there are also reasons to use stand-alone high-side switches:

(1) *Fail-safe-behavior*: If there is a failure causing a short-to-ground from the switching node to ground, the short bypasses the load, and the load (e.g., a motor) does not get activated. The same failure on the low-side switch would unintentionally trigger the load. See also Section 6.5 for details on short-to-ground and how this condition can be detected.

(2) *Easier interconnection*: If electronic modules are built using a ground-level chassis or a PCB ground plane, all loads can be tied to this ground connection. One V_{sw} connection will be required. For the low side, two wires are needed per load.

A half-bridge is formed by combining a low-side and a high-side switch, as shown in Fig. 2.1c), again with the option for a p-type and n-type transistor at the high side.

Two half-bridges form a full bridge, also called H-bridge. The high-side switches can be n-channel devices, as shown in Fig. 2.1d) or p-channel transistors.

There are countless application areas for power transistors, including control of motors and drives, valves, solenoids and relays, switched-mode power supplies, and LED drivers. Higher power levels require discrete power transistors placed externally, off-chip (see Chapter 3).

2.2 On-Resistance and Dropout

If used as a switch, the transistor operates in its triode region, characterized by its fundamental I–V characteristics:

$$I_D = \beta \left((V_{GS} - V_{th}) V_{DS} - \frac{V_{DS}^2}{2} \right)$$

$$\approx \beta (V_{GS} - V_{th}) V_{DS} \quad \text{with} \quad \beta = \frac{W}{L} \mu_o C_{ox} \tag{2.1}$$

The approximation assumes that V_{DS} is minimal if the transistor is fully switched-on and the square of a small value is even smaller. Hence, the whole term $\frac{V_{DS}^2}{2}$ is usually negligible. Equation (2.1) allows to derive the on-resistance R_{DSon} of the transistor. In the case of the n-channel transistor:

$$R_{DSon} = \frac{V_{DS}}{I_D} = \frac{1}{\beta \left(V_{GS} - V_{th} \right)} = \frac{1}{\frac{W}{L} \mu_o C_{ox} \left(V_{GS} - V_{th} \right)} \tag{2.2}$$

Two parameters can be modified by design to achieve a specified R_{DSon}, the W/L-ratio and the gate–source voltage V_{GS}, also called gate overdrive. All other parameters are technology-specific and cannot be changed by design (μ_o, C_{ox}, V_{th}). In analog circuit design, the term ($V_{GS} - V_{th}$) is called overdrive, which is, in the true sense, also the gate overdrive in switch operation. As V_{GS} of the switch is usually much larger than V_{th}, V_{GS} itself is referred to as overdrive.

The gate overdrive should be as large as possible. The typical value for on-chip power stages is 5 V, but it depends on the available supply domains in the design. Technologies at the 0.35 μm node and above also provide devices suitable for 8 and 12 V gate voltage. Sub-130 nm technologies are often restricted to 3.3 V. Discrete power transistors usually require more than 10 V gate overdrive to provide the nominal on-resistance.

Once the gate–source voltage is chosen, the required W/L-ratio can be determined from Eqn. (2.2):

$$\frac{W}{L} \geq \frac{1}{R_{DSon} \mu_o C_{ox} \left(V_{GS} - V_{th} \right)} \tag{2.3}$$

Equation (2.3) needs to be fulfilled under all conditions, including worst-case parameters for V_{GS} and V_{th}. The transistor length should be the minimum possible limit L_{min} as given in the design manual. The width of the transistor is then $W = \frac{W}{L} L_{min}$. The on-resistance is indirectly proportional to the transistor's active area $W \cdot L$. Low resistance can lead to very large transistors that occupy a significant die area.

The design target is usually the overall resistance that can be measured between the pin of the switching node V_{sw} and the ground or supply pin. The resistance R_{met} of the metal interconnections needs to be included in R_{DSon}. On-chip metallization, bond wire, lead frame resistance, chip-scale package interconnection, etc., contribute to R_{met}. See Chapter 12 on physical implementation, in particular, the electrical model in Section 12.2.3 and in Fig. 12.5. R_{met} can be higher than the actual on-resistance of the power transistor.

Why is the on-resistance important? The fundamental reason is the resistive conduction power losses P_{cond} in the switch, also referred to as static losses or DC losses:

$$P_{cond} = I_D^2 R_{DSon} \tag{2.4}$$

Note that R_{DSon} includes the metallization resistance R_{met}. The smaller the on-resistance, the lower the losses at a given current.

Another reason for low R_{DSon} is the dropout voltage. The typical example is a linear regulator with a p-channel power transistor, shown in Fig. 2.2a) (see also Section 7.2). Dropout V_{do} is the voltage drop across the transistor (between drain and source) at a given I_D for worst-case gate overdrive V_{GS}. If the battery gets discharged over time, a small dropout virtually extends the battery time as illustrated in Fig. 2.2b). If the dropout is low, the regulator's output voltage will stay constant for a longer time until the battery needs to be recharged or replaced.

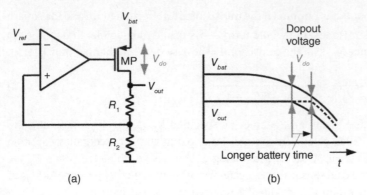

(a)   (b)

Figure 2.2 a) Definition of the dropout as the minimum source–drain voltage in a linear regulator; b) importance of low dropout for longer battery time.

2.3 Parasitic Capacitances

Earlier in this chapter, we have seen that power transistors can get large in area. Large dimensions are associated with large parasitic capacitances at the transistor, mainly oxide capacitances at the gate and junction capacitances at the drain and source regions. All capacitances are lumped into the equivalent circuit shown in Fig. 2.3. Integrated transistors, including their Spice models, consider the capacitances C_{GS}, C_{GD}, and C_{DS} between the corresponding terminals as indicated in Fig. 2.3. In contrast, the capacitances in discrete power switches are modeled by the input capacitance C_{iss}, the output capacitance C_{oss}, and the reverse capacitance C_{rss}. All three capacitances are shown in Fig. 2.3. C_{iss}, and C_{oss} are defined for $V_{DS} = 0$ and $V_{GS} = 0$, respectively. Based on Fig. 2.3, we can derive a correspondence between C_{GS}, C_{GD}, C_{DS} and C_{iss}, C_{oss}, C_{rss}:

$$C_{in} = C_{iss} = C_{GS} + C_{GD} \quad (V_{DS} = 0) \tag{2.5}$$

$$C_{out} = C_{oss} = C_{GD} + C_{DS} \quad (V_{GS} = 0) \tag{2.6}$$

$$C_{rss} = C_{GD} \tag{2.7}$$

$$C_{GS} = C_{iss} - C_{rss} \tag{2.8}$$

$$C_{DS} = C_{oss} - C_{rss} \tag{2.9}$$

$$C_{GD} = C_{rss} \tag{2.10}$$

Figure 2.3 Parasitic capacitances in a power transistor.

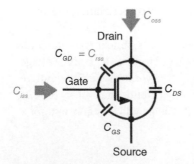

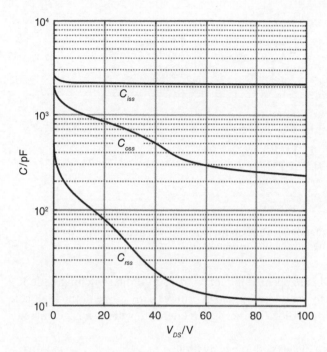

Figure 2.4 Transistor capacitances versus drain–source voltage V_{DS} for a discrete 100 V power transistor (BSC070N10NS5, Infineon Technologies, values extracted from data sheet).

Example 2.1 *The datasheet for a discrete power transistor lists C_{iss} = 2100 pF, C_{oss} = 340 pF, C_{rss} = 16 pF. Determine the corresponding values for C_{GS}, C_{GD}, C_{DS}.*

$$C_{GS} = C_{iss} - C_{rss} = 2100\,\text{pF} - 16\,\text{pF} = 2084\,\text{pF} \sim 2.1\,\text{nF}$$

$$C_{DS} = C_{oss} - C_{rss} = 340\,\text{pF} - 16\,\text{pF} = 324\,\text{pF} \tag{2.11}$$

$$C_{GD} = C_{rss} = 16\,\text{pF} \tag{2.12}$$

Figure 2.4 shows an example of how the capacitances depend on the drain–source voltage V_{DS}. C_{oss} and C_{rss} incorporate the junction capacitance at drain and source. Hence, they are highly non-linear dependent on V_{DS}. Higher voltages result in lower capacitance as the depletion region gets wider.

Various losses are associated with the transistor capacitances; see Section 2.7. Section 2.6 discusses the correlation with gate charge.

2.4 The Body Diode

Due to the PN-junctions at the source and drain, Complementary Metal-Oxide-Semiconductor (CMOS) power transistors have a body diode parallel to their channel. Figure 2.5 shows the location of the body diode for the two half-bridge configurations of Fig. 2.1c). This diode is always present and cannot be avoided. Even if the symbol of the switch does not show the diode like in Fig. 2.3, we need to imagine the diode to be present at any time.

Conventional circuit design always assumes body diodes to be reverse biased, i.e., they are not considered (except for leakage and junction capacitance). In contrast, it is commonplace that body

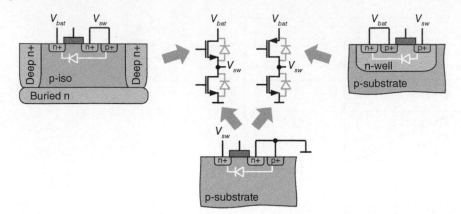

Figure 2.5 Body diodes in half-bridge configurations. The body diode appears between each power transistor's drain and source node. It is formed by the drain-to-bulk junction.

diodes in power stages get activated. This happens if inductive loads pull the switching node V_{sw} below ground (low-side body diode in forward bias) or above supply (body diode of the high-side switch conducting). In such a case, the body diode can act as a free-wheeling diode for the inductor current.

Nevertheless, there are drawbacks associated with the body diode:

- The power switch may not get turned off. Even if there is no conducting channel anymore (controlled by the gate voltage), the body diode may still conduct in forward bias. It would be the case if V_{sw} exceeds V_{bat} and can be prevented by a back-to-back combination of two switches, as discussed in Section 2.11.
- The PN-junction is the base-emitter junction of a parasitic bipolar transistor. If activated, unwanted currents into the substrate, coupling to sensitive circuits, and latch-up may occur in addition to higher power dissipation ($V_F I_{load}$). We discuss these effects in Section 3.3.1.
- The reverse recovery effect draws reverse recovery charge Q_{rr}, which impacts switching behavior and associated losses. We will get back to this effect shortly.

There are benefits, too. Besides the purpose of a free-wheeling diode, the body diode can be utilized as a general-purpose power diode. Its current capability and voltage ratings are identical to the actual power transistor. Imagine the gate terminal of the isolated n-channel transistor to the left in Fig. 2.5 shorted to its source–bulk node. This way, the actual transistor will always remain off while the power diode is present between the drain and the source. The isolation avoids the anode and cathode being tied to the ground or supply.

Reverse recovery happens if the body diode goes from forward bias to a blocking state. Figure 2.6a) shows the scenario of a half-bridge during a switching transition with an inductive load and $I_{load} > 0$. We start with the case that both transistors are off, usually during a dead time between switching transitions (see Section 2.5.5). In this situation, the body diode of the low-side switch delivers the load current into the inductor. As the body diode is in forward bias, the switching node potential is one diode forward voltage below ground, i.e., $V_{sw} = -V_F$. Now the high-side switch turns on and pulls V_{sw} toward V_{bat}. At the same time, the high-side switch takes over the current from the low-side body diode. The diode current goes down, starting from I_{load}, as illustrated in Fig. 2.6b). The diode current I does not directly approach zero. It instead flows in the negative direction for a limited time until the diode finally blocks with zero current.

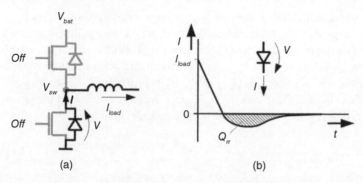

Figure 2.6 a) Half-bridge with an inductive load indicating the body diode with its current and voltage. b) Reverse recovery charge Q_{rr} flows if the voltage V across the body diode changes to negative values. The diode current I goes negative for a limited time until the diode blocks in reverse bias with zero current.

Reverse recovery charge Q_{rr} flows into the depletion region of the body diode to block the diode:

$$Q_{rr} = \int_0^\infty I\,dt. \tag{2.13}$$

Modeling the reverse recovery charge Q_{rr} is complex because it depends on various parameters [4]. It is mainly determined by the dead time and the current amount I. The dead time between low-side turn-off and high-side turn-on determines the body diode conduction time in forward bias. Q_{rr} also scales with the amount current I that flows through the body diode during that time. The current I is equal to the current in the inductor, which, on average, is equal to the load current I_{load}. During circuit design, Q_{rr} can be extracted from a transistor-level simulation using device models that include the reverse recovery effect [5].

As Q_{rr} is delivered by the high-side switch from V_{bat}, reverse recovery losses $P_{rr} = V_{bat}Q_{rr}f_{sw}$ occur, see Eqn. (2.41) in Section 2.7.

For a load current with opposite polarity than shown in Fig. 2.6, body diode conduction and reverse recovery losses would occur at the high side.

2.5 Switching Behavior

We will now investigate the switching behavior of a power stage. The gate current and the transistor capacitances determine the timing. Since they are not constant during the transition and are rather voltage-dependent, the below equations are an estimation. It is instructive to derive the major mechanism that determines the turn-on and turn-off transition of power switches.

2.5.1 Resistive Load

We consider the low-side switch of Fig. 2.1a) with a purely ohmic load resistance R_{load}. Figure 2.7a) shows the turn-on switching behavior. The drain current I_D starts to flow when the gate–source voltage V_{GS} reaches the threshold voltage V_{th} at time t_1. Until t_1 the gate–source capacitance C_{GS} gets charged (due to the large drain voltage, we neglect the gate–drain capacitance), and T_1 can be approximated by

$$T_1 = \frac{V_{th}C_{GS}}{I_{gate}}. \tag{2.14}$$

T_1 is sometimes called the precharge phase.

At this moment, the drain current, which flows through R_{load}, causes a voltage drop across R_{load}. It decreases the drain–source voltage V_{DS}. The drain current builds up as the voltage drop increases. At t_2, the drain–source voltage reaches nearly zero, and the drain current is at its maximum, which is the steady-state load current I_{load}, i.e., $I_{D,max} = I_{load}$. As the switch has a finite on-resistance, the remaining voltage drop across the switch at t_2 is ($I_{load} \cdot R_{DSon}(t_2)$). Typically, $R_{DSon} \ll R_{load}$ and $I_{load} \approx V_{bat}/R_{load}$. $R_{DSon}(t_2)$ is given by Eqn. (2.2) with $V_{GS} = V_{GS}(t_2)$. We can assume that the power transistor operates in the saturation region for most of the transition time, hence

$$V_{GS}(t_2) = V_{th} + \sqrt{\frac{2I_{load}}{\beta_{MP}}} = V_{th} + \frac{2I_{load}}{g_m}. \tag{2.15}$$

Equation (2.15) shows two ways to calculate $V_{GS}(t_2)$. The first one is based on the transfer curve of the MOS transistor in its saturation region. The right-hand term in Eqn. (2.15) uses the small-signal transconductance parameter g_m.

The V_{GS} slope within T_2 is much smaller than during T_1, as shown in Fig. 2.7a). This is because as soon as the V_{DS} transition starts, also C_{GD} gets discharged in parallel to the charging of C_{GS}. The current into C_{GD} can be approximated based on the switching node transition (V_{DS} transition),

$$I_{GD} = C_{GD}\frac{dV_{GD}}{dt} \approx -C_{GD}\frac{dV_{DS}}{dt}. \tag{2.16}$$

For a first-order assessment, we neglect that C_{GD} depends on V_{DS}. With a steep slope dV_{DS}/dt at the switching node, the current I_{GD} will be large. As the gate driver provides roughly a constant gate current I_{gate}, most of the gate current will contribute to I_{GD}. Consequently, compared to the time

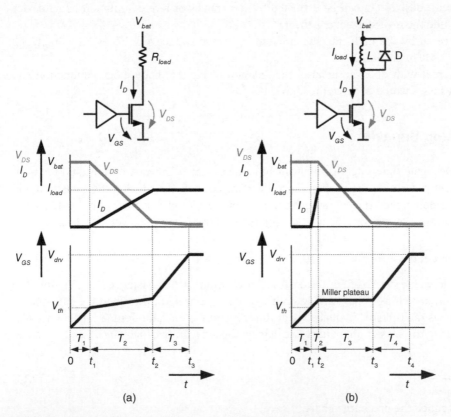

Figure 2.7 Turn-on switching behavior: a) resistive load; b) inductive load.

before t_1 where the full gate current flows into C_{GS}, a significantly smaller current will remain to charge C_{GS}, resulting in a smaller slope during T_2.

Miller Effect

The much larger current that flows into C_{GD} during T_2 can be explained by the Miller effect. The Miller effect is known from circuit theory. Even though the power stage is switching, it can be considered an (inverting) analog amplifier during the transition. The input voltage is V_{GS}, and the output is V_{DS}. A capacitance C between the input and the output of an inverting amplifier with a voltage gain of $A = dV_{DS}/dV_{GS} < 0$ can be transferred into an equivalent capacitance $C_{eq} = |A| \cdot C$ at the input, referred to ground. If we assume that I_{GD} of Eqn. (2.16) is the input current of the "amplifier," the equivalent input capacitance is

$$C_{eq} = \frac{I_{in}}{dV_{in}/dt} = \frac{-C_{GD}dV_{out}/dt}{dV_{in}/dt} = |A| \cdot C_{GD}. \tag{2.17}$$

Equation (2.17) assumes that the transition of V_{DS} is approximately equal to V_{DG}, which than both correspond dV_{out}/dt. As a result, C_{eq} appears in parallel to C_{GS}, which is much smaller and practically negligible. The large value of C_{eq} explains why there is only a small change of V_{GS} during T_2.

The duration T_2 can be estimated by the discharge time of C_{GD} if we assume a full swing of V_{bat}:

$$T_2 = \frac{C_{GD}V_{bat}}{I_{gate}} = \frac{Q_{GD}}{I_{gate}} \tag{2.18}$$

As the capacitances depend on the actual terminal voltages, it is more precise and convenient to use the gate charge Q_{GD} to calculate T_2. If the switching node capacitance (C_{DS}) is large, it will also impact the transition time as it can be considered in parallel to C_{DG}. Placing an additional capacitor in parallel to C_{GD} is an easy way to reduce the transition speed at the switching node to reduce noise, ringing, and electromagnetic interference (EMI). Section 12.3 explains how the transition slope influences the EMI noise.

After the transition of V_{DS} has finished, the Miller effect is not present anymore. The gate current I_{gate} charges both the gate–source and the gate–drain capacitance (which appear in parallel as C_{GD} has its drain node nearly at ground). The gate transition is finished if V_{GS} reaches its maximum, which is the gate drive voltage V_{drv}, determined by the gate driver, see Fig. 2.7a). During T_3, the on-resistance of the switch reaches its target value according to Eqn. (2.2),

$$T_3 = (C_{GS} + C_{GD})\frac{V_{drv} - V_{GS}(t_2)}{I_{gate}} \tag{2.19}$$

with $V_{GS}(t_2)$ given in Eqn. (2.15).

Figure 2.8a) shows the switching curves at turn-off. The behavior is similar to the turn-on case, except that the sequencing occurs in reversed order. The phases T_1 to T_3 can be calculated using the same expressions in this section.

2.5.2 Inductive Load

Most power management applications do not have purely resistive loads. Instead, inductive loads are common. Real applications will consist of a mix of resistive, capacitive, and inductive load components. As the inductive load component dominates the switching, we study the switching behavior for that case.

Figure 2.7b) depicts the turn-on transition for a purely inductive load L. The inductor conducts a load current I_{load}, which is assumed to be constant in the considered time frame. The transistor is off for $t < 0$, and the diode D acts as a free-wheeling diode that carries the inductor current I_{load}.

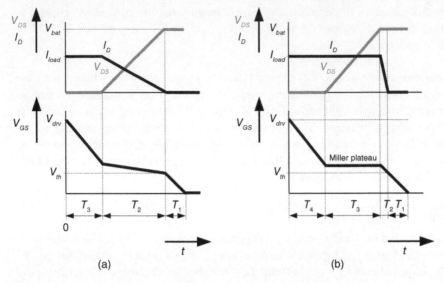

Figure 2.8 Turn-off switching behavior: a) resistive load; b) inductive load.

The diode can be a dedicated free-wheeling diode or, in a half-bridge configuration, the body diode of the high-side switch. The situation for $t < 0$ is similar to Fig. 2.6a), just that the inductor current has the opposite polarity.

Why do we need the diode D? Without the diode, the inductor current would not see a conducting path when the transistor is off. The current would flow into an infinite impedance, and the switching node voltage would theoretically rise to infinity. In practical cases, a finite series resistance of the inductor would limit the voltage spike, but it would still reach values far above V_{bat}. The drain–source voltage would exceed its maximum ratings and destroy the power transistor.

The diode D clamps the drain voltage one diode voltage V_F above V_{bat}. We neglect V_F for simplification, which is a valid approximation as we usually have $V_{bat} \gg V_F$. Hence, the transition starts at $V_{DS} = V_{bat}$. We also assume that the diode has zero reverse recovery charge. We will cover the case for $Q_{rr} \neq 0$ shortly. After a few switching cycles, a load current I_{load} is present in the system, stored in the inductor. This fact represents the main difference to a purely resistive load. It causes the differences in switching behavior outlined below.

During T_1, the gate–source voltage rises identically to the case of a resistive load shown in Fig. 2.7a). T_1 is given by Eqn. (2.14). At time t_1 V_{GS} reaches the power transistor's threshold voltage V_{th}. The transistor starts to conduct. It will pull more and more current out of the inductor. At the same time, this will reduce the current in the free-wheeling diode. At t_2, the transistor will carry the full load current, i.e., $I_D = I_{load}$. We note that V_{DS} remains at V_{bat} during T_2. T_2 is called the current transition or di/dt phase during switching. The drain–source voltage does not change before the drain current carries the full load current I_{load}. This condition is reached at t_2 when the gate–source voltage reaches the value $V_{GS}(t_2)$ required for a drain current $I_D = I_{load}$. During T_2 the power transistor's gate voltage swing is $V_{GS}(t_2) - V_{th}$ with $V_{GS}(t_2)$ given by Eqn. (2.15), which can be derived from the drain current. T_2 is the time required to charge C_{GS} by the voltage swing $V_{GS}(t_2) - V_{th}$. It corresponds to the gate charge Q_{ct} needed during the current transition, resulting in

$$T_2 = \frac{C_{gate}}{I_{gate}} \left(V_{GS}(t_2) - V_{th} \right) = \frac{Q_{ct}}{I_{gate}}. \tag{2.20}$$

Q_{ct} can be determined from the gate charge diagram of the power transistor (see Section 2.6). This diagram is provided in the datasheet of discrete devices or can be generated from a simulation for integrated power stages with low effort.

During T_3 the drain–source voltage changes from V_{bat} to $I_{load} \cdot R_{DSon}(t_3)$, which is close to zero. $R_{DSon}(t_3) = R_{DSon}(t_2)$ is given by Eqn. (2.2) with $V_{GS} = V_{GS}(t_2)$ from Eqn. (2.15). T_3 is called the voltage transition (dV/dt) phase or the Miller phase. As outlined in Section 2.5.1, the Miller effect occurs during the V_{DS} transition. As the full load current has already developed in the power switch, the voltage gain is larger than for a resistive load. Hence, the Miller effect is so strong that the gate–source voltage stays virtually flat at the so-called Miller plateau, indicated in Fig. 2.7b). Figure 5.19 (Gate Drivers) shows a transistor-level simulation with a noticeable Miller effect.

The duration T_3 of the Miller phase is identical to T_2 of Eqn. (2.18) for the resistive load case, described in Section 2.5.1:

$$T_3 = \frac{C_{GD} V_{bat}}{I_{gate}} = \frac{Q_{GD}}{I_{gate}} \tag{2.21}$$

As the gate–drain charge Q_{GD} is usually more significant than the charge Q_{ct}, the current transition is typically faster than the duration of the voltage transition (Miller).

The final phase T_4 is identical to T_3 for the case of a resistive load. Similar to Eqn. (2.19), T_4 can be calculated:

$$T_4 = (C_{GS} + C_{GD}) \frac{V_{drv} - V_{GS}(t_2)}{I_{gate}} = \frac{Q_{cond}}{I_{gate}} \tag{2.22}$$

Q_{cond} is the gate charge for the transistor to go into full conduction with its targeted on-resistance.

Like for a resistive load, the turn-off transition in Fig. 2.7b) happens in reversed order. The phases T_1 to T_4 can be calculated using the expressions above for the turn-on case. After turning off the switch, the inductor keeps the load current flowing and pulls the switching node above V_{bat}. This transition happens during phase T_2. The free-wheeling diode gets forward biased and clamps V_{DS} at $V_{bat} + V_F$, one forward voltage above supply. For simplicity, the forward voltage drop is not shown in Fig. 2.7. The diode conducts the total load current. The inductor current is said to commutate into the free-wheeling diode (from the power switch). Without a free-wheeling diode, the drain voltage would approach high values, ideally infinity. This mechanism also nicely explains why the diode got its name *free-wheeling* diode. Eventually, the energy in the inductor will be dissipated such that V_{DS} reaches V_{bat}. However, many power management architectures take advantage of the fact the inductor keeps the current during the off-time of the switch.

2.5.3 Reverse Recovery and Switching Node Capacitance

We now consider the case for finite reverse recovery charge Q_{rr} of the body diode. Section 2.4 explains that reverse recovery occurs if the body diode changes from forward to reverse polarity. Hence, the turn-off behavior does not change in the presence of Q_{rr}. The turn-on behavior with reverse recovery, is shown in Fig. 2.9. In comparison to Fig. 2.7b), the drain current exceeds the load current by the maximum reverse recovery current I_{rr} as defined in Fig. 2.6b). A current peak I_{pk} occurs while the charge Q_{rr} is delivered until the drain current approaches I_{load}. The gate voltage and the times t_1 to t_4 remain approximately as shown in Fig. 2.7b). During turn-on, the switching node capacitance gets discharged, mainly the transistor's drain–source capacitance $C_{DS} = C_{oss}$. The discharge current adds to the reverse recovery current and the current peak I_{pk}. Both I_{rr} and I_{oss} are dissipated through the transistor's conducting channel and contribute to the switching losses.

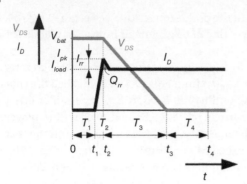

Figure 2.9 Turn-on switching behavior for an inductive load including reverse recovery charge Q_{rr} of the body diode.

In a half-bridge, the switching node sees a rising transition of the switching node voltage when the high-side device turns on. The high-side switch delivers I_{rr} into the body diode of the low-side transistor (or into the free-wheeling diode). It also supplies the current to charge the switching node capacitance. Both currents are delivered from V_{bat} but do not contribute to the load current. Hence, the reverse recovery charge (Q_{rr}) and the switching node capacitance (C_{oss}) cause losses.

2.5.4 Power and Gate Loop Inductance

Figure 2.10 shows the power stage with an inductive load and major parasitic components. The distributed R-L-C network can be approximated sufficiently by lumped components. The parasitic inductances $L_{D'}$, $L_{S'}$, and $L_{G'}$ account for the transistor package. They are often negligible for fully integrated power stages due to very short interconnections. See also Section 12.2 on IC packages. There is also an inherent gate resistance $R_{G'}$. In addition, resistive connections at the drain and source contribute to the on-resistance (not shown in Fig. 2.10 for clarity). The connections outside the power transistor are also associated with parasitic inductances, as indicated in Fig. 2.10. There are more inductance effects like the connections of diode D (or another power switch, e.g., the

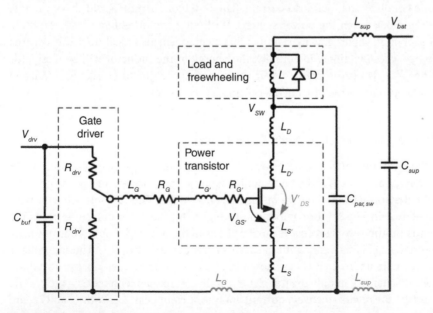

Figure 2.10 Power stage with parasitics.

high-side transistor, see Section 2.5.5), which have been omitted for clarity. The switching node capacitance $C_{par,sw}$ comprises the junction capacitance of diode D, the parasitic capacitance of the inductor L, and the parasitic capacitance between the switching node and ground. As it occurs inside the transistor, the drain–source capacitance $C_{DS} = C_{oss}$ is not part of $C_{par,sw}$, but combining them is often a suitable approximation.

Power Loop Inductance
The parasitic L-C components in the power path form multiple resonant tanks. According to Fig. 2.10, this includes the supply buffer capacitor C_{sup}, the parasitic switching node capacitance $C_{par,sw}$ and, not shown, the drain–source capacitance of the transistor. Oscillations are triggered at each switching transition. Regarding the switching node, $C_{par,sw}$ form a series resonant tank with the equivalent power loop inductance $L_p = L_{S'} + L_{D'} + L_S + L_D$. No surprise, L_p influences the switching behavior. It sees a voltage drop at turn-on while the drain current builds up. It reduces the inner drain–source voltage $V_{DS'}$, discharging the drain–gate capacitance C_{DG}. Consequently, part of the gate current recharges C_{DG} and does not flow into C_{GS}. Therefore, the current transition during T_2 in Fig. 2.7b) slows down. The current in L_p at turn-off results in a voltage overshoot of the drain–source voltage initiated during T_2, as shown in Fig. 2.11. According to $V_L = L\,di/dt$, the peak voltage increases with larger power loop inductance and faster current transitions. The transition speed increases for larger load currents if the switching times remain the same.

The induced voltage can exceed the maximum drain–source voltage ratings of the transistor. The blocking-voltage rating of the transistor may have to be 2–3 times larger than V_{bat}. This requirement is why fast-switching power stages demand considerable effort to achieve a low-inductive design. As L_p forms a resonant tank with the switching node capacitance $C_{par,sw}$, the voltage overshoot is usually followed by ringing, damped by the parasitic resistances in the power loop. This ringing also contributes to EMI noise (see Section 12.3). The voltage peak can be reduced by lowering the driving strength of the gate driver by inserting a gate resistor R_G or by controlling the gate current (Section 5.9).

Gate Loop Inductance
For integrated power stages with the power switch and the driver on the same die, gate loop inductance does not play a role (but be aware of the connection to the external buffer capacitor C_{buf}. However, in discrete power electronics, the driver can be at a significant distance from the power transistor. The interconnections between them and the wiring within the driver and the transistor result in a gate loop inductance $L_{loop} = L_{G'} + L_G$. There is also an inner parasitic gate resistance $R_{G'}$ in addition to an external gate resistor. R_G may be placed to reduce the transition speed at the switching node to reduce EMI, $V_{DS'}$ voltage overshoot, and ringing (at the expense of higher switching losses). All L-C-R components along the gate loop form a resonant tank. So the gate voltage will see a similar behavior, as shown in Fig. 2.11 for $V_{DS'}$. If the gate resistor is appropriately chosen, it prevents the gate voltage from exceeding its maximum rating. Details are covered in the gate driver chapter; see Sections 5.9 and 5.10.

Because L_{loop} slows down the current transition time, we can redefine Eqn. (2.20) by adding a delay due to the inductive current build-up in L_{gate},

$$T_2 = \frac{Q_{ct}}{I_{gate}} + L_{loop}\frac{I_{gate}}{V_{drv} - V_{th}}. \tag{2.23}$$

Equation (2.23) assumes that L_{loop} sees a voltage drop of $(V_{drv} - V_{th})$ during the current transition (phase T_2 in Figs. 2.7 and 2.11).

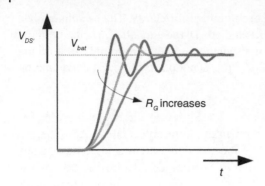

Figure 2.11 Turn-off switching behavior including power loop parasitics for a varying gate resistor R_G.

2.5.5 Half-Bridge with Inductive Load

A half-bridge allows to accomplish pulse-width modulation as outlined in Section 1.3.2 and, in more detail, in Chapter 10. We modify the essential power stage of Fig. 2.7b) to derive the switching behavior by adding a high-side transistor to the existing low-side transistor. The configuration is shown in Fig. 2.12a). We define the load current to flow out of the power stage. We assume the output voltage to be constant and buffered by an output capacitor C_{out}. Both switches control the switching node voltage V_{sw}. For $V_{sw} = V_{bat}$, the inductor sees a positive voltage drop V_L, so the inductor current ramps up. Vice versa, if V_{sw} is at low potential, the current decreases. In steady state, the mean value of the inductor current I_L is equal to the load current I_{load}.

A dead time prevents cross-conduction at both transition directions between low-side and high-side conduction. During each dead time, the low-side and the high-side switch are off. Dead times are required to avoid cross currents at large drain–source voltage, which exceed the safe operating area (SOA, see Section 3.4) and lead to damage to the power transistor. In addition, cross-conduction would also add excessive losses. Achieving the optimum dead time is challenging because it depends on various parameters, including the supply voltage and load current. Control loops can be installed as explained in Section 2.8.

For $I_L > 0$, the body diode of the low-side switch carries the load current during the dead time. Therefore, the body diode forms the free-wheeling path for the inductor current, pulling the switching node one forward voltage drop below ground potential, resulting in $V_{sw} = -V_F$. During the dead time, body diode conduction losses $P_{bd} = V_F I_{load}$ occur. After the dead time t_{dHL} in Fig. 2.12b), the low-side switch turns on. It shorts the body diode, resulting in a very low, but still negative, voltage drop between drain and source ($R_{DSon} \cdot I_L$). While I_L decreases, the voltage drop gets smaller

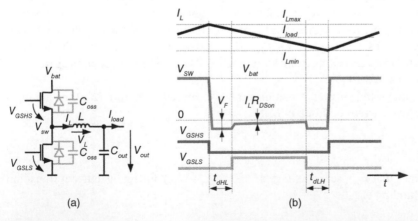

        (a)   (b)

Figure 2.12 Switching behavior of a half-bridge with an inductive load: a) schematic; b) waveforms.

as indicated in Fig. 2.12b). If the low-side switch turns off, the inductor current commutates again into the low-side body diode during dead time t_{dLH} and causes losses P_{bd}.

The on-time of the high-side transistor defines the on-time t_{on} of the pulse-width modulation (PWM) cycle defined in Fig. 1.4b) in Section 1.3.2. This way, the half-bridge in Fig. 2.12a) may constitute a fundamental step-down DC–DC converter as covered in Chapter 10. If conducting, the high-side switch brings energy into the system. Hence, the duty cycle is defined as $D = t_{on}/T$ with the PWM switching period T (see also Fig. 1.4). Often the duty cycle is defined as a digital control signal with $D = 1$ during t_{on} (high-side on) and $D = 0$ during t_{off} (high-side off). Setting the on- and off-times allows controlling the inductor current, for instance, in DC–DC converters but also to vary the passage in solenoid valves.

2.5.6 Switching Trajectories

If we plot the drain current over the drain–source voltage, we can mark two essential points, the on- and off-state of the switch. As indicated in Fig. 2.13a), these points correspond to the

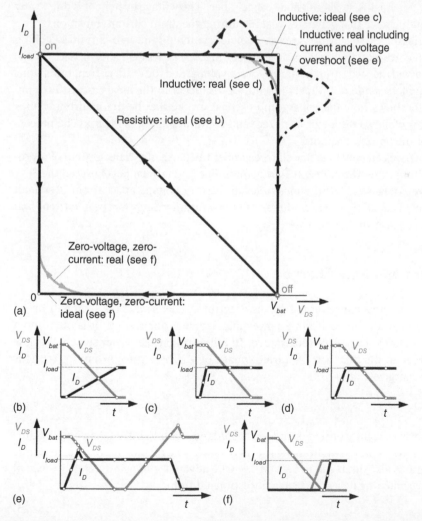

Figure 2.13 a) Switching trajectories for b) resistive switching, c) ideal and d) real inductive switching, e) inductive switching including current and voltage overshoot, and f) zero-voltage switching (ZVS).

axis intersections $(V_{DS} = 0, I_D = I_{load})$ (on) and $(V_{DS} = V_{bat}, I_D = 0)$ (off) in the ideal case with zero on-resistance. For simplicity, we neglect the finite on-resistance, which would move the drain–source voltage in on-state to $V_{DS} = I_{load}R_{DSon}$. The relationship between the drain current and the drain–source voltage transients during switching can be plotted as trajectories in the I–V diagram.

The resistive load case, repeated for convenience in Fig. 2.13b), leads to a straight connection between the on- and off-state (in reality to a transition within a guard band around the shown diagonal).

For an inductive load, the trajectory runs along the outer boundary. Starting in the off-state, the current builds up while the drain–source voltage stays constant at $V_{DS} = V_{bat}$. At $I_D = I_{load}$, the voltage drops down to zero (or $I_{load}R_{DSon}$), which marks the on-state. There is no difference between the off-on and on-off trajectories in this ideal scenario corresponding to Fig. 2.13c). In an actual design, the current and voltage transients may show some overlap, as indicated in Fig. 2.13d). Hence, the trajectories may do touch the outer point in the top-right corner $(V_{DS} = V_{bat}, I_D = I_{load})$. If we also consider current and voltage peak, as descrived in Sections 2.5.3 and 2.5.4, we can draw the trajectories based on Fig. 2.13e). The current overshoot accounts for the reverse recovery charge and additional charge due to parasitic switching node capacitance. The voltage overshoot is due to any parasitic power loop inductance that generates peaking during the drain current transition.

Finally, the trajectories for the so-called case of zero-voltage switching and zero-current switching (ZVS/ZCS) are shown, corresponding to the waveforms in Fig. 2.13f). As described in Section 2.9, these concepts aim to avoid the overlap of the drain–source voltage and the drain current transitions such that the switching losses due to I–V overlap entirely disappear in the ideal case. At turn-on, for instance, Fig. 2.13f) shows how the voltage approaches zero before the drain current ramps up. Hence, the trajectory follows the x-axis until zero and continues along the y-axis. The process happens vice versa for the on-off transition.

Observing Fig. 2.13a), we can conclude that the area under the particular trajectory curve is proportional to the switching losses (Section 2.7). Any point in Fig. 2.13a) must never exceed the SOA of the particular power transistor. For the most relevant case of voltage and current overshoot (Fig. 2.13e)), there must be sufficient margin toward the maximum ratings for drain current and drain–source voltage, see SOA in Section 3.4.

2.6 Gate Current and Gate Charge

Section 2.3 shows that the capacitances at the transistor terminals determine the switching speed and the switching behavior of the device in a power management application. It is often more convenient to consider charge instead of capacitance. To fully turn on the power switch, the gate capacitance gets charged to the final gate overdrive voltage of V_{drv} by a gate current $I_{gate}(t)$. This corresponds to a gate charge of

$$Q_{gate} = C_{gate} \cdot V_{drv} = \int_0^{t_4} I_{gate}dt, \tag{2.24}$$

assuming that the transistor is fully turned on at t_4, in accordance with Fig. 2.7. Likewise, the same gate charge Q_{gate} needs to be withdrawn from the gate at turn-off. The relation between gate charge and gate driver strength is also discussed in Section 5.5. One advantage of using charge instead of capacitance is that the switching delay can be readily estimated from

$$t_4 = Q_{gate}/I_{gate}. \tag{2.25}$$

Table 2.1 Typical gate charge values.

Device type	Gate charge
Integrated power transistor	100 pC...<1 nC
Discrete power transistor	40 nC
GaN	<10 nC
IGBT (1200 V)	50...100 nC

Table 2.1 provides an overview of gate charge values for typical power devices. While discrete power devices have a gate charge of at least 10 nC up to micro coulombs, integrated power transistors reach values of a few hundred pico coulombs. Fast-switching GaN transistors achieve very low gate charge values. Chapter 3 covers more details on gate charges for different types of power devices.

Example 2.2 *Calculate the switching delay for a transistor with a gate charge of 10 nC. Assume a constant gate current of 1 mA.*

$$t = Q_{gate}/I_{gate} = 10\,nC/1\,mA = 10\,\mu s \tag{2.26}$$

How does the delay change for $I_{gate} = 1$ A?

$$t = Q_{gate}/I_{gate} = 10\,nC/1\,A = 10\,ns \tag{2.27}$$

From Example 2.2, we keep in mind that typical gate currents are in the order of $I_{gate} = 1$ A, which leads to appropriate transition delays in the order of tens of nanoseconds. We return to this value if we discuss gate drivers in Chapter 5. Most gate drivers are hard-switching, which connects the gate to V_{drv} by a switch with a finite resistance. As the voltage across this resistance decreases if V_{gate} goes up, the gate current, provided by the driver, decreases. This results in a current peak, as illustrated in Fig. 2.14, which shows the switching transition from Fig. 2.1 with the gate current

Figure 2.14 Turn-on transition with gate currents for a hard-switching and a current-source gate driver (current mode).

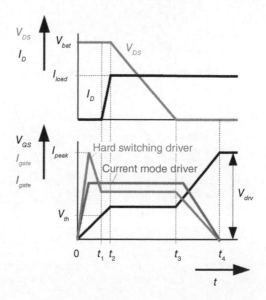

added. A second type of driver, a current-source gate driver, avoids this current peak as the driver output stage operates as a current source. In both driver cases, the area under the curve is equal and identical to the total gate charge (see the integral in Eqn. (2.24)), see also Section 5.5.

If we provide a constant gate current I_{gate}, the circuit in Fig. 2.15a) allows to extract the gate charge Q_{gate} in correlation to the gate voltage $V_{GS}(t)$. The inductive load, as shown in Fig. 2.7b), is modeled by a constant current source I_{load}, while the battery voltage determines V_{DS} as a parameter. Based on the relationship $Q_{gate} = I_{gate} \cdot t$, the time axis of the $V_{GS}(t)$ plot in Fig. 2.7 (or Fig. 2.8 in case of turn-off) can be converted into charge values as depicted in Fig. 2.15b). Moreover, we can assign portions of gate charge to the different phases of the switching transition. From the start until the voltage transition at the switching node, the gate charge mainly flows into C_{GS}, corresponding to a quantity Q_{GS}. During the voltage transition (Miller plateau), the gate charge flows in first order into C_{GD}, hence referred to as Q_{GD}. The remaining gate charge is delivered in the precharge phase after the Miller plateau until the gate reaches its final voltage V_{drv}. The gate charge portions Q_{GS}, Q_{GD} and the total gate charge Q_{gate} (or Q_{tot}) are usually listed in the power transistor's datasheet.

The setup in Fig. 2.15a) can be used for bench characterization as well as to extract Q_{gate} from simulation. Simulation is helpful in the case of integrated power transistors, which usually do not have any gate charge values provided in the design manual. For multilevel gate driving, the current source can also be tied to negative gate potential voltage such that V_{GS} ramps up with respect to the grounded source terminal.

Figure 2.15c) provides an example plot of V_{GS} versus Q_{gate} for an actual discrete power transistor. There is a small dependency on the drain–source voltage due to the voltage dependency of the parasitic transistor capacitances (see Fig. 2.4).

Example 2.3 *Extract the values of Q_{GS}, Q_{DG}, and Q_{tot} for the transistor in Fig. 2.15c) at $V_{DS} = 50$ V. Based on the definition of Fig. 2.15b), we find $Q_{GS} = 10$ nC, $Q_{DG} = 7$ nC, and a total gate charge of $Q_{tot} = 30$ nC, which is reached at $V_{GS} = 10$ V.*

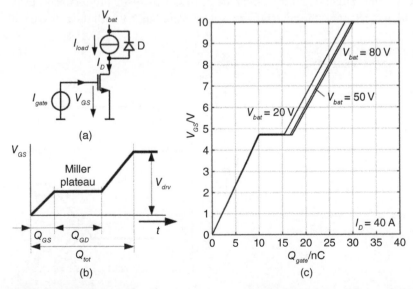

Figure 2.15 a) Test and simulation setup for gate charge extraction; b) gate–source voltage V_{GS} as a function of the gate charge; c) $V_{GS}(Q_{gate})$ for a discrete 100 V power transistor (BSC070N10NS5, Infineon).

2.7 Losses

There are various power losses in a power stage, mainly associated with the power transistor, its parasitics, and the finite resistance of interconnections. Figure 2.16 shows the switching waveforms over one switching cycle and the associated losses.

2.7.1 Conduction Losses

In the introduction, we have found that switched-mode operation ideally has no losses (see Section 1.3.2). A real switch has a finite on-resistance R_{DSon}, which causes resistive conduction losses P_{cond} during the on-state. We can refine Eqn. (2.4) if the switch has a certain on-time t_{on},

$$P_{cond} = \frac{1}{T} \int_0^T I_D^2 R_{DSon} dt = I_{load}^2 R_{DSon} f_{sw} t_{on} \tag{2.28}$$

with the switching frequency f_{sw} and the duty cycle $D = t_{on}/T = t_{on} f_{sw}$, defined in Eqn. (1.2) in Section 1.3.2. For better visibility, in Fig. 2.16, the transition times are exaggerated concerning the on and off times. Instead of power, the loss energy may be convenient to use, as it is independent of the switching frequency f_{sw}:

$$E_{cond} = I_{load}^2 R_{DSon} t_{on} \tag{2.29}$$

Concerning the switching waveforms of Figs. 2.7 and 2.8, conduction losses occur mainly during T_3 and T_4, for the resistive and inductive load case, respectively.

2.7.2 Conduction Losses in Case of Current Ripple

In many power stages, the power switch connects to an inductor L and changes the voltage V_L across the inductor. Consequently, the inductor current will ramp up and down depending on the polarity of V_L. A typical scenario is shown in Fig. 2.12 for a half-bridge. If the high-side switch gets activated for an on-time t_{on} a positive inductor voltage $V_L = V_{in} - V_{out}$ causes a positive slope of the inductor current, as shown in Fig. 2.17. The resulting current ripple is described by

$$\Delta I_L = L \frac{V_L}{\Delta t}. \tag{2.30}$$

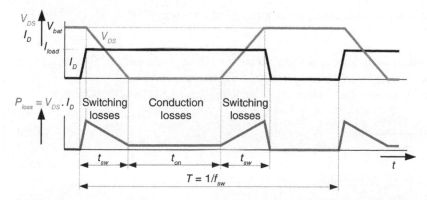

Figure 2.16 Conduction and switching losses at a power transistor for an inductive load during one switching cycle.

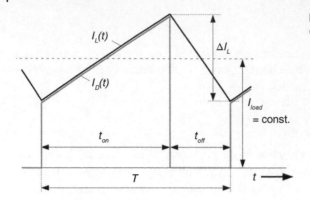

Figure 2.17 Current ripple ΔI_L increases the conduction losses in a power transistor.

The total drain current in the high-side switch becomes

$$I_D = I_{load} + \Delta I_L \left(\frac{t}{t_{on}} - \frac{1}{2} \right). \tag{2.31}$$

The expression in Eqn. (2.31) is only valid during t_{on}, else the drain current is zero. In fact, during t_{off}, the low-side switch will carry the inductor current. For more details, refer to the fundamentals of inductive DC–DC converters in Chapter 10 (Section 10.1).

The current ripple increases the root-mean-square (RMS) value of the total current in the power transistor, and greater conduction losses are generated compared to the case if only a constant load current flows in the switch. The additional losses must be addressed for power management designs with large current ripple. If we consider Eqn. (2.31) as the sum of two terms A and B, we find

$$I_D^2 = A^2 + 2AB + B^2. \tag{2.32}$$

We can determine the conduction losses by inserting Eqn. (2.32) into Eqn. (2.28):

$$P_{cond} = R_{DSon} f_{sw} \int_0^{t_{on}} \left(A^2 + 2AB + B^2 \right) dt. \tag{2.33}$$

The term $(2AB)$ is symmetrical with respect to the horizontal axis. Hence, the integral will be zero, and we only need to consider the terms A^2 and B^2. Solving the integral yields

$$P_{cond} = \left(I_{load}^2 + \frac{\Delta I_L^2}{12} \right) R_{DSon} f_{sw} t_{on}. \tag{2.34}$$

Compared to the case without a current ripple in Eqn. (2.28), the current ripple ΔI_L increases the conduction losses.

Example 2.4 *How much do the conduction losses increase if the ripple is set to $\Delta I_L = 0.4 I_{load}$, and what is the maximum increase for $\Delta I_L = 2 I_{load}$ (boundary conduction)?*

If we denote the conduction losses with ripple $P_{cond,r}$, dividing Eqns. (2.34) and (2.28) gives

$$\frac{P_{cond,r}}{P_{cond}} = \frac{I_{load}^2 + \frac{(0.4 I_{load})^2}{12}}{I_{load}^2} = 1 + \frac{0.16}{12} = 1.013. \tag{2.35}$$

In boundary conduction with $\Delta I_L = 2 I_{load}$, the inductor current will reach zero and move up to reach a maximum of $I_{Lmax} = 2 I_{load}$. In this case,

$$\frac{P_{cond,r}}{P_{cond}} = \frac{I_{load}^2 + \frac{(2 I_{load})^2}{12}}{I_{load}^2} = 1 + \frac{4}{12} = 1.333. \tag{2.36}$$

While in the first case, the losses are only 1.3% higher, the boundary conduction case sees a significant loss increase by 33%.

The ripple will be significant, especially for small inductance values such as for on-chip inductors (see Section 4.2), and the impact on conduction losses needs to be considered. Large ripple also supports fast transient response in DC–DC converters at the expense of higher conduction losses. A particular application is multi-phase DC–DC converters, as discussed in Section 10.14.

2.7.3 Dynamic Losses

Several losses scale dependent on the switching frequenccy and, hence, are called dynamic or switching losses.

Transition Losses

The switching curves in Fig. 2.16 as well as the switching trajectories in Section 2.5.6 reveal that the transistor voltage V_{DS} and current I_D are nonzero at the same time. Consider the inductive load case. During T_2 and T_3 (di/dt and dv/dt phase, see Figs. 2.7 and 2.8), switching losses P_{sw} occur, also referred to as transition losses or I–V overlap losses, shown in Fig. 2.16. To determine the switching losses, we multiply $V_{DS}(t)$ and $I_D(t)$, which yields the instantaneous power $P_{sw}(t)$. We can derive the switching loss energy by integrating $P_{sw}(t)$, resulting in the turn-off/turn-on energy

$$E_{sw} = E_{turn\text{-}on/off} = \int_{t_1}^{t_3} P_{loss}(t). \tag{2.37}$$

The times t_1 and t_3 set the start and end point of the transition, as shown in Figs. 2.7 and 2.8 for turn-on and turn-off, respectively. The transition losses span a triangle with a transition duration of $t_{tr} = T_2 + T_3$. Hence, the turn-off energy can be approximated by the area of the triangle:

$$E_{sw} = E_{turn\text{-}on/off} = \frac{1}{2}V_{bat}I_{load}t_{tr,on/off} \tag{2.38}$$

The transition time will be different between on and off transitions. The turn-on energy can be 5–10 times larger than the loss at turn-off.

Adding the on and off contributions according to Eqn. (2.38) and dividing by $T = 1/f_{sw}$ results in the power loss related to the switching transition with nonzero drain current and drain–source voltage:

$$P_{sw} = \frac{1}{2}V_{bat}I_{load}(t_{tr,on} + t_{tr,off})f_{sw} \tag{2.39}$$

There are also losses P_{oss} related to the switching node capacitance. While these losses are associated with the power transistor itself, there are more loss components in the setup of Fig. 2.7. To account for significant additional switching losses, we include reverse recovery losses P_{rr}, body diode conduction losses P_{bd}, and gate driver losses P_{gate} as described later.

Switching Node Capacitance Losses

If we just consider the transistor itself, the output capacitance C_{oss} represents the switching node capacitance. In the configuration of Fig. 2.7b), it gets charged by the inductor's load current while the transistor is off, which can be considered lossless. At turn-on, the switching node capacitance gets discharged through the transistor channel. The energy E_{oss} stored in the capacitance C_{oss} is converted into heat and contributes losses:

$$P_{oss} = E_{oss}f_{sw} = \frac{C_{oss}}{2}V_{bat}^2 f_{sw} \tag{2.40}$$

Reverse Recovery Losses

These losses are due to the additional charge Q_{rr} that is required to bring the diode D from conduction (during freewheeling) to the blocking state at turn-on of the power transistor in Fig. 2.7b). In a half-bridge configuration with a typical scenario of a load current that flows out of the switching node, as shown in Fig. 2.6, the body diode of the low-side switch shows a reverse recovery effect. In either case, the losses can be calculated by

$$P_{rr} = V_{bat}Q_{rr}f_{sw}.$$ (2.41)

Body Diode Conduction Losses

Let us again refer to Fig. 2.7b). If the transistor is off, the inductor current will cause the switching node to rise and exceed the battery voltage V_{bar}. This way, the diode D gets forward-biased. We have seen for a half-bridge configuration that diode D will be the body diode of a high-side switch, which behaves as a regular PN-diode. The inductor current flows through the diode, which acts as a freewheeling diode. In typical designs, the current will be so large that it keeps flowing as long as the power transistor is off. With a typical forward voltage $V_F = 0.7\,\text{V}$, this may contribute significant losses. If we assume the transistor to be off during $t_{off} = (T - t_{on})$ (see Fig. 1.4b)), the body diode conduction losses are

$$P_{bd} = I_{load}V_F t_{off} f_{sw}.$$ (2.42)

Gate Driver Losses

The gate driver must deliver the gate charge Q_{gate} to turn on the power transistor. At turn-off, the gate charge is withdrawn and dissipated. Hence, the gate charge causes an additional loss component, according to

$$P_{gate} = Q_{gate}V_{drv}f_{sw} = C_{gate}V_{drv}^2 f_{sw},$$ (2.43)

and similarly

$$E_{gate} = Q_{gate}V_{drv} = C_{gate}V_{drv}^2,$$ (2.44)

where V_{drv} is the final gate–source voltage at turn-on. The term $Q_{gate}f_{sw}$ accounts for the average current delivered by the gate supply. As the gate driver is basically a CMOS inverter, its cross current will add further frequency-dependent losses. More details on gate drivers and associated losses are covered in Chapter 5.

2.7.4 Supply Current Related Losses

Some circuit blocks require a continuous bias current, for example, a level shifter (Section 5.12) or analog circuits in a control loop of a DC–DC converter (see Chapter 10, e.g., Fig. 10.5). The losses are easily calculated,

$$P_{bias} = V_{DD}I_{bias}$$ (2.45)

where V_{DD} is the internal supply voltage that delivers the bias current I_{bias}. At light load, if the output power and other losses are low, P_{bias} dominates and will impact the power efficiency of a switched-mode power supply.

2.7.5 Total Losses

Adding up all losses discussed above yields a generic loss definition:

$$P_{loss} = P_{cond} + P_{sw} + P_{oss} + P_{rr} + P_{bd} + P_{gate} + P_{bias}$$ (2.46)

Table 2.2 Parameters of Example 2.5.

V_{bat}	I_{load}	$R_{DS,on}$	t_{tr}	C_{oss}	Q_{rr}	Q_{gate}	V_{drv}
Discrete							
100 V	10 A	60 mΩ	10 ns	23 nF	89 nC	30 nC	10 V
Integrated							
20 V	500 mA	500 mΩ	5 ns	100 pF	1 nC	1 nC	5 V

Except for the static losses (conduction losses) $P_{cond} + P_{bias}$, all losses have in common that they are frequency dependent. As they are related to the switching transition, they are often referred to as switching losses. This way, switching losses is not a unique term as it sometimes only means transition losses, any other loss component, or the sum of all (switching) frequency-dependent losses.

Equation (2.46) is not even complete. Passive components and interconnects contribute additional losses (see Sections 4.1 and 4.2 in Chapter 4). Also, losses due to parasitic bipolar structures and capacitive coupling as outlined in Section 3.3 may occur and not be negligible.

Example 2.5 *Calculate the loss contribution and the total losses for the parameters given in Table 2.2 for a duty cycle of $D = 25\%$. We consider a discrete and an integrated power transistor. We assume equal transition times t_{tr} for turn-on and turn-off. The switching frequency is 100 kHz and 1 MHz for the discrete and the integrated device, respectively. One of the major advantages of integrated power transistors is their higher operating frequency. For this reason, we will also calculate the losses if we further increase the switching frequency to 10 MHz. We assume that the body diode of the transistor forms the freewheeling path. The bias current losses P_{bias} can be neglected because of the large load current I_{load}.*

We need to derive the on- and off-times to calculate the losses. With the period time $T = 1/f_{sw}$, we get $t_{on} = D/f_{sw}$ and $t_{off} = T - t_{on}$. For the discrete transistor, the times are $t_{on} = 2.5$ μs, $t_{off} = 7.5$ μs. Because of the higher switching frequencies, the integrated switch values are ten and a hundred times smaller, respectively.

Table 2.3 lists the calculated losses. It nicely demonstrates the different orders of magnitude for discrete and integrated transistors. While the power level is much higher for discrete devices, the smaller dimensions of the integrated transistors allow higher switching frequencies. The integrated switch still yields lower absolute losses, even for 10 MHz. However, this needs to be seen in relation to absolute power. Power converters express this by the power efficiency; see Eqn. (1.1). In our case, the power levels ($V_{bat} \cdot I_{load}$) are much different, 1 kW and 2 W for the discrete and integrated transistor, respectively. Despite the higher losses, the discrete design has higher efficiency.

2.7.6 Switch Sizing for Minimum Losses

Integrated devices allow the designer to tailor their size to achieve minimum losses. While the conduction losses P_{cond} according to Eqn. (2.28) reduce with increasing $\frac{W}{L}$-ratio of the power transistor, its gate charge losses P_{gate} get more significant because the gate capacitance increases. The same applies if we consider the energy (Eqns. (2.29) and (2.44)) instead of power because the expressions get independent of the switching frequency. A typical design aims to balance both loss contributions

Table 2.3 Losses of Example 2.5.

P_{cond}	P_{sw}	P_{oss}	P_{rr}	P_{bd}	P_{gate}	P_{tot}
Discrete ($f_{sw} = 100$ kHz)						
150 mW	1 W	115 mW	890 mW	5.25 W	30 mW	7.44 W
Integrated ($f_{sw} = 1$ MHz)						
31.3 mW	50 mW	20 mW	20 mW	263 mW	5 mW	389 mW
Integrated ($f_{sw} = 10$ MHz)						
31.3 mW	500 mW	200 mW	200 mW	263 mW	50 mW	1.24 W

to achieve minimum total loss. Hence, the design goal is to find the optimal $\frac{W}{L}$-ratio. With $L = L_{min}$, the optimum transistor width W must be found. For this purpose, both loss energies can be plotted versus $\frac{W}{L}$. Since $\frac{W}{L}$ relates to the on-resistance R_{DSon} (see Eqn. (2.2)), it can be convenient to draw the losses as a function of R_{DSon}, as shown in Fig. 2.18. The $1/x$–relationship suggests a double-log scale.

For the given parameter set in Fig. 2.18, the loss minimum corresponds to approximately $R_{DSon} = 100$ mΩ with a total loss energy of $E_{tot} = E_{cond} + E_{gate} = 2 \times 10^2$ pJ. For $f_{sw} = 2$ MHz, the total power loss calculates to $P_{tot} = E_{tot}f_{sw} = 400$ µW. Equations (2.2) and (2.3) allow to derive the transistor width $W = 1136$ µm that corresponds to the optimum R_{DSon}.

The above consideration is a simplified approach. It neglects the influence of the gate–drain and drain–source capacitances. Another uncertainty is the nonlinear dependence of the gate capacitance on the transistor voltages. Nevertheless, for most designs, this approach points to an approximate sizing for minimum losses, which simulations can further optimize.

Example 2.6 *Derive an analytical expression for the optimum on-resistance and calculate the exact value of the optimum R_{DSon} and W for the parameters given in Fig. 2.18.*

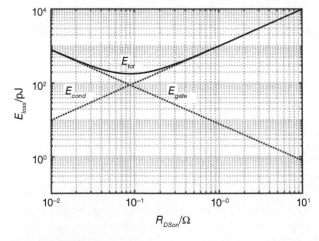

Figure 2.18 Loss energy E_{loss} versus on-resistance. The plot shows the loss components E_{cond}, E_{gate} and the total loss $E_{tot} = E_{cond} + E_{gate}$. Parameters: $V_{GS} = 5$ V, $V_{th} = 0.6$ V, $\mu_o = 500$ cm^2/(Vs), $L = 300$ nm, $I_D = 100$ mA, $t_{on} = 100$ ns, $C_{ox} = 1.5$ fF µm^{-2}

The total loss energy $E_{tot} = E_{cond} + E_{gate}$ follows an expression in the general form $E_{tot} = A \cdot R_{DSon} + B/R_{DSon}$ with $A = I_D^2 t_{on}$ and $B/R_{DSon} = W(R_{DSon}) \cdot L \cdot C_{ox} \cdot V_{GS}^2$. From Eqn. (2.2) we get $W = L/(\mu_o C_{ox} R_{DSon}(V_{GS} - V_{th}))$ and, finally, $B = L^2 V_{GS}^2/(\mu_o(V_{GS} - V_{th}))$.

The optimum corresponds to the on-resistance for which the first derivative gets zero:

$$\frac{\partial E_{tot}}{\partial R_{DSon}} = A - B \cdot R_{DSon}^{-2} = 0 \rightarrow R_{DSon} = \sqrt{\frac{B}{A}}$$

Assuming $V_{GS} - V_{th} \approx V_{GS}$ yields

$$R_{DSon} = \frac{L}{I_D}\sqrt{\frac{V_{GS}}{\mu_o t_{on}}}.$$

In Fig. 2.18, the value of the optimum on-resistance calculates to $R_{DSon} = 94.9$ m$\Omega \approx 95$ mΩ. This corresponds to a transistor width $W = 9583$ µm ≈ 9600 µm and $W/L = 31942 \approx 32000$.

2.7.7 Losses in a Half-Bridge

Equation (2.46) applies to a single transistor power stage. In the case of a half-bridge configuration, we find two power devices, which result in additional losses. Figure 2.12 shows the switching waveforms for a half-bridge. The pulse-width modulation (PWM) duty cycle defines the ratio between on-time t_{on} (high-side transistor turned on) and switching period T. Hence, $t_{off} = T - t_{on}$. This relationship neglects the dead time at each transition, which is usually significantly shorter than T. For the conduction losses, we obtain

$$P_{cond} = P_{cond,HS} + P_{cond,LS} \tag{2.47}$$

with

$$\begin{aligned} P_{cond,HS} &= I_{load}^2 R_{DSon,HS} D \\ P_{cond,LS} &= I_{load}^2 R_{DSon,LS}(1 - D) \end{aligned} \tag{2.48}$$

If both transistors have equal resistance, the total conduction loss is given by Eqn. (2.4), repeated here for convenience:

$$P_{cond} = I_{load}^2 R_{DSon} \tag{2.49}$$

Although the relationship between the high-side and low-side transistors is not affected by the duty cycle, their individual losses are impacted. If our design has a specific duty cycle that falls within a narrow range, we can use Eqn. (2.48) to adjust the on-resistance of both power devices. This will result in similar conduction loss and offer more flexibility in thermal design, PCB, and IC top-level layout.

If the current ripple is not negligible, we can apply the relationship of Eqn. (2.34), valid for a single switch, and expand Eqn. (2.49):

$$P_{cond} = \left(I_{load}^2 + \frac{\Delta I_L^2}{12}\right) R_{DSon} \tag{2.50}$$

For simplicity, let us keep the assumption of identical transistors with equal on-resistance, equal capacitances, and identical switching transition times $t_{tr} = t_{tr,on} = t_{roff}$. As each transistor undergoes one on- and one off-transition, the switching losses (see Eqn. (2.39)) are

$$P_{sw} = 2V_{bat}I_{load}t_{tr}f_{sw}. \tag{2.51}$$

The output capacitances of both transistors can approximate the switching node capacitance. They get charged and discharged during one switching cycle. One transition incorporates lossless charging via the power inductor, contributing to the load current (e.g., the charging current of $C_{oss,HS}$). The other transition results in hard-switching with corresponding losses, hence,

$$P_{oss} = \frac{C_{oss,HS} + C_{oss,LS}}{2} V_{bat}^2 f_{sw} \sim Q_{oss} V_{bat} f_{sw} \tag{2.52}$$

with the total output charge $Q_{oss} = C_{oss} V_{bat}$ that accounts for both devices.

Since we have two power devices, the gate charge nearly doubles compared to a single power transistor. Assuming identical devices with equal gate charge yields

$$P_{gate} = 2Q_{gate} V_{drv} f_{sw}. \tag{2.53}$$

The gate charge for the high-side transistor is usually delivered from a bootstrap supply or from a charge pump, which adds additional losses (see Sections 5.11 and 8.9).

From Fig. 2.12, we can conclude that only the body diode of the low-side transistor contributes reverse recovery and forward conduction losses. The body diode conducts during the dead times at both transitions per cycle. Hence, the reverse recovery losses and body diode conduction loss of Eqns. (2.41) and (2.42) count twice:

$$P_{rr} = 2Q_{rr} V_{bat} f_{sw}$$

$$P_{bd} = I_{load} V_F (t_{dHL} + t_{dLH}) f_{sw} \tag{2.54}$$

Not only the body-diode losses P_{bd} but also the reverse recovery losses P_{rr} depend on the dead time (see Section 2.4). Therefore, both losses are also known as dead time-related losses.

Example 2.7 *Calculate the losses in a half-bridge for both device parameter sets listed in Table 2.2 (Example 2.5). We assume the same duty cycle of D = 25%, i.e., one transistor conducts 25% of the period, the other device during the remaining time. We further assume equal dead times t_{dead} of 20 ns (discrete) and 10 ns (integrated) in between each transition. Body diode conduction losses occur only during this short dead time. For this reason, we can expect much smaller losses due to the body diode.*

Table 2.4 shows the calculated results. We can observe how the body diode losses during freewheeling (off-time) for the single transistor case corresponds to higher conduction losses in the half-bridge because one of the transistors always conducts over the full period (except during the short dead time). The half-bridge brings lower losses if the on-resistance is small enough to overcompensate the forward voltage drop of the body diode. For example, the half-bride with discrete transistors has more than hundred times lower body diode conduction losses, while the conduction losses increase by a factor of four.

Table 2.4 Losses of Example 2.7.

P_{cond}	P_{sw}	P_{oss}	P_{rr}	P_{bd}	P_{gate}	P_{tot}
Discrete (f_{sw} = 100 kHz)						
600 mW	2 W	230 mW	1.78 W	28 mW	60 mW	4.70 W
Integrated (f_{sw} = 1 MHz)						
125 mW	100 mW	40 mW	40 mW	7 mW	10 mW	322 mW
Integrated (f_{sw} = 10 MHz)						
125 mW	1 W	400 mW	400 mW	70 mW	100 mW	2.10 W

2.8 Dead Time Generation

Equation (2.54) shows that dead time-related losses in the power stage are a considerable loss contribution to the power stage. In a bridge configuration, the low-side and the high-side devices must never be turned on simultaneously. Losses are not desired, but there is even more of a risk that cross-conduction may violate the SOA of the switches (see Section 3.4) and destroy the devices. Adding a dead time ensures that one device gets turned off before the other turns on, Fig. 2.19. Typical dead times are 10 ns for switched-mode power supplies with up to 1 MHz switching frequency. Motors, which run at up to 20 kHz pulse-width modulation (PWM) frequency, have a typical dead time of 1 μs with a trend toward shorter dead times to reduce losses. With larger dead time the losses increase because, during the dead time, the low-side body diode usually conducts the load current. Also, the body diode's reverse recovery losses increase with longer dead time.

The optimum dead time depends on the operating conditions, mainly defined by V_{bat} and the load current I_{load} delivered by the power stage, and the parasitic components such as the switching node capacitance, determined by the capacitances of the transistors and the interconnection traces (IC, PCB). This relationship is explored analytically in [5]. Therefore, several dead time generation concepts have been proposed. The most popular ones are (1) fixed dead time, (2) adaptive dead time, and (3) predictive dead time generation.

2.8.1 Fixed Dead Time

This simple method utilizes a fixed delay to ensure a dead time if there is an output transition. Figure 2.20 shows two circuit examples in which a chain of digital inverters generates a fixed delay. While this method is straightforward to implement, it suffers from variations, including the inverter delay process variation. Consequently, there needs to be a worst-case delay margin, which leads to a considerable dead time in the nominal case, i.e., to a nonoptimized switching behavior with increased switching losses and limitations on the switching frequency (minimum duty cycle). Trimming can be applied to reduce the delay margin. It can be installed by adding multiplexers to the inverter delay chain and selecting the appropriate multiplexer output.

Figure 2.19 Gate signals with dead time. The active device turns off before the other turns on with a dead time t_{dead} in between.

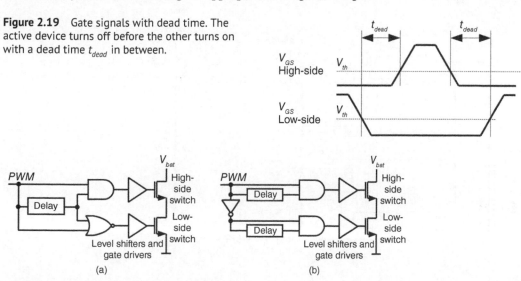

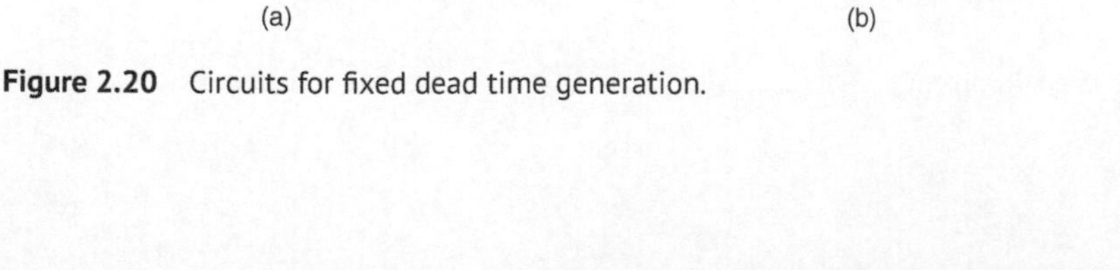

Figure 2.20 Circuits for fixed dead time generation.

2.8.2 Adaptive Dead Time

The method is similar to the break-before-make circuit of Section 5.8.2 (see Fig. 5.23). Figure 2.21 shows how the off-state is sensed, and the respective power transistor is not turned on before a valid off-state signal is detected. As for the fixed dead time case, the dead time can be adapted to a minimal delay with a much shorter delay margin. However, the dead time is determined by the propagation delay through the cascaded driver stages, which limits the minimum achievable dead time. The high-side and low-side sensing should be taken directly at the gate of the power devices. However, to achieve shorter dead times, the turn-off state can be sensed along the cascaded driver stages (see Section 5.8) at one of the first stages. This approach allows a shorter dead time because the signal does not need to propagate through the entire driver cascade.

On the other hand, this approach entails the risk that the device may not be fully turned off when the other transistor is switched on. For this reason, adaptive dead time circuits need to have a maximum gate capacitance specified, especially for large power switches (large gate) like DC–DC converters with external power transistors.

A simple inverter can take care of the low-side sensing. A level-down shifter (see Section 5.12.4) is needed for high-side off-state sensing. A resistor-based level-down shifter can be used, as shown in Fig. 5.41a). Since all signals are available in the low-side or logic domain, the entire control logic can also be implemented in these domains.

2.8.3 Predictive Dead Time

This method comprises a control loop to regulate the dead time cycle-by-cycle, as shown in Fig. 2.22a). A variable-delay circuit can accomplish that, which is adjusted based on switching node sensing. V_{sw} is typically sensed at the turn-on instant of the low-side switch. If the dead time is too large, like in the first part of Fig. 2.22b), the body diode conducts, and there will be reverse recovery losses (with the next rising transition of the switching node, see Sections 2.4 and 2.7.7). The predictive dead time control will decrease the delay in the next cycle to obtain a shorter dead time. If dead time is too small, shown on the left of Fig. 2.22b), the switching node reaches a level

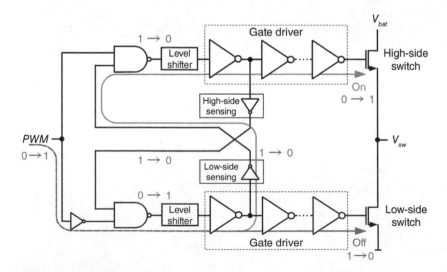

Figure 2.21 Adaptive dead time control.

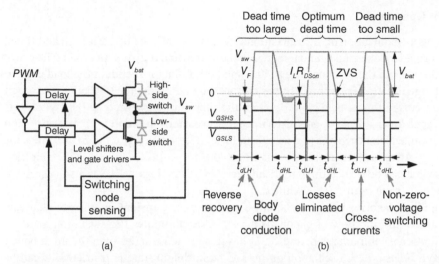

Figure 2.22 Predictive dead time control: a) block diagram; b) low-side/high-side gate signals V_{GSLS}, V_{GSHS} and the switching node voltage V_{sw} if the dead time is too large (left), at the optimum (center), and too small (right).

significantly greater than zero if the low-side switch turns on. Hence, the delay needs to increase. In the optimum case, shown in the center of Fig. 2.22b), the body diode never conducts, and the low-side switch can even benefit from ZVS as described in Section 2.9.

Compared to adaptive dead time generation, the predictive dead time tracks process variations and operating point changes, which may require a modified dead time (e.g., if the load current changes). The propagation delay in the driver and, partially, further delay associated with the sensing cancels out. Hence, predictive dead time control achieves an excellent performance but at the expense of increased complexity. Nevertheless, the cycle-by-cycle operation relaxes the speed requirements of the circuits. The switching node sensing circuits either detect the low-side body diode conduction or the zero crossing of the switching node. See Section 6.7 for suitable circuits. Loss reduction in the order of 30% has been reported, including ZVS [6, 7].

2.9 Soft-Switching

Soft-switching aims to reduce the switching losses due to the loss triangle described by Eqn. (2.39). At turn-on, Fig. 2.16 indicates that the losses are due to the fact that first, the drain current builds up before the drain voltage approaches zero and vice versa for the turn-off case. Figure 2.23a) repeats the switching circuit and the waveforms. Soft-switching adds some control measures, often employing additional circuitry, that allows first to bring down the drain–source voltage V_{DS} to zero before the current ramps up. Because the condition $V_{DS} = 0$ is reached before the actual switching event, the concept is called zero-voltage switching (ZVS). Likewise, the alternative approach of ZCS assumes $I_D = 0$. ZVS and ZCS are subcategories of soft-switching. In contrast, the conventional switching in Figs. 2.16 and 2.23a) is referred to as hard-switching. Even though actual designs often do not fully meet the conditions $V_{DS} = 0$ and $I_D = 0$, respectively, the terms ZVS and ZCS are used, or more general, soft-switching. Interestingly, there is no Miller phase during the turn-on and turn-off switching sequence because the switch turns on at $V_{DS} \sim 0$. Two widely used concepts for achieving soft-switching are (1) resonant operation and (2) dead time control.

2.9.1 Resonant Operation

Figure 2.23b) shows a modification of the conventional switching stage of Fig. 2.23a). An additional resonant tank is formed by inserting L_r and C_r between the switch and the inductive load. The waveforms in Fig. 2.23b) start in the on-state. The switch and both inductors conduct the load current I_{load}, and C_r is charged to V_{bat}. For simplicity, we assume the main inductor L to be large enough such that the load current has a negligible ripple in the time frame. As soon as the switch gets turned off, the I_D goes to zero while C_r gets discharged by the current stored in L_r. This way, L_r and C_r start an oscillation, which brings the transistor's drain–source voltage $V_{DS} = V_r$ to two times the battery voltage V_{bat}. Afterward, this node swings down to zero. At this point, the ZVS condition is met, and the switch can be turned on with ideally zero switching losses. ZVS is also ensured at turn-off because the current ramps down while still $V_{DS} = V_r = 0$.

Comparing the V_{sw} waveforms in Fig. 2.23a,b), we see that their transients are nearly identical. Both concepts can control the flow of energy. However, ZVS eliminates the switching losses at the expense of additional components L_r and C_r. L_r has to be much smaller than the main power inductor L. As ZVS relies on the oscillation of the L-C tank, the on-time is restricted. It can be approximated by $t_{off} = 2\pi\sqrt{L_rC_r}$. The time until the oscillation reaches the maximum depends on I_{load}, V_{bat}, and on the tolerances of the resonant circuit.

Advanced designs sense the value of V_{DS} at turn-on and adjust the on-time, for instance, as part of predictive control. Also, C_r can be implemented as a binary weighted array with a mixed-signal control. As the on-time can be freely chosen, the resonant approach allows for a duty cycle control of the power stage. If one oscillation cycle is too short for a desired on-time, the transistor can be turned on after multiple oscillations.

2.9.2 Dead Time Control

Concepts for dead time control are described in Section 2.8. In a half-bridge configuration, if the dead time is set correctly, we not only eliminate the dead time-related losses but can also achieve

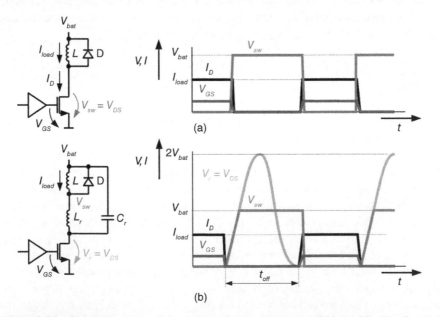

Figure 2.23 Switching losses of a) a hard-switching setup are eliminated by b) soft-switching by inserting a resonant tank, consisting of L_r and C_r.

Table 2.5 Losses of Example 2.8.

P_{cond}	P_{sw}	P_{oss}	P_{rr}	P_{bd}	P_{gate}	P_{tot}
Discrete ($f_{sw} = 100$ kHz)						
600 mW	0	0	0	0	60 mW	660 mW
Integrated ($f_{sw} = 1$ MHz)						
125 mW	0	0	0	0	10 mW	135 mW
Integrated ($f_{sw} = 10$ MHz)						
125 mW	0	0	0	0	100 mW	225 mW

ZVS (see the center part of Fig. 2.22b)). Let us examine the waveforms in Fig. 2.12. After the high-side switch turns off, the inductor current (i.e., the load current) will pull down the switching node. This process is lossless since the discharge current equals the load current, contributing to the output power. ZVS can be achieved if we keep the dead time short enough that the low-side device turns on exactly when V_{sw} reaches zero.

ZVS can also be achieved on the high side, which requires the inductor current to change polarity. If the low-side transistor turns off after a small load current flows into the switching node, this current will pull up the switching node during the dead time before the high-side switch turns on. The corresponding dead time must be controlled such that the turn-on instant of the high-side meets ZVS, i.e., $V_{sw} = V_{bat}$.

Example 2.8 *Repeat Example 2.7 in case of ideal ZVS.*

The results in Table 2.5 confirm that most switching losses vanish and the overall losses become much smaller compared to the hard-switching case in Example 2.7. The loss reduction potential is in the order of 80%. In reality, ZVS will never be perfect, so the gain in power efficiency will be reduced but still significant. Actual designs with integrated power stages typically achieve 20–30% lower losses.

2.10 Switch Stacking

The overview in Fig. 1.8 shows that various application use higher system voltages. These applications require power switches with higher voltage-blocking capabilities. Regular thick-oxide MOS transistors (n- and p-channel), as well as Double-Diffused Metal-Oxide-Semiconductor (DMOS) and drain-extended (DEMOS) devices, can be used as switch transistors (see Fig. 2.1). However, many low-voltage deep sub-micrometer CMOS technologies do not offer these devices. In that case, stacking two or more low-voltage transistors can form a high-voltage switch. Even for power management circuits designed in larger process nodes, which offer single high-voltage transistors, it can be beneficial to implement a stacked option with thin-oxide low-voltage devices, as shown in Fig. 2.24a) [8]. Stacking usually results in lower losses and layout area than single switches. There are various examples of PMICs with stacked power switches. Early implementations are presented in [9–11]. The fully integrated voltage regulator in [12] uses two stacked low-voltage transistors in a synchronous inductive step-down converter (see Chapter 10).

In the stacking circuit with two low-voltage transistors in Fig. 2.24a), the gate driver controls the lower transistor M1 directly. The upper transistor M2 operates as a cascode with its gate connected to V_{drv}, the gate supply. The bulk and source of each transistor are shorted. If the control signal V_{ctrl} transitions from low to high, first, transistor M1 discharges the node V_1 to V_{low}, which can be global power ground (in case of a low-side switch) or a floating high-side ground level (high-side switch). Subsequently, the transistor M2 turns on. Both transistors M1 and M2 operate in the triode region. V_{ctrl} turns low to turn the switch stack off. Transistor M1 turns off, and the intermediate node rises to $V_1 = V_{drv} - V_{th}$. It may then slowly rise to V_{high} due to the subthreshold current of M2. This way, the transistor M2 cuts off as well. At the beginning of the turn-off transient, the upper switch M2 sees a drain–source voltage V_{DS2} larger than V_{drv}. The loop equation results in $V_{DS2} = V_{blk} - V_{DS1}$ with the off-state blocking voltage $V_{blk} = 2V_{drv}$ and $V_{DS1} = V_{drv} - V_{th}$. Hence, $V_{DS2} = V_{drv} + V_{th}$. V_{DS2} needs to stay below the maximum allowed drain–source voltage $V_{DS,max}$ of the switches. The typical assumption of $V_{DS,max} \sim 1.5V_{drv} > V_{DS2}$ results in $V_{drv} > 2V_{th}$, which is well fulfilled in most designs.

If the gate supply V_{drv} is provided by a charge pump, the losses due to the parasitic transistor capacitances are scaled up by the charge pump efficiency (see Section 8.5), which can typically be at 50% [8]. However, as a particular advantage of the switch stacking circuit in Fig. 2.24a), only the parasitic capacitors C_{GS1} and C_{GD1} of the transistor M1 have to be charged by the charge pump. The loss can be significantly reduced since all other parasitic capacitances are directly charged from the input (V_{high}).

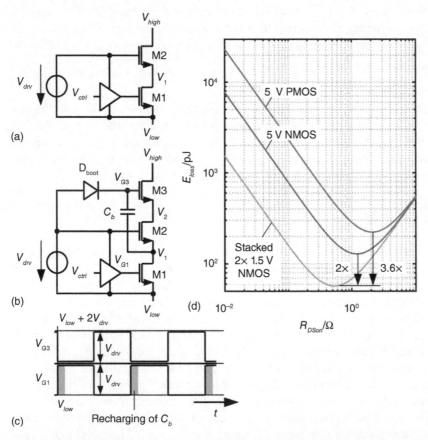

Figure 2.24 Switch stacking of a) two and b) three low-voltage transistors; c) timing for b); d) loss comparison versus on-resistance.

Figure 2.24b,c) shows a configuration for stacking of three switches along with the timing [8]. If the gate supply V_{drv} comes from a charge pump, the circuit tolerates arbitrary flying voltage levels at node V_{low} and does not require to return to ground. The bulk and source of each switch are shorted. Compared to the stack of two switches in Fig. 2.24a), no additional level shifters and gate drivers are required. The gate voltage V_{G3} is provided by a bootstrap capacitor C_b (typically, a MIM type), which is referred to node V_1 (see also Section 5.11 on bootstrapping). The bootstrap capacitance C_b is recharged from the gate drive supply via the diode D_{boot} during the on-phase of the control signal V_{ctrl}. C_b is designed to be 5–10 times greater than the gate capacitance of transistor M3. Neglecting the forward voltage drop of D_{boot}, V_{G3} automatically switches between $V_{low} + 2V_{drv}$ and $V_{low} + V_{drv}$, as shown in Fig. 2.24c). D_{boot} can be replaced by a switch that turns on for a few nanoseconds according to the timing in Fig. 2.24c). All transistors are sized identically and relatively large, preventing excessive voltage stress due to global and local variations.

The loss energy versus R_{DSon} can be plotted similarly to Fig. 2.18 to identify the optimum on-resistance of the switches. The scenario in Fig. 2.24d) achieves minimum losses for an on-resistance of about 500 mΩ. For single n-channel (NMOS) and p-channel (PMOS) switches with 5 V ratings, the loss minimum occurs at a different value R_{DSon}. However, Fig. 2.24d) confirms that the losses are significantly lower for switch stacking (2 and 3.6 times, in this example). Also, area-wise, switch stacking can bring a considerable advantage [8]. On the downside, switch stacking makes the power stage more sensitive to noise coupling. Glitches may occur and impact the switching performance. There is a remaining risk that the maximum voltage ratings of the switches get violated. Especially during start-up, this needs to be ensured by careful design.

2.11 Back-to-Back Configuration

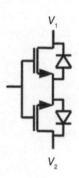

In off-state, an n-type transistor switch can only block the voltage across it if the drain potential keeps above the source (shorted to the bulk in a typical power transistor). Otherwise, the body diode may get forward biased and establish a current flow even though the transistor channel is turned off. Assume a power switch connected with its drain to a battery input. If the source voltage gets pulled up by the load (e.g., inductive load, capacitive charge pump level above battery, and different supply domains), unwanted back-supply into the battery may occur. Suppose the input voltage at the drain of the switch comes from a linear regulator instead of a battery. In that case, higher source potential may pull up the regulator's output voltage as it usually cannot sink any current. Devices connected to the regulator may get damaged if the output exceeds their maximum ratings.

Figure 2.25 Back-to-back configuration.

The back-to-back configuration, shown in Fig. 2.25, prevents body diode conduction. As the main drawback, two transistors are required. The stacking of two transistors means that the total channel length is doubled. Their width also needs to be doubled to keep the same on-resistance as of a single switch. Hence, the back-to-back configuration requires four times the area of a regular transistor. It will, therefore, only be applied if necessary.

References

1 Mohan, N., Undeland, T.M., and Robbins, W.P. (2003) *Power Electronics. Converters, Applications and Design*, John Wiley & Sons, Inc., 3rd edn.

2 Murari, B., Bertotti, F., and Vignola, G. (2002) *Smart Power ICs*, Springer, Berlin, Heidelberg, New York, 2nd edn.

3 Erickson, R.W. and Maksimovic, D. (2001) *Fundamentals of Power Electronics*, Springer.

4 Benda, H. and Spenke, E. (1967) Reverse recovery processes in silicon power rectifiers. *Proceedings of the IEEE*, 55 (12), 2159–2159, doi: 10.1109/PROC.1967.6093.

5 Wittmann, J., Barner, A., Rosahl, T., and Wicht, B. (2016) An 18 V input 10 MHz buck converter with 125 ps mixed-signal dead time control. *IEEE Journal of Solid-State Circuits*, 51 (7), 1705–1715, doi: 10.1109/JSSC.2016.2550498.

6 Maderbacher, G., Jackum, T., Pribyl, W., and Sandner, C. (2010) A sensor concept for minimizing body diode conduction losses in DC/DC converters, in *2010 Proceedings of the ESSCIRC*, pp. 442–445, doi: 10.1109/ESSCIRC.2010.5619738. URL http://ieeexplore.ieee.org/stamp/stamp.jsp?arnumber=5619738.

7 Wittmann, J., Barner, A., Rosahl, T., and Wicht, B. (2015) A 12 V 10 MHz buck converter with dead time control based on a 125 ps differential delay chain, in *ESSCIRC 2015 - 41st European Solid-State Circuits Conference (ESSCIRC)*, pp. 184–187, doi: 10.1109/ESSCIRC.2015.7313859. URL http://ieeexplore.ieee.org/stamp/stamp.jsp?arnumber=7313859.

8 Renz, P., Kaufmann, M., Lueders, M., and Wicht, B. (2021) Switch stacking in power management ICs. *IEEE Journal of Emerging and Selected Topics in Power Electronics*, 9, 3735–3743, doi: 10.1109/JESTPE.2020.3012813.

9 Xiao, J., Peterchev, A., Zhang, J., and Sanders, S. (2004) A 4 μA-quiescent-current dual-mode buck converter IC for cellular phone applications, in *2004 IEEE International Solid-State Circuits Conference (IEEE Cat. No.04CH37519)*, pp. 280–528, Vol. 1, doi: 10.1109/ISSCC.2004.1332703.

10 Kursun, V., Narendra, S., De, V., and Friedman, E. (2004) High input voltage step-down DC-DC converters for integration in a low voltage CMOS process, in *International Symposium on Signals, Circuits and Systems. Proceedings, SCS 2003. (Cat. No.03EX720)*, pp. 517–521, doi: 10.1109/ISQED.2004.1283725.

11 Serneels, B., Piessens, T., Stepert, M., and Dehaene, W. (2004) A high-voltage output driver in a standard 2.5 V 0.25 /spl mu/m CMOS technology, in *2004 IEEE International Solid-State Circuits Conference (IEEE Cat. No.04CH37519)*, pp. 146–518, Vol. 1, doi: 10.1109/ISSCC.2004.1332636.

12 Burton, E.A., Schrom, G., Paillet, F., Douglas, J., Lambert, W.J., Radhakrishnan, K., and Hill, M.J. (2014) FIVR - Fully integrated voltage regulators on 4th generation Intel(R) Core(TM) SoCs, in *2014 IEEE Applied Power Electronics Conference and Exposition - APEC 2014*, pp. 432–439, doi: 10.1109/APEC.2014.6803344.

3

Semiconductor Devices

In this chapter, we take a more detailed look at various types of power transistors and diodes. All material is presented from a circuit designer's point of view. The contents are not intended to replace a comprehensive book on power semiconductors such as [1–3].

We distinguish between discrete power devices and transistors available in technologies for integrated circuits. Figure 3.1a) shows a transistor in a D2PAK package (Double Decawatt Package, also classified as TO-263), suitable for currents of 100A. The on-resistance of discrete power transistors is in the <10 mΩ range. Typical drain currents are in the range of 10–100 A. Voltage classes reach from below 50 V to more than 1200 V. See also Fig. 1.17 in the introduction chapter.

Figure 3.1b) shows an example of integrated power transistors in an automotive power management integrated circuit (PMIC). The IC contains a variety of power management blocks like linear regulators, buck, boost, and buck-boost converters in addition to several monitoring and safety functions on approximately 20 mm^2 die size. The right part of the IC includes several power transistors. A thick-copper metallization layer connects them. Each switch can carry currents in the order of 1A from a 12 V car battery, which can reach up to 60 V in some instances (see Section 1.5.7).

While the full integration onto one single chip brings several advantages, including reduced board space, and increased reliability (fewer parasitics), there are some restrictions: (1) on-resistance ≥ 100 mΩ, (2) maximum drain current of 1–5 A, and (3) drain-source voltage up to 100 V. These values are intended as a general overview. There are plenty of technology options that exceed these limits. The achievable drain current also depends on the actual power dissipation. If the current is drawn out of a low-voltage supply, even >10 A may be possible. Dedicated high-voltage technologies (bulk Complementary Metal-Oxide-Semiconductor (CMOS) or silicon-on-insulator (SOI)) support 350 and 600 V classes.

3.1 Discrete Power Transistors

While the bipolar junction transistor (BJT) was used as an early power switch, in the mid-1970s, the power metal-oxide-semiconductor field-effect transistor (MOSFET) entered the market with superior switching characteristics. The insulated-gate bipolar transistor (IGBT) came up in 1985 with a higher voltage and current capability, combined with an isolated gate structure like the power MOSFET. It still keeps improving due to new designs and new materials, resulting in a promising market forecast. In 1998, the superjunction MOSFET was commercialized, which can switch significantly faster than previous devices and exhibits lower forward resistance compared to conventional power MOSFETs at high breakdown voltages. Recently, transistors based on wide-bandgap materials such

Design of Power Management Integrated Circuits, First Edition. Bernhard Wicht.
© 2024 John Wiley & Sons Ltd. Published 2024 by John Wiley & Sons Ltd.

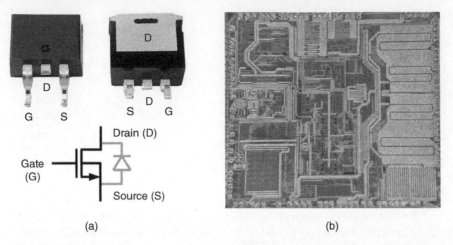

(a)   (b)

Figure 3.1 Discrete and integrated power transistors: a) discrete power transistor in a D2PAK package with lead assignment; b) power management IC with integrated power transistors. Source: courtesy of Bosch.

as gallium nitride (GaN) and silicon carbide (SiC) are on the rise because they offer inherently better characteristics resulting in low conduction and dynamic losses and better thermal behavior. GaN transistors achieve extremely low gate and output charge and show no reverse recovery losses. Beyond 650 V, SiC shows superior switching performance compared to the IGBT. Nevertheless, advanced superjunction transistors remain very competitive. Figure 1.17 in Chapter 1 maps the different transistor types with respect to power level and switching frequency.

Table 3.1 gives an overview of key parameters for silicon and GaN transistors in the 100 V class. A widely used metric (figure of merit (FOM)) that incorporates both the conduction losses and the switching losses of a power transistor is the product of its on-resistance R_{DSon} and the gate charge Q_{gate} that needs to be delivered to turn the device on. The expression $R_{DSon} \cdot Q_{gate}$ is listed in the last row of Table 3.1. A low FOM value reflects how reduced parasitic device capacitances at a certain on-state resistance lead to lower dynamic losses, faster switching transitions with higher dV/dt, and lower gate driver losses. Among the components in Table 3.1, the GaN device, by far, shows the lowest FOM of 31.2 pV/s. With FOM numbers of 180 and 235 pV/s, the silicon MOSFETs are at least six times worse.

Table 3.1 Key parameters of discrete 100 V power transistors.

	Si MOSFET Infineon OptiMOS5 BSC070N10NS5	Si MOSFET TI NexFET CSD19533KCS	GaN EPC EPC2022
$V_{DS,max}$	100 V	100 V	100 V
$I_{D,max}$	80 A	86 A	90 A
R_{DSon}	6 mΩ	8.7 mΩ	2.4 mΩ
$V_{GS,nom}$	10 V	10 V	5 V
$V_{th,nom}$	3.0 V	2.8 V	1.4 V
Q_{gate}	30 nC	27 nC	13 nC
Q_{rr}	89 nC	211 nC	0
C_{oss}	340 pF	395 pF	840 pF
$R_{DSon} \cdot Q_{gate}$	180 pV/s	235 pV/s	31.2 pV/s

The product $R_{DSon} \cdot Q_{oss}$ is another FOM related to the output capacitance (not shown in Table 3.1). It is also called the hard-switching FOM because in hard-switching power stages, Q_{oss} contributes a significant loss. In general, wide-bandgap devices such as GaN and SiC have a lower FOM than traditional silicon devices. This means that they can switch faster and handle higher power levels with lower losses.

3.1.1 The Silicon Power MOSFET

Figure 3.2a) shows the cross section of the fundamental silicon power MOSFET. It exhibits a planar construction with vertical current flow. The source and gate connection is on the top side of the semiconductor substrate, and the drain connection is on the bottom side. Vertical current flow is common for discrete power devices as it enables the current to flow through a maximum cross section resulting in low on-resistance as well as good electrical and thermal connection to the package. In comparison to lateral devices, the power losses are dissipated across the entire volume. This way, vertical transistors show a robust behavior when handling transient load current peaks.

The device in Fig. 3.2a) is also referred to as double-diffused MOS transistor (DMOS) or, more specifically, as vertical DMOS (VDMOS). The name indicates the sequence in which the p-backgate is diffused first, followed by highly doped n+ source diffusion.

A gate voltage above the device threshold voltage forms a conducting n-channel from the drain contact and its n-region at the bottom through the n-area, further along the p+ doped region under the gate toward the n-region at the source terminal. The threshold voltage is typically around 3 V, and the target on-resistance is reached at a gate voltage of 7–10 V, while up to 15 V can be applied to achieve a minimum resistance.

The high-voltage capability is determined by the thickness of the lightly doped n-region (n-), the so-called drift zone. Unfortunately, the on-resistance increases with the thickness of the drift zone. This trade-off is essential for many power devices. Hence, the achievable specific on-resistance of a silicon power MOSFET is limited for a given voltage class. Silicon power transistors are typically available with voltage ratings up to 200 V. The trench gate technology in advanced MOSFETs places the polysilicon of the gate and the channel vertically. As a result, the channel density is maximized, providing an improved resistance. In addition, low conduction losses per unit area allow the chip size to be reduced, improving switching losses.

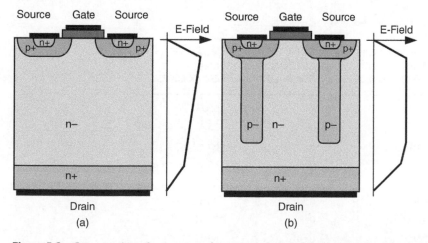

Figure 3.2 Cross section of power transistors and electrical field strength: a) fundamental planar silicon power MOSFET; b) superjunction MOSFET.

The on-resistance has a positive temperature coefficient. Hence, several devices can be placed in parallel. Figure 3.2a) shows only one unit transistor, called finger, which is connected in parallel in large numbers to achieve the final low switch-on resistance. There is no risk of thermal runaway. Thermal runaway occurs if the device has a negative temperature coefficient, like in a BJT. In that case, the finger with the lowest resistance takes most of the current. As this heats up the finger locally, its on-resistance decreases, and more and more current flows through this one finger until overheating destroys the device. As the silicon power MOSFET is not prone to a thermal runaway failure, even several power devices can be installed in parallel to form a power module that supports very large currents and voltages.

For $V_{DS} > 0$, all PN junctions are reverse biased. The junction of the p and n-regions forms the body diode of the transistor. The switching behavior is determined by the parasitic capacitances. Especially the drain-gate and drain-source capacitances are strongly nonlinear with respect to the terminal voltages, see Fig. 2.4.

3.1.2 The Superjunction MOSFET

The introduction of the superjunction transistor, shown in Fig. 3.2b), allowed to break the inverse relationship between the on-resistance and the blocking capability of the conventional silicon power transistor. This is achieved by adding p-doped trenches, which form a compensation depletion region at the junction – the *superjunction* – toward the n-drift zone. The doping of the trenches is adjusted exactly such that the n-doping is compensated. This results in a very low effective doping. The vertical electrical field distribution changes from triangular to rather rectangular, resulting in higher breakdown voltage. Thanks to the compensation structure, the n-region can be doped much more heavily. This way, superjunction transistors achieve low on-resistance together with high-voltage capability. Superjunction transistors operate mainly in the range of 200–900 V. Table 3.2 lists a typical transistor rated for 600 V in comparison to GaN and SiC devices.

3.1.3 The Insulated-Gate Bipolar Transistor (IGBT)

The IGBT supports very high currents and low on-resistances. Figure 3.3a) shows its construction, which is very similar to the DMOS. The main difference is a p-region at its drain, which forms the emitter of a PNP bipolar transistor, as indicated.

Table 3.2 Key parameters of discrete 600 V power transistors.

	Si Superjunction Infineon CoolMOS IPDD60R105CFD7	GaN GaN Systems GS66508B	SiC MOSFET Cree C3M0060065J
$V_{DS,max}$	600 V	650 V	650 V
$I_{D,max}$	31 A	28 A	36 A
R_{DSon}	88 mΩ	50 mΩ	60 mΩ
$V_{GS,nom}$	10 V	6 V	15 V
$V_{th,nom}$	4.0 V	1.7 V	2.3 V
Q_{gate}	36 nC	6.1 nC	46 nC
Q_{rr}	410 nC	0	75 nC
C_{oss}	28 pF	65 pF	80 pF
$R_{DSon} \cdot Q_{gate}$	3.2 nV/s	0.3 nV/s	2.8 nV/s

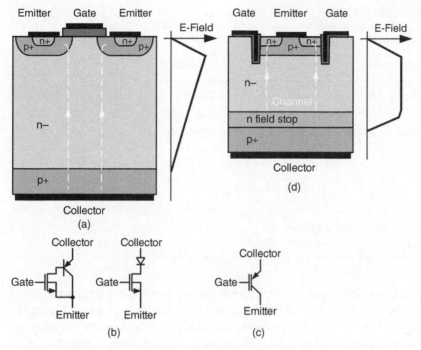

Figure 3.3 The IGBT: a) Cross section of a conventional IGBT; b) its equivalent circuits; c) IGBT symbol; d) cross section of a trench-gate field-stop IGBT.

Together with the rest of the structure, the equivalent circuit consists of the PNP transistor and an MOS structure, as shown in Fig. 3.3b). It can be considered an MOS transistor with a series diode, representing the PNP transistor's emitter-base junction. In on-state, this diode adds a forward voltage drop. Consequently, the IGBT exhibits a significant voltage drop for low collector currents. It corresponds to more considerable conduction losses compared to the power MOSFET, where no forward-biased PN junction is present. The series diode also indicates that the IGBT can only conduct uni-polar currents. An external freewheeling diode is usually added to conduct the reverse current (in case of inductive switching).

The three terminals of the IGBT are a mix of the bipolar transistor's emitter and collector and a gate terminal. This is reflected in the IGBT symbol shown in Fig. 3.3c). Because of the insulated gate, the IGBT as a whole remains voltage controlled.

The typical range of the gate-emitter voltage V_{GE} is −8 to 15 V. If V_{GE} exceeds the threshold voltage V_{th}, an n-channel is created with the electrons flowing to the collector. The emitter-base junction of the PNP structure gets forward-biased and injects minority carriers into the n-drift zone (the base). The increased charge carrier density reduces the resistance of the drift zone. This fundamental mechanism in the IGBT is called conductivity modulation (the conductivity of the n-drift zone is modulated). It is responsible for very low on-resistance, even at very high voltages. The turn-on mechanism is identical to the behavior of a DMOS transistor with a similar rise time of the current, fall time of the voltage, influence of capacitances, and gate drive current. In on-state, the IGBT consists of an on-resistance, similar to the MOSFET, in series with a forward-biased junction, reflected in Fig. 3.3b).

The turn-off process is significantly different because it is determined by minority carriers in the n-drift region (i.e., holes). When the gate-emitter voltage turns to zero or, in the case of bipolar gate control, to negative values, the gate capacitance of the IGBT is discharged, and the channel current is interrupted. The current slowly decays in proportion to the minority charge through

Table 3.3 Key parameters of discrete 1200 V power transistors.

	IGBT (Silicon) ST STGQ15H120DF2	IGBT (Silicon) Infineon IKW15N120BH6	SiC MOSFET Rohm SCT3080KW7
$V_{DS,max}$	1200 V	1200 V	1200 V
$I_{D,max}$	30 A	30 A	30 A
R_{DSon}	140 mΩ	126 mΩ	80 mΩ
$V_{GS,nom}$	15 V	15 V	18 V
$V_{th,nom}$	6.0 V	5.7 V	4.5 V
Q_{gate}	67 nC	92 nC	60 nC
Q_{rr}	720 nC	830 nC	261 nC
C_{oss}	105 pF	60 pF	75 pF
$R_{DSon} \cdot Q_{gate}$	9.4 nV/s	11.6 nV/s	4.8 nV/s

recombination. For long carrier lifetimes, this can take a few microseconds. Thus, it leads to a slow turn-off speed and, consequently, high turn-off loss. The recombination determines the relatively slow switching frequency of IGBTs, which is in a range of 1 to 20 kHz, although recent advances in device construction have pushed effective switching frequency toward 100 kHz. During the presence of the tail current, the gate driver can no longer control the device. In this period, the IGBT still sees nearly the entire input voltage across its collector-emitter path. Therefore, a large portion of the IGBT's switching losses is due to its tail current.

The cross section in Fig. 3.3a) reveals that the parasitic body diode is lacking in an IGBT, so an external freewheeling diode is required in the power stage.

Despite the slower switching speed, IGBTs are attractive for many applications in the voltage range of 400 to 1700 V. Table 3.3 lists the key parameters of typical 1200 V devices. The lower on-state voltage drop of the IGBT compared to the DMOS reduces the conduction losses. Furthermore, higher power ratings are reached with a smaller die, which improves cost-effectiveness compared to a power MOSFET. Assuming that the positive temperature coefficient of the on-resistance dominates, the IGBT can be paralleled in IGBT modules up to 6500 V and 3600 A.

Advanced IGBT technologies, Fig. 3.3d), improve the long recombination times by two measures:

(1) The thickness of the drift zone is reduced, and an n-zone of increased doping is added next to the p-collector zone. Adding the field-stop layer changes the electric field from a triangular to a trapezoidal distribution. With a total chip thickness of approximately 100 μm and less, it is possible to use a very-lightly doped collector region and still achieve low resistance without affecting the voltage capability.
(2) By introducing the trench gate concept, the MOS channel of the device is rotated by 90° compared with a traditional planar IGBT structure. It allows to optimize the carrier profile to reduce charge within the drift zone that must be removed during the turn-off phase.

3.1.4 The Gallium-Nitride Transistor

As a wide-bandgap material, gallium nitride (GaN) enables transistors with low on-resistance and parasitic device capacitances for high-voltage fast-switching applications. State-of-the-art enhancement mode (e-mode) GaN transistors are lateral devices, as shown in Fig. 3.4. Silicon

Figure 3.4 Cross section of a GaN power transistor.

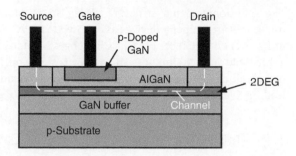

wafers are typically used as mechanical carriers due to their low cost and high availability. For this reason, the GaN technology is also referred to as GaN-on-Si. The layers of gallium nitride (GaN) and aluminum-doped GaN (AlGaN) create a heterojunction. In GaN heterostructures, electrons are generated as carriers at the interface between GaN and AlGaN. These electrons form a lateral two-dimensional electron gas (2DEG) with low specific on-resistance due to the high carrier density and high electron mobility. That's why the GaN transistor is also called a high electron mobility transistor (HEMT) or GaN HEMT. The gate is implemented by removing the AlGaN layer and applying a new thinner AlGaN layer. The natural AlGaN/GaN interface forms a normally on depletion mode (d-mode) device. However, most applications prefer enhancement (e-mode or normally off) transistors without current flow if no gate-source voltage is present. One of the most popular ways to achieve e-mode behavior is to grow p-doped GaN (p-GaN) or p-AlGaN on top of the AlGaN, which depletes the 2DEG channel below. If a positive gate-source voltage is applied, the electrons are reaccumulated, and the transistor turns on. Unlike discrete DMOS, IGBT, and SiC devices, GaN has a low threshold voltage below 2 V. Its target on-resistance is typically reached for $V_{GS} = 5$ V. The maximum rating slightly above 5 V makes the gate drive challenging, especially in the presence of parasitics during fast switching transitions. An exception are gate-injection GaN transistors (GIT), which show a "self-clamping" behavior at the gate. GaN transistors up to 600 V with current capabilities of 10 A and above are available. There is a possibility that GaN reaches good performances up to 1200 V. Likewise, there is also promising research on low-voltage GaN-based switches (5 V and below), which demonstrates more than four times better $R_{DSon} \cdot Q_{gate}$ FOM compared to thin-gate silicon transistors [4].

The performance benefits of GaN transistors for high-voltage switching applications are mainly due to the higher critical electrical field. It reduces the drift region significantly, leading to a low area-specific on-resistance. However, the 2DEG exhibits a significant temperature coefficient of ~9000 ppm/K. This leads to larger variations of transistor parameters over temperature compared to state-of-the-art silicon technologies.

A primary benefit of the GaN process for integrating power converters is the lack of p/n-junctions. Hence, there is no body diode, and GaN does not suffer from reverse recovery losses since no junction depletion region needs to be formed to stop conducting. However, GaN transistors show a quasi-body-diode behavior (third quadrant conduction) when they are in reverse NMOS-diode configuration with gate and source shorted and at a higher potential than the drain. The voltage drop across the GaN transistor in third quadrant conduction is considerably higher than the forward voltage of a silicon body diode.

3.1.5 The Silicon-Carbide Transistor

Silicon Carbide (SiC) devices have emerged as the most viable candidate for next-generation, low-loss semiconductors in the voltage range above 600 V, see Table 3.3 and Fig. 1.17.

The traditional planar structure of Fig. 3.2 is adopted for SiC power transistors. Hence, SiC transistors are unipolar devices as opposed to the IGBT. State-of-the-art SiC transistor technologies utilize trench gate technologies. Challenges include handling high field strength and maintaining a threshold voltage similar to silicon devices. The much higher breakdown field strength and thermal conductivity of SiC power devices offer significant advantages over their silicon counterparts, leading to a longer lifetime and higher reliability.

3.2 Power Transistors in Integrated Circuits

Many state-of-the-art integrated circuit technologies offer high-voltage MOS transistors suitable for power applications. Unlike their discrete counterparts, they are designed as a lateral device, which can be placed beside analog and digital components as part of a microchip. They can be contacted at the IC surface.

Monolithic integration of high-voltage power transistors together with digital, analog, and mixed-signal functions enables all kinds of integrated power circuits, like switch-mode power supplies and power amplifiers. This way, one IC can replace entire electronic modules, which may even contain additional monitoring and safety functions. Besides innovation through new functions, these highly integrated solutions enable cost reduction, smaller size, and increased reliability. The conventional power transistors in IC technologies are the drain-extended MOS transistor and the lateral DMOS. Some recent semiconductor technologies offer lateral IGBT devices to be used in power management ICs [5, 6]. Even monolithic GaN technologies are available that enable the design of power management circuits in GaN [7] based on lateral GaN HEMT devices.

The designer can size all power devices individually with variable widths as a primary advantage of the on-chip integration. It offers a wide degree of freedom, and the design can be optimized for a suitable trade-off between on-resistance and parasitic capacitances (switching losses).

3.2.1 Drain-Extended Transistors

The drain-extended MOS (DEMOS) transistor comprises a lightly doped drain extension that increases the drain breakdown voltage by reducing the electric field under the gate at the drain end of the transistor. Figure 3.5 shows how the DEMOS transistors are built with an extended n- or p-well as the drain region without added process complexity. This way, n- and p-channel transistor types are available, as shown in Fig. 3.5. DEMOS transistors can operate at higher drain voltages up to more than 60 V. The threshold voltage is similar to regular CMOS transistors. Their on-resistance is relatively large, so DEMOS transistors are high-voltage components but not real power devices. The width, as well as the length, can be freely chosen to meet any design requirement.

3.2.2 Lateral DMOS Transistors

The double-diffused MOS transistor (DMOS) is the standard power device in integrated power technologies. It offers low on-resistance but requires more process complexity. Its name refers to the *double diffusion* that forms the n+ source and p-doped body. Figure 3.6 shows the cross section of a lateral DMOS transistor (LDMOS). The current flows in the lateral direction and within a confined area. It allows co-integration with various other components like low-voltage analog transistors. Therefore, LDMOS transistors are available in Bipolar-CMOS-DMOS (BCD) technologies used for power management IC designs (see also Section 1.8.1).

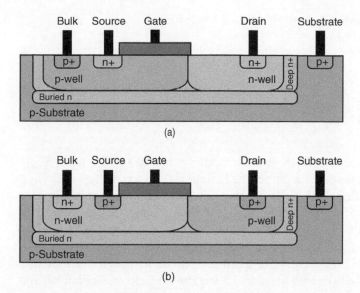

(a)

(b)

Figure 3.5 Cross section of a drain-extended MOSFET (DEMOS): a) n-type and b) p-type DEMOS.

LDMOS Cross Section

Figure 3.6 shows the source diffusion in the center. It has an integrated p-type back-gate, i.e., no separate back-gate contact exists. A widely spaced drain area surrounds the whole source area. The threshold voltage is higher than for normal CMOS transistors. Two key points can be observed from the LDMOS cross section:

- *High V_{DS}*: The separation of the drain contact enables large maximum drain-source voltages. The distance between the drain and the source is called the drift region. Power technologies support maximum ratings of typically 100 V (50–120 V).
- *Low R_{DSon}*: The current flows through a large cross section between drain and source, resulting in low channel resistance if the switch is turned on. Values in the order of 100 mΩ can be achieved in PMICs. In addition, currents in the order of single-digit amperes can be handled on-chip.

There are two parasitic PN junctions, the body diode between drain and source and a substrate-diode from drain to substrate. Both diodes are responsible for unwanted substrate coupling during switching transitions. Figure 3.7 shows the equivalent circuit of the LDMOS with both parasitic diodes included. The diodes are always present and must be carefully considered during design. They can even be utilized as a freewheeling or power diode. We also note that the source contact is fully isolated. Hence, the source terminal can go below ground potential. The drain must stay above ground potential to keep its PN junction to the p-substrate in reverse bias.

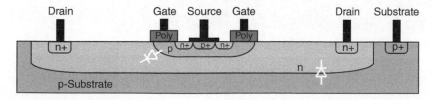

Figure 3.6 Lateral DMOS transistor cross section.

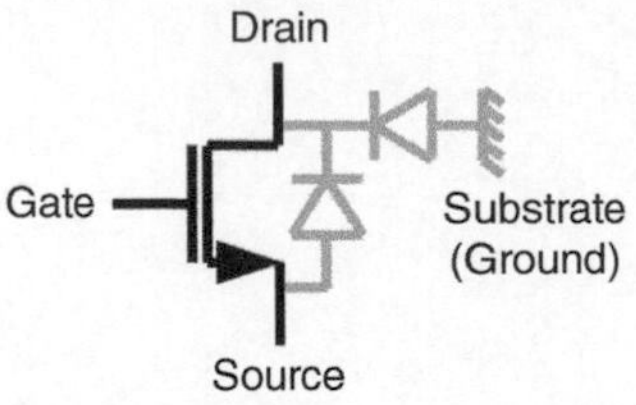

Drain

Gate ——

Source

Substrate
(Ground)

Figure 3.7 DMOS transistor equivalent circuit.

DMOS Transistor Layout

Multiple unit transistors are usually placed in parallel, as shown in Fig. 3.8 to achieve a low on-resistance. The unit device is also called a finger. The cross section in Fig. 3.6 represents one LDMOS finger. Its symmetrical shape allows overlapping adjacent drain areas, resulting in a compact and regular layout of the overall power transistor as indicated in Fig. 3.8. The technology fixes the channel length for a specific blocking voltage, and it cannot be modified by design.

The tapered metal connections of the drain and source aim to equalize the current density across the entire transistor. The total drain current enters the transistor from the left. Therefore, the metal width is maximized on the left and decreases finger by finger. The source metallization has the opposite layout because the current adds up from left to right with the entire current flowing on the outer right edge of the transistor.

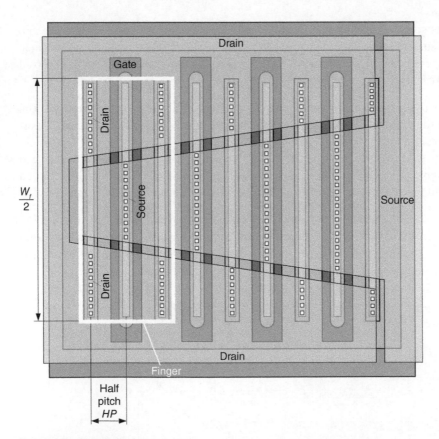

Figure 3.8 DMOS transistor layout.

Due to the symmetry in each finger, the drain current flows from both sides to the source in the center. Therefore, the width of the finger in the layout corresponds to half of the finger width W_f as depicted in Fig. 3.8. For n fingers in parallel, the total width is

$$W = n \cdot W_f. \tag{3.1}$$

Typical switches implement $n = 1000$ parallel fingers and more.

DMOS Transistor Sizing

The design task is to determine the appropriate W/L-value for a targeted on-resistance R_{DSon}. In general, as this is a switch, the transistor operates in the triode region and follows its basic characteristics:

$$I_D = \beta \left((V_{GS} - V_{th})V_{DS} - \frac{V_{DS}^2}{2} \right)$$

$$\rightarrow R_{DSon} = \frac{V_{DS}}{I_D} \approx \frac{1}{\beta \left(V_{GS} - V_{th} \right)} = \frac{1}{\frac{W}{L} \mu_o C_{ox} \left(V_{GS} - V_{th} \right)} \tag{3.2}$$

The approximation assumes that V_{DS}^2 is negligible (small drain-source voltage in on-state). There are only two variables to be defined by design, the gate-source voltage and the W/L-ratio. The gate-source voltage $V_{GS} = V_{drv}$ should be set to the maximum possible value. The maximum gate-source voltage of the transistor limits it. It also depends on the voltage level, which is available as a gate supply (see Section 5.11 in the chapter on gate drivers). Once V_{drv} is defined, the W/L-ratio can be derived.

Area-Specific on-Resistance

The area-specific on-resistance R_{sp} is a useful parameter that supports the power transistor sizing. It is a measure of the achievable on-resistance per die area,

$$R_{sp} = R_{DSon} \cdot n \cdot \frac{W_f}{2} \cdot HP = R_{DSon} \cdot W \cdot HP \rightarrow W = \frac{R_{sp}}{R_{DSon} \cdot HP}, \tag{3.3}$$

where the half-pitch HP corresponds to half the distance from drain to drain within the finger, see Fig. 3.8. Table 3.4 provides typical parameter values.

The area consumption is critical as the LDMOS is large. Moreover, the high-voltage (drift region) and high-current (cross section) capability covers far more layout space than the active area defined by W and L. For this reason, Eqn. (3.2) does not suit very well as a sizing equation. The specific on-resistance R_{sp} along with the half-pitch parameter HP is provided in the design manual for each LDMOS type. Besides the usefulness for the actual transistor sizing, R_{sp} also allows for a quick die size estimation in a given technology. It also acts as a benchmark to compare different technologies (and to decide on the more suitable one if the LDMOS dominates the die area).

The power transistor must guarantee its target on-resistance over the entire temperature range. The specific on-resistance is characterized by its first- and second-order temperature coefficients resulting in the expression

$$R_{spo} = R_{sp,27\,°C} + TC_{Rsp,lin} \cdot (T - 27\,°C) + TC_{Rsp,quad} \cdot (T - 27\,°C)^2. \tag{3.4}$$

R_{sp} has positive temperature coefficients; see also Table 3.4 for typical values. It applies in general. Hence, R_{sp} increases with temperature. We discuss this characteristic of the power transistor related to the safe operating area in Section 3.4 (Fig. 3.17).

Table 3.4 Typical data sheet for an integrated LDMOS.

Parameter	Symbol	Value	Unit	Note
Channel length	L	1.5	μm	
Half pitch	HP	5.29	μm	
Oxide thickness	t_{ox}	150	Å	1Å = 0.1 nm
Specific on-resistance	R_{sp}	$6.4 \cdot 10^{-2}$	Ωmm²	at 27 °C, $V_{GS} = 5$ V
Linear temperature coefficient	$TC_{Rsp,lin}$	$2.7 \cdot 10^{-4}$	$\frac{\Omega mm^2}{K}$	
Quadratic temperature coefficient	$TC_{Rsp,quad}$	$6.0 \cdot 10^{-7}$	$\frac{\Omega mm^2}{K^2}$	
Scaling factor for V_{GS}	F_{GS}	1.27		$V_{GS} = 2.5$ V
		1.17		$V_{GS} = 3.0$ V
		1.09		$V_{GS} = 3.5$ V
		1.05		$V_{GS} = 4.0$ V
		1.02		$V_{GS} = 4.5$ V
		1.00		$V_{GS} = 5.0$ V

For the transistor sizing, we must first determine the worst case R_{sp} at the maximum junction temperature. The actual IC may have a significant power dissipation, mainly due to the losses in the power stage as outlined in Section 2.7. Hence, the junction temperature may be much higher than the ambient temperature. In the second design step, we must consider that the on-resistance is only valid for the nominal gate-source voltage. Different values of V_{GS} are taken into account by a scaling factor F_{GS} such that

$$R_{sp} = F_{GS} \cdot R_{spo}. \tag{3.5}$$

The factor F_{GS} is provided in the design manual for various values of V_{GS}.

Example 3.1 *A 50 V LDMOS shall be sized to achieve a total on-resistance of no more than $R_{DSon} = 600\,m\Omega$ over a temperature range $T = -40 \dots 150\,°C$ with the parameters given in Table 3.4. For contact resistances and metallization, $100\,m\Omega$ should be considered. The DMOS is driven with a minimum gate-source voltage of 3 V.*

(a) *Calculate the specific on-resistance R_{sp} for the given parameters (worst case).*
 Subtracting $100\,m\Omega$ for contact resistances and metallization leaves $R_{DSon} = 500\,m\Omega$ for the LDMOS. This value has to be achieved at 150 °C and $V_{GS} = 3V$ (worst case). With $F_{GS} = 1.17$ for $V_{GS} = 3$ V from Table 3.4 and Eqns. (3.4) and (3.5), we calculate $R_{sp} = 124\,m\Omega/mm^2$. This value is almost twice as large as the given original value for 27 °C and $V_{GS} = 5$ V.

(b) *Calculate the required width of the transistor.*
 $W = R_{sp}/(R_{DSon}HP) = 124\,m\Omega/mm^2/(500\,m\Omega \cdot 5.29\,\mu m) = 46.88\,mm \approx 47\,mm$

(c) *What on-resistance results for the width determined in (b) at room temperature ($T = 27\,°C$)?*
 $R_{DSon}(T = 27\,°C) = F_{GS}R_{sp,27\,°C}/(W \cdot HP) = 301.1\,m\Omega$

(d) *Calculate what chip area the transistor occupies.*
$A = R_{sp}/R_{DSon}$ with R_{sp} from (a) and $R_{DSon} = 500\,m\Omega$
$\rightarrow A = 124\,m\Omega/mm^2/500\,m\Omega = 0.249 \times 10^6\,\mu m^2 = 0.249\,mm^2$

(e) *Calculate the number of transistor fingers n if a square area is desired. Round the result because n can only be an integer.*
Due to the ring structure, for $n = 1$, one dimension of the transistor extends to $W/2$. For a square area, this dimension is to be reduced to $\sqrt{A}$ by dividing it into parallel fingers: $n = W/(2\sqrt{A}) = 47.09 \rightarrow$ chosen: $n = 47$

(f) *Calculate the width W_f of a finger from the number of fingers n.*
$W_f = W/n = 47\,mm/47 = 1\,mm$
Thus, the dimension of the whole transistor is $W_f/2 = 0.5\,mm$.

(g) *Specify the geometric dimensions (length L_T and width W_T) that the transistor occupies on the chip.*
$W_T = W_f/2 \approx 0.5\,mm, L_T = A/(W_f/2) = n \cdot 2HP = 0.5\,mm$

(h) *Estimate the expected gate capacitance C_{gate} based on the oxide capacitance. The permittivity of the oxide is $\varepsilon_{ox} = 0.34\,pF/cm$.*
$C_{gate} = C_{ox} = \frac{\varepsilon_{ox}}{t_{ox}}(W \cdot L) = \frac{0.34\,pF\,cm^{-1}}{15\,nm}(47\,mm \cdot 1.5\,\mu m) = 159.8\,pF$

(i) *Estimate how much an IC would cost to build (cost of building, COB), consisting of a half-bridge with two LDMOS devices and requiring an additional $0.5\,mm^2$ of chip area for isolation (guard-rings), wiring, and control. The IC will be fabricated on 200 mm wafers (32,000 mm² total area per wafer) with a wafer cost of 2000 USD per wafer. For the cost estimation, yield losses, development, and test costs shall not be considered.*

Total area $= A_{IC} = 2 \cdot$ transistor area $+ 0.5\,mm^2 \approx 1\,mm^2$
Cost $= COB = 2000\,USD \cdot A_{IC} / 32000\,mm^2 = 0.625\,USD \approx 63\,Cent$

It is interesting to note that one LDMOS transistor costs approximately 15.6 Cent. Please note that this is a very simplistic approach. For a more accurate cost estimate, yield losses, development and test costs, and package/bonding must also be considered, which can increase the cost by at least a factor of 2.

3.2.3 Silicon-on-Insulator Technologies (SOI)

Recent SOI technologies with power and high-voltage capabilities offer excellent isolation at competitive cost [8]. SOI provides oxide isolation instead of junction isolation. Each active component is placed in its own buried oxide (BOX), as shown in Fig. 3.9. Parasitics like bipolar effects and leakage currents are negligible in SOI, which eases design and improves efficiency. However, losses caused by parasitic capacitances such as at drain-source and toward substrate are challenging in fast-switching high-voltage converters for low-power applications [9].

Figure 3.9 Power transistor in an SOI technology, isolated by deep-trench isolation (DTI) and buried oxide (BOX).

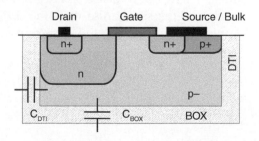

3.2.4 Monolithic GaN Integration

Contrary to vertical transistor structures (silicon superjunction, silicon carbide), the lateral GaN technology enables monolithic integration of the high-voltage power transistor with analog and digital functions [7]. As one of the advantages, the gate loop parasitics are minimized. The lack of a suitable p-type device is a significant challenge for achieving monolithic integration in GaN. While p-type transistors can be manufactured in GaN technology, their performance is far inferior to n-type GaN transistors due to a lower carrier density in the two-dimensional hole gas (2DHG) and far lower carrier mobility of holes in GaN. Another challenge for monolithic integration is a more significant device mismatch compared to modern silicon technologies since the GaN technology still needs to mature. Random crystal defect rates between 5 and 10 per μm^2 strongly affect the electron mobility and, thereby, the transconductance matching of adjacent transistors.

3.3 Parasitic Effects

3.3.1 Parasitic Bipolar Junction Transistor

Assume a half-bridge with an inductive load and a load current flowing out of the switching node. As soon as the high-side switch turns off, the current commutates into the body diode of the low-side device. The low-side transistor keeps turned off for a finite dead time. The switching node gets pulled below ground by one diode drop, limited by the substrate junction of the DMOS transistor (see Section 2.5.5). The substrate diode turns on and forms the base-emitter junction of a parasitic NPN structure, as shown in Fig. 3.10a). The substrate becomes the base of a lateral NPN transistor, and any other n-well of the substrate forms the collector. The NPN transistor injects electrons into the substrate, which can cause the malfunction of sensitive circuits. It may even trigger a silicon-controlled rectifier (SCR) and cause a latch-up (see the section below).

Guard Rings

The purpose of guard rings is to reduce the injection of minority carriers to prevent interference with other devices. For the case of Fig. 3.10a), electrons are the minority carriers injected into the substrate. As depicted in Fig. 3.10b), a sacrificial n-type collector forms a collector that takes most of the injected electron current. The effectiveness of guard rings depends on several points, most importantly (1) the bias voltage of the guard ring, (2) the number of guard rings, and (3) the structure/doping concentration of the guard ring. The table in Fig. 3.11 shows typical guard ring data [10]. If a single guard ring is biased at 12 V, 160 μA of substrate current flow at the n-area of the sensitive circuits with respect to 500 mA pulled out of the LDMOS drain (first table entry).

Suppose the guard ring bias is kept at ground level ($V_{GR} = 0$ V, the second entry of the table in Fig. 3.11), the 500 mA disturbance is reduced to 450 μA. This result is still a reduction of 1000. The ground-referred zero voltage bias of guard rings is usually preferred as it does not need a dedicated guard ring supply. Also, the power dissipation may force a ground bias instead of $V_{GR} \gg 0$ V. Assume an emitter current of $I_D = 500$ mA, a guard ring bias of $V_{GR} = 12$ V, and a worst case base-emitter forward voltage of $V_F = 1$ V. The associated power dissipation is $P_{diss} = (V_{GR} + V_F) \cdot (-I_D) = (12 \text{ V} + 1 \text{V}) \cdot 500$ mA $= 6.5$ W. It exceeds the maximum power dissipation of an IC in a standard power package (see Section 12.2.4). For $V_{GR} = 0$ V, the power dissipation reduces to an acceptable value of $P_{diss} = V_F \cdot (-I_D) = 500$ mW.

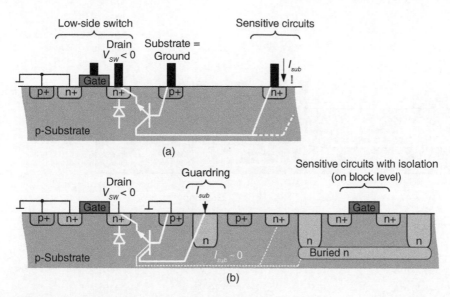

Figure 3.10 MOS transistor with parasitic bipolar NPN transistor in case of negative drain potential: a) any n-area forms a collector from which parasitic substrate currents are drawn; b) a guard ring is inserted to act as a collector and to prevent substrate currents at other n-area.

The third table entry in Fig. 3.11 gives the parameters for a double guard ring. It achieves much higher efficiency than a single guard ring at the expense of significant area consumption (remember that high-voltage capabilities require layout space). Nevertheless, if a sensitive circuit with only a few microamperes of bias current needs to be protected against electron injection, only the double guard ring provides sufficient protection. Up to three guard rings are used in practical designs. Concerning power dissipation, the first guard ring is usually biased at ground level, while any further guard ring can be at a higher bias voltage as the current is minimal.

Figure 3.12 shows the cross section for two typical guard ring structures [11]. The structure in Fig. 3.12a) can be biased with high voltages up to 80 V and above and achieves very good efficiency in terms of remaining substrate current. However, it covers a large area (it surrounds the entire power transistor, including necessary spacing). The guard ring type of Fig. 3.12b) usually represents a good trade-off between efficiency and layout area.

If the inductive load at the switching node sees a current that flows into the power stage (inductor connected to supply V_{bat}), the switching node gets pulled one (body) diode drop above the supply. The body diode and the substrate diode of the high-side device together form a parasitic PNP transistor as illustrated in Fig. 3.13. A hole current is injected into the substrate. A highly doped n+ buried layer surrounding the drain of the LDMOS reduces the hole injection and, hence, the efficiency of the PNP structure significantly. The current gain reduces to < 0.1%.

Latch-Up

The cross section of a Complementary Metal-Oxide-Semiconductor (CMOS) inverter in Fig. 3.14 reveals two parasitic BJTs, a PNP and an NPN device, which form a silicon-controlled rectifier (SCR) structure. As the n and p-regions of the CMOS transistors have a finite (nonzero) resistance, any parasitic current may forward bias the base-emitter junctions of the PNP and NPN device, respectively. For $I \cdot R_{p,n} > 0.7\,\mathrm{V}$ both transistors start conducting. It also means that the product

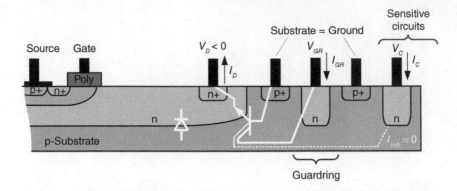

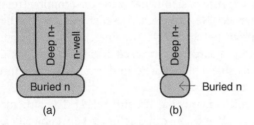

Guard ring type	Injection current I_D	Guard ring bias V_{GR}	Collector bias V_C	Collector current I_C
Single	−500 mA	12 V	12 V	160 μA
Single	−500 mA	0 V	12 V	450 μA
Double	−500 mA	12 V	12 V	6 μA

Figure 3.11 Guard ring characteristics. Source: Table adapted from Efland et al. [10].

Figure 3.12 Guard ring types: a) guard ring with high efficiency, b) guard ring with reduced area.

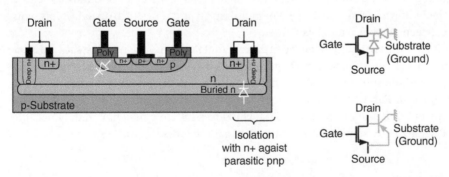

Figure 3.13 DMOS transistor with parasitic bipolar PNP transistor. The equivalent circuits are shown on the right.

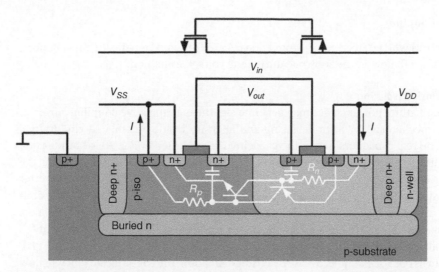

Figure 3.14 Cross section with parasitic PNP and NPN structures that can cause latch-up.

of the current gain of both transistors equals unity, i.e., $\beta_p \cdot \beta_n > 1$. This condition is defined as a latch-up. The root cause for a parasitic current may be leakage current, capacitive coupling, transient voltage undershoot and overshoot, electro-static discharge, electromagnetic interference, and particle upset. The case of capacitive coupling is explained in Section 3.3.2.

Let us assume that the PNP transistor carries a collector current; that current is supplied from V_{DD} and flows through R_n. If the voltage drop across R_n exceeds 0.7 V, the parasitic PNP device turns on. The PNP's collector current maintains the voltage drop across R_p and keeps the parasitic NPN transistor turned on, even though the root cause for the NPN turn-on may have disappeared. Consequently, the large parasitic current may result in local hotspots with permanent damage. Latch-up can only be terminated by interrupting the power supply. Hence, latch-up presents a severe risk if it occurs in the final application. There are dedicated latch-up tests, such as the JEDEC standard JESD78. In these tests, pin by pin, a test current is pulled out or sourced into each IO pin. Any excessive increase in current at the supply pin(s) or other pins indicates a latch-up risk.

These essential rules help to prevent latch-up:

- Ensure a low-resistive connection of n and p-areas by placing sufficient substrate and well contacts. Carriers will be diverted via these contacts in short distances.
- Place guard rings around p- and n-wells with low-resistive connections.
- Larger lateral spacing will increase the base width and reduce the current gain. Likewise, a highly doped n+ buried layer will degrade the PNP transistor.

Isolation of Circuit Blocks

In addition to adding guard rings to protect sensitive blocks, those sensitive circuit parts can also be isolated, as illustrated in the right-hand part of Fig. 3.10b). As a general rule, all circuit blocks should be isolated this way. The only limitation arises from the fact that some devices use the isolation layers as part of their layout. For example, a vertical NPN BJT in a bandgap reference may utilize a buried n+ layer. In any case, sensitive blocks should be placed at a distance from any noisy device because the recombination of minority carriers follows a 1/distance relation. Silicon-on-insulator technologies (SOI) may be a suitable alternative as they provide oxide isolation instead of junction isolation, as shown in Fig. 3.9. Each active component is isolated by its own BOX, and there is no parasitic bipolar effect.

3.3.2 Capacitive Coupling

Several active structures will see steep voltage transitions. As they represent parasitic capacitances, displacement currents will flow in the substrate during the voltage transition.

Capacitive Coupling Inside Isolation

Due to the switching nature, power stages may generate significant capacitive coupling currents. It may be a trigger mechanism for local debiasing and latch-up inside an isolated circuit area. The root cause can be real capacitors like in the capacitive level shifter in Fig. 3.15a) or parasitic node capacitances like in gate driver stages. Figure 3.15b) shows the equivalent structure for the lower floating inverter of the level shifter in Fig. 3.15a). Depending on the polarity of the inverter output V_2, the transistor channel will form another parallel path, which is not shown for visibility. In any case, C_c causes a capacitive (dis)charge current if the floating inverters experience a common-mode transition. A rising edge at node V_2 leads to a positive backgate transient current I_{bp}, while a falling transition results in I_{bn} to flow in a positive direction. In consequence, either the PNP or the NPN device turns on. For example, in case of a rising transition, the coupling current will flow through R_p and may turn on the NPN device. Its collector current, in turn, flows via R_n such that the PNP can turn on and triggers latch-up. The latch-up rules of Section 3.3.1 can prevent these effects.

Substrate Coupling Due to Fast dv/dt Transitions

As they are huge, the depletion capacitance of the isolation structures and the buried layer may cause extreme substrate currents. A large amount of charge is injected into the substrate within a very short time. Figure 3.16 shows this scenario for a high-side power switch where the gate driver

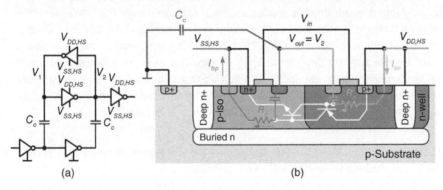

Figure 3.15 Capacitive coupling inside isolation for a) a capacitive level shifter and b) cross section with parasitic structures that can cause latch-up.

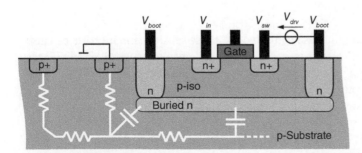

Figure 3.16 Parasitic substrate coupling due to parasitic capacitances of the isolation structure.

is referred to the switching node. Its isolation is connected to the high-side bootstrap supply V_{boot} (see Section 5.11), which is one drive voltage level V_{drv} (typically 5 V) higher than V_{sw}.

For slopes greater than 10 V/ns (see Table 5.3 in Section 5.10), capacitances in the range of a few picofarad cause substrate current in the order of 100 mA. The current spreads across the substrate and is typically diverted through substrate contacts and bond wires to the system ground. Due to the finite resistance of the substrate, local debiasing of a few volts can be observed and may lead to the malfunction of analog circuits and digital designs (bit flips).

Countermeasures include p-guard rings, conducting trench isolation, and back-side metallization (BSM). A p-guard ring contains a p+-buried layer under a p+-doped deep well, but with the typical mask set of a standard process, it does not extend down to the p-substrate. P-guard rings achieve a suppression of the coupling current by >60%. The conducting trench close to the high-side isolation well diverts nearly all the coupling currents. However, only a few technologies offer conducting trenches, as they require additional process steps. Back-side metallization comprises a conducting layer at the back side (bottom in Fig. 3.10). It diverts any substrate current directly underneath the power devices toward the system ground. BSM is a nonstandard step during backend production associated with additional cost and reliability requirements.

Capacitive coupling is also present in SOI technologies due to each device's oxide isolation (BOX). Due to the oxide isolation, this may not lead to malfunctions but can contribute to significant charging losses with a negative impact on the power conversion efficiency at large input voltages above 100 V. High-resistively biased wells provide an effective loss reduction. These techniques are described for integrated capacitors in Section 4.1. A high-voltage-biased junction below the BOX is one of the most effective countermeasures against capacitive coupling in SOI, which is demonstrated to improve the efficiency of a power converter by approximately 30% [9].

3.4 Safe Operating Area (SOA)

Each power transistor has a maximum rating for its drain source voltage V_{DS} and its drain current I_D. The relation between V_{DS} and I_D forms the safe operating area (SOA), as shown in Fig. 3.17a) for a typical power transistor. Any voltage and current peaks in switching operation, as considered in Section 2.5.6, must stay within the SOA. The SOA graph generally assumes an ambient temperature of 25 °C with a specific junction temperature, usually below 175 °C. The SOA is plotted in a double-log scale. This way, the SOA can be approximated by straight lines. Within these limit lines, the transistor can be safely operated. The relevant mechanisms are as follows, indicated in Fig. 3.17a).

(1) The on-resistance limit line is not a real limit for safe operation, but the on-resistance R_{DSon} rather prevents getting beyond $I_D = V_{DS}/R_{DSon}$ at a given voltage V_{DS}. The slope of the line is simply the maximum R_{DSon} of the transistor at maximum junction temperature (175 °C) for a given gate-source voltage V_{GS}. The on-resistance limit line will be lower for lower V_{GS} because R_{DSon} increases.
(2) The voltage limit line is defined by the breakdown voltage V_{DSmax} of the device. It depends on the junction temperature T_j.
(3) The current limit line I_{Dmax} is typically defined by the maximum current the package can handle. Also, electromigration may determine the current limit.
(4) Maximum steady-state power dissipation P_{max} defines the limit $I_D = P_{max}/V_{DS}$. Because the SOA graph generally assumes a specific junction temperature, the power dissipation-related

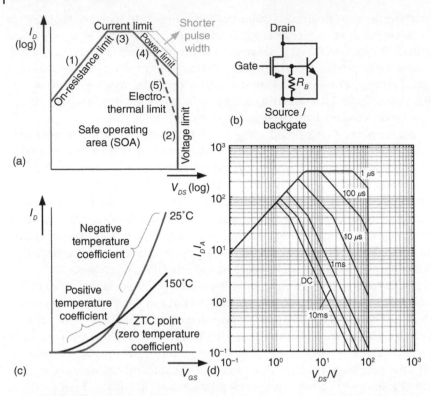

Figure 3.17 Safe operating area (SOA): a) SOA graph; b) power transistor equivalent circuit with parasitic bipolar junction transistor; c) transfer characteristic over temperature; d) SOA graph for OptiMOS5 BSC070N10NS5 (Infineon, extracted from data sheet).

boundary is given for the DC case and pulses of varying duration. These lines are constant power lines representing the maximum power the system is allowed to generate to raise the device junction temperature to the maximum rated junction temperature (e.g., 150 °C) while maintaining a constant 25 °C case temperature (ideal cooling). The pulsed lines allow more power as the die can absorb a certain amount of energy.

(5) Limit due to electrothermal interactions in the power transistor, often referred to as thermal SOA and electrical SOA. These effects may cause undesired turn-on of the parasitic bipolar device that exists in parallel with the actual power transistor, as shown in Fig. 3.17b). Due to the parasitic bipolar effect, the drain current shows a positive temperature coefficient, which causes thermal runaway in this region of the SOA plot.

The definition of the SOA limits according to items (1) to (4) gives a good general approximation. It is based on purely thermal considerations and neglects further limitations in (5) related to electrothermal interactions, as not all transistors show these limits. The latter combines two effects, electrical SOA and thermal SOA. Both impact ionization and thermal generation create excess carriers within the device, which provide currents that the gate voltage cannot control.

The electrical SOA boundary corresponds to the point in the output characteristic of the transistor, where the device begins to exhibit negative resistance. This triggering of the parasitic BJT is also called snapback. Snapback occurs within tens of pico-seconds. A large current causes local heating that may destroy the transistor. Higher gate bias V_{GS} (i.e., higher drain current I_D) causes a reduction in V_{DS} rating because a higher I_D accelerates the electron–hole pair generation that can forward bias the bipolar transistor's base-emitter junction (via the parasitic base resistance R_B, Fig. 3.17b)).

The second effect, thermal SOA, is related to thermal instability. It occurs if the power generation P_{gen} rises faster than the power dissipation P_{diss} over temperature, i.e., $\partial P_{gen}/\partial T = V_{DS}\partial I_D/\partial T > \partial P_{diss}/\partial T$. If the transistor is fully turned on, the temperature coefficient $\partial I_D/\partial T$ is negative, and the transistor is immune to thermal instability. This is confirmed by the transfer characteristic of the power transistor in Fig. 3.17c). However, the current will rise over temperature for lower gate-source voltage, indicating a positive temperature coefficient. This effect is because the threshold voltage's negative temperature coefficient dominates over the positive temperature coefficient of the on-resistance.

Consequently, local hot spots will draw more current as they heat up. It leads to increased local power dissipation and further heating. Accordingly, this results in thermal runaway and destruction of the chip.

Figure 3.17d) shows the SOA graph for an exemplary discrete power transistor (OptiMOS^{TM}, BSC070N10NS5, Infineon). We can go through some example values based on this graph. The transistor is rated for $V_{DS} = 100\,V$, which sets the voltage limit. The on-resistance at $V_{GS} = 10\,V$ and $T_j = 150\,°C$ is $R_{DC,on} = 12.5\,m\Omega$. For the DC-case, the maximum current of 80 A is reached at $V_{DS} = 1\,V$. The current limit increases to about 300A for very short pulses of $1\,\mu s$. This is linked to the power limit line, which supports about 83 W DC power (such as $2\,V \cdot 40\,A$) and as much as $15\,kW$ for a $1\,\mu s$-pulse. The positive temperature coefficient kicks in for currents below approximately 40 mA, which marks the upper onset of the thermal SOA limit.

3.5 Integrated Diodes

Diodes are widely used in power management. Examples include power "switches" in a charge pump and drain-gate clamps for overvoltage protection of the power stage. Two options exist, defined by the underlying PN junction: a shallow p-area inside a deep n-well and, vice versa, a shallow n-area inside a p-well. Both diode options come along with a parasitic bipolar transistor. The design challenge is to keep the bipolar effect at a minimum.

3.5.1 Diodes with a Parasitic PNP Transistor

The first diode option is a shallow p-area placed inside a deep n-well, as shown in Fig. 3.18. The p-area is usually the one that forms a drain or source of a low-voltage MOS transistor. It corresponds to the anode of the diode, while the n-well forms the cathode. Figure 3.18 indicates that this structure forms a parasitic PNP device. Its current gain may be as large as 50, i.e., without any precautions, most of the diode current will be lost toward the substrate. This setup nicely illustrates how critical it is to take the bipolar effect into account. If we treat the diode as a black box with its two terminals, we may wonder why Kirchhoff's current law is not fulfilled. There may be a current of 100 mA flowing into the anode of the forward-biased diode, while only 50 mA arrive at the cathode. Only by recognizing that the missing 50 mA flow toward the "third" terminal, the p-substrate (the collector), everything makes sense again. Unfortunately, this bipolar effect is not always modeled in a given technology.

How to eliminate the parasitic PNP? Figure 3.18 shows how deep n-wells and a buried n-layer can encapsulate the diode structure. By adequately connecting all areas to the anode and cathode terminals of the diode, the PNP current gain reduces to below 1% [11]. A PN diode without any

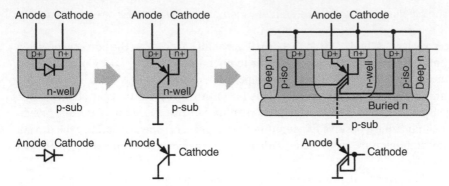

Figure 3.18   A PN diode formed by a shallow p-area inside a deep n-well. This structure is associated with a parasitic PNP bipolar transistor. Isolation reduces its effect significantly.

isolation may cause circuits to fail. Charge pumps are a typical example, as covered in Chapter 8, see Section 8.4 and Fig. 8.8.

3.5.2   Diodes with a Parasitic NPN Transistor

The other option to implement a PN diode is to place a shallow n-area (cathode) inside an isolated p-area (anode), as shown in Fig. 3.19a). Figure 3.19b) indicates that a parasitic NPN transistor is formed. However, in this case, the parasitic transistor contributes to the main diode current. Therefore, this diode has inherently lower parasitic substrate currents and should be preferred. Moreover, if Zener diodes are available in the technology, they readily offer all required structures, as shown in Fig. 3.19c). In the forward direction, they can be used as regular diodes with nearly no impact from the parasitic bipolar effect. Nevertheless, at high temperatures (above 100 °C), junction leakage toward the substrate will be present.

3.5.3   The DMOS Transistor As a Power Diode

By shorting the gate-source terminals, the body diode of a lateral DMOS transistor forms a power diode, as shown in Fig. 3.20. The current capability is large and similar to the actual DMOS transistor. The body diode forms a parasitic PNP transistor with the substrate diode. Without any precautions, as much as half of the diode current may flow toward the substrate. The n+ isolation

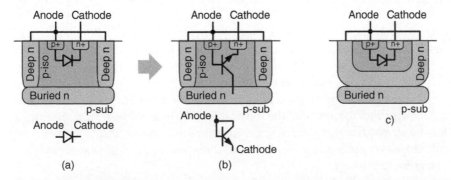

Figure 3.19   a) A PN diode formed by a shallow n-area inside an isolated p-area; b) a parasitic NPN bipolar structure occurs, contributing to the main diode current. c) A Zener diode in forward bias can be used as a PN diode.

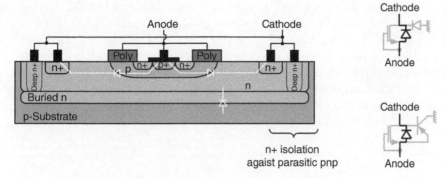

Figure 3.20 DMOS transistor forming a power diode.

shown in Fig. 3.20, consisting of a deep n-well and a buried n layer, reduces the PNP current gain to approximately 0.1%. Still, a diode current of 10 mA may be associated with a substrate current of $\sim 10\,\mu A$.

3.5.4 Zener Diodes

Several technologies offer Zener diodes, which can be used as clamps and voltage references. Zener diodes utilize the breakdown voltage of the base-emitter junction of a vertical BJT, as shown in Fig. 3.21a). The emitter of the NPN transistors forms the cathode, while the base acts as the anode. The symbol is shown in Fig. 3.21b). The collector acts as the tank that provides isolation from the p-substrate such that the cathode and anode are separated from the ground.

Two options for connecting the tank form a C-B and C-E shorted version of a Zener diode, Fig. 3.21c,d). Both options are significantly different. The C-B shorted Zener diode has a parasitic junction toward the substrate. Hence, it should only be used if the anode (base) stays above 0 V to ensure that the junction remains reverse-biased. If this is fulfilled, we should prefer the C-B option because the C-E option incorporates a parasitic substrate PNP transistor as indicated in Fig. 3.21d). Hence, a significant emitter current flows toward the substrate in the forward direction, increasing the latch-up risk, as the NPN and PNP of Fig. 3.21a) form an SCR (see Section 3.3.1). A highly doped buried layer may reduce the current gain of the substrate PNP. Typical Zener voltages are in the range of 6 V. This way, Zener diodes can nicely be used as clamps that protect the maximum gate-source voltage of (power) transistors.

As a rule of thumb, these Zener diodes can handle 10 to 100 μA per micron of emitter periphery (up to 500 μA for short events like in electro-static discharge (ESD) protection cells). For higher current capabilities, a Zener diode can be combined with a power transistor to form an active Zener

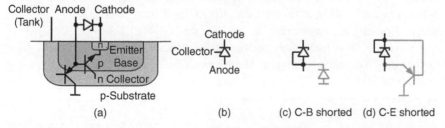

Figure 3.21 Integrated Zener diodes: a) cross section; b) symbol; c) collector-base shorted Zener diode; d) collector-emitter shorted Zener diode.

circuit as described in Section 6.1.2 (Fig. 6.3). For breakdown voltages of 6 V and above, the conduction mechanism is avalanche multiplication, which shows a positive temperature behavior. The typical temperature coefficient *TC* of the Zener voltage is 2 to 40 mV/°C (the higher V_z the larger the *TC*). This way, partial temperature compensation can be achieved by combining one or more Zener diodes with forward-biased PN diodes (*TC* = −2 mV/°C at room temperature). C-E shorted Zener diodes can also be stacked to achieve high clamping voltages. In that case, the tank can be connected to the cathode of one or two diodes above.

References

1 Lutz, J., Schlangenotto, H., Scheuermann, U., and Doncker, R.D. (2011) *Semiconductor Power Devices*, Springer-Verlag, Berlin, Heidelberg.

2 Baliga, B.J. (2018) *Fundamentals of Power Semiconductor Devices*, Springer, Cham, doi: 10.1007/978-3-319-93988-9.

3 Mohan, N., Undeland, T.M., and Robbins, W.P. (2003) *Power Electronics. Converters, Applications and Design*, John Wiley & Sons, Inc., 3rd edn.

4 Then, H.W., Radosavljevic, M., Desai, N., Ehlert, R., Hadagali, V., Jun, K., Koirala, P., Minutillo, N., Kotlyar, R., Oni, A., Qayyum, M., Rode, J., Sandford, J., Talukdar, T., Thomas, N., Vora, H., Wallace, P., Weiss, M., Weng, X., and Fischer, P. (2020) Advances in research on 300mm gallium nitride-on-Si(111) NMOS transistor and silicon CMOS integration, in *2020 IEEE International Electron Devices Meeting (IEDM)*, pp. 27.3.1–27.3.4, doi: 10.1109/IEDM13553.2020.9371977.

5 Tee, E.K.C., Hoelke, A., Pilkington, S., Pal, D.K., Antoniou, M., Udrea, F., Abidin, W.A.B.W.Z., and Yew, N.L. (2013) 200V superjunction lateral IGBT fabricated on partial SOI, in *2013 25th International Symposium on Power Semiconductor Devices & IC's (ISPSD)*, pp. 389–392, doi: 10.1109/ISPSD.2013.6694427.

6 Rindfleisch, C., Wicht, B., Tee, E.K.C., and Hölke, A. (2021) The on-chip lateral super-junction IGBT in integrated high-voltage low-power converters, in *2021 33rd International Symposium on Power Semiconductor Devices and ICs (ISPSD)*, pp. 51–54, doi: 10.23919/ISPSD50666.2021.9452214.

7 Kaufmann, M. and Wicht, B. (2022) *Monolithic Integration in E-Mode GaN Technology*, Springer International Publishing, doi: 10.1007/978-3-031-15625-0.

8 Hölke, A., Pal, D.K., Hao, Y., Yaw, K.K., Kho, E., Kittler, G., Kuniss, U., and Gessner, J. (2010) A 200V partial SOI 0.18 μm CMOS technology, in *2010 22nd International Symposium on Power Semiconductor Devices & IC's (ISPSD)*, pp. 257–260.

9 Rindfleisch, C. and Wicht, B. (2021) A resonant one-step 325 V to 3.3-10 V DC-DC converter with integrated power stage benefiting from high-voltage loss-reduction techniques. *IEEE Journal of Solid-State Circuits*, 56 (11), 3511–3520, doi: 10.1109/JSSC.2021.3098751.

10 Efland, T., Devore, J., Hastings, A., Pendharkar, S., and Teggatz, R. (2000) Bipolar issues in advanced power BiCMOS technology, in *Proceedings of the IEEE 2000 BIPOLAR/BiCMOS Circuits and Technology Meeting*, IEEE, Minneapolis, MN, USA, pp. 20–27, doi: 10.1109/BIPOL.2000.886166.

11 Hastings, A. (2006) *The Art of Analog Layout*, Prentice-Hall.

4

Integrated Passives

Passive components, capacitors, and inductors are essential for high conversion efficiency in switched-mode converters. They also help reduce the current and voltage ripple as they store energy. Operating as a buffer, they are crucial in minimizing the droop in case of fast load transients. In this chapter, we cover the on-chip integration of capacitors and inductors. In general, integrated passives are not comparable with their discrete counterparts. However, they are key to supporting the ongoing demand for miniaturization.

4.1 Capacitors

This section reviews various types of integrated capacitors along with their pros and cons. Capacitors are widely used in power management integrated circuits. They can be applied as a buffer for the input and output voltage, also known as bypass or decoupling capacitor (decap). In switched-capacitor (SC) and hybrid DC–DC converters, capacitors form an essential part of the power path. Key parameters include capacitance density (capacitance per area) and voltage ratings. Also, parasitics, mainly the parasitic top- and bottom-plate capacitances toward substrate (C_{tp}, C_{bp}), have to be considered. Figure 4.1a) shows a capacitor C with added parasitic capacitances C_{tp} and C_{bp}. Both parasitic capacitances can be expressed as a fraction of the actual capacitor value C,

$$C_{bp} = \gamma C, \tag{4.1}$$

with a coupling factor γ. A similar definition applies to C_{tp}. Sometimes, γ is called the capacitor's quality factor. In that case, it must be noted that high quality factor values correspond to larger parasitics [1]. The key parameters of the common integrated capacitor types are summarized in Table 4.1.

4.1.1 Metal-Oxide-Semiconductor Capacitors

Referred to as the MOS capacitor, this capacitor corresponds to the oxide capacitance of the MOS transistor. Hence, this capacitor is available in any baseline Complementary Metal-Oxide-Semiconductor (CMOS) technology. It is formed between the gate (capacitor top-plate) and the shorted terminals source, drain, and back gate (bottom-plate). Drain and source areas are not needed, as shown by the cross-section in Fig. 4.1b). Relatively large capacitor values can be implemented with typical densities of 2–5 fF/μm^2, related to the thickness and dielectric constant of the gate oxide. The dielectric strength is identical to the gate-source voltage ratings of the underlying

Design of Power Management Integrated Circuits, First Edition. Bernhard Wicht.
© 2024 John Wiley & Sons Ltd. Published 2024 by John Wiley & Sons Ltd.

Table 4.1 Key parameters of integrated capacitors.

	MOS	MOM	MIM
Capacitance density	2–5 fF/μm^2	0.5 fF/μm^2	1 fF/μm^2
Voltage rating	$\leq$ 5 V	$\geq$ 100 V	10 V
Coupling factor	4–6%	1%	1–2%
Voltage dependence, nonlinearity	High	Low	Low

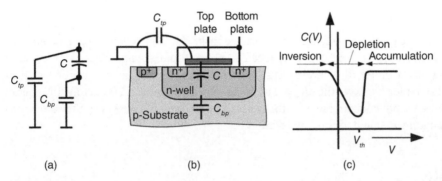

(a)   (b)   (c)

Figure 4.1 MOS capacitor: a) Symbol with parasitic top- and bottom-plate capacitance; b) PMOS transistor as a MOM capacitor with parasitic capacitance; c) voltage dependency of the MOS capacitor.

MOS transistor (typ. $\leq$ 5 V). Both NMOS and PMOS transistors can be used. The bottom plate is formed by an n-doped region (NMOS) or p-doped region (PMOS). For this reason, the MOS capacitance value strongly depends on the actual voltage across the capacitor. For designs that require an isolated bottom plate, either a p-type MOS transistor with its n-well connected to the bottom plate or an isolated n-type transistor is needed. An example would be flying capacitors in switched-capacitor converters. Due to the additional junctions, a parasitic bottom-plate capacitance C_{bp} occurs. The p-type MOS transistor, shown in Fig. 4.1, results in lower values of C_{bp} compared to an isolated NMOS because of the low doping concentration of the n-well [2]. Nevertheless, C_{bp} depends on the bottom-plate voltage. The coupling factor γ reaches typical values of 4–6% [3].

The highly nonlinear voltage dependency of the MOS capacitor has to do with the fact that it can be operated in inversion and accumulation. Figure 4.1c) shows how the capacitance C depends on the capacitor voltage (between the top and bottom plate). For a positive voltage at the gate of the p-type transistor in Fig. 4.1b), electrons will accumulate under the gate oxide. The gate oxide acts as an insulating medium between the gate and the accumulated electrons, forming a capacitor. In opposite polarity, with gate voltage (the top-plate potential) being lower than the bottom-plate potential by more than the transistor's threshold voltage V_{th}, an inversion layer appears at the semiconductor surface, which would be the channel in regular transistor operation. Again, a capacitor is formed between the gate and the inversion layer with similar capacitance values like in accumulation. In between accumulation and inversion, i.e., around zero bias, the device can be considered an insulator with large thickness (depletion). Consequently, the effective capacitance will decrease significantly.

The relatively large capacitance density, in combination with the strong voltage dependency, makes the MOS capacitor a preferred choice for an output buffer or bypass capacitor.

The non-linearity can be accepted since the output voltage range is usually small. Also, the bottom plate is directly connected to the grounded p-substrate ($C_{bp} = 0$).

4.1.2 Metal-Oxide-Metal Capacitors

Metal-oxide-metal capacitors (MOM capacitors) utilize the capacitance between adjacent metal lines, as illustrated in Fig. 4.2. Interdigitated structures improve the capacitor density. For the same reason, multiple metal layers are used. Consequently, MOM capacitors easily integrate on-chip without any additional mask layer. They also show negligible voltage dependence due to their structure. Advanced process technologies achieve higher capacitor density as the metal spacing shrinks to lower values. In addition, more metal layers can be utilized to improve the unit area size. Nevertheless, the metal spacing is still relatively large, and the achievable capacitance densities are usually low. Typical values are below $0.5\,\mathrm{fF/\mu m^2}$. Due to the metal spacing, the dielectric strength is relatively high with voltage ratings of more than 100 V.

Using lower layers in the CMOS process, such as polysilicon or metal 1, increases the parasitic coupling to the substrate. The parasitic bottom-plate capacitance can reach 10% of the actual capacitance. It can be reduced by placing wells underneath as described for MIM capacitors in Section 4.1.3 (see also Fig. 4.3).

4.1.3 Metal-Insulator-Metal Capacitors

This capacitor type, also called a MIM capacitor, uses an *insulator* with a much smaller dielectric thickness and achieves much higher capacitor density. Figure 4.3a) shows the cross-section. The dielectric material with high permittivity is referred to as *high-k material* with ~4× higher permittivity k than silicon dioxide. The MIM capacitor requires additional mask steps. MIM capacitors are mainly composed of two top metal layers with a special metal layer (capacitor top metal) in between, forming a parallel plate capacitor as illustrated in Fig. 4.3a). Using the upper metal layers results in very low parasitic coupling to the substrate. Parasitic capacitances are in the order of 1–2% of the actual capacitance. That is why MIM capacitors are often used as flying capacitors in SC converters. The capacitor densities are in the order of $1\,\mathrm{fF/\mu m^2}$, typically more than three times better than MOM capacitors. The dielectric strength is in the order of 10 V.

In advanced technologies, the MIM structure can be repeated between lower metal layers, and all sub-capacitors are connected in parallel, increasing the parasitic capacitances. These capacitances

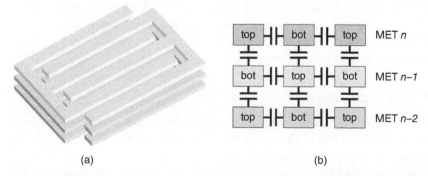

(a)   (b)

Figure 4.2 MOM capacitor: a) Typical interdigitated multi-layer structure; b) cross-section with various metal–metal capacitances that are connected in parallel between the bottom and top electrodes (bot, top) of the MOM capacitor. MET n, MET $n-1$, and MET $n-2$ indicate the metal layers.

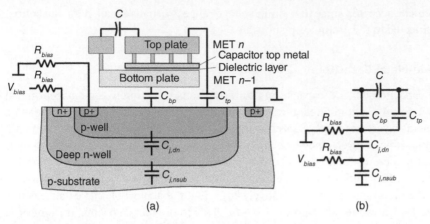

Figure 4.3 MIM capacitor: a) Capacitor cross-section with p-well and deep n-well underneath to minimize the effective parasitic top- and bottom-plate capacitances. The wells are high-resistively biased via R_{bias}; b) equivalent circuit.

can be reduced by placing wells with suitable biasing underneath the MIM capacitor, Fig. 4.3a). Additional junction capacitances appear in series, reducing the overall effective bottom-plate and top-plate capacitance. This effect is reflected in the equivalent circuit of Fig. 4.3b). A common technique is to place a single n-well [4]. If the technology offers a deep n-well, an additional high-resistively biased p-well at ground potential can be inserted, as shown in Fig. 4.3a) [5]. It adds a series capacitance $C_{j,dn}$ between the p-well and deep n-well. To minimize the junction capacitance $C_{j,nsub}$ toward the p-substrate, the n-well bias voltage V_{bias} should be large, preferably connected to the input supply voltage. Connecting V_{bias} via a large resistor R_{bias} is beneficial. It ensures proper DC bias of the well but avoids charging losses due to the capacitances. The combination of n- and p-wells can reduce the coupling factor γ from 2% (no wells) to $\gamma \sim 1.3\%$ (at $V_{bias} \sim 4\,\text{V}$) [3]. The bias resistance R_{bias} provides DC bias of the wells but prevents fast recharging of the capacitances. R_{bias} should be highly resistive with typical values in the order of $1\,\text{M}\Omega$. R_{bias} can be implemented by pseudo resistors [6] at the expense of one additional voltage drop with respect to V_{bias}. If available, a high-sheet resistor can be used instead.

For SC converters, the impact of parasitic capacitances is discussed in Chapter 9 (see Section 9.9.3). Usually, only one capacitance, C_{bp} or C_{tp}, contributes to the charge-sharing losses, like in the 2:1 converter cell, as explained in Fig. 9.11. For MIM capacitors and most other capacitor types, $C_{bp} > C_{tp}$. Therefore, the loss contributions from parasitic capacitances can be reduced by connecting the flying capacitor in reverse polarity. Applied to the scenario in Fig. 9.11, C_{bp} will then contribute to the load current while only the smaller top-plate capacitance C_{tp} causes losses. This approach is often used in SC converters.

4.1.4 Trench and Ferroelectric Capacitors

These capacitor types achieve very high area density along with low bottom-plate para-sitics. However, they are not available in standard CMOS technologies and require expensive post-processing. Etching out the high aspect ratio trenches increases the cost significantly. The maximum voltage ratings are low, typically $< 1.8\,\text{V}$. Andersen et al. [7] demonstrates a SC DC–DC converter in 32 nm silicon-on-insulator (SOI) CMOS, which utilizes deep-trench capaci-tors with a total $C_{fly} = 1\,\text{nF}$ and $V_{in} = 1.8\,\text{V}$. A multi-ratio SC DC–DC converter with ferroelectric capacitors in 130 nm CMOS with input voltages of $1.5\,\text{V}$ is implemented in [8].

4.2 Inductors

The integration of inductors is very challenging. In a standard CMOS technology, only planar designs that utilize the available metalization layers are possible. Alternatively, bond wires can be used as an inductor. The achievable inductance values are low. Inductances in the range of a few nH lead to remarkably high switching frequencies in DC–DC converters. Therefore, high-frequency phenomena like the skin effect need to be considered.

Post-processing techniques for on-chip inductors with magnetic cores achieve higher inductance and quality factor values. However, the inductance values are still small. In addition, the magnetic material introduces core losses. Parasitic resistance degrades the inductor's quality much more compared to discrete inductors. Quality factors of integrated inductors are in the range of 5–25.

This section explores air-cored bond wire and planar on-chip inductors. We briefly review implementation examples of inductors with magnetic filling, which can be processed on-die or on a separate silicon interposer. The concepts can be expanded to integrated transformers [9].

4.2.1 Ampere's Law and Inductance

The electrical current generates a magnetic field $\vec{H}$. The relationship is described by Ampere's Law as illustrated in Fig. 4.4a). For a single wire with current I:

$$\oint_C \vec{H}\, d\vec{l} = I \tag{4.2}$$

The line integral of the magnetic field $\vec{H}$ along path elements $d\vec{l}$ that follow a closed curve C is equal to the electric current I enclosed by the loop C. We use that wire to build a solenoid with N windings, as shown in Fig. 4.4b). This setup represents the helical coil as the widely used inductive component. If the length-to-diameter ratio is large, we can assume that the internal field $\vec{H}_{int}$ is homogeneous while the external field component $\vec{H}_{ext}$ is negligible. Hence, the left-hand part of Eqn. (4.2) becomes simply $(H \cdot l)$ with where H is the absolute value of $\vec{H}$ and l the length of the solenoid. Inside the closed path C, the current I flows in each of the N windings such that the right-hand part of Eqn. (4.2) corresponds to $N \cdot I$. Consequently, we can calculate the magnetic field strength H from the geometry parameters N and l and the current I. Knowing H, we can further determine the inductance L of the solenoid. We first need to find the magnetic flux Φ as the surface integral of the flux density $\vec{B} = \mu\vec{H}$ passing through a surface A,

$$\Phi = \iint_A \vec{B}\, d\vec{A} = \mu \iint_A \vec{H}\, d\vec{A}. \tag{4.3}$$

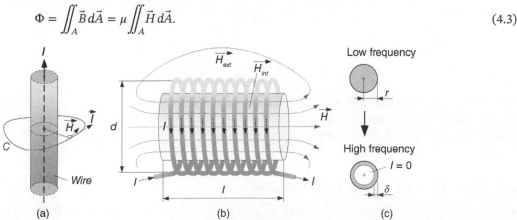

Figure 4.4 a) Ampere's law applied to a cylindrical wire; b) solenoid inductor; c) skin effect.

The permeability μ reflects the material dependency. In air (vacuum) μ is equal to the magnetic field constant $\mu_o = 4\pi 1 \times 10^7$ Vs/(Am).

For the solenoid in Fig. 4.4b), the surface corresponds to the circular area $A = \frac{\pi d^2}{4}$ of the solenoid's diameter d. This area has to be counted N-times because the flux goes through N circular windings. Hence,

$$\Phi = \frac{\mu N^2 I}{l} \frac{\pi d^2}{4}. \tag{4.4}$$

According to Eqn. (4.4), the magnetic flux Φ and the current I are proportional. The (self) inductance L is defined as the proportionality constant. For the solenoid we obtain

$$L = \frac{\Phi}{I} = \frac{\mu N^2 \pi d^2}{4l}. \tag{4.5}$$

Besides the magnetic core material (μ), the inductance L is defined by the geometry, specifically by the number of windings, their diameter, and the solenoid length.

Series Resistance and Quality Factor

We can also calculate the wire series resistance R from its length l_w and its cross-section area A. The length is equal to the solenoid's perimeter times the number of windings (turns), i.e., $l_w = \pi d N$. The wire radius r defines the area $A = \pi r^2$. With the specific resistance ρ of the wire material:

$$R = \rho \frac{l_w}{A} = \rho \frac{d N}{r^2} \tag{4.6}$$

The quality factor Q of the inductor relates the inductance L to its series resistance R. Q is defined by the quality factor of a resonator comprising L, R, and an ideal capacitor C. At the resonance frequency $\omega_o = 1/2\pi \sqrt{LC}$ the quality factor corresponds to the ratio between the inductive reactance and the (lossy) resistance,

$$Q = \frac{\omega_o L}{R}. \tag{4.7}$$

Skin Effect

At high frequencies, the resistance R increases due to the skin effect. Figure 4.4c) illustrates how the electrons flow at the outer diameter at the *skin* of the wire. The effective cross-sectional area reduces from $A = \pi r^2$ to

$$A_{eff} = \pi r^2 - \pi(r - \delta)^2 = 2\pi r \delta - \pi \delta^2 \tag{4.8}$$

with the skin depth

$$\delta = \sqrt{\frac{\rho}{\pi \mu_o f}} \tag{4.9}$$

at frequency f.

Example 4.1 *At $f = 50$ MHz, the skin depth δ in a golden bond wire ($\rho = 2.44 \times 10^{-8}$ Ωm) reaches ~11.1 μm, which is approximately equal to the radius of a standard bond wire (25 μm diameter). Consequently, the skin effect will impact the effective area of a bond wire for frequencies above 50 MHz.*

4.2.2 Bond-Wire and Package-Layer Inductors

Bond wires are available as part of standard IC packaging. Straight segments of equal length l_s can be combined to form one turn of the bond wire inductor. Bond pads connect the segments.

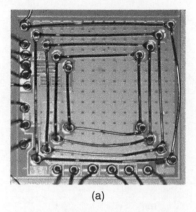 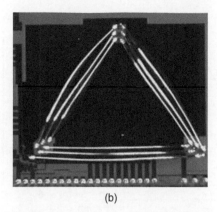

(a)   (b)

Figure 4.5 Bondwire inductors: a) Square shape, $L = 18$ nH, four turns, outer segment length $l_s = 1.4$ mm. Source: Mike Wens et al. [10]/from IEEE; b) Triangular shape, $L = 21$ nH, three turns, outer segment length $l_s = 2.3$ mm. Source: Gerard Villar and Eduard Alarcon [11]/from IEEE.

Multiple turns can be implemented as a spiral inductor. Figure 4.5 shows two early examples of bond wire inductors, the square shape implemented by Wens et al. in 2007 [10] and the triangular shape presented by Villar and Alarcón in 2008 [11]. The square shape hollow spiral inductor in Fig. 4.5a) has four turns with a segment length $l_s = 1.4$ mm of the outer turn. The total inductance is $L = 18$ nH. Villar and Alarcón [12] investigate various shapes of bond wire inductors and finds that the triangular implementation offers the highest inductance per area among various shapes (triangle, square, pentagon, hexagon, and octagon). The triangular spiral inductor in Fig. 4.5b) achieves a total inductance of $L = 21$ nH with three turns and an outer side length of $l_s = 2.3$ mm.

The general inductance equation for the solenoid inductor, Eqn. (4.5), is not valid anymore because the field distribution is not uniform. Also, the assumption that the coil diameter is much smaller than the length of the solenoid does not apply. Instead, the inductance can be calculated based on Grover's inductance model [13]. For one bond wire segment of length l_s with radius r, we get

$$L_s = \frac{\mu_0}{2\pi} \cdot l_s \cdot \left(\ln\left(\frac{2\,l_s}{r}\right) - 0.75 \right) = \frac{l_s}{5 \cdot 10^6} \cdot \left(\ln\left(\frac{l_s}{r}\right) - 0.0569 \right). \tag{4.10}$$

From this model, we can determine the inductance of 1 mm bond wire. For a typical bondwire with 25 μm diameter, Eqn. (4.10) results in 0.87 nH. As a rule-of-thumb, 1 mm bond wire corresponds to an inductance of approximately 1nH. This result is useful for estimating parasitics due to long bond wires. See Chapter 12 where Section 12.2.2 provides more details on bond wires.

Returning to our bond wire inductor, we can also determine the resistance R_s of the segment. This is given by Eqn. (4.6) (for $l_w = l_s$). The bond pad resistance must be added and can dominate the overall resistance.

The total inductance L and the resistance R of one turn correspond to the sum of all segment values. Consequently, we need to multiply L_s of Eqn. (4.10) and R_s of Eqn. (4.6) by 3 or 4 for the triangular and the square shape inductor, respectively. Figure 4.6a) plots the inductance L as a function of the segment length l_s. The inductance L increases with l_s, but it results also in larger resistance R as shown in Fig. 4.6b) for a typical gold bond wire with 25 μm diameter ($r = 12.5$ μm). The square shape gives larger L for the same segment length because it encloses a larger area with a larger diameter d. It aligns with the general inductance expression in Eqn. (4.5). Since the square also has one more leg than the triangle, its resistance is larger.

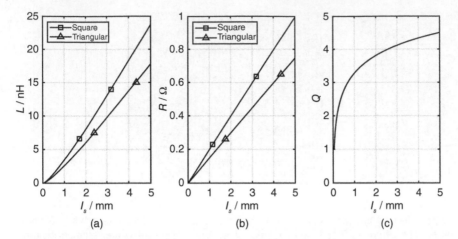

Figure 4.6 Bondwire inductor in triangular and square shape: a) inductance L; b) total wire resistance R, and c) quality factor at $f = 30\,\text{MHz}$ as a function of the segment length l_s for one turn of a gold bond wire with 25 μm diameter. The quality factor is independent of the inductor shape.

Finally, we can analyze the quality factor of the bond wire inductor as defined by Eqn. (4.7) and plotted in Fig. 4.6c) at $f = 30\,\text{MHz}$. Interestingly, Q is independent of the shape because of the inductance and the resistance scale with the number of segments and cancel out in the expression for Q. Q increases with l_s, but it remains low in general.

The inductance can be further increased by implementing multiple turns. The positive mutual inductance M (see Section 4.2.3) can improve the total inductance L. However, due to the spacing between the wires, M is relatively low. We can estimate the total inductance by multiplying the inductance values of Fig. 4.6a) by the number of turns N. However, this requires the use of a mean segment length because of the spiral shape with decreasing diameter toward its center.

Example 4.2 *Let us verify the inductance values of the two bond wire inductors in Fig. 4.5 through Eqn. (4.10). We assume gold bondwire with radius $r = 12.5\,\mu\text{m}$. The magnetic field constant is $\mu_o = 4\pi \cdot 1 \times 10^7\,\text{Vs/(Am)}$.*

Figure 4.5a): We know the outer segment length 1.4 mm and we can estimate a mean segment length of 1.1 mm. For one segment Eqn. (4.10) yields

$$L_s = \frac{l_s}{5 \cdot 10^6} \left(\ln\left(\frac{1.0\,\text{mm}}{12.5\,\mu\text{m}}\right) - 0.0569 \right) = 0.97\,\text{nH}. \tag{4.11}$$

This value needs to be multiplied by the number of segments (4) and $N = 4$ turns,

$$L = 4 \cdot 4 \cdot L_s = 16 \cdot 0.97\,\text{nH} \sim 16\,\text{nH}. \tag{4.12}$$

We would get the same result by extracting the inductance of one turn for the square shape inductor in Fig. 4.6a), which is roughly 4 nH, and multiplying it by $N = 4$ turns.

Figure 4.5b): Similar to Eqn. (4.11), we obtain $l_s \sim 2.1\,\text{mm}$ and $L_s = 2.1\,\text{nH}$ per segment. The total inductance will be $L = 3 \cdot 3 \cdot L_s \sim 18.9\,\text{nH}$.

In series production, there will be several challenges related to bond wire inductors. Sufficient spacing between the wires must be ensured to prevent shorts, which may impact reliability and reproducibility. Overall, wire bonding is increasingly being replaced by flip-chip assembly of the die.

(a)   (b)

Figure 4.7 Package layer inductor. a) The bottom of an Intel® 4th generation Core™ microprocessor LGA package and corresponding die. A group of eight FIVR inductors is pulled off to the side. b) Enlarged 3-D view of two FIVR inductors. Source: Burton et al. [14]/IEEE.

Therefore, a promising alternative is the implementation of inductors using the bottom metal layers of a standard flip-chip package. Figure 4.7 shows the approach presented by Burton et al. [14] for a land grid array package (LGA). The nonmagnetic inductors are used in a fully integrated voltage regulator (FIVR) with up to 360 phases. The current density is 31 A/mm². An LGA package with four CPUs fits 59 inductors on 10 different voltage rails in a 20 mm × 20 mm area. The package design allows the inductors to be customized per rail to meet efficiency, ripple, and transient response requirements.

4.2.3 Planar Metal-Layer Inductors

Using the available metal layers, the solenoid inductor of Fig. 4.4b) can be implemented as an integrated planar spiral inductor (without magnetics, i.e., air core). Circular spirals are usually not possible because the layout design rules restrict the possible angles to 90°, 45°, and perhaps 30°. Therefore, on-chip inductors are implemented as polygonal spirals in square, hexagonal, and octagonal shapes, as shown in Fig. 4.8. Another metal layer can connect the inner winding end. A direct bond connection and flip-chip bonding can achieve lower resistance.

Basic Inductor Model
The planar on-chip inductor is characterized by its inductance L and its series resistance R. Due to the short distance, the parasitic capacitance C_{sub} toward the substrate must be considered. The basic lumped model is shown in Fig. 4.9. The model does not contain a coupling capacitance between the two terminals since the interwinding capacitances are small. Also, there is no return path under the inductor, which would introduce significant coupling between the terminals.

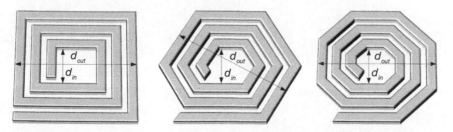

Figure 4.8 On-chip spiral inductor in the square, hexagonal, and octagonal shape.

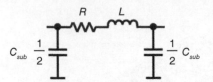

Figure 4.9 Equivalent circuit of a planar metal-layer inductor.

Substrate losses due to the eddy currents induced by the coil's magnetic field are not included in this simple inductor model.

Inductor Implementation

One of the first power management designs with an on-chip metal-layer inductor is presented in [15]. The inductor is implemented as a 1.5-turn coil on the top metal layer resulting in $L = 1.5\,\text{nH}$ and $Q = 4.4$ (area: $0.25\,\text{mm}^2$). Figure 4.10 shows an example of a square shape inductor, published in 2019 [16]. Multiple metal layers are connected in parallel to minimize the series resistance R. An ultra-thick top metal layer forms the upper level (metal 7). This layer has a typical thickness of 3 µm and may be available in aluminum (sheet resistance: $>10\,\text{m}\Omega/\square$) or copper ($2$–$5\,\text{m}\Omega/\square$). The second top layer (metal 6) is made of thick aluminum (with a typical thickness of 900 nm). The lower metal layers have a standard height (typ. 400 nm). The inductor design in Fig. 4.10 achieves $L = 9\,\text{nH}$, $R = 280\,\text{m}\Omega$, and $Q = 6$ at $f = 35\,\text{MHz}$.

Compact Expressions

Unlike the solenoid inductor, the magnetic field is not homogeneous, and the inductance cannot be calculated directly. Fortunately, compact expressions are available. Mohan et al. [17] provides various formulas, including a modified version of the Wheeler formula [18] for multiple shapes of on-chip inductor realizations:

$$L = K_1\,\mu_o\,\frac{N^2\,d_{avg}}{1 + K_2\,d_f} \tag{4.13}$$

This is a handy equation that requires only two geometry parameters, the average diameter d_{avg} and the fill ratio d_f. The coefficients K_1 and K_2 depend on the shape format given in Table 4.2. Knowing the inner and outer diameter of the spiral inductor, d_{in} and d_{out}, indicated in Fig. 4.8, the geometry parameters are defined as

$$d_{avg} = \frac{d_{out} + d_{in}}{2}, \tag{4.14}$$

and

$$d_f = \frac{d_{out} - d_{in}}{d_{out} + d_{in}}. \tag{4.15}$$

Table 4.2 Coefficients for the compact expression in Eqn. (4.13).

Layout	K_1	K_2
Square	2.34	2.75
Hexagonal	2.33	3.82
Octogonal	2.25	3.55

A small fill ratio d_f represents a hollow inductor ($d_{out} \approx d_{in}$) while a higher value of d_f describes a full inductor ($d_{out} \gg d_{in}$). The fill ratio appears in the denominator of Eqn. (4.13). Consequently, higher fill ratios result in lower inductance because the inner turns are so close to the center of the spiral that they will contribute more negative mutual inductance and less positive mutual inductance. Since the metal trace will be shorter, hollow coils will also have lower resistance R, resulting in higher quality factor Q.

Example 4.3 *Verify the square spiral inductor of Fig.* 4.10 *using the compact expression of Eqn.* (4.13).

From Fig. 4.10 we know $d_{out} = 1500\,\mu m$, $N = 3$ turns and we assume $d_{in} = 300\,\mu m$. Equations (4.14) and (4.15) yield an average diameter of $d_{avg} = 900\,\mu m$ and a fill factor $d_f = 0.667$. For a square shape, the coefficients are $K_1 = 2.34$ and $K_2 = 2.75$. Inserting these parameters into Eqn. (4.13):

$$L = K_1\,\mu_0\,\frac{N^2\,d_{avg}}{1 + K_2\,d_f}$$

$$= 2.34 \cdot 4\pi \cdot 1 \times 10^7\,\text{Vs/(Am)}\,\frac{3^2 \cdot 900\,\mu m}{1 + 2.75 \cdot 0.667} = 8.4\,\text{nH} \tag{4.16}$$

The result is slightly lower than the value of approximately 9 nH in [16].

Could we further increase the inductance? By enlarging the inner diameter d_{in}, we get a higher fill ratio d_f. This way, we could easily double the inductance. However, the series resistance R will increase significantly as the trace width decreases.

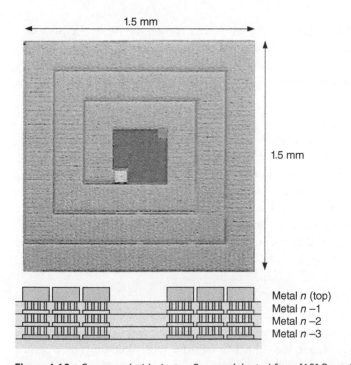

Figure 4.10 Square spiral inductor. Source: Adapted from [15] Renz, P et al. 2019. Top view and cross-section showing the implementation with various metal layers for low series resistance.

Series Resistance

The resistance of the planar inductor can be calculated similarly to Eqn. (4.6) for a rectangular cross-sectional area $A = w \cdot t_w$ of the winding with a total length l_w:

$$R = \rho \frac{l_w}{A} = \rho \frac{l_w}{w \cdot t_w} \tag{4.17}$$

The resistance is impacted by two high-frequency phenomena: the skin and proximity effects. For planar inductors with the windings on the same plane, [19] shows that the proximity effect is negligible. The skin effect considers that the current flows within a thin surface of the inductor winding. With the skin depth δ given by Eqn. (4.9) the effective winding thickness reduces to [19]

$$t_{w,eff} = \delta \cdot \left(1 - e^{-t_w/\delta}\right). \tag{4.18}$$

The skin depth reduces with frequency (see Eqn. (4.9)), and so does the effective thickness $t_{w,eff}$. Hence, the series resistance increases toward higher frequencies. It is critical for fast switching and resonant DC–DC converters operating in the multi-megahertz range. At $f = 50$ MHz the effective thickness of a 3 μm copper winding ($\rho = 1.71 \times 10^{-8}$ Ωm, $t = 3$ μm) is already only $t_{w,eff} = 2.6$ μm.

Substrate Capacitance

The parasitic capacitance toward the substrate limits the achievable quality factor Q. It can be approximated as a planar plate capacitor with an electrode distance given by the oxide thickness t_{ox}. The plate area is equal to the area of the winding trace. Hence,

$$C_{sub} = l_w \cdot w \cdot \frac{\epsilon_{ox}}{t_{ox}}. \tag{4.19}$$

The shorter the width w and length l_w (related to the number of turns), the lower the parasitic capacitance. However, smaller windings degrade the series resistance R. A proper inductor design usually requires finding a trade-off, which gives maximum inductance and quality factor at minimum resistance and substrate capacitance.

Greenhouse Method

Based on Grover's inductance model [13], Greenhouse [20] derived a systematic and easy-to-use design method for planar rectangular on-chip inductors. Unlike the compact model, this method allows for various inductor aspect ratios, and the influence of individual parameters can be comprehensively analyzed. Figure 4.11 defines the geometry of such an inductor, including the outer diameter d_{out}, winding width w, and the spacing s. All segments are assumed to be shortened at each connecting end by half the winding width w. Hence, the length of segment 2 is $l_2 = d_{out} - w$. The dash-dot line in Fig. 4.11 represents the dimensions of each segment and the entire spiral.

As the key idea presented by Greenhouse, the total inductance L is calculated by

$$L = L_0 + M_+ - M_- \tag{4.20}$$

with the self-inductance L_0, the sum of all positive mutual inductances M_+ and the sum of all negative mutual inductances M_-. M_+ results from the coupling between two segments with currents in the same directions like $M_{1,5}$, $M_{1,9}$ in Fig. 4.11. The negative mutual conductance corresponds to segments where the current flows in opposite directions, such as $M_{1,3}$, $M_{1,7}$, and $M_{1,11}$ in Fig. 4.11.

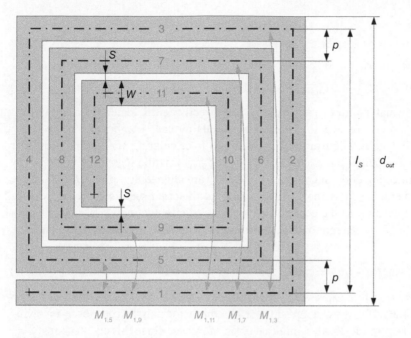

Figure 4.11 Square planar coil with dimensions and mutual inductances $M_{j,m}$ according to Greenhouse [20].

Equation (4.10) defines the self-inductance for a circular cross-section based on Grover's inductance model. For a rectangular cross-section of width w, thickness t a segment of length l_s has a self-inductance of [13]

$$L_s = \frac{\mu_0}{2\pi} \cdot l_s \left(\ln \left(\frac{2 l_s}{w+t} \right) - 0.50049 + \frac{w+t}{3 l_s} \right). \tag{4.21}$$

The total self-inductance L_o is the sum of all segment self-inductances:

$$L_o = \sum_{s=1}^{n} L_s \tag{4.22}$$

For the three-turn inductor of Fig. 4.11, $L_o = L_1 + L_2 + \cdots + L_{12}$.

The mutual inductance between two parallel tracks of length λ is given by

$$M(\lambda) = \frac{\mu_0}{2\pi} \cdot \lambda \left(\ln \left(\frac{\lambda}{GMD} + \sqrt{1 + \left(\frac{\lambda}{GMD} \right)^2} \right) - \sqrt{1 + \left(\frac{GMD}{\lambda} \right)^2} + \frac{GMD}{\lambda} \right) \tag{4.23}$$

where GMD denotes the geometric mean distance between the two tracks, which is approximately equal to the distance between the track centers. According to Fig. 4.11, $GMD \approx w + s$ with track width w and spacing s. Refer to [20] for a more exact equation for GMD.

Consider the mutual inductance between segments 1 and 5 in Fig. 4.11. A component $M_{1,5}$ is caused by the current flowing in segment 1, as well as a component $M_{5,1}$ due to the current flowing in segment 5. Therefore, the total mutual inductance linking segments 1 and 5 is $M_{1,5} + M_{5,1}$. The same relationship exists between all other parallel segments. If the current in both tracks flows in the same direction, the mutual inductances contribute to the total positive mutual inductance M_+ of the planar coil. Likewise, mutual inductances between segments with opposite current flow add to the total negative mutual inductance M_-. For the square inductor in Fig. 4.11,

$$\begin{aligned} M_+ = 2 \, (&M_{1,5} + M_{1,9} + M_{5,9} + M_{2,6} + M_{2,10} + M_{6,10} \\ &+ M_{3,7} + M_{3,11} + M_{7,11} + M_{4,8} + M_{4,12} + M_{8,12}) \end{aligned} \tag{4.24}$$

and

$$M_- = 2\left(M_{1,3} + M_{1,7} + M_{1,11} + M_{5,3} + M_{5,7} + M_{5,11} \right.$$
$$+ M_{9,3} + M_{9,7} + M_{9,11} + M_{2,4} + M_{2,8} + M_{2,12}$$
$$\left. + M_{6,4} + M_{6,8} + M_{6,12} + M_{10,4} + M_{10,8} + M_{10,12}\right). \tag{4.25}$$

If the segments have equal length l_s, the term $M_{j,m}$ can directly be calculated from Eqn. (4.23) for $\lambda = l_s$. However, Fig. 4.11 indicates that the two segments for each component in the above equations are unequal in length. There are two cases related to calculating M_+ and M_-. Consider segments 2 and 6 (M_+). Segment 2 has a length of $l_s = l_2 = d_{out} - w$ while segment 6 is shorter by a distance p at both connecting ends. Segments 2 and 4 (M_-) are aligned at the upper end but are different by p at the lower end. The same scenario applies to the other segments for the positive and negative mutual inductance. In the general case, we always have two segments, one of length l_s and the other of length $l_s - \Delta l$. Greenhouse approximates the mutual inductance between these segments by

$$M_{j,m} = M(\lambda = l_s - \Delta l/2) - M(\lambda = \Delta l/2). \tag{4.26}$$

Greenhouse suggests using this expression to calculate the values for all individual terms in Equations (4.24) and (4.25). It gives exact results for all conductor pairs except those involving segment 1. However, this introduces only a minimal error. Moreover, since this error occurs in M_+ and M_-, it will somewhat cancel out.

Example 4.4 *We want to calculate the inductance and the series resistance of a square spiral inductor according to Fig. 4.11 using the Greenhouse method. The top-metal copper layer has a thickness of $t = 3\ \mu m$ (specific resistance of copper: $\rho = 1.71 \times 10^{-8}\ \Omega m$) with a spacing $s = 3\ \mu m$. The outer diameter should be $d_{out} = 1.5\ mm$. We use four different winding widths $w = 50, 100, 200, 300\ \mu m$ and the number of turns should be $N = 2$ and 3.*

We first determine the length of each segment and the corresponding DC resistance and self-inductance according to Equations (4.17) and (4.21), respectively. Summing up the values of all segments gives the total resistance R and self-inductance L_o. We calculate all mutual inductance components as defined in Equations (4.24) and (4.25) from Eqn. (4.26) with the use of Eqn. (4.23). It requires determining the geometry parameters l_s, Δl, and GMD for each considered segment pair.

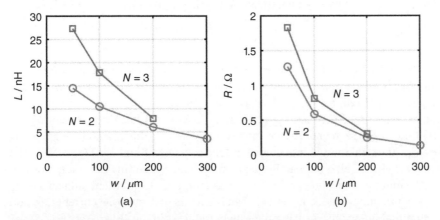

Figure 4.12 a) Inductance L and b) series DC resistance R versus winding width w for a square spiral inductor with 1.5 mm outer diameter for $N = 2$ and 3 turns.

For two turns, there are only eight segments. Hence, we can omit the terms with index nine and higher in Equations (4.24) and (4.25). The results are plotted in Fig. 4.12. The lowest winding width of 50 μm gives the highest inductance of L = 27.3 nH for N = 3, which reduces by half for two turns. However, the DC resistance of more than 1 Ω will not be acceptable for power management applications. On the other end, the maximum width of 300 μm achieves R = 131 mΩ but only L = 3.5 nH. Also, only two turns are possible within the given outer boundary of 1.5 mm. A width of 200 μm represents a good trade-off. It has been chosen for the inductor in Fig. 4.10 [16].

4.2.4 Inductors with Magnetic Core

The above sections cover *air-core* inductors with inductance values in the order of 10 nH. The series resistance is relatively large, resulting in single-digit quality factors Q. Introducing a magnetic core increases the inductance per unit area by more than one order of magnitude with quality factors higher than 15. Today's integrated inductors with magnetic cores reach typical densities of 125 nH/mm^2. Another key parameter is the ratio between inductance and series resistance with typical values of a few hundred nH/Ω. This metric measures how much the inductor structure contributes to the loss.

There are two categories: fully integrated inductors and inductors fabricated on a separate interposer die. On-chip solenoid inductors with magnetic cores are available in CMOS fabrication as a nonstandard backend-of-line processing step. The windings and the via connections are usually implemented by electroplating.

As one of the early examples [21] introduces an integrated solenoid inductor that achieves $Q \geq 13$ at 100 MHz, Fig. 4.13a). The inductor is a vertical solenoid structure formed by multiple vertical windings around a planar magnetic core. A two-turn inductor reaches $L \sim 3$ nH and a four-turn version achieves $L \sim 6.5$ nH on an area of 0.25 and 0.5 mm^2, respectively. The inductors can handle currents up to 1 A. The copper windings have a width of 200 μm.

The vertical solenoid structure became the most popular inductor design. An example layout from [22] is shown in Fig. 4.13b). This design implements five vertical turns utilizing two thick-top metal layers of 6 and 12 μm. The inductance is reported to be $L \sim 1$ nH and the series resistance is $\sim$300 mΩ at 100 MHz at currents up to 300 mA. The measured DC resistance is 75 mΩ. One inductor covers roughly 0.15 mm^2.

The inductor implemented in [23], shown in Fig. 4.13c), measures an inductance of 5.86 nH at 10 MHz with an area footprint of 0.0525 mm^2, which leads to a total inductance density of 111 nH/mm^2. Interestingly, without magnetics, the same inductor is measured to reach about 2.2 nH. The AC series resistance is 2.7 Ω.

The work in [25] presents a silicon-integrated thin-film inductor technology with improved magnetic core laminations to reduce eddy current loss and improve high-frequency behavior. Three inductor designs are investigated with inductance values of approximately 20–85 nH. The corresponding peak-quality factor ranges from 21 to 15 over a frequency span of 25–150 MHz. The inductor saturation current reaches up to 400 mA. The maximum inductance density is 127 nH/mm^2 and the ratio L/R is 425 nH/Ω.

Thin-film inductors that integrate on a separate interposer achieve higher inductance and quality factors. Also, magnetic-core transformers can be implemented [26, 27]. The early work by Sturcken et al. [24] integrates the inductor on a silicon interposer with a multi-phase buck converter IC by 2.5-D chip stacking. The preferred design with eight single-turn coupled inductors provides 12.5 nH with 270 mΩ DC resistance, which supports more than 5 A of load current, Fig. 4.13d). The winding thickness is 5 μm. Eddy currents are induced in the magnetic core, which causes the inductance to

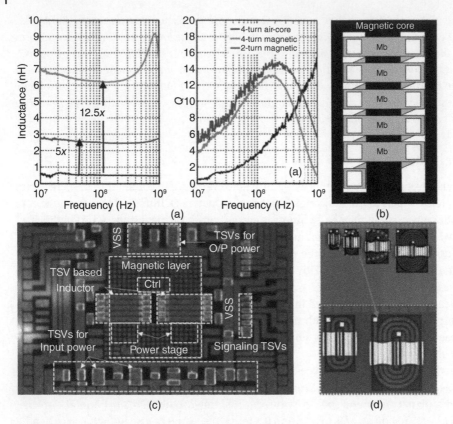

Figure 4.13 Inductors with magnetic core: a) Inductance and quality factor plots of inductor designs. Source: Wang et al. [21]/IEEE. b) Typical vertical solenoid structure. Source: Krishnamurthy et al. [22]/IEEE. c) Die photo of a fully integrated voltage regulator with 3-D through-silicon-via (TSV) based on-die solenoid inductor. Source: Krishnamurthy et al. [23]/from IEEE. d) Thin-film inductors. Source: Sturcken et al. [24]/from IEEE.

fall off, starting in the tens of MHz range. Over the same frequency range, skin depth and proximity effect increase the winding resistance.

Le et al. [28] provides a comprehensive review of state-of-the-art technologies for inductor fabrication and discusses the challenges and opportunities of integrated inductors for power management applications.

References

1 Breussegem, T.V. and Steyaert, M. (2013) *CMOS Integrated Capacitive DC-DC Converters*, Springer Science+Business Media.
2 Jiang, J., Lu, Y., Huang, C., Ki, W.H., and Mok, P.K.T. (2015) 20.5 A 2-/3-phase fully integrated switched-capacitor DC-DC converter in bulk CMOS for energy-efficient digital circuits with 14% efficiency improvement, in *2015 IEEE International Solid-State Circuits Conference - (ISSCC) Digest of Technical Papers*, pp. 1–3, doi: 10.1109/ISSCC.2015.7063078.
3 Renz, P. and Wicht, B. (2021) *Integrated Hybrid Resonant DCDC Converters*, Springer International Publishing, doi: 10.1007/978-3-030-63944-0.

4 Lutz, D., Renz, P., and Wicht, B. (2016) 12.4 A 10 mW fully integrated 2-to-13V-input buck-boost SC converter with 81.5% peak efficiency, in *2016 IEEE International Solid-State Circuits Conference (ISSCC)*, pp. 224–225, doi: 10.1109/ISSCC.2016.7417988.

5 Renz, P., Kaufmann, M., Lueders, M., and Wicht, B. (2019) A 3-ratio 85% efficient resonant SC converter with on-chip coil for Li-ion battery operation. *IEEE Solid-State Circuits Letters*, 2, 236–239, doi: 10.1109/LSSC.2019.2927131.

6 Butzen, N. and Steyaert, M. (2017) 10.1 A 1.1 W/mm²-power-density 82%-efficiency fully integrated 3:1 switched-capacitor DC-DC converter in baseline 28 nm CMOS using stage outphasing and multi-phase soft-charging, in *2017 IEEE International Solid-State Circuits Conference (ISSCC)*, pp. 178–179, doi: 10.1109/ISSCC.2017.7870319.

7 Andersen, T.M., Krismer, F., Kolar, J.W., Toifl, T., Menolfi, C., Kull, L., Morf, T., Kossel, M., Brändli, M., Buchmann, P., and Francese, P.A. (2014) 4.7 A sub-ns response on-chip switched-capacitor DC-DC voltage regulator delivering 3.7 W/mm² at 90% efficiency using deep-trench capacitors in 32 nm SOI CMOS, in *2014 IEEE International Solid-State Circuits Conference Digest of Technical Papers (ISSCC)*, pp. 90–91, doi: 10.1109/ISSCC.2014.6757351.

8 El-Damak, D., Bandyopadhyay, S., and Chandrakasan, A.P. (2013) A 93% efficiency reconfigurable switched-capacitor DC-DC converter using on-chip ferroelectric capacitors, in *2013 IEEE International Solid-State Circuits Conference Digest of Technical Papers*, pp. 374–375, doi: 10.1109/ISSCC.2013.6487776.

9 Bevilacqua, A. (2020) Fundamentals of integrated transformers: From principles to applications. *IEEE Solid-State Circuits Magazine*, 12 (4), 86–100, doi: 10.1109/MSSC.2020.3021844.

10 Wens, M., Cornelissens, K., and Steyaert, M. (2007) A fully-integrated 0.18 μm CMOS DC-DC step-up converter, using a bondwire spiral inductor, in *ESSCIRC 2007 - 33rd European Solid-State Circuits Conference*, pp. 268–271, doi: 10.1109/ESSCIRC.2007.4430295.

11 Villar, G. and Alarcon, E. (2008) Monolithic integration of a 3-level DCM-operated low-floating-capacitor buck converter for DC-DC step-down donversion in standard CMOS, in *2008 IEEE Power Electronics Specialists Conference*, pp. 4229–4235, doi: 10.1109/PESC.2008.4592620.

12 Piqué, G.V. and Alarcón, E. (2011) *CMOS Integrated Switching Power Converters - A Structured Design Approach*, Springer, New York, Dordrecht, Heidelberg, London.

13 Grover, F.W. (1946) *Inductance Calculations - Working Formulas and Tables*, D. van Nostrand Company, New York.

14 Burton, E.A., Schrom, G., Paillet, F., Douglas, J., Lambert, W.J., Radhakrishnan, K., and Hill, M.J. (2014) FIVR—Fully integrated voltage regulators on 4th generation Intel(R) Core(TM) SoCs, in *2014 IEEE Applied Power Electronics Conference and Exposition - APEC 2014*, pp. 432–439, doi: 10.1109/APEC.2014.6803344.

15 Krishnamurthy, H.K., Vaidya, V.A., Kumar, P., Matthew, G.E., Weng, S., Thiruvengadam, B., Proefrock, W., Ravichandran, K., and De, V. (2014) A 500 MHz, 68% efficient, fully on-die digitally controlled buck Voltage Regulator on 22 nm Tri-Gate CMOS, in *2014 Symposium on VLSI Circuits Digest of Technical Papers*, pp. 1–2, doi: 10.1109/VLSIC.2014.6858438.

16 Renz, P., Kaufmann, M., Lueders, M., and Wicht, B. (2019) 8.6 A fully integrated 85%-peak-efficiency hybrid multi-ratio resonant DC-DC converter with 3.0-to-4.5 V input and 500 μA-to-120mA load range, in *2019 IEEE International Solid- State Circuits Conference (ISSCC)*, IEEE, San Francisco, CA, USA, pp. 156–158, doi: 10.1109/ISSCC.2019.8662491.

17 Mohan, S., del Mar Hershenson, M., Boyd, S., and Lee, T. (1999) Simple accurate expressions for planar spiral inductances. *IEEE Journal of Solid-State Circuits*, 34 (10), 1419–1424, doi: 10.1109/4.792620.

18 Wheeler, H. (1928) Simple inductance formulas for radio coils. *Proceedings of the Institute of Radio Engineers*, 16 (10), 1398–1400, doi: 10.1109/JRPROC.1928.221309.

19 Yue, C. and Wong, S. (2000) Physical modeling of spiral inductors on silicon. *IEEE Transactions on Electron Devices*, 47 (3), 560–568, doi: 10.1109/16.824729.

20 Greenhouse, H. (1974) Design of planar rectangular microelectronic inductors. *IEEE Transactions on Parts, Hybrids, and Packaging*, 10 (2), 101–109, doi: 10.1109/TPHP.1974.1134841.

21 Wang, N., Doris, B.B., Shehata, A.B., O'Sullivan, E.J., Brown, S.L., Rossnagel, S., Ott, J., Gignac, L., Massouras, M., Romankiw, L.T., and Deligianni, H.L. (2016) High-Q magnetic inductors for high efficiency on-chip power conversion, in *2016 IEEE International Electron Devices Meeting (IEDM)*, pp. 35.3.1–35.3.4, doi: 10.1109/IEDM.2016.7838547.

22 Krishnamurthy, H.K., Vaidya, V., Kumar, P., Jain, R., Weng, S., Kim, S.T., Matthew, G.E., Desai, N., Liu, X., Ravichandran, K., Tschanz, J.W., and De, V. (2018) A digitally controlled fully integrated voltage regulator with on-die solenoid inductor with planar magnetic core in 14-nm tri-gate CMOS. *IEEE Journal of Solid-State Circuits*, 53 (1), 8–19, doi: 10.1109/JSSC.2017.2759117.

23 Krishnamurthy, H.K., Weng, S., Mathew, G.E., Desai, N., Saraswat, R., Ravichandran, K., Tschanz, J.W., and De, V. (2018) A digitally controlled fully integrated voltage regulator with 3-D-TSV-based on-die solenoid inductor with a planar magnetic core for 3-D-stacked die applications in 14-nm tri-gate CMOS. *IEEE Journal of Solid-State Circuits*, 53 (4), 1038–1048, doi: 10.1109/JSSC.2017.2773637.

24 Sturcken, N., O'Sullivan, E.J., Wang, N., Herget, P., Webb, B.C., Romankiw, L.T., Petracca, M., Davies, R., Fontana, R.E., Decad, G.M., Kymissis, I., Peterchev, A.V., Carloni, L.P., Gallagher, W.J., and Shepard, K.L. (2013) A 2.5D integrated voltage regulator using coupled-magnetic-core inductors on silicon interposer. *IEEE Journal of Solid-State Circuits*, 48 (1), 244–254, doi: 10.1109/JSSC.2012.2221237.

25 Raju, S., Soh, S., Lulu, P., Disney, D., Ho, D., Jien, C.K., Stenger-Koob, M., Wrona, J., Landmann, M., Ocker, B., Langer, J., Selvaraj, S.L., Annamalai, M.A., and Singh, R.P. (2020) Thin-film magnetic inductors on silicon for integrated power converters, in *IECON 2020 The 46th Annual Conference of the IEEE Industrial Electronics Society*, pp. 2292–2295, doi: 10.1109/IECON43393.2020.9255388.

26 Wu, H., Lekas, M., Davies, R., Shepard, K.L., and Sturcken, N. (2016) Integrated transformers with magnetic thin films. *IEEE Transactions on Magnetics*, 52 (7), 1–4, doi: 10.1109/TMAG.2016.2515501.

27 Dinulovic, D., Shousha, M., Haug, M., Beringer, S., and Wurz, M.C. (2017) Comparative study of microfabricated inductors/transformers for high-frequency power applications. *IEEE Transactions on Magnetics*, 53 (11), 1–7, doi: 10.1109/TMAG.2017.2734878.

28 Le, H.T., Haque, R.I., Ouyang, Z., Lee, S.W., Fried, S.I., Zhao, D., Qiu, M., and Han, A. (2021) MEMS inductor fabrication and emerging applications in power electronics and neurotechnologies. *Microsystems & Nanoengineering*, 7, 59, doi: 10.1038/s41378-021-00275-w.

5

Gate Drivers and Level Shifters

5.1 Introduction

The gate driver turns the power transistor on and off with sufficient gate overdrive. The gate overdrive equals the driver voltage V_{drv}. Figure 5.1a) shows the circuit principle of a low-side gate driver. Its gate connects to V_{drv} to turn on the power transistor, while its gate is pulled to the ground to turn the power switch off. This way, the driver provides either a pull-up path toward V_{drv} (turn-on) or a pull-down path toward the ground (or, more precisely, to the source potential of the power transistor). A simple Complementary Metal-Oxide-Semiconductor (CMOS) inverter can be used as a gate driver as shown in Fig. 5.1b). The turn-on/turn-off speed of the power transistor is determined by the driver strength, which indicates how much current (the gate current I_{gate}) the driver can deliver in a given time. This characteristic also relates to its equivalent output resistance R_{drv}.

A complete driver design includes several circuit blocks, as shown in the overall driver block diagram in Fig. 5.2:

- *Power transistor(s) including protection*: One or more transistors form the power stage. The transistors can be in low-side, high-side, or bridge configurations. The protection prevents damage to the power transistors in case of over-voltage, over-current, and over-temperature.
- *Gate driver*: Depending on the switch location, the driver can be a low-side or a high-side driver.
- *Level shifter*: A level shifter converts the driver control signal V_{ctrl} from the low-voltage domain into the V_{drv} voltage domain. Usually, the control voltage is lower than the gate overdrive voltage V_{drv}. For the high-side driver, the level shifter may shift the control signal up to the input supply voltage V_{bat} of the power stage or, in case of an n-channel high-side switch, even higher $(V_{bat} + V_{drv})$.
- *Gate supply*: This block provides the gate overdrive voltage V_{drv}. Many gate drivers utilize a linear or shunt regulator from V_{bat}. For an n-type high-side power transistor, a gate supply larger than V_{bat} must be available to turn the power transistor on. It can be accomplished with a bootstrap supply, a charge pump, or a boost converter.
- *Dead-time control*: If the power stage is configured as a half-bridge, a dead-time control is mandatory to avoid damage at the power stage due to cross-conduction.

The control logic delivers the turn-on or turn-off signal for the power stage, usually in a pulse-width modulated fashion.

Design of Power Management Integrated Circuits, First Edition. Bernhard Wicht.
© 2024 John Wiley & Sons Ltd. Published 2024 by John Wiley & Sons Ltd.

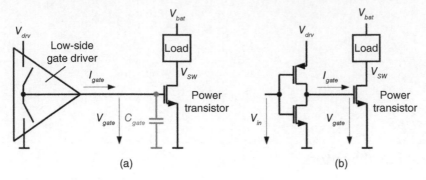

(a)   (b)

Figure 5.1   Gate driver: a) circuit principle; b) using a CMOS inverter.

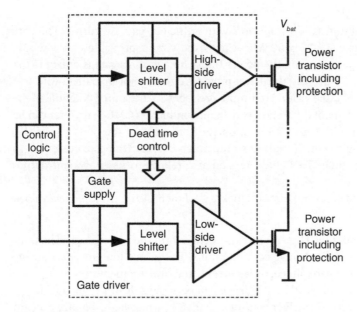

Figure 5.2   Gate driver block diagram.

5.2   Gate Driver Configurations

Figure 5.3 shows the commonly used gate driver configurations. The upper two circuits use p-channel high-side switches, while the two configurations at the bottom use n-channel devices at both the high side and low side. Hence, Fig. 5.3a,b) are rather for lower output currents (output power), while the lower on-resistance of the n-channel high-side devices in Fig. 5.3c,d) makes them suitable for higher output currents (output power).

A different classification can be applied to the left column of circuits, Fig. 5.3a,c), and the right-hand column, Fig. 5.3b,d). The gate driver circuits on the left are used for low supply voltages V_{bat}, compatible with the maximum gate–source voltage of both power transistors, typically 5 V and below. For instance, these configurations fit power management designs that run from a Li-Ion battery. The gate drivers in Fig. 5.3b,d) are suitable for higher input voltages, up to 100 V and above. These circuits can be found in DC–DC converters for automotive and compute power applications, supplied from 12 or 48 V (see application examples in Section 1.5).

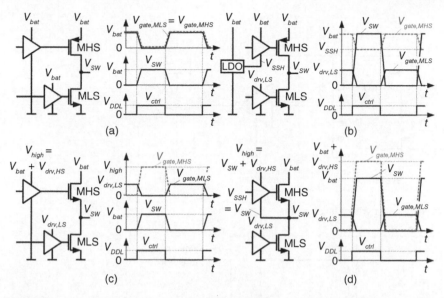

Figure 5.3 Gate driver configurations: a) and b) with p-channel high-side switch; c) and d) with n-channel high-side switch; a) and c) are suitable for low V_{bat} and b) and d) for high V_{bat}.

Let us discuss the specifics of each configuration separately. Figure 5.3a) is the most simple design. The power stage forms a CMOS inverter. The gate terminals of the switches are driven with identical signals. A control signal V_{ctrl} initiates the switching transition. The control logic (not shown) runs from a low-voltage supply V_{DDL}. In most applications, V_{ctrl} is a pulse-width modulated signal *PWM*. Due to the inverting nature of the p-channel transistor, one switch is on at a time. The switching node voltage V_{sw} remains zero if the low-side device is turned on. V_{sw} reaches V_{bat} if the high-side switch is activated. For simplification, the timing diagrams in Fig. 5.3 do not include a dead time in between transitions (see Section 2.8).

Figure 5.3b) is very similar, except that the high-side gate driver's ground V_{SSH} is provided by a linear regulator (LDO). V_{SSH} is set such that $V_{SSH} = V_{bat} - V_{drv,HS}$ where $V_{drv,HS}$ sets the desired source–gate voltage of the high-side switch MHS. It ensures that the maximum gate voltage ratings of MHS are never exceeded, even for large battery voltages V_{bat}.

If we replace MHS of Fig. 5.3a) by an n-channel transistor, we obtain Fig. 5.3c). A complementary gate control switches V_{sw} between zero and V_{bat}. The essential difference to Fig. 5.3a) is the increased high-side gate driver supply V_{high}, which is larger than V_{bat} by the gate overdrive $V_{drv,HS}$ of MHS. Figure 5.3c) can be modified to Fig. 5.3d) if the high-side driver's ground gets connected to the switching node V_{sw}. If V_{sw} undergoes a rising transition, the high-side driver "moves" simultaneously. The gate–source voltage of MHS is defined by $V_{drv,HS}$. It is usually provided by bootstrapping (see Section 5.11), by a charge pump (see Section 8.9), or by a boost converter (see Section 10.10).

The most commonly used configurations are Fig. 5.3a) for low-voltage supplies ($V_{bat} \leq 5$ V) and Fig. 5.3d) for larger voltages ($V_{bat} \geq 5$ V). Using n-type power switches with low on-resistance, Fig. 5.3d) supports higher load currents (output power). For the same reason, it is also the standard driver configuration with external power switches, which are almost exclusively n-type transistors.

Figure 5.4 shows an implementation example of a low-side gate driver with a level shifter. Such a setup could be used in any of the four cases shown in Fig. 5.3. A strong inverter acts as a gate driver, which provides the gate overdrive voltage V_{drv} for the low-side power transistor MLS. A cross-coupled level shifter converts the digital input signal V_{ctrl} from a low-voltage domain V_{DDL} into a V_{drv} voltage level. The level shifter will be discussed later in this chapter (Section 5.12).

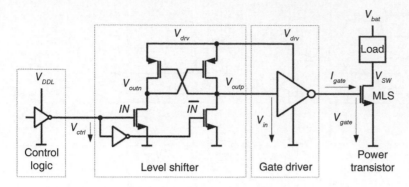

Figure 5.4 Low-side gate driver with a level shifter.

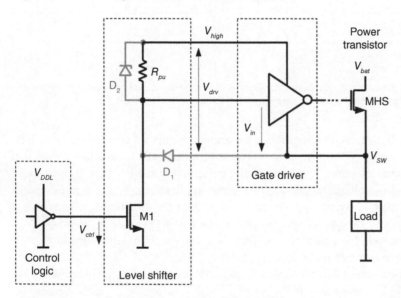

Figure 5.5 High-side gate driver with a level shifter.

A typical signal transfer at turn-on of MLS is: $V_{ctrl} \to 0$ ($IN \to 0$, $\overline{IN} \to 1$) $\Rightarrow V_{outn} = V_{in} \to 0 \Rightarrow V_{gate} \to V_{drv} \Rightarrow$ MLS turns on.

Figure 5.5 shows a high-side driver example in which the gate driver ground is referred to the switching node V_{sw}. This circuit corresponds to the high-side driver in the configuration of Fig. 5.3d). For $V_{ctrl} \to 1$, the resistor-based level shifter provides a "0" level at the gate driver input ($V_{in} \to 0$). The driver pulls the gate node of the power transistor to the gate supply V_{high}. Hence the power transistor turns on with $V_{GS} = V_{drv}$. The level shifter will be discussed later in this chapter (Section 5.12.1).

5.3 Driver Circuits

We have seen that a CMOS inverter is a well-suitable gate driver. Figure 5.6 shows alternative driver circuits along with the CMOS inverter, depicted in Fig. 5.6a).

If external power transistors with $V_{gate} > 10$ V have to be driven, the driver output transistors need to have the corresponding V_{DS} voltage ratings. The output stage can become huge since

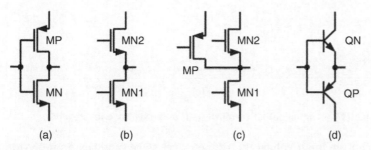

Figure 5.6 Gate driver circuits: a) CMOS inverter, b) two NMOS transistors, c) NMOS pull-down and NMOS/PMOS pull-up transistors, and d) bipolar transistor stage.

such a high-voltage p-type (PMOS) transistor is typically about three times larger than an n-type (NMOS) transistor with the same on-state resistance. Furthermore, many technologies do not offer area-efficient high-voltage p-type devices. In that case, the driver circuit of Fig. 5.6b) with two NMOS transistors can be used. The drawback of this two-NMOS driver is that the pull-up device needs sufficient gate overdrive, which requires some auxiliary supply (local charge pump or bootstrapping).

Figure 5.6c) shows a way to avoid a bootstrap circuit by placing a p-channel transistor parallel to the NMOS pull-up. During turn-on, the gate node of MN2 is connected to V_{drv}, carrying the large drive current at the beginning of the switching phase. When the switching node approaches V_{drv}, the gate–source voltage of MN2 decreases, and MN2 finally turns off. MP pulls the output node fully to V_{drv}. Due to MN2, MP does not need to be as strong as the p-channel transistor in the CMOS inverter of Fig. 5.6a).

Figure 5.6d) shows a driver stage with bipolar transistors. This kind of driver has been the standard circuit in early driver designs. For legacy reasons, many commercially available driver ICs, especially high-voltage drivers, are still based on Fig. 5.6d). The bipolar transistor driver is still popular for external discrete drivers, added as a booster if the driver IC does not provide sufficient driving capability (too little peak current and too small driver resistance). However, in state-of-the-art integrated drivers, the MOS transistor circuits of Fig. 5.6a–c) are used instead of their bipolar counterparts.

The most suitable output stage configuration choice depends on technology and the power transistor characteristics. For most driver applications, the simple solution of an integrated CMOS inverter according to Fig. 5.6a) achieves sufficiently good performance. For this reason, we will focus on this circuit in most parts of this chapter. Nevertheless, many aspects apply to alternative driver circuits as well.

5.4 DC Characteristics

As the power transistor represents a capacitive load equal to the equivalent gate capacitance C_{gate} (see Fig. 5.1a)), the driving strength of the inverter determines the turn-on and, respectively, the turn-off speed of the power transistor. The larger the W/L-ratios of the NMOS and PMOS switches in the inverter, the larger the gate current I_{gate} and the faster is the rate of change of the gate voltage V_{gate}.

Consider a CMOS inverter to be used as a gate driver. Figure 5.7a,b) shows the symbol and the circuit, respectively. If we assume that the power transistor is turned on, we have $V_{out} = V_{gate} = V_{drv}$.

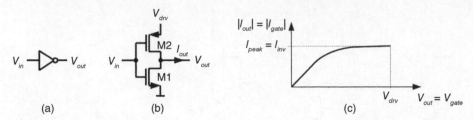

Figure 5.7 CMOS inverter: a) Symbol; b) circuit; c) output current versus output voltage for $V_{in} = V_{drv}$.

The power switch turns off when the input voltage V_{in} of the driver stage switches from 0 V to V_{drv}. In this case, the characteristic output curve of the inverter's pull-down NMOS transistor M1 defines the relation between $V_{out} = V_{gate}$ and $I_{out} = I_{gate}$ as shown in Fig. 5.7c). Since the current flows into the driver toward the ground, it has a negative sign. For convenience, the absolute value $|I_{out}| = |I_{gate}|$ is drawn in Fig. 5.7c). The transition starts at $V_{gate} = V_{drv}$ and $I_{gate} = 0$, reaches the peak current I_{gate} after some delay, and goes back to $I_{gate} = 0$ after the transition has finished.

The maximum gate current $I_{peak} = I_{inv}$, as shown in the diagram, cannot be achieved in practical designs due to the finite delay when (dis)charging the power transistor's gate. The gate driver, along with its control path, may have some delay, which causes $I_{peak} < I_{gate,max}$. Parasitic gate loop resistance and inductance, as explained in Section 2.5.4, further increase the delay. We discuss this relationship in Section 5.5 and neglect these effects for now. The turn-off transition ends if $V_{gate} = 0$ and $I_{gate} = 0$, corresponding to the origin of the graph in Fig. 5.7c).

Figure 5.8 shows a more general graph of the DC characteristics. If the x-axis is chosen as indicated in the graph, the curve is valid for both the turn-on and turn-off cases. Two basic parameters can be extracted, the peak current I_{peak} and the driver resistance R_{drv}. R_{drv} is usually specified at a given gate current I_{gate1},

$$R_{drv} = \frac{V_1}{I_{gate1}}. \tag{5.1}$$

As the transistor operates in the triode region, R_1 represents the slope of the curve, which remains constant versus V_{gate} in the first order. Typical values for R_{drv} are in the range of 1–10 Ω. The peak current I_{peak} allows us to estimate the driving speed for a power transistor with a given gate capacitance. Typical values are in the range of 1 to 5 A. Depending on the application, the peak current

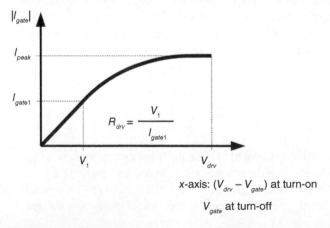

x-axis: $(V_{drv} - V_{gate})$ at turn-on

V_{gate} at turn-off

Figure 5.8 Gate driver DC characteristics with the two basic driver parameters I_{peak} and R_{drv} highlighted.

may be lower or much higher. R_{drv} is used to calculate the current capability of the driver in a more general way. Especially for external power transistors, an additional outer gate resistor is inserted in the gate path to reduce the transition slope and to improve the electromagnetic interference (EMI, see Section 12.3).

Example 5.1 *What is the time t required to drive a power transistor with* $C_{gate} = 200$ pF, $V_{drv} = 5$ V, $I_{peak} = 1$ A?

$$t = \frac{Q_{gate}}{I_{peak}} = \frac{C_{gate}V_{drv}}{I_{peak}} = \frac{200 \text{ pF} \cdot 5 \text{ V}}{1 \text{ A}} = 1 \text{ ns} \tag{5.2}$$

How does the time change if I_{peak} *is reduced to* 10 mA?

$$t = \frac{200 \text{ pF} \cdot 5 \text{ V}}{10 \text{ mA}} = 100 \text{ ns} \tag{5.3}$$

In conclusion, the peak current value directly influences the power transistor's switching speed. The peak current corresponds to the driving strength of the gate driver. We will investigate this in Section 5.5.

5.5 Driving Strength

We consider the configuration in Fig. 5.9a) to investigate the influence of the inverter driving strength. To keep the diagrams in Fig. 5.9a–e) applicable to both the turn-on and the turn-off behavior of the power transistor, the plots show the absolute value $|I_{gate}|$ of the gate current. At turn on, $I_{gate} > 0$ flows out of the driver, delivering the gate charge onto the gate (from the gate supply). At turn-off, the gate charge will be removed and flow via the gate driver's pull-down path to its ground.

Let us first assume a minimum-size inverter in the driver configuration of Fig. 5.9a). The inverter consists of transistors with small W/L-ratios. Therefore, the achievable peak current is low, as indicated in Fig. 5.9b). The x-axis depends on the direction of the transition as defined in Fig. 5.8.

The turn-off transition of the power transistor, where the x-axis is defined as V_{gate}, is initiated by switching V_{in} from 0 to V_{drv}. Using, instead, $V_{drv} - V_{gate}$ for the x-axis, the same procedure describes the turn-on behavior, starting with V_{in} set to 0.

The first part of the transition corresponds to the time before t_1 in the transient curve in Fig. 5.9c). It will happen almost instantaneously for a minimum-sized inverter because the input capacitance is minimal. The area under the curve represents the gate charge Q_{gate}, which corresponds to the integral of I_{gate} over time as defined in Eqn. (2.24) (Section 2.6).

In the DC diagram in Fig. 5.9b), t_1 is the time to go from point A to point B. Due to the small and decreasing gate current, the discharge of the gate capacitance of the power transistor in the second part of the transition, from t_1 to t_2, takes much longer. It corresponds to the delay from point B to point C in Fig. 5.9b). As this increases the switching rise-fall time of the power transistor, its switching losses can become very large (see Eqn. (2.39)). At t_2, the transition is finished, corresponding to point C. The driving strength needs to be increased to achieve reasonable switching losses of the power transistor at acceptable turn-on or turn-off times.

We now use a strong inverter, which requires transistors with large W/L-ratios. For $V_{in} = V_{drv}$, the dark line in Fig. 5.9e) represents the DC behavior. The turn-off transition starts again at point A. Due to the large transistors in the inverter, the input capacitance of the driver is large. Hence, V_{in}

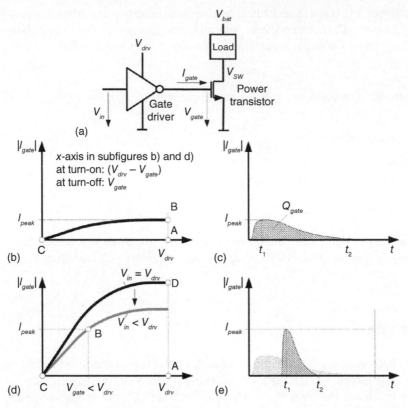

Figure 5.9 Influence of the driver strength: a) setup; b) DC and c) transient behavior for a minimum-size inverter as the gate driver; d) and e) corresponding diagrams for an inverter with large driving strength.

will not immediately reach V_{drv}. The transition of V_{gate} will be finished before V_{in} approaches V_{drv}. In consequence, the maximum possible gate current (I_{inv} in Fig. 5.7c)), indicated by point D, cannot be reached. Instead, the gray curve gets valid since $V_{in} < V_{drv}$, which is similar to the output characteristics of the NMOS transistor for different V_{GS}. So, the actual transition goes from point A to point B, corresponding to time t_1 in Fig. 5.9e). Compared to the weak driver in Fig. 5.9c), this phase takes much longer. The benefit of the strong driver takes place in the second phase of the transition because the peak current I_{peak} is much larger than the one in Fig. 5.9b). The transition from point B to point C is finished at time t_2. Figure 5.9e) shows the transient behavior of the gate current (the absolute value $|I_{gate}|$) in comparison to the curve from Fig. 5.9c). The power transistor is identical in both cases, and so is the gate charge Q_{gate}. Hence, the area under the curve is the same for both scenarios. Finding the optimum sizing of the transistors in the gate driver is a key design task covered throughout this chapter.

5.6 The CMOS Inverter as a Gate Driver

The CMOS inverter, as shown in Fig. 5.7, is the essential circuit in gate drivers. This section will revisit inverter fundamentals to understand its influence on gate driver speed and power dissipation. Further details can be found in textbooks on digital design like [1–3]. We assume that the transistors have equal minimum lengths, as this would give a minimum driver area for a specified gate drive current. A minimum-size inverter corresponds to a given technology's standard logic inverter cell with minimum driver strength.

5.6.1 Input and Output Capacitance

The inverter capacitances are depicted in Fig. 5.10a). The input capacitance is dominated by the gate oxide capacitance of both the NMOS and the PMOS device of the inverter, C_{ox1}, C_{ox2} (neglecting the parasitic capacitance of the gate metal connection). As they appear in parallel, they can be combined to obtain an equivalent circuit, as indicated in Fig. 5.10b). We will use the term C_{inv} for the resulting input capacitance $C_{inv} = C_{ox} = C_{ox1} + C_{ox2}$ as illustrated in Fig. 5.10b,c). The output capacitance accounts for the drain–bulk capacitance $C_{db} = C_{db1} + C_{db2}$ of each inverter device. If the inverter has to drive an identical inverter, Fig. 5.10d), its output capacitance is the sum of C_{inv} and C_{db}. Typical values of C_{inv} are in the range of 1–20 fF and $C_{db} \ll C_{inv}$. In gate driver circuits, the inverter has to drive a much larger capacitive load, either a stronger inverter or the gate of a power transistor. In that case, as shown in Fig. 5.10e), we neglect the contribution of C_{db} to the output capacitance and approximate it by αC_{inv} where α accounts for the scaling of I2 with respect to I1, also called tapering factor.

5.6.2 Output Current

At the beginning of each inverter output transition, the corresponding transistor operates in its saturation region related to the $I\text{--}V$ curve in Fig. 5.7c). We keep in mind that I_{inv} flows out of the inverter if the PMOS transistor is enabled and, vice versa, I_{inv} is the pull-down current that flows into the inverter if the NMOS device is turned on. The latter case is shown in Fig. 5.7c). Hence the absolute output current can be approximated by

$$I_{inv} = \frac{\beta_{inv}}{2}(V_{drv} - V_{th})^2, \tag{5.4}$$

which applies to NMOS as well as the PMOS transistor of the inverter if their driving strength is set equal:

$$\beta_{inv} = \left(\frac{W_1}{L_1}\right)\mu_{on}C_{ox} = \left(\frac{W_2}{L_2}\right)\mu_{op}C_{ox} \tag{5.5}$$

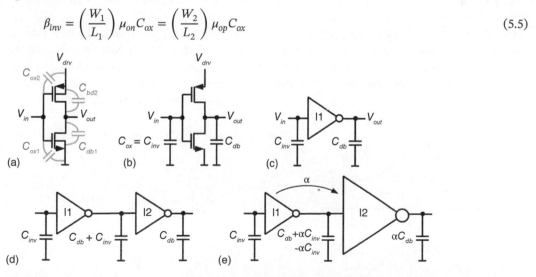

Figure 5.10 Inverter stage a) with parasitic capacitances that are merged into equivalent input and output capacitances, shown b) in the schematic and c) along with the symbol; d) inverter I1 with equally sized inverter I2 as a load; e) inverter I1 connected to an α-times larger inverter I2.

The widths of both transistors have to be sized such that $W_2/W_1 = \mu_{on}/\mu_{op}$, i.e., proportional to the carrier mobilities with $\mu_{on} = 2 \dots 3\mu_{op}$.

The inverter may be larger than a minimum-size inverter by scaling the factor α to increase its driver strength. For example, inverter I2 in Fig. 5.10e) is α times larger. Its maximum output current is equal to

$$I_{peak} = \alpha I_{inv} = \alpha \frac{\beta_{inv}}{2}(V_{drv} - V_{th})^2. \tag{5.6}$$

See also Fig. 5.7c). This condition holds until the output reaches $\sim V_{drv}/2$.

Example 5.2 *What is the maximum inverter output current I_{inv1} if $V_{drv} = 5\,\text{V}$, $V_{th} = 0.7\,\text{V}$, $\beta_{inv} = 80\,\mu\text{A}/\text{V}^2$?*

$$I_{inv1} = \frac{\beta_{inv}}{2}(V_{drv} - V_{th})^2 = \frac{80\,\mu\text{A}/\text{V}^2}{2}(5\,\text{V} - 0.7\,\text{V})^2 = 0.74\,\text{mA} \tag{5.7}$$

Note that I_{inv1} flows only in a very short moment during the output transition of the inverter (peak current).

Calculate the output current I_{inv2} if the inverter gets scaled by a factor of $\alpha = 1000$.

$$I_{inv2} = \alpha I_{inv1} = 1000 \cdot I_{inv1} = 0.74\,\text{A} \tag{5.8}$$

As indicated by Eqn. (5.6), that current can be used to drive the gate of a power transistor connected to the inverter ($I_{peak} = I_{inv2}$).

5.6.3 Rise-Fall Time

With equal transconductances, the rise time and fall time of the inverter's output voltage will be identical, referred to as a 10–90% swing. A rise-fall time t_{rf} can account for both the rise and the fall time. We need to consider that inverters in gate drivers have usually higher voltage ratings (e.g., 5 V) with intrinsic rise-fall times that are much larger than for advanced CMOS technologies, for instance, 1 ns.

For the scenario in Fig. 5.10e) where the second inverter I2 is α-times stronger than a minimum-size inverter, inverter I1 sees a load capacitance of αC_{inv}. I1 needs to deliver a charge of αQ_{inv} where $Q_{inv} = C_{inv}V_{drv}$ defines the input charge of a minimum-size inverter supplied by V_{drv}. We assume that the inverter input charge Q_{inv} is transferred during t_{rf} by a mean inverter output current of $I_{inv}/2$, as depicted in Fig. 5.11. I_{inv} is the maximum output current of the inverter

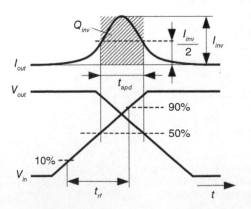

Figure 5.11 Rise-fall time t_{rf} and average propagation delay t_{apd}.

according to Fig. 5.7c), hence

$$\alpha Q_{inv} = \int_0^{t_{rf}} I_{out}(t)\, dt \approx \frac{I_{inv}}{2} t_{rf} \tag{5.9}$$

and

$$t_{rf} = \frac{2\alpha Q_{inv}}{I_{inv}}. \tag{5.10}$$

Example 5.3 *Calculate the rise-fall time t_{rf} if an inverter with $Q_{inv} = 50$ fC, $\beta_{inv} = 80\,\mu A/V^2$, $V_{drv} = 5\,V$ drives an identical inverter and a five times stronger inverter ($\alpha = 5$).*

From Example 5.2 we know that $I_{inv} = 0.74$ mA. With I_{inv}, we can determine the value of the rise-fall time for a minimum-size inverter as a load ($\alpha = 1$):

$$t_{rf} = \frac{2\alpha Q_{inv}}{I_{inv}} = \frac{2 \cdot 50\ \text{fC}}{0.74\ \text{mA}} = 135.1\ \text{ps} \approx 135\ \text{ps}$$

For the larger load with $\alpha = 5$ we simply need to multiply the above result by α:

$$t_{rf} = 5 \cdot 135\ \text{ps} = 685\ \text{ps} \tag{5.11}$$

5.6.4 Average Propagation Delay

Similarly to the rise-fall time, the average propagation delay t_{apd} can be derived. As indicated in Fig. 5.11, t_{apd} is the delay from the input to the output transition if the inverter drives an identical inverter, referred to the 50%-level. From

$$Q_{inv} = \int_0^{t_{apd}} I_{out}(t)\, dt \approx I_{inv} t_{apd} \tag{5.12}$$

we get

$$t_{apd} = \frac{Q_{inv}}{I_{inv}}. \tag{5.13}$$

This expression is a rough estimation but helps obtain an optimized gate driver design.

5.6.5 Power Dissipation

Besides speed, the power dissipation in gate drivers is important. It is even more true as modern CMOS technology can operate at switching frequencies >100 MHz. In integrated power converters, supporting circuits, including the gate driver, significantly influence efficiency.

The inverter has three kinds of loss components:

(1) Static power dissipation due to leakage currents, mainly associated with junction leakage at drain and source and channel leakage. Static power can often be neglected for gate drivers because they typically use transistors with higher voltage ratings or I/O-level devices, which have a reasonable leakage performance.
(2) Short-circuit energy dissipation E_{sc} due to cross-conduction between both transistors in the inverter. E_{sc} causes a dynamic power dissipation that scales with the frequency of switching cycles. The losses are caused by a short-circuit current I_{sc} during the input transition when the input is in the range of $V_{drv}/2$ where both transistors are conducting.
(3) Dynamic energy dissipation E_{dyn} associated with the charging and discharging of a capacitive load at the output of the inverter.

Short Circuit Losses

During the input transition, both inverter transistors conduct at input voltages around $V_{drv}/2$. A short-circuit current I_{sc} flows out of the driver supply V_{drv} toward the ground. Figure 5.12a) shows that a current peak arises roughly in the middle of the V_{in} transition. The magnitude of current depends mainly on two parameters: (1) the input rise-fall time and (2) the driving strength of the inverter expressed by β_{inv}. For a rise-fall time t_{rf2} larger than t_{rf1}, the average short-circuit current is larger. I_{sc} further increases with β_{inv}. The switching energy E_{sc} associated with short-circuit losses can be estimated if the input voltage $V_{in}(t)$ is modeled by a linear ramp function, as shown in Fig. 5.12b) [4]:

$$E_{sc} = V_{drv} \int_{t_1}^{t_3} I_{sc}(t) \, dt = 2V_{drv} \int_{t_1}^{t_2} I_{sc}(t) \, dt \tag{5.14}$$

If we assume both transistors to be in the saturation region, the short-circuit current $I_{sc}(t)$ can be calculated similarly to Eqn. (5.4),

$$I_{sc}(t) = \frac{\beta_{inv}}{2}(V_{in}(t) - V_{th})^2, \tag{5.15}$$

where

$$V_{in}(t) = \frac{V_{drv}}{t_{rf}}t \tag{5.16}$$

and

$$t_1 = \frac{V_{th}}{V_{drv}}t_{rf}, \qquad t_2 = \frac{t_{rf}}{2}. \tag{5.17}$$

Equation (5.17) assumes $V_{thn} = V_{thp} = V_{th}$. Substituting Eqns. (5.15), (5.16) and (5.17) into Eqn. (5.14) yields

$$E_{sc} = \frac{\beta_{inv}}{24}(V_{drv} - 2V_{th})^3 t_{rf} \tag{5.18}$$

in accordance with the expression in [4].

Equation (5.18) accounts for one single switching transition. With two transitions per switching period, the losses of Eqn. (5.14) must be multiplied by two to obtain the loss energy per period.

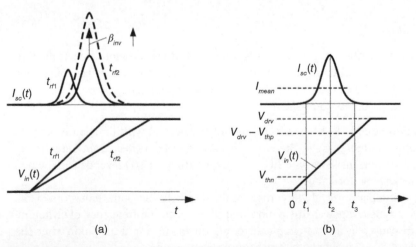

(a)   (b)

Figure 5.12 Short-circuit losses: a) short-circuit current versus rise-fall time; b) short-circuit current model.

E_{sc} increases with the W/L-ratio of the inverter and with the rise-fall time. The faster the input transition, the lower the losses. The losses also decrease for lower drive voltage V_{drv}, which is contrary to the requirement for low static losses in the power transistor, which requires its $V_{GS} = V_{drv}$ to be high.

Example 5.4 *Calculate the short-circuit losses E_{sc} for a full switching period if $\beta_{inv} = 80\,\mu A/V^2$, $V_{drv} = 5\,V$, $V_{th} = 0.7\,V$. Take the rise-fall time t_{rf} from Example 5.3 for $\alpha = 1$, i.e., $t_{rf} = 135$ ps.*

$$E_{sc} = 2\frac{\beta_{inv}}{24}(V_{drv} - 2V_{th})^3 t_{rf} = \frac{\beta_{inv}}{12}(V_{drv} - 2V_{th})^3 t_{rf}$$

$$= \frac{80\,\mu A/V^2}{12}(5\,V - 2 \cdot 0.7\,V)^3 \cdot 135\,ps = 41.99\,fJ \sim 42\,fJ \tag{5.19}$$

Let us do one more exercise. Assume we scale the minimum-size inverter by a factor of 1000 to drive a large power transistor at a reasonable speed. What is the short-circuit loss if, in addition, the input rise-fall time increases by a factor of 5?

If the inverter size increases by 1000, its transconductance parameter gets $1000 \cdot \beta_{inv}$ (where β_{inv} accounts for the minimum-size inverter). From Eqn. (5.18) we can calculate the losses by multiplying E_{sc} from above by $1000 \cdot 5 = 5 \cdot 10^3$:

$$E_{sc} = 5 \cdot 10^3 \cdot 42\,fJ = 210\,pJ \tag{5.20}$$

Dynamic Losses

The dynamic losses correspond to the energy required to charge and discharge the load capacitance C_{out}, as shown in Fig. 5.13a).

As depicted in the equivalent circuit of Fig. 5.13b), the energy E_{drv} delivered from V_{drv} during the first switching transition is

$$E_{drv} = Q_{out}V_{drv} = C_{out}V_{drv}^2 \tag{5.21}$$

with the charge Q_{out} stored by the output capacitor C_{out}. During the second transition within period T, Fig. 5.13c), the capacitor C_{out} gets discharged, and it releases the energy, which was stored on C_{out}, to ground. Hence, over the full period T the dynamic losses are equal to E_{drv}:

$$E_{dyn} = E_{drv} = Q_{out}V_{drv} = C_{out}V_{drv}^2 \tag{5.22}$$

The timing diagram of Fig. 5.13d) shows the output current and its direction for both switching transitions. The result in Eqn. (5.22) is identical to the gate charge loss introduced in Eqn. (2.44).

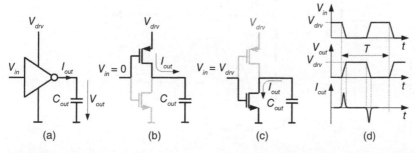

Figure 5.13 Dynamic losses: a) loss scenario with C_{out}; equivalent circuits during b) rising and c) falling transition at the output; d) timing diagram.

In actual designs, C_{out} represents the input capacitance of a subsequent inverter stage or a power transistor. Therefore, C_{out} will always show a voltage dependence. It is adequate to use Q_{out}; see also the details on gate charge in Section 2.6. Q_{out} also includes the gate–drain charge Q_{GD} of the power transistor, delivered during the Miller phase, as defined in Fig. 2.15.

Example 5.5 *Calculate the dynamic energy dissipation E_{dyn} for a full switching period if the load is a minimum-size inverter with $C_{inv} = 10$ fF and for the case $C_{out} = 100$ pF. The drivers supply voltage is $V_{drv} = 5$ V.*

$$E_{dyn,inv} = C_{inv}V_{drv}^2 = 10 \text{ fF} \cdot (5 \text{ V})^2 = 10 \cdot 10^{-15}\frac{\text{As}}{\text{V}} \cdot 25 \text{ V}^2 = 250 \text{ fJ} \tag{5.23}$$

$$E_{dyn} = C_{out}V_{drv}^2 = 100 \text{ pF} \cdot (5 \text{ V})^2 = 100 \cdot 10^{-12}\frac{\text{As}}{\text{V}} \cdot 25 \text{ V}^2 = 2.5 \text{ nJ} \tag{5.24}$$

5.7 Gate Driver with a Single-Stage Inverter

For some applications, a single inverter with sufficient driving strength is suitable as a gate driver. We consider the configuration of Fig. 5.14a) with a minimum-size inverter I1 as an input stage and the inverter I2 as the actual driver. The design task is to identify the required driving strength of inverter I2. The driving strength relates to the maximum available gate current, which is proportional to the W/L-ratio of the transistors in I2. If we assume I2 as the gate driver to be α times stronger than I1, the configuration of I1 and I2 is exactly like in Fig. 5.10e). Hence, the factor α determines the driving strength of I2 with respect to I1, which is a minimum-size inverter.

5.7.1 Speed

The total propagation delay t_{gd} of the driver and power transistor configuration of Fig. 5.14a) can be estimated by

$$t_{gd} = t_{in} + t_{out}, \tag{5.25}$$

with two delay components, as indicated in Fig. 5.14b).

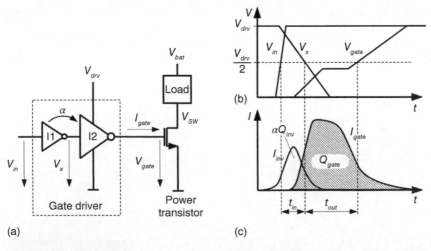

Figure 5.14 Single stage driver transient behavior: a) setup; b) voltage, and c) current waveforms.

Note that the level shifter controls V_{in}. Hence, the total delay from the control logic signal V_{ctrl} to the switching transition of the power transistor is the sum of t_{gd} and the propagation delay of the level shifter.

The propagation delay t_{in} is the time from the transition of V_{in} until the output V_x of I1 reaches $V_{drv}/2$:

$$t_{in} \sim \frac{\alpha Q_{inv}}{I_{inv}} \tag{5.26}$$

This approximation assumes that I2's input gate charge αQ_{inv} is transferred within t_{in} by the maximum inverter current I_{inv} of I1, which is defined in Eqn. (5.4). It is similar to the definition of the average propagation delay t_{apd} in Fig. 5.11, except that the load is not a minimum-size inverter, but α times larger. In fact, referring to Eqn. (5.13), we find $t_{in} = \alpha t_{apd}$.

The second delay component, t_{out}, is the delay related to delivering or withdrawing the gate charge Q_{gate} of the power transistor,

$$t_{out} \approx \frac{Q_{gate}}{I_{gate}} = \frac{Q_{gate}}{\alpha I_{inv}}. \tag{5.27}$$

The gate current I_{gate} is α times larger than the output current I_{inv} of a minimum-size inverter, according to Eqn. (5.4). Hence, Eqn. (5.25) can be rewritten in the form

$$t_{gd} = \alpha \frac{Q_{inv}}{I_{inv}} + \frac{1}{\alpha} \frac{Q_{gate}}{I_{inv}} = \frac{1}{I_{inv}} \left(\alpha Q_{inv} + \frac{1}{\alpha} Q_{gate} \right). \tag{5.28}$$

As the driver I2 gets larger, t_{in} scales proportionally to α, while t_{out} decreases because I_{gate} increases with α according to Eqn. (5.6).

Figure 5.15 demonstrates this relation for the example parameters given in Table 5.1 and I_{inv} from Example 5.2. The diagram shows the gate driver delay according to Eqn. (5.28) as a function of the scaling factor α. Due to the opposed influence of Q_{inv} and Q_{gate}, a minimum delay occurs. We can

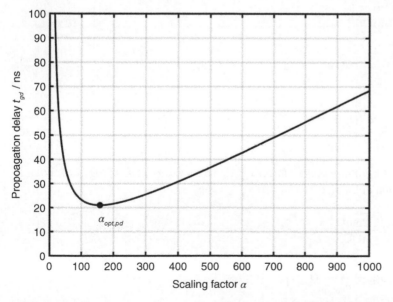

Figure 5.15 Propagation delay t_{gd} of a single-stage gate driver versus scaling factor α (Q_{inv} = 50 fC, Q_{gate} = 1.25 nC, I_{inv} = 0.74 mA).

Table 5.1 Gate driver design parameters. Values for external (discrete) power transistors are marked by *. Unless otherwise noted, the graphs and examples in this chapter use the parameter values in the rightmost column.

Parameter	Symbol	Typical value	Value used in examples
Gate drive voltage	V_{drv}	3.3 V, 5 V	5 V
		10–15 V*	
Inverter parameters:			
Intrinsic rise-fall time	$t_{rf,i}$	10 ps	10 ps
Input capacitance	C_{inv}	1–20 fF	10 fF
Input charge	Q_{inv}	3–100 fC	50 fC
Transconductance	β_{inv}	100 μA/V^2	80 μA/V^2
Power transistor parameters:			
Gate capacitance	C_{gate}	100 pF–1 nF	250 pF
		1–10 nF*	
Gate charge	Q_{gate}	1–5 nC	1 nC
		5–100 nC*	
Gate–drain capacitance	C_{gd}		70 pF
Transconductance	β_{MP}	>1 A/V^2	1 A/V^2

find the optimum scaling factor $\alpha_{opt,pd}$ for a minimum delay t_{gd} from

$$\frac{\partial t_{gd}}{\partial \alpha} = \frac{1}{I_{inv}}\left(Q_{inv} - \frac{Q_{gate}}{\alpha^2_{opt,pd}} \right) = 0,$$

resulting in

$$\alpha_{opt,pd} = \sqrt{\frac{Q_{gate}}{Q_{inv}}}, \tag{5.29}$$

or alternatively

$$\alpha_{opt,pd} = \sqrt{\frac{C_{gate}V_{drv}}{C_{inv}V_{drv}}} = \sqrt{\frac{C_{gate}}{C_{inv}}}. \tag{5.30}$$

Example 5.6 *Calculate the optimum scaling factor $\alpha_{opt,pd}$ for the parameters of Fig. 5.15. What is the corresponding propagation delay t_{gd}?*

$$\alpha_{opt,pd} = \sqrt{\frac{Q_{gate}}{Q_{inv}}} = \sqrt{\frac{1.25 \text{ nC}}{50 \text{ fC}}} = \sqrt{25000} = 158 \tag{5.31}$$

$$t_{gd} = \frac{1}{I_{inv}}\left(\alpha_{opt,pd}Q_{inv} + \frac{1}{\alpha_{opt,pd}}Q_{gate} \right)$$

$$= \frac{1}{0.74 \text{ mA}}\left(158 \cdot 50 \text{ fC} + \frac{1}{158}1.25 \text{ nC} \right) = 21.4 \text{ ns} \tag{5.32}$$

If we consider that the input capacitance of the gate driver stage I2 is $C_{in} = \alpha C_{inv}$, Eqn. (5.30) reveals that the minimum gate driver delay is achieved for the following scaling of the input and load capacitances,

$$C_{gate} = \alpha C_{in} = \alpha^2 C_{inv}. \tag{5.33}$$

This conclusion confirms that scaling by a fixed factor α is essential to achieve minimum delay. It is in line with the original approach presented in [5]. This relation was also found in [6] for optimizing a single-stage digital buffer delay. In consequence of Eqn. (5.33), the optimum delay can be achieved by cascading several inverter stages, while the optimum factor α can be satisfied from stage to stage. Section 5.8 further discusses cascaded inverter stages.

Example 5.7 *Calculate the sizing of the transistors in the driver stage for $\alpha_{opt,pd} = 158$ as derived in Example 5.6. The minimum-size inverter for a 5 V supply consists of an n-channel transistor with $W_{min,n} = 1.0\,\mu m$ and a p-channel device with $W_{min,p} = 2.4\,\mu m$. The minimum length for both devices is 0.6 μm.*

$$\text{NMOS}: W_n = \alpha_{opt,pd} \cdot W_{min,n} = 158\,\mu m$$

$$\text{PMOS}: W_p = \alpha_{opt,pd} \cdot W_{min,p} = 379\,\mu m$$

The length of all devices is kept at the minimum dimension of 0.6 μm.

5.7.2 Loss Energy

Losses in the power stage are covered in Chapter 2, Section 2.7. The gate driver influences several loss components, such as the switching losses and the gate charge losses. The gate driver optimization for speed does not necessarily correspond to the optimum driver transistor sizing for minimum energy loss. While the switching losses in the power transistor decrease with α (faster transition times $t_{tr,on/off}$), the dynamic losses in the driver increase. The switching losses E_{sw} from Eqn. (2.38) can be rewritten in the form

$$E_{sw} = \frac{E_{sw,inv}}{\alpha} \tag{5.34}$$

where $E_{sw,inv}$ corresponds to the switching losses for the case that the power transistor is driven by a minimum-size inverter ($I_{gate} = I_{inv}$ in Eqns. (2.20) and (2.21)). The larger α (larger driving strength), the lower the switching losses in the power transistor.

Example 5.8 *Determine the gate charge loss energy E_{gate} and the switching loss energy $E_{sw,inv}$, if a minimum-size inverter drives the power transistor. Calculate the losses for one full switching period. Use the parameters given in Table 5.1.*

From Eqns. (2.44) and (2.38):

$$E_{gate} = C_{gate}V_{drv}^2 = 250\,\text{pF} \cdot (5\,\text{V})^2 = 6.25\,\text{nJ} \tag{5.35}$$

$E_{sw,inv}$ requires to calculate the switching times T_2, T_3 (Eqns. (2.20) and (2.21)) for $I_{gate} = I_{inv}$ (I_{inv} can be taken from Example 5.2):

$$T_2 = \frac{C_{gate}}{I_{gate}} \sqrt{\frac{2I_{load}}{\beta_{MP}}} = \frac{250\,\text{pF}}{0.74\,\text{mA}} \sqrt{\frac{2 \cdot 1\,\text{A}}{1\,\text{A/V}^2}} = 477.8\,\text{ns} \tag{5.36}$$

$$T_3 = \frac{C_{GD}V_{bat}}{I_{gate}} = \frac{70\,\text{pF} \cdot 12\,\text{V}}{0.74\,\text{mA}} = 1.135\,\mu s \tag{5.37}$$

The expression in Eqn. (2.38) needs to be multiplied by two to cover a full switching period, resulting in

$$E_{sw,inv} = V_{bat}I_{load}(T_2 + T_3)$$
$$= 12\,\text{V} \cdot 1\,\text{A} \cdot (477.8\,\text{ns} + 1.135\,\mu\text{s}) = 19.35\,\mu\text{J}. \tag{5.38}$$

The losses in the driver stage consist of three contributions:

(1) *Gate charge losses of the power transistor*: The gate charge losses E_{gate}, defined in Section 2.7 (see Eqn. (2.44)), are independent of α for a given power transistor size.

(2) *Short-circuit losses E_{sc}* in the actual driver inverter, defined by Eqn. (5.18). With the short-circuit losses $E_{sc,inv}$ of a minimum-size inverter as the baseline, the short-circuit losses increase if α scales up. Since both β_{inv} ($\propto \frac{W}{L}$) and t_{rf} (see Eqn. (5.10)) are proportional to α, the short-circuit losses in the gate driver scale with α^2.

(3) *Gate charge losses E_{dyn}* at the input of the driving stage (inverter) as defined by Eqn. (5.22). These dynamic losses occur in the control circuit, which sees αC_{inv} as a load. If $E_{dyn,inv}$ represents the dynamic losses for a minimum-size inverter as a gate driver ($\alpha = 1$), the gate charge losses are determined by $E_{dyn} = \alpha E_{dyn,inv}$. In the most straightforward case, the output stage of the control circuit can be another inverter.

Hence, the driver losses can be estimated by

$$E_{driver} = E_{dyn} + E_{sc} + E_{gate}$$
$$= \underbrace{\alpha E_{dyn,inv}}_{\text{Control Logic}} + \underbrace{\alpha^2 E_{sc,inv}}_{\text{Gate Driver}} + \underbrace{E_{gate}}_{\text{Gate Charge}} \tag{5.39}$$

with the losses $E_{sc,inv}$ and $E_{dyn,inv}$ of a minimum-size inverter according to Eqns. (5.18) and (5.22), respectively. Equation (5.39) assumes that the output capacitances of the control logic and the gate driver are much smaller than the corresponding load capacitance of the next stage, i.e., $\alpha \gg 1$. Equation (5.39) also omits the dynamic short-circuit losses in the control logic because they are negligible.

The switching losses E_{sw} of the power transistor depend on the transition times of its drain–source voltage and its drain current (see Eqn. (2.38)), which, in turn, are a function of the gate current delivered by the gate driver. Therefore, it is useful to consider the combined losses E_{gd} as the sum of E_{sw} and E_{driver}. Adding up Eqns. (5.34) and (5.39) yields

$$E_{gd} = \frac{1}{\alpha}E_{sw,inv} + \alpha E_{dyn,inv} + \alpha^2 E_{sc,inv} + E_{gate}. \tag{5.40}$$

We do not consider the conduction losses of the power transistor because they do not depend on the gate driver, see Eqn. (2.28).

If the size of the power transistor is fixed, the gate charge losses E_{gate} can be determined by a simple calculation according to Eqn. (2.44). If we, in addition, characterize a minimum-size inverter by its loss quantities $E_{sw,inv}$, $E_{dyn,inv}$, and $E_{sc,inv}$, we can predict and optimize the overall gate driver losses E_{gd}. For practical designs with reasonable speed and power losses, α will reach large values ($\alpha > 100$), and the driver will suffer from large short-circuit currents because V_{in} will remain for some time in the range of $V_{drv}/2$ where both transistors conduct. Hence, the short-circuit losses E_{sc} dominate over the dynamic losses E_{dyn} in the driver stage, which leads to the approximation

$$E_{gd} = \frac{1}{\alpha}E_{sw,inv} + \alpha^2 E_{sc,inv} + E_{gate}. \tag{5.41}$$

It allows to estimate the optimum scaling factor $\alpha_{opt,loss}$ for minimum losses:

$$\frac{\partial E_{gd}}{\partial \alpha} = -\frac{E_{sw,inv}}{\alpha_{opt,loss}^2} + 2E_{sc,inv}\alpha_{opt,loss} = 0$$

This yields

$$\alpha_{opt,loss} = \sqrt[3]{\frac{E_{sw,inv}}{2E_{sc,inv}}}. \tag{5.42}$$

Figure 5.16 shows the combined driver losses E_{gd} versus scaling factor α. The losses are taken from Examples 5.4 ($E_{sc,inv}$) and 5.8 ($E_{sw,inv}$, E_{gate}). The minimum losses can easily be identified. While the run of the curve is similar to the delay plot of Fig. 5.15, the optimum scaling factor is different.

Example 5.9 *Calculate the optimum scaling factor $\alpha_{opt,loss}$ for the parameters of Fig. 5.16. What is the corresponding loss energy E_{gd} and the average power dissipation P_{gd} if the transistor switches at $f_{sw} = 1$ MHz?*

$$\alpha_{opt,loss} = \sqrt[3]{\frac{E_{sw,inv}}{2E_{sc,inv}}} = \sqrt[3]{\frac{19.35\,\mu J}{2\cdot42\,fJ}} = 613 \tag{5.43}$$

$$E_{gd} = \frac{E_{sw,inv}}{\alpha_{opt,loss}} + \alpha_{opt,loss}^2 E_{sc,inv} + E_{gate} \tag{5.44}$$

$$= \frac{19.35\,\mu J}{613} + (613)^2 \cdot 42\,fJ + 6.25\,nJ = 53.6\,nJ$$

$$P_{gd} = E_{gd} \cdot f_{sw} = 53.6\,nJ \cdot 1\,MHz = 53.6\,mW \tag{5.45}$$

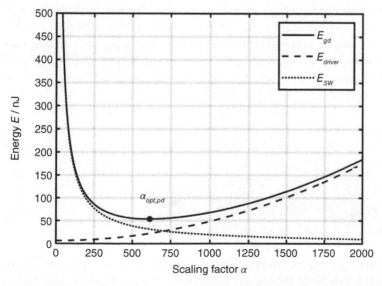

Figure 5.16 Combined driver and switching loss E_{gd} of a single-stage gate driver versus scaling factor α ($E_{sw,inv} = 19.35\,\mu J$, $E_{sc,inv} = 42\,fJ$, $E_{gate} = 6.25\,nJ$, $V_{bat} = 12$ V, all parameters from Table 5.1).

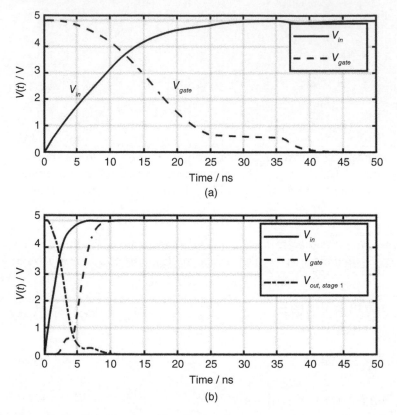

Figure 5.17 Switching transition for a) a single-stage driver according to Fig. 5.14 and b) a two-stage driver.

Even for $\alpha = \alpha_{opt,pd}$, the driver's speed (propagation and transition slope) is still relatively slow. Consequently, such a driver will suffer from large shoot-through currents because V_{in} will remain in the $V_{drv}/2$ range where both transistors are conducting. Excessive loading of the V_{drv} supply and poor power efficiency are the consequence. The disadvantages regarding limited peak current and slow propagation delay can be circumvented by cascading several inverters in series. Already adding one more driver stage gives a much lower propagation delay, as illustrated by circuit simulation in Fig. 5.17. This plot also shows how the rise and fall times of each stage improve by adding more stages. Due to the inverting behavior, the direction of the gate voltage transition changes if only one stage is added. It can be addressed in the control logic. The performance can be significantly improved by adding multiple inverter stages. We will explore the concept of cascaded gate drivers in the following section.

5.8 Cascaded Gate Drivers

Cascaded inverters, as shown in Fig. 5.18, give faster propagation delay than one large single stage. This approach was used originally for digital CMOS buffers to drive large off-chip (capacitive) loads, e.g., for I/O buffers. The driver is designed to increase the W/L-ratio of the transistors from stage to stage. The most common approach is based on a fixed scaling factor per stage, introduced in [5]. Lin and Linholm [5] derive a delay model based on the propagation delay between two minimum-size

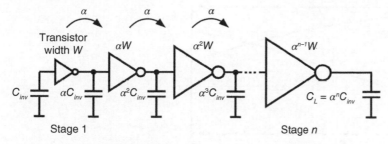

Figure 5.18 Cascaded gate driver.

inverters. Based on that model, [7] found the optimum scaling factor for minimum propagation delay. We have already seen by Eqn. (5.33) in Section 5.7.1 that a fixed scaling factor α results in minimum propagation delay. The design task is to find the optimum scaling factor α and the number of stages n.

With the known average propagation delay t_{apd} of a minimum-size inverter according to Eqn. (5.13) its propagation delay for any capacitive load C_{gate} can be derived:

$$t_{direct} = \frac{Q_{gate}}{I_{inv}} = t_{apd}\frac{Q_{gate}}{Q_{inv}} = t_{apd}\frac{C_{gate}}{C_{inv}} \tag{5.46}$$

t_{direct} is the propagation delay for the case that a single-stage inverter had to directly drive a power transistor with a gate capacitance of C_{gate} like in Fig. 5.1.

Example 5.10 *Calculate* t_{direct} *for* $t_{apd} = t_{rf}/2 = 67.5$ *ps (Example 5.3),* $C_{inv} = 10$ *fF,* $C_{gate} =$ 250 *pF.*

$$t_{direct} = t_{apd}\frac{C_{gate}}{C_{inv}} = 67.5\ \text{ps} \cdot \frac{250\ \text{pF}}{10\ \text{fF}} = 1.69\ \mu\text{s} \tag{5.47}$$

This represents a huge delay. We will investigate how the delay can be reduced by a cascaded driver approach.

If the transistor width is scaled by a constant scaling factor α from stage to stage, the sizing of the cascaded driver follows Fig. 5.18. The transistor length is kept at the minimum value for the used device type in a given technology.

Similar to Fig. 5.17, the transistor-level simulation in Fig. 5.19 shows how the transient behavior of the driver improves with an increasing number of cascaded stages. We can observe that independent of the number of stages, the rise-fall times and the propagation delay per stage are identical throughout the whole driver. Figure 5.19 illustrates how the propagation delay of the entire gate driver can be significantly improved by cascading multiple stages. While a two-stage driver does not even reach a complete transition after 15 ns, with a six-stage driver, the power transistor is already turned-off at $t \sim 3$ ns. The output voltage of the last stage corresponds to the power transistor's gate voltage.

The simulations in Fig. 5.19 also show the Miller plateau during the switching node transition of the power stage (i.e., the drain voltage of the power transistor) as described in Section 2.5.

Since each driver stage's propagation delay is identical, Eqn. (5.46) can be used to calculate the propagation delay t_{stage} per stage. According to Fig. 5.18, the load capacitance of each inverter stage in the driver is α times larger than its input capacitance. Hence,

$$t_{stage} = t_{apd}\frac{\alpha^{i}C_{inv}}{\alpha^{i-1}C_{inv}} = \alpha t_{apd}, \tag{5.48}$$

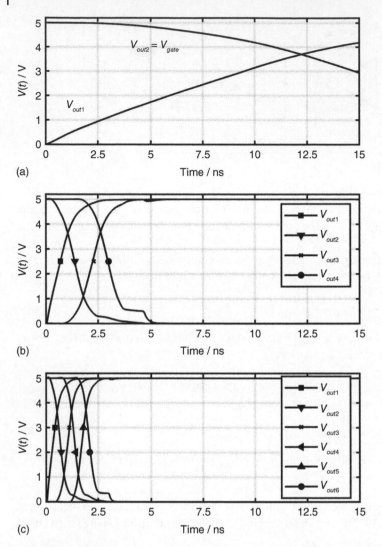

Figure 5.19 Switching transitions for a) two, b) four, and c) six driver stages (circuit simulation for $C_{gate}/C_{inv} = 25{,}000$ and $V_{bat} = 12$ V).

where i corresponds to the stage number. With a total number of n stages, the cascaded driver can be characterized by an overall propagation delay $t_{cascade}$:

$$t_{cascade} = n\alpha t_{apd}. \tag{5.49}$$

5.8.1 Optimization for Speed

Even though a cascaded driver brings a significant speed advantage, we need to determine if there is an optimum set of parameters α and n. From Fig. 5.18 we can derive

$$\alpha^n = \frac{C_{gate}}{C_{inv}}, \tag{5.50}$$

which results in expressions for α and n,

$$\alpha = \sqrt[n]{\frac{C_{gate}}{C_{inv}}} \tag{5.51}$$

and

$$n = \frac{\ln \frac{C_{gate}}{C_{inv}}}{\ln \alpha}.$$
(5.52)

Equation (5.52) was obtained from Eqn. (5.50) by forming the natural logarithm on both sides of the equation. For $n = 2$, Eqn. (5.51) is identical to the expression in Eqn. (5.30) that was found for a single-stage driver (which is a two-stage driver if we include the initial minimum-size inverter, hence $n = 2$).

Our goal is to have the cascaded delay as small as possible with respect to t_{direct} of a single driver stage. In other words, the ratio of Eqns. (5.49) and (5.46) needs to be as small as possible:

$$\frac{t_{cascade}}{t_{direct}} = \frac{n\,\alpha}{\frac{C_{gate}}{C_{inv}}}$$
(5.53)

We note that Eqn. (5.53) does not depend on t_{apd}. Substituting n by Eqn. (5.52) we get

$$\frac{t_{cascade}}{t_{direct}} = \frac{\ln \frac{C_{gate}}{C_{inv}}}{\frac{C_{gate}}{C_{inv}}} \cdot \frac{\alpha}{\ln \alpha}.$$
(5.54)

Since the capacitance values C_{inv} and C_{gate} are fixed for a given design, optimization can be achieved by adjusting the scaling factor α. The α-dependent term of Eqn. (5.54), $\alpha / \ln \alpha$, is plotted in Fig. 5.20a). The graph indicates that the minimum of the expression occurs if α is equal to Euler's number $e \sim 2.72$. Choosing $\alpha \sim 2.72$, we achieve the smallest propagation delay with respect to a single-stage driver. Since the circuit model of the cascaded driver is based on several simplifications (like neglecting the output capacitance of the inverters), the optimum value of α is typically between 3 and 6.

Figure 5.20b) shows the number of stages n as a function of the capacitance ratio C_{gate}/C_{inv} for various values of α according to Eqn. (5.52). The number of stages strongly influences the size and, hence, the layout area of the whole driver. However, the propagation delay in Fig. 5.20a) shows low sensitivity to the actual value of α. Larger α leads to fewer stages n in compliance with Eqn. (5.52).

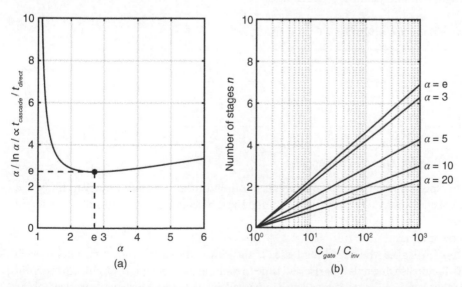

Figure 5.20 Optimization for speed: a) ratio $\alpha/\ln \alpha$ as a function of the scaling factor α; b) number of stages n versus capacitance ratio C_{gate}/C_{inv} for various values of α.

Layout area can be saved if the application can afford a slightly larger delay. This can also be seen in Fig. 5.20b) by the varying slope for different values of α. While α has a low influence on the delay, it can significantly change the number of stages.

Example 5.11 *Determine the minimum possible propagation delay $t_{cascade,min}$ for the values given in Example 5.10. Compare the result with t_{direct} from Example 5.10. In the second step, choose $\alpha = 6$ and recalculate $t_{cascade}$. How does the number of stages n change?*

$$n = \frac{\ln \frac{C_{gate}}{C_{inv}}}{\ln \alpha} = \frac{\ln \frac{C_{gate}}{C_{inv}}}{\ln e} = \ln \frac{250\,\text{pF}}{10\,\text{fF}} = \ln 25000 = 10.13 \sim 10 \tag{5.55}$$

$$t_{stage} = \alpha\, t_{apd} = 2.78 \cdot 67.5\,\text{ps} = 187.7\,\text{ps} \tag{5.56}$$

$$t_{cascade,min} = n\,\alpha\, t_{apd} = n t_{stage} = 10 \cdot 187.7\,\text{ps} = 1.88\,\text{ns} \tag{5.57}$$

The propagation delay $t_{cascade}$ of the cascaded driver is nearly 1000 times smaller than the delay $t_{direct} = 1.69\,\mu s$ for a single-stage driver, which is calculated in Example 5.10.
For $\alpha = 6$:

$$n = \frac{\ln \frac{C_{gate}}{C_{inv}}}{\ln \alpha} = \frac{\ln \frac{250\,\text{pF}}{10\,\text{fF}}}{\ln 6} = \frac{\ln 25000}{1.792} = 5.65 \sim 6 \tag{5.58}$$

$$t_{cascade} = n\,\alpha\, t_{apd} = 6 \cdot 6 \cdot 67.5\,\text{ps} = 2.43\,\text{ns} \tag{5.59}$$

The number of stages n reduces from 10 to 6 while the delay has increased approximately by 0.6 ns. If speed is not extremely critical, the latter case will result in a much smaller area and energy dissipation, as seen below.

Example 5.12 *Calculate the sizing of the driver transistors in each stage for $\alpha = 6$ and $n = 6$ as derived in Example 5.11. The minimum-size inverter for a 5V supply consists of an n-channel transistor with $\frac{W}{L} = \frac{1.0\,\mu m}{0.6\,\mu m}$ and a p-channel device with $\frac{W}{L} = \frac{2.4\,\mu m}{0.6\,\mu m}$.*
The sizing of each stage can be calculated according to Fig. 5.18, listed in Table 5.2. Only widths are shown as the length remains identical for all devices.

Table 5.2 Driver sizing for $\alpha = 6$ and $n = 6$ (Example 5.12).

Stage:	1	2	3	4	5	6
Width PMOS/μm	2.4	14.4	86	518	3110	18662
Width NMOS/μm	1.0	6.0	36	216	1296	7776

5.8.2 Optimization for Energy Efficiency

Optimization for speed does not necessarily result in minimum driver loss. Similar to the speed optimization in Section 5.8.1, the losses can be minimized by choosing the optimum number of stages. The short-circuit current and dynamic gate charge losses in each inverter stage contribute to the overall power losses.

If there are too few stages, the switching losses of the power transistor and transistors in each inverter stage will increase due to larger rise/fall times, resulting in larger short-circuit currents out of V_{drv}. On the other hand, too many stages cause increased driver losses. Moreover, large switching current peaks will be present at the driver supply V_{drv} as the rise-fall times get smaller and the gate charge is delivered during a shorter time. It increases the requirements on the driver supply regarding switching noise rejection and low IR drop. Gate drivers must aim to optimize the power transistor's switching losses concurrently with the driver losses themselves. We will investigate energy optimization based on an energy model to determine the energy consumption of a cascaded gate driver proposed in [8–10].

Let us consider once again the scenario of Fig. 5.10e) with a minimum-size inverter that drives an α times larger inverter (α times larger $\frac{W}{L}$-ratios of both the NMOS and the PMOS transistor). As the rise-fall time increases with the scaling factor α (Eqn. (5.10)), also the short-circuit losses in the minimum-size inverter increase according to Eqn. (5.18). Assuming the minimum-size inverter is the initial gate driver stage, the short-circuit losses are

$$E_{sc,stage} = \alpha E_{sc,inv}, \tag{5.60}$$

in which $\alpha E_{sc,inv}$ accounts for the losses due to the α times larger rise-fall time. It should be noted that this energy does not fully contribute to the losses in the commonly used high-side driver configuration of Fig. 5.3d), also shown in Fig. 5.5. As the driver ground is connected to the switching node, the short-circuit current contributes to the load current. Nevertheless, the short-circuit current still causes losses in the driver stage and loads the high-side driver supply (e.g., a bootstrapping capacitor or a charge pump).

For the same scenario, the gate charge losses associated with the α times larger inverter are proportional to α. Counting them as dynamic losses $E_{dyn,stage}$ in the initial driver stage with a minimum-size inverter according to Eqn. (5.22), we can derive

$$E_{dyn,stage} = \alpha E_{dyn,inv}. \tag{5.61}$$

The gate charge losses $E_{dyn,inv}$ of the minimum-size inverter are multiplied by α to account for an α times larger inverter at its output.

Adding up Eqns. (5.60) and (5.61) yields an expression for the overall losses in a minimum-size driver stage in relation to the short-circuit and dynamic losses of a minimum-size inverter:

$$E_{driver,stage} = \alpha(E_{sc,inv} + E_{dyn,inv}) \tag{5.62}$$

Example 5.13 *Take the short-circuit losses of Example 5.4 and the dynamic losses of Example 5.5 ($E_{sc,inv} = 42$ fJ, $E_{dyn,inv} = 250$ fJ) to calculate the losses $E_{driver,stage}$ of one driver stage for $\alpha = 10$.*

$$E_{driver,stage} = \alpha(E_{sc,inv} + E_{dyn,inv}) = 10 \cdot 292 \text{ fJ} = 2.92 \text{ nJ} \tag{5.63}$$

For a cascaded driver with n stages, the gate charge losses get α time larger from stage to stage. The rise-fall time, given by Eqn. (5.10), will not change between stages because I_{inv} (the denominator in Eqn. (5.10)) gets α times larger and cancels the scaling effect αQ_{inv} of the gate charge (the numerator in Eqn. (5.10)). However, the transconductance parameter β_{inv} in Eqn. (5.18) and, concurrently, E_{sc} scale proportionally to α. In conclusion, E_{sc} and E_{dyn} increase by α from stage to stage. Hence, the total energy consumption is

$$\begin{aligned} E_{driver} &= E_{driver,stage\,1} + E_{driver,stage\,2} + \cdots + E_{driver,stage\,n} \\ &= E_{driver,stage} \sum_{i=0}^{n-1} \alpha^i \\ &= E_{driver,stage} \frac{\alpha^n - 1}{\alpha - 1} \approx E_{driver,stage} \alpha^{n-1}. \end{aligned} \tag{5.64}$$

According to Eqn. (5.52), n increases as α scales to lower values. Hence, with an increasing number of stages, the losses per stage $E_{driver,stage}$ get smaller, but the overall driver losses E_{driver} increase due to the larger number of stages. Equation (5.64) confirms this relationship because the exponential dependency on n dominates.

This finding would indicate that a single-stage driver would be the optimum solution for minimum power loss. While it is true for the losses of the gate driver itself, the conclusion is wrong as we also have to consider switching losses of the power transistor. For this reason, the switching losses depend on the driving strength. The combined driver losses are

$$E_{gd} = E_{driver} + E_{sw}. \tag{5.65}$$

From Eqns. (2.20), (2.21), and (2.38), we know that

$$E_{sw} \propto (T_2 + T_3) \propto \frac{Q_{gate}}{I_{gate}} = \frac{\alpha^n Q_{inv}}{\alpha^{(n-1)} I_{inv}} = \alpha \frac{Q_{inv}}{I_{inv}} \propto \alpha. \tag{5.66}$$

While the driver losses E_{driver} decrease with α, the switching losses of the power transistor increase proportional to α. It again indicates that a minimum of the combined losses can be found.

Figure 5.21 shows the total driver loss E_{gd} as a function of the number of driving stages n. The loss minimum can be obtained analytically based on the given equations or by simulation. If the switching losses in the power transistor get smaller, the optimum number of stages also reduces. Figure 5.21a) has larger switching losses and $n_{opt,loss} = 7$ because the power stage supply V_{bat} is at

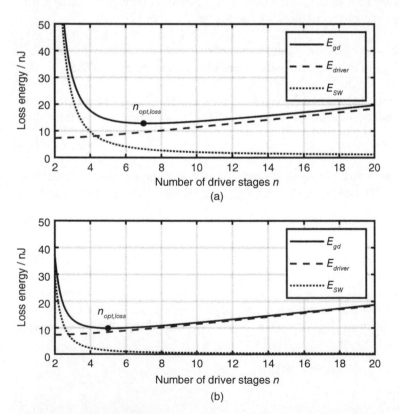

Figure 5.21 Combined driver and switching loss E_{gd} of a cascaded gate driver versus number of driver stages n for a) $V_{bat} = 12$ V and b) $V_{bat} = 5$ V.

12 V. For $V_{bat} = 5$ V, the switching losses decrease and result in a lower number of stages $n_{opt,loss} = 5$. In conclusion, the number of stages n should not be too large, as a rule-of-thumb, in the range of 4–5.

Short-Circuit Losses in the Last Driver Stage

Even for an optimized number of stages, losses will be due to the short-circuit currents in each inverter (shoot-through from supply to ground). Based on Eqns. (5.60), (5.62) and (5.65) the total short-circuit losses can be derived:

$$E_{sc} = \alpha E_{sc,inv} \frac{\alpha^n - 1}{\alpha - 1} \sim \alpha^n E_{sc,inv}, \tag{5.67}$$

where $E_{sc,inv}$ accounts for the short-circuit losses of a minimum-size inverter in a chain of minimum-size inverters. The approximation in Eqn. (5.67) is valid for $\alpha \gg 1$. Assuming a ratio $C_{gate}/C_{inv} = 25000$, Eqn. (5.52) yields $n = 6.3$ for $\alpha = 5$. If we accept $\alpha = 5 \gg 1$ as a boundary condition, the approximation holds for $n \leq 6$ driver stages.

As the shoot-through current scales with the inverter size, i.e., with its driving strength, defined by its W/L-ratio, the last stage is of major influence. Its short-circuit loss is given by

$$E_{sc,n} = \alpha^n E_{sc,inv}. \tag{5.68}$$

A comparison of Eqn. (5.68) with the approximation in Eqn. (5.67) indicates that the last stage accounts for nearly all short-circuit losses. The expressions are identical due to the approximation. To estimate how much these losses contribute to the overall gate driver losses E_{driver}, we put Eqn. (5.68) in relation to Eqn. (5.65). Substituting $E_{driver,stage}$ by Eqn. (5.62):

$$\frac{E_{sc,n}}{E_{driver}} = \frac{\alpha^n E_{sc,inv}}{\alpha^n (E_{sc,inv} + E_{dyn,inv})} = \frac{1}{1 + \frac{E_{dyn,inv}}{E_{sc,inv}}} \tag{5.69}$$

The approximation is valid in practical designs with $\alpha \gg 1$. Especially if the cascaded driver gets optimized for minimum power loss, the scaling factor α gets large, and we can assume $\alpha \gg 1$.

The knowledge about the loss components $E_{sc,inv}$ and $E_{dyn,inv}$ in the minimum-size inverter allows us to predict the short-circuit loss contribution of the last driver stage. It depends on the ratio between the dynamic losses and the short-circuit losses. For the realistic case of $E_{dyn,inv} = (1 \ldots 10)E_{sc,inv}$ the short-circuit losses in the last stage account for approximately $10 \ldots 50\%$ of the total gate driver losses. This conclusion includes the short-circuit and the gate charge losses.

Example 5.14 *Calculate the percentage of the short-circuit losses referred to the total driver losses for $E_{dyn,inv}$ from Example 5.5 and $E_{sc,inv}$ from Example 5.4, i.e., $E_{dyn,inv} = E_{dyn,e} = 250$ fJ and $E_{sc,inv} = E_{sc,e} = 42$ fJ.*

$$\frac{E_{sc,n}}{E_{driver}} = \frac{1}{1 + \frac{E_{dyn,inv}}{E_{sc,inv}}} = \frac{1}{1 + \frac{250 \text{ fJ}}{42 \text{ fJ}}} = \frac{1}{1 + 5.95} = 14.4\% \tag{5.70}$$

The pull-up and pull-down transistor of the last stage can be controlled by separated branches, leaving a dead time when changing between the on/off states of the driver. If the dead time is properly chosen, there is never a conducting path between supply and ground. Thus, cross-conduction in the last driver stage can be eliminated.

Split-Path Gate Driver

Figure 5.22 shows how a driver can be modified to achieve a separate control for each transistor in the last driver stage. The idea is that the pull-up and the pull-down transistor of the final stage (MP3

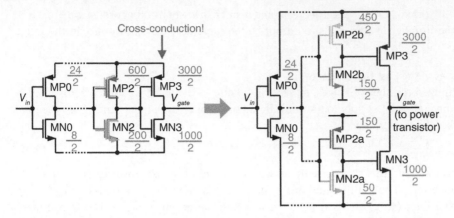

Figure 5.22 Splitting up the gate driver into separate branches.

and MN3, respectively) get a separate cascade of intermediate driver stages. This way, the transistors get activated with a dead time in between to eliminate cross-current causing short-circuit losses E_{sc}. The concept is known as split-path gate driver [11, 12]. Starting from the initial cascaded driver (Fig. 5.22 left), each inverter stage gets split into two parts, one for the pull-up transistor cascade and one for the pull-down transistor cascade. The size of these inverters scales proportionally to the size of the corresponding transistor of the last stage. For the example in Fig. 5.22, the output transistors have a W/L-ratio of 3000/2 (pull-up path, PMOS-transistor) and 1000/2 (pull-down path, NMOS-transistor). Hence, the ratio is 3:1. As the size of the PMOS-transistor MP2 of the second to last stage of the original driver is 600/2, it gets split up into two devices, MP2b and MP2a, by a ratio of 3:1. This results in a W/L-ratio of 450/2 and 150/2, respectively. The other preceding driver stages are modified in the same way. The first stage remains unchanged. There may be more common stages in large drivers with many stages until the driver chain splits into two branches. The split-path approach has no area penalty except for some wiring and spacing overhead. Based on this two-branch driver architecture, two main methods exist to reduce cross-conduction.

Break-Before-Make Circuit
This method detects the gate voltage of each transistor of the last driver stage, Fig. 5.23. Only if the corresponding transistor of the last stage is turned off the other transistor gets turned on. It is achieved by detecting the driver output (feedback inverters in Fig. 5.23) and incorporating a break-before-make logic. In the case of a high-side driver, these circuits can be implemented with the actual driver fully in the high-side domain. As explained above, the driver stages are partitioned and scaled according to Fig. 5.22.

The disadvantage is that this circuit results in a relatively large dead time and response time because the switching signal has to propagate through both driver branches. Moreover, as the dead time is defined by the propagation delay through one driver branch, it suffers from process and corner variations. Consequently, some delay margin is required, which sets a limit for high-speed gate drivers.

Asymmetrical Sizing
This method is depicted in Fig. 5.24. Unlike the break-before-make concept, this method needs no feedback. The idea is to make the devices in each inverter stronger or weaker with respect to the original size obtained based on the theory presented in Sections 5.8.1 and 5.8.2. The scaling is done asymmetrically, as indicated in Fig. 5.24 at each transistor.

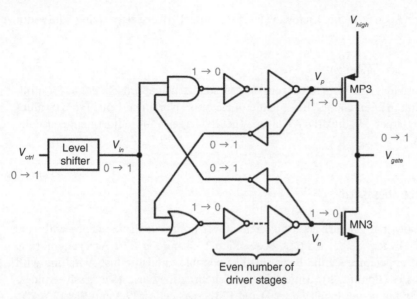

Figure 5.23 Gate driver with a break-before-make circuit.

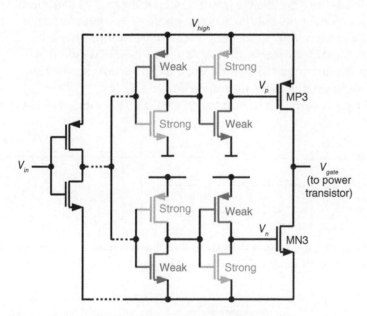

Figure 5.24 Gate driver with asymmetrical device sizing.

To avoid cross-conduction at the last stage, the corresponding transistor MN3 must be turned off before activating the complementary output transistor MP3. Therefore, the NMOS transistor is driven by an inverter with a strong NMOS device, while a strong PMOS device controls the PMOS transistor. The same approach is continued for all preceding stages, i.e., the turn-off signal propagation is emphasized by using strong devices while weak devices drive the turn-on signal. This way, almost automatically, a dead time is achieved. As no feedback is involved, process variations have less influence compared to the break-before-make circuit.

An asymmetry factor AF can be defined, allowing for systematical driver device sizing. The width is given by

$$W_{strong,\,weak} = W\left(1 \pm \frac{AF}{100\%}\right), \tag{5.71}$$

where W refers to the original width obtained from the cascaded driver approach, either for minimum delay or for minimum energy loss. The asymmetry factor is given in percent. Typical values are $AF = 20\dots30\%$. For larger AF, the driver delay may get too large. A fine adjustment is usually achieved by simulation.

5.9 External Gate Resistor

From Chapter 2, we know that gate drivers for discrete power transistors have to cope with gate loop parasitics expressed by R_{loop}, L_{loop}, and C_{loop}, see Section 2.5.4 and Fig. 2.10. As a general rule, the corresponding loop impedance should be as small as possible to ensure fast switching with minimum switching losses (Eqn. (2.39)). However, fast switching may lead to excessive voltage overshoot and ringing in the power loop (Fig. 2.11) and EMI (see Section 12.3 and Fig. 12.7c,d)). For a given gate driver, an external gate resistor R_G can be added to adjust the driving strength and, hence, the transition speed to optimally drive the discrete power transistor. Figure 5.25 illustrates how the gate resistor reduces the gate current and the drain current rise time. The same influence applies to the fall time when the switch turns off. R_G basically adds to the driver resistance R_{drv}. A trade-off between losses (large R_G) and over-voltage (small R_G) can be achieved. We recall how the gate driving strength influences the current and voltage transition speed. For an inductive load, Eqns. (2.20) and (2.21) show that transition times are both proportional to $1/I_{gate}$.

The R-L-C components along the gate loop form a resonant tank, see Fig. 2.10. Its impedance is given by

$$Z = R_{loop} + j\omega L_{loop} + \frac{1}{j\omega C_{loop}} = R_{loop}\left(1 + j\underbrace{\frac{1}{R_{loop}}\sqrt{\frac{L_{loop}}{C_{loop}}}}_{Q}\left(\frac{\omega}{\omega_o} - \frac{\omega_o}{\omega}\right)\right), \tag{5.72}$$

where Q denotes the quality factor and ω_o the natural frequency. The step response is a damped oscillation at $f_o = \omega_o/(2\pi)$ with a damping factor $D = 2/Q$. The overshoot of the gate voltage may

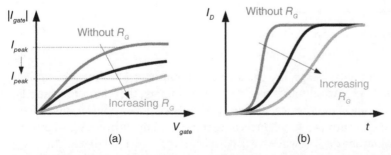

Figure 5.25 Influence of an external gate resistor R_G: a) gate current versus gate voltage; b) drain current transition.

exceed the maximum gate–source voltage. For this reason, the damping coefficient D of the series R-L-C resonant tank needs to be sufficiently large.

$$D = \frac{R_{loop}}{2} \cdot \sqrt{\frac{C_{loop}}{L_{loop}}} \rightarrow R_{loop} > 2 \cdot D \cdot \sqrt{\frac{L_{loop}}{C_{loop}}}. \tag{5.73}$$

R_{loop} is the sum of all resistance components, which includes $R_G + R_{G'}$ and the driver resistance R_{drv} (see Section 5.4). $D = 1$ corresponds to critical damping. However, $D = 0.7$ is usually a good trade-off, resulting in only a small overshoot at smaller values of R_{loop}. Even though they are not shown in Fig. 2.10, the finite resistance of the source terminal and the remaining interconnections must be considered. The loop capacitance is formed by the series connection of the transistor's gate–source capacitance C_{GS} and the driver buffer capacitor C_{buf}. Since C_{buf} is typically in the order of 100 nF, the total loop capacitance can be approximated by C_{GS}, which is only a few nano-farads. C_{buf} can also be a bootstrap capacitor (see Section 5.11). Finally, $L_{loop} = L_G + L_{G'} + L_S + L_{S'}$ represents the total parasitic loop inductance.

Example 5.15 *What is the damping factor D for $R_{loop} = 2\,\Omega$, $L_{loop} = 18$ nH, $C_{gs} = 2$ nF, $C_{buf} = 100$ nF?*

With $C_{loop} \sim C_{gs}$:

$$D = \frac{R_{loop}}{2} \cdot \sqrt{\frac{C_{loop}}{L_{loop}}} = \frac{2\,\Omega}{2} \cdot \sqrt{\frac{2\text{ nAs/V}}{18\text{ nVs/A}}} = 0.33 \tag{5.74}$$

The damping can be improved by increasing the gate resistor R_G. Which value of R_{loop} results in critical damping ($D = 1$)?

$$R_{loop} = 2\sqrt{\frac{L_{loop}}{C_{loop}}} = 2\sqrt{\frac{18\text{ nVs/A}}{2\text{ nAs/V}}} = 6\,\Omega \tag{5.75}$$

5.10 *dv/dt* Triggered Turn-On

In Sections 5.8 and 5.9, we explored the design space for gate drivers regarding speed and power efficiency. One more criterion that any driver must fulfill is the immunity against turn-on at steep voltage slopes at the switching node, Fig. 5.26a). The driver needs to be strong enough to keep the power transistor off at fast dv/dt transients across the drain–source of the power switch. Table 5.3

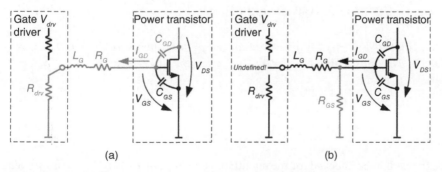

(a) (b)

Figure 5.26 dv/dt triggered turn-on: a) general case; b) at start-up.

Table 5.3 Overview of typical transition slopes of the switching node for various applications.

Application	Slope in V/ns (= kV/μs)
Standard integrated SMPS $f_{sw} \sim 1$ MHz	≤ 1
Fast-switching integrated SMPS	10–80
Motors and drives (MOS, IGBT)	1–10 (max. 50)
Off-line converters (SJ MOSFETs, SiC)	100–150 and above
Fast-switching power MOSFETs, GaN	Up to 200–250

SMPS: Switched-mode power supply.

provides an overview of the typical slopes dV_{DS}/dt. In a particular application, the slope depends on the output node impedance (R, L, C), which includes the load, and on the driving strength. Typical rise times are 1–10 V/ns in integrated designs and ten times larger for discrete power stages.

Section 2.5.4 shows how the parasitic gate loop inductance and resistance add to the driver resistance. Assume that the low-side switch in Fig. 5.26a) has just turned off. Now, the drain node will be pulled from $V_{DS} \approx 0$ toward supply by either the load or a high-side switch. The gate–drain capacitance C_{GD} gets charged during this transition. The corresponding current I_{GD} flows from the drain via C_{GD} into the pull-down path of the driver. Due to the finite pull-down resistance of the driver and the gate loop parasitics, the gate–source voltage will increase. To prevent the power transistor from getting turned on and to avoid cross-conduction from the supply, the gate–source voltage must not exceed the power transistor's threshold voltage V_{th}. As the gate loop parasitics are significant for discrete power transistors, it makes sense to consider integrated and discrete power stages separately.

5.10.1 Integrated Power Stages

For integrated power stages, we can neglect the loop inductance ($L_{loop} = 0$). Hence,

$$R_{drv} + R_G \leq \frac{V_{th}}{C_{GD} \frac{\partial V_{DS}}{\partial t}}. \tag{5.76}$$

This relationship assumes that the transition rates of V_{DS} and V_{DG} are approximately equal (see Section 2.5 on switching behavior and Eqn. (2.16)). The driver resistance R_{drv} is defined according to Fig. 5.8. Figure 5.26a) shows the scenario on the low-side gate. The same effect can be observed on the high side, and Eqn. (5.76) is also valid for the high-side driver.

Example 5.16 *Calculate the maximum driver resistance R_{drv} for a power transistor with $C_{GD} = 70$ pF and $V_{th} = 0.5$ V. The transition slope is 2 V/ns. We neglect the gate resistance, i.e., $R_G = 0$.*
For $R_G = 0$, Eqn. (5.76) yields

$$R_{drv,max} = \frac{V_{th}}{C_{GD} \frac{\partial V_{DS}}{\partial t}} = \frac{0.5 \text{ V}}{70 \text{ pF} \cdot 2 \text{ V/ns}} = 3.57 \text{ k}\Omega \sim 3.6 \text{ k}\Omega. \tag{5.77}$$

Any driver design needs to be checked for the condition given in Eqn. (5.76). If the requirement of Eqn. (5.76) is not fulfilled for an existing driver design, R_{drv} needs to be reduced by increasing the W/L-ratio (i.e., the width W because the length L is kept constant at minimum) of the driver output stage until the condition is met.

If a cascaded driver of Fig. 5.18 requires to increase the width W_n in the last stage, the scaling factor α recalculates to

$$\alpha = \sqrt[n-1]{\frac{W_n}{W_1}}, \tag{5.78}$$

where W_1 is the width of the first-stage inverter (generically named W in Fig. 5.18). The width can be referred to the NMOS or the PMOS device in the corresponding driver stage.

Alternatively, some designs use a second pull-down switch parallel to the gate driver's output stage that is only active during the switching transition to ensure sufficiently low pull-down resistance.

Example 5.17 *The device sizing determined in Example 5.12, listed in Table 5.2, gives a driver resistance of $R_{drv} = 5.5\ \Omega$. Recalculate the required width of the NMOS transistor in the last driver stage as well as the new scaling factor α to fulfill the condition $R_{drv,max} = 3.6\ \Omega$ from Example 5.16.*

In Example 5.12, the calculated width of the NMOS transistor in the last driver stage is 7776. This width corresponds to $R_{drv} = 5.5\ \Omega$. Hence, the new target $R_{drv} = 3.6\ \Omega$ requires to scale the width of the last stage to $W_n = W_6 = \frac{5.5\ \Omega}{3.6\ \Omega} \cdot 7776 = 11880$. Inserting into Eqn. (5.78) yields the new scaling factor,

$$\alpha = \sqrt[n-1]{\frac{W_n}{W_1}} = \sqrt[5]{\frac{W_6}{W_1}} = \sqrt[5]{\frac{11880\ \mu m}{1\ \mu m}} = 6.53 \sim 6.5. \tag{5.79}$$

5.10.2 Discrete Power Stages

In a discrete design with significant loop inductance, Eqn. (5.76) allows estimating the driver impedance in first order, assuming zero loop inductance. Equation (5.76) may conflict with the minimum loop resistance according to Eqn. (5.73) to suppress excessive overshoot of the gate–source voltage. It can only be solved by improving the gate loop design such that L_{loop} becomes lower (see Eqn. (5.73)). There are, of course, physical limits. Therefore, some designs increase the gate–source capacitance by placing an external capacitor as close as possible to the power switch. According to Eqn. (5.73), increasing $C_{GS} \sim C_{loop}$ yields a lower minimum gate loop resistance. However, larger C_{GS} destroys all effort for a good gate driver that achieves fast switching and minimizes switching losses. As a third solution, we can use a bipolar gate driver that applies a negative gate–source voltage $V_{GS,off}$ to keep the transistor safely turned off. Hence, we can replace V_{th} in Eqn. (5.76) by the term $V_{th} + |V_{GS,off}|$. This way, the design can tolerate a much larger gate resistance value, which may allow it to fulfill Eqn. (5.73) concurrently.

It is instructive to estimate the maximum allowed inductance by neglecting the resistive components in the gate loop. Solving the energy equation for an L-C-resonant tank formed by $C_{loop} \sim C_{GS}$ and L_{loop}, we obtain the maximum value of L_{loop} that ensures $V_{GS} < V_{th}$:

$$\frac{L_{loop}}{2}I_{dg}^2 = \frac{C_{loop}}{2}V_{th}^2 \rightarrow L_{loop} < \frac{C_{GS} \cdot V_{th}^2}{\left(C_{GD}\frac{\partial V_{DS}}{\partial t}\right)^2} \tag{5.80}$$

Example 5.18 *What is the maximum acceptable value of L_{loop} for $R_{loop} = 0$, $C_{GS} = 4\ nF$, $C_{GD} = 100\ pF$, $V_{th} = 2\ V$? The transition slope is $20\ V/ns$.*

From Eqn. (5.80):

$$L_{loop} < \frac{C_{GS} \cdot V_{th}^2}{\left(C_{GD}\frac{\partial V_{DS}}{\partial t}\right)^2} = \frac{4 \text{ nF} \cdot (2 \text{ V})^2}{(100 \text{ pF} \cdot 20 \text{ V/ns})^2} = 4 \text{ nH}$$

5.10.3 Save Start-Up

A particular case for dv/dt triggered turn-on is the start-up when the power stage settles, and the gate voltage is undefined, Fig. 5.26b). At start-up, the slope is determined by the rate at which the input supply voltage of the power stage rises. The slopes are usually significantly lower than in regular operation, typically in the order of a few V/µs. To ensure that the power transistor remains off in that case, a pull-down resistor R_{GS} can be added, as shown in Fig. 5.26b). While a low R_{GS} gives the best protection against dv/dt triggered turn-on, it causes additional loading of the driver of its supply V_{drv}. Hence, its value should not be too low. 100 kΩ is usually a good trade-off.

5.11 Bootstrap Gate Supply

High-side drivers for n-type high-side transistors must be supplied with a voltage larger than the input supply voltage of the power stage V_{bat} to ensure sufficient gate overdrive. If the high side is turned on, the switching node V_{sw} reaches nearly V_{bat}. If the gate–source voltage of the high-side switch is equal to V_{drv}, the driver supply needs to be $(V_{bat} + V_{drv})$. A charge pump or an inductive boost converter can supply this voltage. See Section 8.9 for suitable charge pump circuits and Section 10.10 for boost converters.

A bootstrap supply provides an easily implemented gate supply for applications with a steadily repeating pulse pattern like pulse width modulation (PWM). Bootstrapping is common for switched-mode converters, as covered in Chapters 10 and 11. In contrast, motor bridge drivers are typically not suitable because the PWM is not steadily repeating (e.g., during breaking).

5.11.1 General Operation

The idea is to store the required energy in a capacitor while the high side is off and to supply the driver from that capacitor at turn-on. If V_{sw} is at zero (PWM = 0, high-side off, low-side on), the bootstrap capacitor C_{boot} in Fig. 5.27a) gets charged from V_{DD} through the bootstrap diode

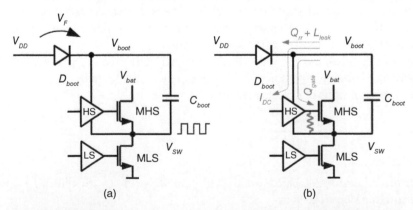

(a)   (b)

Figure 5.27 a) Bootstrap gate supply; b) with currents and charge supplied by C_{boot}.

D_{boot}. The voltage across C_{boot} is the driver voltage $V_{drv} = V_{DD} - V_F$, which provides the high-side gate supply voltage. The high-side switch pulls V_{sw} toward V_{bat} at turn-on. The diode gets reverse biased and blocks the path to V_{DD} (which would otherwise be back supplied). The configuration in Fig. 5.27a) can be seen as the first stage of a charge pump. If the high-side is fully turned on, the top plate potential of C_{boot} reaches $V_{boot} = V_{bat} + V_{drv}$ (referred to ground). The bootstrap capacitor becomes the only source of energy for the high-side driver. For this reason, bootstrapping is only suitable in applications where the PWM ensures frequent recharge. The driver sees only V_{drv} and can be implemented with transistors of lower voltage ratings.

5.11.2 Charge Balance and Bootstrap Capacitor Sizing

The voltage drop across D_{boot} can be eliminated by an active bootstrap circuit, as described in Section 5.11.4. Nevertheless, suppose the half-bridge drives an inductive load, for instance, in an inductive DC–DC converter. In that case, the switching node gets pulled to $-V_F$ during the dead time as shown in Fig. 2.12b) (for the typical case of a positive inductor current I_L). When C_{boot} recharges during the dead time, the voltage drop across the bootstrap diode somewhat compensates.

As illustrated in Fig. 5.27b), the bootstrap capacitor needs to deliver:

- the total gate charge Q_{gate} to turn on the high-side transistor
- the reverse recovery charge Q_{rr} of the bootstrap diode D_{boot}
- the leakage current of the bootstrap diode D_{boot}
- the quiescent current of the gate driver and level shifter
- gate–source leakage current (including pull-down resistor according to Fig. 5.26b), if applicable)

The net current comprised of the latter three items will be referred to as I_{boot}. I_{boot} is only present if PWM = 1 (high-side on, low-side off), which corresponds to a maximum duty cycle D_{max}, i.e., to a duration of $D_{max}T_{sw} = D_{max}/f_{sw}$. These charge and current components cause a charge loss ΔQ_{boot} at the bootstrap capacitor per switching cycle:

$$\Delta Q_{boot} = C_{boot}\Delta V_{boot} = Q_{gate} + Q_{rr} + I_{boot}\frac{D_{max}}{f_{sw}} \tag{5.81}$$

The voltage droop ΔV_{boot} has to be small enough to ensure proper power transistor switching. As a rule-of-thumb, ΔV_{boot} should be around 10% of V_{drv}.

Solving Eqn. (5.81) for C_{boot} yields a sizing equation for the bootstrap capacitor:

$$C_{boot} \geq \frac{Q_{gate} + Q_{rr} + I_{boot}\frac{D_{max}}{f_{sw}}}{\Delta V_{boot}} \tag{5.82}$$

Typical values of C_{boot} are in the order of 100 nF. It is, therefore, implemented as a discrete off-chip component because its value is typically far too large for on-chip integration.

Example 5.19 *What value of C_{boot} is required for $Q_{gate} = 10$ nC, $C_{rr} = 1$ nC, $I_{boot} = 500\,\mu A$, and $f_{sw} = 500$ kHz? We go for the worst case of 100% duty cycle, i.e., $D_{max} = 1$ and allow a voltage droop of $\Delta V_{boot} = 0.1$ V.*

Inserting all values into Eqn. (5.82) yields

$$C_{boot} \geq \frac{Q_{gate} + Q_{rr} + I_{boot}\frac{D_{max}}{f_{sw}}}{\Delta V_{boot}}$$

$$\geq \frac{10\,\text{nC} + 1\,\text{nC} + 500\,\mu A\,\frac{1}{500\,\text{kHz}}}{0.1\,\text{V}} = 120\,\text{nF}. \tag{5.83}$$

This result is indeed a typical value.

The gate charge Q_{gate} is usually the dominating portion of the charge balance. If we neglect all other losses in Eqn. (5.82) except for Q_{gate}, we can estimate C_{boot} in relation to the actual gate capacitance $C_{gate} = Q_{gate}/V_{drv}$:

$$C_{boot} \geq C_{gate} \frac{V_{drv}}{\Delta V_{boot}} \tag{5.84}$$

Also, this relationship confirms that C_{boot} is usually too large to be integrated.

Example 5.20   *For $V_{drv} = 5$ V and $\Delta V_{boot} = 0.1$ V, Eqn. (5.84) gives $C_{boot} \geq 50\, C_{gate}$.*

5.11.3   Practical Aspects of Bootstrapping

There are several issues with bootstrap supplies, mainly related to insufficient charging of the bootstrap capacitor. The entire power stage may fail to operate if the switching node does not reach zero at start-up. Some solutions are presented in Fig. 5.28.

The circuit in Fig. 5.28a) adds an additional charging path for C_{boot} from the battery voltage via R_1. The zener diode D_1 protects against overcharging. R_1 should not be too low to keep the static current consumption at an acceptable level (typically some tens of microamperes, i.e., $R_1 \sim 100$ kΩ).

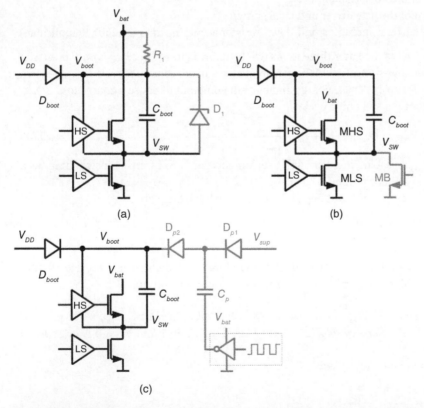

(a)   (b)

(c)

Figure 5.28   Bootstrap supply with a) recharge path from V_{bat}, b) pull-down transistor MB, and c) charge pump to recharge C_{boot}.

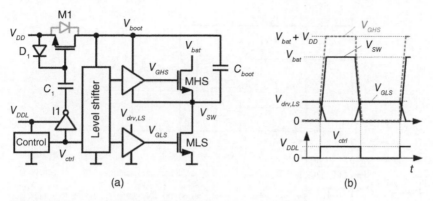

Figure 5.29 Active bootstrapping: a) circuit; b) timing diagram.

Other designs use an additional pull-down switch parallel to the low-side power transistor as shown in Fig. 5.28b). Transistor MB can be much smaller than the power transistor. Controlled by a power-on-reset signal (see Section 6.9), MB will initially pull down V_{sw} to zero.

Bootstrapping cannot support long on-times of the high-side transistor since C_{boot} gets eventually discharged by any static current as shown in Fig. 5.27b). Adding a charge pump can recharge C_{boot} without requiring the switching node to return to zero. In Fig. 5.28c), two diodes and a pumping capacitor are added to form a charge pump that re-charges C_{boot}. The charge pump clock frequency (at the input of the inverter), the size of C_p, and the voltage drop on C_{boot} determine the amount of charge delivered onto C_{boot}. For more details on charge pumps, refer to Chapter 8. In that chapter, more elaborate gate supply circuits are covered; see Section 8.9.

5.11.4 Active Bootstrapping

Some applications cannot tolerate the voltage drop by the bootstrap diode. The concept of active bootstrapping replaces the diode with a transistor, as shown in Fig. 5.29. The NMOS-transistor M1 replaces the bootstrap diode. It gets activated in sync with the power stage's gate control signal V_{ctrl}. The inverter I1, capacitor C1, and diode D_1 implement local bootstrapping to turn on the transistor M1. The polarity of M1 is critical. Since V_{boot} reaches $V_{bat} + V_{drv}$, it needs to be connected to the drain of M1. M1 has to be a high-voltage device rated at V_{boot}. As long as M1 is off, its body diode precharges C_{boot} already up to $V_{DD} - V_F$.

5.12 Level Shifters

Most gate control circuits utilize level shifters as a linking element between low-voltage control and higher-voltage gate drive domains (V_{drv}). We have seen two representative level shifter examples in Figs. 5.4 and 5.5, the resistor-based level shifter and the cross-coupled level shifter. Another widely used class of level shifters are capacitive level shifters. There exist various implementations for all types of level shifters. We will explore the most essential concepts in this section.

5.12.1 Resistor-Based Level Shifters

The resistor-based circuit of Fig. 5.30a) is the most basic implementation of a level shifter. If the input signal is "0," M1 is off, and hence, there is no current flow through the pull-up resistor R_{pu},

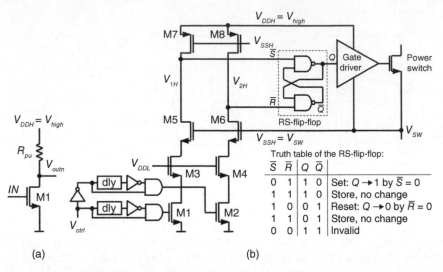

Figure 5.30 Resistor-based level shifters: a) basic configuration; b) pulsed level shifter with transistors M7, M8 as resistive pull-ups.

resulting in a logical "1" at the level shifter output. For a "1" input level, M1 turns on and conducts a current I_1 defined by the W/L-Ratio of M1. This current leads to a voltage drop across R_{pu}. The level shifter output reaches $V_{drv} - (I_1 \cdot R_{pu})$, which sets the logical "0" level at the level shifter output. The "0" level may vary depending on von I_1 and R_{pu}. There are two drawbacks associated with this variation, (1) the logic level may be undefined and cause a shoot-through current in the high-side domain logic and gate driver, and (2) the gate–source voltage maximum rations of the gate driver's input inverter and any other transistors on the high-side may be violated. Therefore, resistor-based level shifters comprise clamping diodes to define the output "0" level. A clamping diode D_1 or D_2, as indicated in Fig. 5.5, has to be added. It comes at the expense of additional power losses, i.e., a part of I_1 flows through R_{pu} and contributes to power losses at V_{high}, and the remaining part of I_1 flows through the clamping diode. The clamping option with a regular diode connected to V_{sw} is preferable, as the current flows through the high-side transistor out of V_{bat} and contributes to the overall load current of the power stage. Especially for high-output power, the additional current is negligible. The current drawn out of V_{high} has not only a power loss aspect, but the supply loading itself is also critical. Especially for the widely used bootstrap supply, the excessive current causes a more voltage droop ΔV_{boot} (influence of I_{boot} in Eqn. (5.81)). Consequently, the gate overdrive and the turn-on time of the driven power switch degrade. On the other hand, I_1 must not be too small to ensure a fast response time of the level shifter. In conclusion, I_1 and R_{pu} must be chosen for the best trade-off between speed and losses.

Table 5.4 summarizes the key characteristics of the resistor-based level shifter. It benefits from its simple design. It also provides a default state during system start-up because R_{pu} keeps the output V_{outn} at V_{high} (logical "1"). The drawback is the static power consumption.

A design guideline is as follows:

(1) Choose R_{pu} such that the time constant $R_{pu}C_{par}$ with the parasitic output node capacitance C_{par} of the level shifter meets the required response time τ_{pu}:

$$R_{pu} \leq \frac{\tau_{pu}}{C_{par}} \tag{5.85}$$

Table 5.4 Comparison of three classes of level shifters, resistor-based level shifters, cross-coupled level shifters, and capacitive level shifters.

Resistor-based	Cross-coupled	Capacitive
− static current	+ no static current	+ no static current
+ simple, small area	− slow and large area	+ fast
+ default state	− no default state	− no default state

C_{par} is usually not well defined and can be estimated for simulation. Many designs simply choose $R_{pu} = 100\ \text{k}\Omega$. With that typical value a 1 ns response time can be achieved for a typical C_{par} value of 10 fF. Please note that τ_p applies to the low-high transition at the level shifter output. The high-low transition is determined by the current flow of I_1 into the parallel configuration of R_{pu} and C_{par}. Neglecting R_{pu} in the initial instant of the transition, we get

$$\tau_{pd} \propto C_{par}\frac{V_{drv}}{I_1} \tag{5.86}$$

with the target gate overdrive V_{drv} of the power transistor MP.

(2) Determine the W/L-ratio of M1 to achieve I_1 according to

$$I_1 \cdot R_{pu} \geq V_{drv}. \tag{5.87}$$

Equation (5.87) needs to be fulfilled for worst-case corners. If we assume a gate overdrive of $V_{drv} = 5\ \text{V}$ and $R_{pu} = 100\ \text{k}\Omega$, I_1 needs to be greater than $50\ \mu\text{A}$. For $I_1 = 50\ \mu\text{A}$, Eqn. (5.86) yields $\tau_{pd} = 1\ \text{ns}$ which is identical to τ_{pu} in this case. Usually, even a minimum transistor M1 will deliver rather currents in the milliampere range. The excessive current needs to be withdrawn by the clamping diodes. Therefore M1 can be combined with a current source to save power at the expense of complexity.

(3) Sizing of the clamping diode (see Fig. 5.5) to fulfill the current density rules. The diode D_1 in Fig. 5.5 can be made of a body diode of a transistor with the same voltage ratings as the devices in the high-side voltage domain. The body diode can withstand the same current as the channel of the transistor can deliver. So by sizing the W/L, the current capability of the body diode can be adjusted. For example, if M1 delivers 5 mA, a transistor of nearly minimum size can be used for the diode clamp. See also Section 3.5 on integrated diodes.

Pulsed Level Shifter

There are several variants of resistor-based level shifters that aim to reduce static power consumption. One example is shown in Fig. 5.30b). It implements various improvements of the basic level shifter of Fig. 5.30a). Immediately noticeable, it is a differential implementation. This approach can be found in many level shifters as it helps to improve the robustness, especially in case of coupling effects during fast switching transitions at the power stage (see CMTI, Section 5.13). Moreover, transistors M6 and M7 replace the pull-up resistor R_{pu} of the basic level shifter. Short pulses transfer the signal at the gates of M1 and M2. The gate control signal is stored in an RS flip-flop with cross-coupled NAND gates. Activating M1 or M2 causes V_{1H} and V_{2H}, respectively, to undergo a negative pulse with respect to V_{DDH}. The flip-flop can be set or reset according to the truth table in Fig. 5.30b). The invalid state will be further discussed in the context of common-mode transient immunity (CMTI) in Section 5.13.

M3 and M4 are high-voltage devices that protect M1, M2. This way, thin-oxide devices can be used for M1 and M2 (e.g., 1.5 V rating), which brings several advantages regarding shifting speed and power dissipation [13]. M5 and M6 form a common-gate gain stage on each branch to amplify the signal at the drain of M3 and M4 and enhance the speed of the level shifter [14]. It further reduces the transition energy and the propagation delay. Their body diodes also prevent V_{1H} and V_{2H} from falling below V_{SSH} by more than their forward voltage V_F. It protects M7, M8, and the flip-flop inputs from over-voltage.

Pulse-forming circuits control the input transistors M1 and M2. This way, M1 and M2 conduct only for a short duration. During that time, static current flows out of the high-side supply V_{DDH}. In steady state M1 and M2 are off and both high-side voltages V_{1H} and V_{2H} are pulled to V_{DDH} by M7, M8. Due to the complementary control of M1 and M2, either V_{1H} or V_{2H} experience a negative pulse and set the RS flip-flop accordingly. After the gate control signal is stored, the pulse generation can turn off M1 and M2, respectively, to avoid static power consumption.

Level Shifter with Diode-Connected Pull-Up Transistors

The pull-up resistor can also be implemented by a diode-connected transistor. Consequently, this diode can be the input of a current mirror. The level shifter in Fig. 5.31 implements this approach. It provides a differential signal transfer, which makes the level shifter more robust against coupling (from the switching node) and other parasitic effects. M1 and M2 see a complementary control signal. M3 to M8 form a kind of symmetrical amplifier. For $IN = $ "1," M2 carries the full bias current I_{bias} while M1 is off. The current mirror M5-M6 pulls up the gate driver input V_{in} to turn on the high-side power transistor. The opposite sequence occurs for $IN = $ "0." Optionally, the current source I_{bias} can be omitted. In that case, the driving strength of M1 and M2 determines the current that is transferred to the high side. I_{bias} can also be adaptive such that it provides maximum speed during signal transfer and goes back to a small standby current in a steady state (low power dissipation from V_{high}). A simple inverter delay chain can control it. Typical values of I_{bias} are in the order of a few tens of microampere, which can be reduced to less than 5 µA in standby.

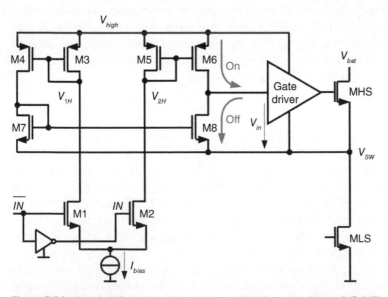

Figure 5.31 Level shifter with diode-connected pull-up transistors (M3, M5).

This level shifter has been demonstrated to operate at switching frequencies of up to 10 MHz [12]. The level shifter's response time and propagation delay can be enhanced by pre-biasing M3 and M5 such that their source–gate voltage is already near their threshold voltage. It can be achieved by pull-down resistors at the drain nodes of M3 and M5. At the same time, the geometry of M3 and M5 should be as small as possible to keep the parasitic capacitance low.

One limitation of this concept is that there is a considerable current through M3 and M5 for fast slopes, such that their gate voltage drops and the transistors may exceed the maximum V_{SG} ratings. Protection diodes may be added similar to D_1 and D_2 in Fig. 5.5. The diodes need to be sized up to handle the currents, they increase the parasitic capacitance at nodes V_{1H} and V_{2H}, which slows down the level shifter. A trade-off between speed and susceptibility on fast switching node transients has to be found.

To further speed up the level shifter, the signal detection on the high side can be done with a comparator that senses the difference between the two branches (V_{1H} and V_{2H}). Instead of waiting for the signal to reach the logic threshold, a difference of ~ 100 mV is sufficient to change the state. This concept is used in [15] for a level shifter that can handle up to 50 V.

5.12.2 Cross-Coupled Level Shifters

The basic cross-coupled level shifter of Fig. 5.32a) is a fully symmetrical circuit that is controlled by the input signal *IN* and its complement $\overline{IN}$. Moghe et al. [16] provides a good overview of the circuit and its design details. There are two complimentary outputs, V_{outp}, V_{outn}. Even if the input voltage level is less than V_{high}, each output sees a full voltage swing $V_{drv} = V_{high}$. As the output swing is equal to the source–gate voltage of the cross-coupled PMOS pair, the basic level shifter can only be used if V_{high} is lower than the maximum rating of the PMOS source–gate voltage. This voltage limit is in the range of 5 V or less. For larger voltages V_{high}, the cross-coupled PMOS pair can be protected by cascode devices, as depicted in Fig. 5.32b). It follows the general protection technique of Fig. 6.2 (Chapter 6). This configuration is commonly used in a high-side application, similar to Fig. 5.5.

The switching happens in two phases. Let us assume high-level at V_{outp}, i.e., $V_{outp} = V_{high}$, $V_{outn} = 0$. To toggle the level shifter state, V_{outp} is pulled down first. As soon as $V_{SG3} = V_{high} - V_{outp}$ reaches $V_{drv} - V_{thp}$, the PMOS transistor M3 turns on, and the first transition phase ends. In the

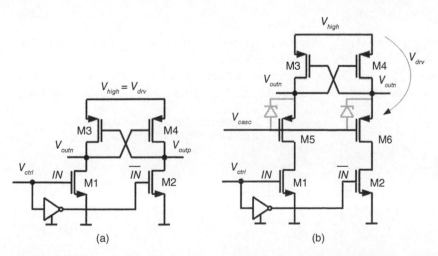

Figure 5.32 Cross-coupled level shifter: a) general circuit; b) high-voltage version with cascodes.

following second phase, M3 pulls up V_{outn} toward V_{high}. At the end of the two phases, the level shifter has fully toggled its state, and we get $V_{outp} = 0$, $V_{outn} = V_{high}$. Due to the two phases, the cross-coupled level shifter is relatively slow. One output signal switches faster than the other one. Its main advantage is that there is no static current consumption because there is always one transistor switched off in each branch, similar to CMOS logic.

Table 5.4 lists the key characteristics of the cross-coupled level shifter compared to other types of level shifters. Compared to the resistor-based level shifter, there is one more disadvantage as discussed below.

Default State

Due to its symmetrical construction, the level shifter suffers from meta-stability, and there is no default state. It is usually a concern at the start-up of a power management system when the input signals are not yet fully defined. In smart power applications, motors may start to run, or solenoids can get activated in an uncontrolled way. Such a failure must be prevented in safety-critical systems. A practical way to force a default "1"-state at one of the level shifter outputs is to place a large pull-up resistor (a few MΩ) between the output and V_{high}. Other designs place a pull-down switch at one of the outputs, controlled by a power-on-reset signal (POR, see Section 6.9). It can be the global POR signal that observes the general start-up, or the POR circuit can be connected directly to V_{high}. An additional resistive level shifter is used to transfer the POR signal to the high-side domain and to release the pull-down switch after the start-up of the IC or if V_{high} has reached its stable value.

Analysis of the DC Switching Condition, Sizing Equation

To initiate the switching transition, the pull-down current of M1 (or M2) has to be greater than the drain current of M3 (or M4). From the equivalent circuit in Fig. 5.33, we can derive that M1 is in its saturation region while M3 is in the triode region because M3 sees the full V_{drv} as its source–gate voltage. Due to the high source–gate voltage, the PMOS devices can deliver high currents. For the drain current I_3 of M3 in the triode region, we can write

$$I_3 = \frac{W_3}{L_3} \mu_{op} C_{ox} \left((V_{drv} - V_{thp})V_{dsx} - \left(\frac{V_{dsx}}{2}\right)^2 \right)$$

$$\approx \frac{W_3}{L_3} \mu_{op} C_{ox} (V_{drv} - V_{thp})V_{dsx}. \tag{5.88}$$

Note that we count V_{thp} positive. As V_{dsx} is small, we neglect the quadratic term for simplicity. The value of V_{dsx} must be V_{thp} or larger. We will discuss this shortly.

Even though M1 and M2 are n-channel devices with inherently higher driving capability, their gate–source voltage is provided by the low-voltage domain V_{DDL}. Hence, M1 and M2 must be strong

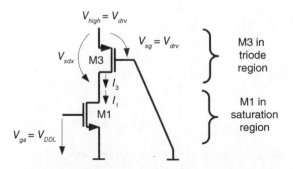

Figure 5.33 Cross-coupled level shifter: equivalent circuit of Fig. 5.32a).

enough to ensure proper level shifter switching across corner variations. The drain current I_1 of M1 in the saturation region is

$$I_1 = \frac{1}{2}\frac{W_1}{L_1}\mu_{on}C_{ox}(V_{DDL} - V_{thn})^2. \tag{5.89}$$

Equating Eqns. (5.88) and (5.89) allows to derive the required relation between the W/L-ratios of M1 and M3,

$$\frac{\left(\frac{W_1}{L_1}\right)}{\left(\frac{W_3}{L_3}\right)} = 2\frac{\mu_{op}}{\mu_{on}}\frac{(V_{drv} - V_{th})V_{sdx}}{(V_{DDL} - V_{th})^2}. \tag{5.90}$$

Sizing of the Cascoded Cross-Coupled Level Shifter

For the cascode version of the cross-coupled level shifter in Fig. 5.32b), the level shifter output is limited by the cascodes M5, M6 to $V_{casc} + V_{th,casc} \propto V_{casc}$. We use the equivalent circuit of Fig. 5.34 to size the devices for proper switching. Similar to the fundamental cross-coupled level shifter, the switching requires that the drain voltage of M3 gets pulled at least one threshold voltage V_{thp} below V_{high} to turn on M4. As there are three devices in series, M1 as well as M5 need to be able to provide the drain current of M3. One drawback of the cascodes is that their driving capability decreases while the drain potential of M3 and, thus, the source–gate voltage of M5 drops. Fast switching requires a large W/L-ratio of the cascodes M5 and M6, which, in turn, increases the parasitic capacitances and deteriorates the switching speed. Figure 5.34 indicates the operating regions of the series transistors M1, M3, M5. The value of V_{dsx} must be V_{thp} or larger. The drain currents of M1 and M3 are again given by Eqns. (5.89) and (5.88) and, consequently, also the sizing ratio by Eqn. (5.90).

We assume that the drain voltage $VD1$ of M1 is immediately pulled to a value less than V_{casc} as soon as M1 turns on (before any switching activity occurs at M3 at all). Hence, M5 is biased in the saturation region,

$$I_5 = \frac{1}{2}\frac{W_5}{L_5}\mu_{op}C_{ox}(V_{drv} - V_{sdx})^2. \tag{5.91}$$

Equating Eqns. (5.88) and (5.91) yields the sizing equation for the W/L-ratios of M3 and M5. To keep simple expressions, we assume all threshold voltages and the oxide capacitance to be equal

Figure 5.34 Cross-coupled level shifter: equivalent circuit of Fig. 5.32b).

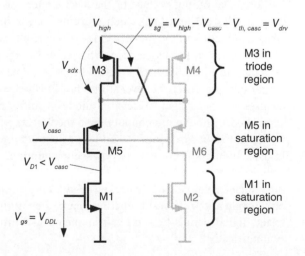

(i.e., $V_{th} = V_{thn} = V_{thp} = V_{th,casc}$). Depending on the different voltage ratings of the devices, this may not always be applicable. However, the simplified expressions will allow for an initial sizing, which can be adjusted by circuit simulation,

$$\frac{\left(\frac{W_5}{L_5}\right)}{\left(\frac{W_3}{L_3}\right)} = 2\frac{(V_{drv} - V_{th})V_{sdx}}{(V_{drv} - V_{sdx})^2}. \tag{5.92}$$

Example 5.21 *Calculate the ratios* $\frac{\left(\frac{W_1}{L_1}\right)}{\left(\frac{W_3}{L_3}\right)}$ *and* $\frac{\left(\frac{W_5}{L_5}\right)}{\left(\frac{W_3}{L_3}\right)}$ *for* $\frac{\mu_{op}}{\mu_{on}} = \frac{1}{3}$, $V_{drv} = 5$ V, $V_{th} = 0.7$ V, $V_{sdx} = 1.2$ V, $V_{DDL} = 2.5$ V.

Equations (5.90) and (5.92) result in $\frac{\left(\frac{W_1}{L_1}\right)}{\left(\frac{W_3}{L_3}\right)} = 1.06$ *and* $\frac{\left(\frac{W_5}{L_5}\right)}{\left(\frac{W_3}{L_3}\right)} = 0.71$.

Interestingly, the n-channel transistor M1 has to be nearly equal in size to the PMOS transistor M3 because M3 sees the large drive level V_{drv} *as its source–gate voltage.*

Design Guideline for Cross-Coupled Level Shifters

Based on the above considerations, we can derive a design guideline for both the basic cross-coupled and the cascoded cross-coupled level shifter.

(1) Use minimum length L for all devices to keep parasitic capacitances small.
(2) Choose minimum widths W_1, W_3 that fulfill sizing equation (5.90) (basic level shifter) for $V_{dsx} = V_{thp}$. For the cascoded type, W_5 needs to fulfill Eqn. (5.92) for the same value of V_{dsx}.
(3) Increase V_{dsx} up to approximately $1.5 \times V_{thp}$ and recalculate W_1, W_3, W_5 to enhance the switching speed. The speed can also be increased by concurrently enlarging W_1, W_3 (and W_5). However, the increase in parasitic node capacitance counteracts the increase in driving capability.

Improvements of the Cross-Coupled Level Shifter

Various modifications of the cross-coupled level shifter have been proposed. Moghe et al. [16] provides an overview of useful modifications. In a high-side configuration, according to Fig. 5.5, the bias V_{casc} at the gate of the cascodes transistors is equal to the switching node voltage. In that case, the low-level output voltage of the level shifter stays above V_{casc} by one source–gate voltage of the cascodes M5 and M6, which prevents a clean logic "0" level at driver input. Consequently, cross-conduction in the first driver stage occurs because its NMOS transistor cannot fully turn off. This issue can be addressed by inserting transistors parallel to the source–gate path of the cascodes as shown in Fig. 5.35. In Fig. 5.35a) M7, M8 are controlled directly by the complementary level shifter outputs. After the level shifter has finished switching, the output is pulled to V_{casc}, which ensures a correct logic "0" level at the level shifter output. This way, the gate driver is properly controlled without cross-conduction in the first stage. Inserting inverters as depicted in Fig. 5.35b) increases the loop gain, and M7 and M8 can be of minimum size. The main benefit of this approach is faster switching and reduced power consumption.

Figure 5.36 is a further extension of Fig. 5.35b). A flip-flop, formed from two NAND gates, solves the issue of asymmetrical switching delays. The timing diagram in Fig. 5.36 shows that the outputs only toggle at the slowest transition phase, the pull-up transition at M3 or M4. Instead of the distorted digital signals V_p and V_n, a simple delay is introduced, and V_{outp}, V_{outn} are generated more symmetrically.

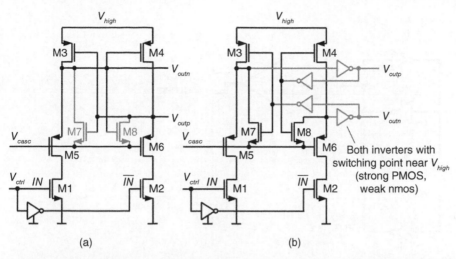

Figure 5.35 Cross-coupled level shifter with full logic level output swing. a) M7 and M8 pull the corresponding level shifter output to V_{casc}; b) increased loop gain by inserting inverters to drive M7, M8.

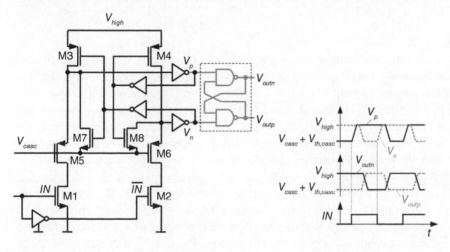

Figure 5.36 Cross-coupled level shifter with full logic level output swing and symmetrical switching delay.

5.12.3 Capacitive Level Shifters

The capacitive level shifter in Fig. 5.37 applies two coupling capacitors C_c to transfer the control signal V_{ctrl} from the low-side to the high-side domain [17]. Voltage pulses V_{ctrl} are buffered by inverter I1. Signal V_{1L} is inverted by I2 to a complementary signal V_{2L}. The cross-coupled inverters I3 and I4 form a latch on the high-side domain. The coupling capacitors C_c transfer V_{1L} and V_{2L} to I3 and I4. If the input capacitance of I3 and I4 is low and their input/output node impedance is high, the voltages V_{1H} and V_{2H} will follow V_{1L} and V_{2L}. If V_{1H} and V_{2H} rise or fall to the switching threshold of I3 and I4, the state of the latch will toggle due to positive feedback. Buffer I5 provides a well-driven signal $V_{in} = \overline{V}_{ctrl}$ at the input of the gate driver. The coupling capacitors C_c need to be able to handle voltages as high as $V_{high} = V_{DDH}$ (equal to V_{bat} plus the gate drive level V_{drv}). In low-power applications, C_c can become as small as 50 fF to lower the shifting energy. However, smaller values of C_c increase the noise sensitivity. The key characteristics of capacitive

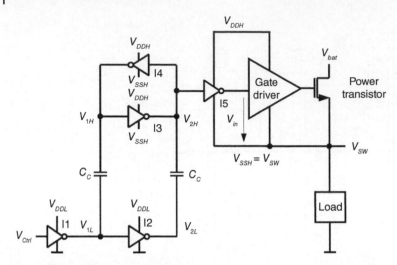

Figure 5.37 Capacitive level shifter.

level shifters are summarized in Table 5.4 compared to resistor-based and cross-coupled level shifters.

Open-Loop Model

While the robustness of this level-shifter benefits from the complementary signal transfer, it may fail to operate if voltages V_{1H} and V_{2H} do not cross the switching thresholds of I3, I4. It requires proper design as investigated in [18]. The approach presented in [18] is based on an open-loop model as shown in Fig. 5.38a). Both sides of I3 and I4 are considered independently without any dynamic effects. For this reason, replicas of both inverters are inserted at both sides of the capacitive level shifter. The node impedance at V_{1H} is only modeled by the input capacitance C_o of I3 (assuming it dominates over the output capacitance of I4) in parallel to the output impedance R_{on} of I4 and, vice versa, at V_{2H}. Due to symmetry, C_o and R_{on} are assumed to be identical on both sides of I3 and I4 (voltages V_{1H} and V_{2H}) with the equivalent circuits according to Fig. 5.38b). Consequently, we assume that the transition V_{1H} is only caused by a rise or fall of V_{1L}. The transition at node V_{2H} is considered the same way. Due to the R-C combination, we can expect an exponential charging and discharging behavior, as shown in Fig. 5.38c). The voltages V_{1L} and V_{2L} are assumed to rise or fall linearly with corresponding rise and fall times t_r, t_f, respectively. Applying Kirchhoff's Current Law, the equivalent circuit at node V_{1H} in Fig. 5.38b) yields

$$C_c \frac{d(V_{1L}(t) - V_{1H}(t))}{dt} = C_o \frac{dV_{1H}(t)}{dt} + \frac{V_{1H} - V_{SSH}}{R_{on}}. \tag{5.93}$$

A similar expression for node V_{2H} can be found. Solving these equations, we obtain:

$$V_H(t) = R_{on} C_c \frac{dV_L(t)}{dt} \left(1 - e^{-\frac{t}{R_{on}(C_c + C_o)}}\right) + V_x \quad 0 \le t \le t_{rf} \tag{5.94}$$

V_H and V_L represent either V_{1H} and V_{1L} or V_{2H} and V_{2L}. t_{rf} corresponds to t_r or t_f. V_x depends on the direction of the transition of V_{1L} and V_{2L}, respectively. For a rising transition of V_L, V_x is equal to V_{SSH} and $\frac{dV_L(t)}{dt} = \frac{V_r}{t_r}$. The falling direction requires $V_x = V_{DDH}$ while the slope gets a negative sign, i.e., $\frac{dV_L(t)}{dt} = -\frac{V_f}{t_f}$.

For $t \ge t_r$, V_{1L} reaches its steady-state. For $\frac{dV_{1L}(t)}{dt} = 0$, Eqn. (5.93) thus becomes

$$-C_c \frac{V_{1H}(t)}{dt} = C_o \frac{dV_{1H}(t)}{dt} + \frac{V_{1H} - V_{SSH}}{R_{on}}. \tag{5.95}$$

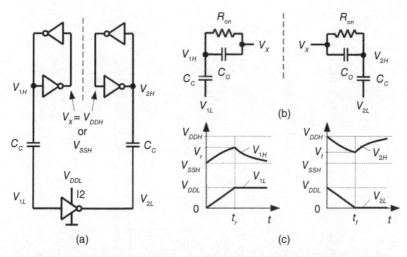

Figure 5.38 a) Open-loop circuit model of the capacitive level shifter with b) the equivalent circuits at nodes V_{1H} and V_{2H} and c) the corresponding charging and discharging behavior.

Likewise, a similar expression can be found for node V_{2H}. Solving these equations yields:

$$V_H(t) = \left(V_{rf} - V_x\right) e^{-\frac{t-t_{rf}}{R_{on}(C_c+C_o)}} + V_x \quad t \geq t_{rf} \tag{5.96}$$

Depending on the rising or falling transition, t_{rf} is either t_r or t_f and V_x is either V_{DDL} or V_{DDH} as discussed above for Eqn. (5.94). V_{rf} corresponds to $V_H(t = t_{rf})$ and can be calculated from Eqn. (5.94).

Equations (5.94) and (5.96) describe the charging and discharging transients as illustrated in Fig. 5.38c). As a design criterion, we can conclude that the correct operation of the level shifter requires

$$V_r \geq V_t. \tag{5.97}$$

This implies that V_{1H} and V_{2H} reach the switching threshold V_t of I3 and I4. It depends on the voltage levels V_{SSH}, V_{DDH}, and the value of both coupling capacitors C_c. Moreover, critical design parameters are the rise and fall times t_r, t_f, the equivalent output resistance R_{on} of inverters I3, I4, as well as their parasitic input capacitance C_o. Also, the propagation delay t_d of inverter I2 is important because it adds a time shift between the points when V_r and V_f are reached.

Figure 5.39 studies the influence of various parameters including t_{rf}, t_d, C_c. The level shifter will correctly operate if there is some enclosed area between the high-side voltages V_{1H} and V_{2H}. While Figure 5.40a,b) are generated from Eqns. (5.94) and (5.96), Figure 5.39c,d) are obtained from circuit simulations. The simulated results demonstrate that the open-loop model reproduces the actual behavior of the level shifter very well, even though the influence of positive feedback between I3 and I4 is not modeled. In Fig. 5.39c) this is clearly visible at $t \sim 70$ ps when the positive feedback kicks in.

Design Guideline

Based on the analysis, the following design procedure is suggested for sizing the inverters I1, I2, I3, and I4 and C_c [18]:

(1) *Determining the size of I3 and I4*: The switching threshold should be in the middle of the floating voltage levels V_{SSH} and V_{DDH}. R_{on} needs to be sufficiently large such that V_{1H} and V_{2H} follow the low side transitions with nearly constant voltage across C_c. However, large R_{on} makes the inverters more sensitive to noise and disturbances. $R_{on} \approx 10$ kΩ is a typical trade-off. There is

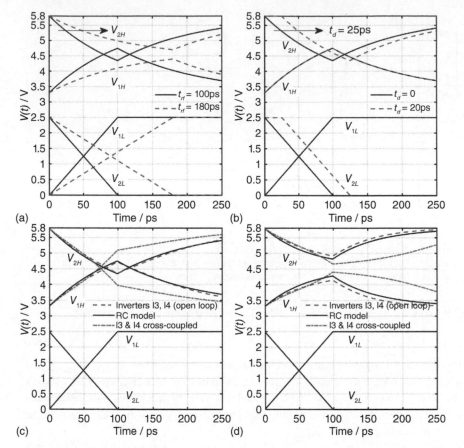

Figure 5.39 Transient behavior of the capacitive level shifter for $V_{DDL} = 2.5$ V, $V_{SSH} = 3.3$ V, $V_{DDH} = 5.8$ V, $C_c = 100$ fF, $C_o = 15$ fF, $R_{on} = 1$ kΩ. a) Influence of the rise/fall times of V_{1L}, V_{2L}: If t_{rf} increases from 100 to 180 ps the high-side domain voltages V_{1H} and V_{2H} do not touch each other anymore and the level shifter fails to operate; b) Influence of the propagation delay t_d of inverter I2: Larger delay will cause to level shifter to fail; c) Circuit simulation: The open-loop circuit of Fig. 5.38a) with I3 and I4 as real inverters behaves similar to the R-C-model of Fig. 5.38a). Both models reproduce the closed-loop behavior of the full-level shifter with cross-coupled inverters I3 and I4. At $t \sim 70$ ps the positive feedback kicks in and accelerates the transition of V_{1H} and V_{2H}; d) Repeated simulation from c) with C_c reduced from 100 to 50 fF, which causes the level shifter to fail. The model matches the actual behavior well and predicts the failure.

also some asymmetry due to I5, which should be considered. Symmetry can be maintained by placing a replica inverter, identical to I5, at the complementary output node (V_{1H}).

(2) *Sizing of C_c*: The coupling capacitor C_c can be determined with the help of Eqn. (5.94) such that V_{1H} and V_{2H} reach the switching threshold of I3 and I4, respectively. C_c should be larger than the parasitic input capacitance of I3 and I4, which is set in step (1).

(3) *Determining the size of I1 and I2*: Strong inverters improve the rise and fall times t_r, t_f as well as the propagation delay t_p of I2. They also determine the transition speed at the high-side voltages V_{1H} and V_{2H}. The concepts of Section 5.7 may be applied for optimizing the inverters.

(4) *Fine-tune the level shifter*: The strength of I1 and I2, I3 and I4 can be simultaneously increased if the rise and fall times of the level shifter output are not satisfying. Increase C_c to minimize the propagation delay.

Capacitive Level Shifter with Pull-Up Resistors and RS Flip-Flop

The above study and, more specific, Eqn. (5.94) indicates that R_{on} plays an important role. The larger R_{on}, the more reliable the operation of the level shifter. Even though R_{on} increases the R-C time constant of the transition, it improves the overall response because it also appears as a scaling factor for $V_H(t)$ in Eqn. (5.94).

In the level shifter of Fig. 5.37 R_{on} cannot be chosen independently because the cross-coupled inverters I3 and I4 need to have some gain to ensure proper latching. Therefore, the capacitive level shifter in Fig. 5.40 uses a dedicated pull-up resistor R_{pu} at each coupling path. Typical values of R_{pu} are tens of kΩ. Instead of cross-coupled inverters, an RS flip-flop stores the state and controls the gate driver accordingly. Two buffers, B1 and B2, transmit the pulses at V_{1H} and V_{2H} to the set and reset inputs of the flip-flop. B1 and B2 are two inverters supplied from V_{DDH} and V_{SSH} at the high side. Two cross-coupled NAND gates form the RS flip-flop. This way, the flip-flop inputs are low-active. A negative pulse at $\overline{S}$ and $\overline{R}$, respectively, sets or resets the state as shown in the truth table in Fig. 5.40. By default, both inputs $\overline{S}$ and $\overline{R}$ are at high level because resistors R_{pu} pull up V_{1H} and V_{2H}. According to the truth table, this keeps the flip-flop state, and no change occurs. Diodes D_1 and D_2 (see also Section 5.12.1) clamp V_{1H} and V_{2H} with respect to V_{SSH} to protect the inputs of B1 and B2 from over-voltage, which is critical in case of fast transients at the switching node $V_{SSH} = V_{sw}$. This topic is taken up again in Section 5.13 concerning CMTI. Also, the invalid state of the RS flip-flop is discussed in this regard.

5.12.4 Level-Down Shifters and Ground Level Shifters

So far, we have considered level-up shifters supporting signal transfer from the low-side to the high-side domain. Some designs require level shifters that accomplish the signal transfer in the opposite direction. Applications include various sensing functions as described in Chapter 6. Another example is the adaptive dead time control shown in Fig. 2.21.

Figure 5.41 shows two level-down shifters. Both circuits are the complementary counterparts of the fundamental resistor-based and cross-coupled level shifters covered in Sections 5.12.1

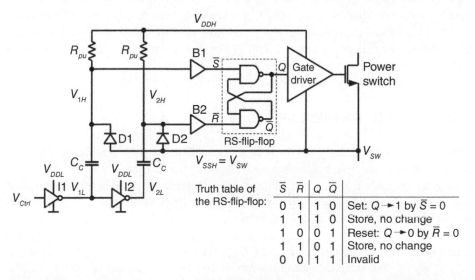

$\overline{S}$	$\overline{R}$	Q	$\overline{Q}$	
0	1	1	0	Set: $Q \rightarrow 1$ by $\overline{S} = 0$
1	1	1	0	Store, no change
1	0	0	1	Reset: $Q \rightarrow 0$ by $\overline{R} = 0$
1	1	0	1	Store, no change
0	0	1	1	Invalid

Figure 5.40 Capacitive level shifter with dedicated pull-up resistors R_{pu} and RS flip-flop to store the control signal on the high side. The truth table shows how negative pulses at $\overline{S}$ and $\overline{R}$ set and reset the flip-flop.

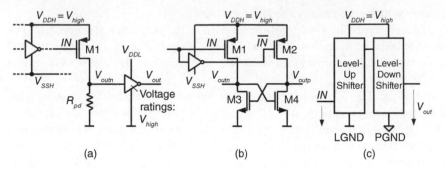

Figure 5.41 a) Resistor-based and b) cross-coupled level-down shifters; c) ground level shifter.

and 5.12.2. In both circuits, the input control signal *IN* is a logic level of the high-side domain, referred to V_{SSH} and the supply level V_{DDH}. For $IN = 1$, the connected PMOS transistor M1 turns off, causing the low-side domain output signal V_{outn} to go to zero. In the cross-coupled circuit of Fig. 5.41b), the second output reaches $V_{outp} = V_{DDH}$.

The output voltage swing in both circuits is between low-side ground and $V_{DDH} = V_{high}$. As the low-side domain usually has a lower supply voltage $V_{DDL} < V_{DDH}$, the circuits connected to V_{outn} and V_{outp} need to be protected. Figure 5.41a) uses an inverter in between V_{outn} and the final output voltage V_{out}, which is designed with a transistor that has suitable ratings to withstand V_{DDH}. Because the inverter is supplied from V_{DDL} its output voltage V_{out} does not exceed V_{DDL} such that all subsequent logic requires only a maximum voltage rating of V_{DDL}. The same inverter can be connected to the outputs of Fig. 5.41b). Alternatively, both circuits can use clamping diodes similar to Fig. 5.5 or high-voltage cascode techniques like in Fig. 5.36 and, more fundamental, in Fig. 6.2 to limit the output swing of the level-down shifter to V_{DDL}.

If a control signal needs to be transferred between different ground domains, a ground level shifter, shown in Fig. 5.41c), will help improve signal integrity. It ensures that the signal remains stable even if a ground shift between logic ground (LGND) and power ground (PGND) occurs, for instance, due to ground bouncing during switching because of the parasitic bond wire inductance. The input part is a level-up shifter. Its input signal *IN* refers to LGND. A level-down shifter forms the second stage and provides the output signal V_{out}, referred to PGND. Both level shifters can be of any type, for instance, a resistor-based output stage of Fig. 5.41b) combined with a cross-coupled input stage from Section 5.12.2.

5.13 Common-Mode Transient Immunity

With the trend toward faster switching and the progress in semiconductor technology, the switching node of a power stage experiences increasing transition speeds with more than 100 V/ns. An overview is given in Table 5.3. Fast transitions come with large coupling currents associated with parasitic capacitance. Let us investigate this effect for the level shifter of Fig. 5.31, which is depicted again in Fig. 5.42 including parasitic capacitances C_{par1} and C_{par2}. Those capacitances are the lumped representation of all parasitic capacitances at V_{1H} and V_{2H}. During a rising transition of the switching node, both capacitances C_{par1} and C_{par2} get recharged from V_{high}. The charging current flows equally on both sides, called common-mode current I_{CM} as indicated in Fig. 5.42. In case of full symmetry, $C_{par1} = C_{par2} = C_{par}$ and

$$I_{CM} = C_{par} \frac{dV_{sw}}{dt}.$$

(5.98)

For $C_{par} = 10\,\text{pF}$ and $\frac{dV_{sw}}{dt} = 10\,\text{V/ns}$, the common-mode current becomes as large as 100 mA. This current is superimposed to the signal current generated from the bias current I_{bias}. With typical bias currents in the order of $10\,\mu\text{A}$, the common-mode current I_{CM} is larger by several orders of magnitude. It explains why single-path level shifters (e.g., the resistor-based type in Fig. 5.30) are unsuitable.

But also, for differential level shifters, this is challenging. Assume there is some mismatch between C_{par1} and C_{par2} as the layout will always have some asymmetry. For 1% mismatch ($C_{par1} = 10\,\text{pF}$, $C_{par2} = 9.9\,\text{pF}$), the common-mode currents will also show 1% difference. For the above example, we can estimate this mismatch to be $\Delta I_{CM} = 1\,\text{mA}$, which is still much higher than $I_{bias} = 10\,\mu\text{A}$. In consequence, false switching may occur even for differential-level shifter designs. Also, oscillations between "on" and "off" can happen. For this reason, level shifters need to provide some *immunity* against those common-mode transients.

Common-mode transient immunity (CMTI) has become a critical parameter for advanced gate drivers. CMTI is specified as the correct operation up to a maximum transition speed given in V/ns or kV/μs. The CMTI scenario is explained for the specific level shifter in Fig. 5.42 but is present in all kinds of level shifters.

Ke and Ma [19] proposes to place damping resistors at the drain of M3 and M5 to limit I_{CM}. However, a standard way to guarantee CMTI is to latch the control signal of the gate driver in a flip-flop. This way, the state is stored before the switching transition happens. As a drawback, there is no way to change the state during transition but this is usually not necessary (except for some emergency "breaking" functions). The capacitive level shifter of Fig. 5.40 would be suitable because an RS flip-flop is already in place. Other differential-level shifters can be adapted similarly.

Let us investigate Fig. 5.40 from a CMTI point of view. If the set and reset happen during on- and off-state ($V_{sw} = 0$ and $V_{sw} = V_{bat}$, respectively), the level shifter will always operate properly. In steady state, both nodes V_{1H} and V_{2H} will be at high-level, i.e., at V_{high}. What happens during the rising transition of the switching node? Keep in mind that parasitic capacitances C_{par1} and C_{par2} are present between V_{1H} and ground and, likewise, between V_{2H} and ground. The pull-up resistors R_{pu} provide the common-mode currents I_{CM} that charge C_{par1} and C_{par2}. Consequently, both nodes V_{1H} and V_{2H} will experience a negative pulse that will result in $\overline{S} = 0$ and $\overline{R} = 0$ at the same time.

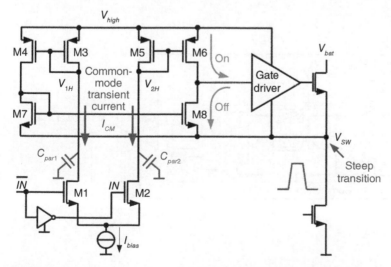

Figure 5.42 Level shifter with diode-connected pull-up transistors (M3, M5).

Clamping diodes D_1 and D_2 kick in to prevent that V_{1H} and V_{2H} do not fall too low with respect to V_{sw} and protect the buffers B1 and B2 from over-voltage. Nevertheless, this is an invalid state for the RS flip-flop, as confirmed by the truth table in Fig. 5.40. The state cannot be stored because outputs Q and $\overline{Q}$ go to the 1-level, and the subsequent state is indeterminate. A blanking circuit needs to be added to prevent this invalid state, as discussed shortly.

But before, let us review the falling transition of the level shifter in Fig. 5.40. In this case, C_{par1} and C_{par2} get discharged. I_{CM} flows through R_{pu} in opposite direction, toward V_{high}, resulting in $\overline{S} = 1$ and $\overline{R} = 1$. Fortunately, this condition is allowed, and no change occurs (see the truth table in Fig. 5.40). However, V_{1H} and V_{2H} may exceed the upper end of the maximum voltage rating (above V_{high}). Consequently, some clamping needs to be installed in addition to the blanking circuit against the invalid state.

Figure 5.43 shows the modified capacitive level shifter of Fig. 5.40 [20–22]. It consists of a common-mode detection and blanking circuit in addition to a pull-up circuit with pull-up transistors MP in parallel to R_p, which are turned off in steady-state (no transition). Both buffers B1 and B2 sense a 0 level during the rising transition and provide it to the OR gate I3. I3 detects the transient and generates two blanking signals, $blk1 = 0$ and subsequently a positive pulse $blk2$ (delay stage dly1, 8 ns in [22]). A delay stage dly2 (3 ns in [22]) is inserted in the signal detection path such that the blanking is always detected before the signal propagates to the RS flip-flop. While $blk2 = 1$ the connected NOR gates I4, I5 each provide 1-level such that $\overline{S} = \overline{R} = 1$ and the flip-flop safely keeps its state. After detecting the rising transition, the output signal $blk1$ of NOR gate I3 generates a pulse pu, which activates both transistors MP and pulls V_{1H} and V_{2H} up to V_{high}. This way, the level shifter reaches its default state quickly and can receive the next signal at maximum speed. Both MP devices turn off at the end of the pulse pu, and the pull-up is only formed by R_{pu}. As outlined in Section 5.12.3 for the level shifter in Fig. 5.40, a fast reaction can be achieved for larger values of R_{pu} (>10 kΩ).

The RS flip-flop can also be implemented by cross-coupled NOR gates. In that case, the pull-up resistors need to be converted to pull-down resistors by connecting them to $V_{SSH} = V_{sw}$. This way, the default state at both the set and the reset input is 0. Positive pulses set and reset the flip-flop. The roles of the rising and falling transition concerning CMTI interchange such that the invalid condition occurs if V_{sw} falls. A similar blanking circuit like in Fig. 5.43 can be implemented as published in [23].

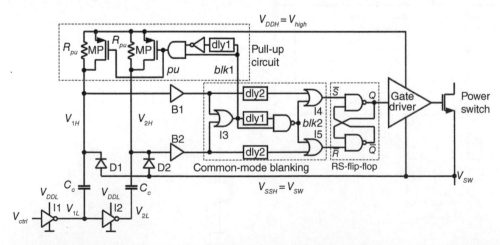

Figure 5.43 A capacitive level shifter with blanking and protection for common-mode transient immunity (CMTI).

References

1 Rabaey, J.M., Chandrakasan, A., and Nikolic, B. (2003) *Digital Integrated Circuits A Design Perspective,* Prentice-Hall.

2 Weste, N.H.E. and Eshragian, K. (1994) *Principles of CMOS VLSI Design A Systems Perspective,* Addison Wesley.

3 Mukherjee, A. (1986) *Introduction to nMOS and CMOS VLSI systems design,* Prentice-Hall.

4 Veendrick, H.J. (1984) Short-circuit dissipation of static CMOS circuitry and its impact on the design of buffer circuits. *IEEE Journal of Solid-State Circuits*, 19 (4), 468–473, doi: 10.1109/JSSC.1984.1052168. URL http://ieeexplore.ieee.org/stamp/stamp.jsp?arnumber=1052168.

5 Lin, H. and Linholm, L. (1975) An optimized output stage for MOS integrated circuits. *IEEE Journal of Solid-State Circuits*, 10 (2), 106–109.

6 Li, N., Haviland, G., and Tuszynski, A. (1990) CMOS tapered buffer. *IEEE Journal of Solid-State Circuits*, 25 (4), 1005–1008, doi: 10.1109/4.58293. URL http://ieeexplore.ieee.org/stamp/stamp.jsp?arnumber=58293.

7 Jaeger, R. (1975) Comments on "an optimized output stage for MOS integrated circuits" [with reply]. *IEEE Journal of Solid-State Circuits*, 10 (3), 185–186, doi: 10.1109/JSSC.1975.1050587. URL http://ieeexplore.ieee.org/stamp/stamp.jsp?arnumber=1050587.

8 Stratakos, A., Sanders, S., and Brodersen, R. (1994) A low-voltage CMOS DC-dc converter for a portable battery-operated system, in *Power Electronics Specialists Conference, PESC '94 Record., 25th Annual IEEE,* pp. 619–626, doi: 10.1109/PESC.1994.349672. URL http://ieeexplore.ieee.org/stamp/stamp.jsp?arnumber=349672.

9 Kursun, V., Narendra, S., De, V., and Friedman, E. (2004) Low-voltage-swing monolithic DC-DC conversion. *IEEE Transactions on Circuits and Systems—Part II: Express Briefs*, 51 (5), 241–248, doi: 10.1109/TCSII.2004.827557. URL http://ieeexplore.ieee.org/stamp/stamp.jsp?arnumber=1299038.

10 Villar, G., Alarcon, E., Madrenas, J., Guinjoan, F., and Poveda, A. (2005) Energy optimization of tapered buffers for CMOS on-chip switching power converters, in *IEEE International Symposium on Circuits and Systems, 2005. ISCAS 2005,* pp. 4453–4456, doi: 10.1109/ISCAS.2005.1465620. URL http://ieeexplore.ieee.org/stamp/stamp.jsp?arnumber=1465620.

11 Hamzaoglu, F. and Stan, M. (2001) Split-path skewed (SPS) CMOS buffer for high performance and low power applications. *IEEE Transactions on Circuits and Systems II: Analog and Digital Signal Processing*, 48 (10), 998–1002, doi: 10.1109/82.974792.

12 Wittmann, J. and Wicht, B. (2013) MHz-converter design for high conversion ratio, in *2013 25th International Symposium on Power Semiconductor Devices IC's (ISPSD),* pp. 127–130, doi: 10.1109/ISPSD.2013.6694445.

13 Renz, P. and Wicht, B. (2021) *Integrated Hybrid Resonant DCDC Converters*, Springer International Publishing, doi: 10.1007/978-3-030-63944-0.

14 Lehmann, T. (2014) Design of fast low-power floating high-voltage level-shifters. *Electronics Letters*, 50 (3), 202–204, doi: 10.1049/el.2013.2270. URL https://ietresearch.onlinelibrary.wiley.com/doi/abs/10.1049/el.2013.2270.

15 Wittmann, J., Rosahl, T., and Wicht, B. (2014) A 50V high-speed level shifter with high dv/dt immunity for multi-MHz DCDC converters, in *ESSCIRC 2014 - 40th European Solid State Circuits Conference (ESSCIRC),* pp. 151–154, doi: 10.1109/ESSCIRC.2014.6942044.

16 Moghe, Y., Lehmann, T., and Piessens, T. (2011) Nanosecond delay floating high voltage level shifters in a 0.35μm HV-CMOS technology. *IEEE Journal of Solid-State Circuits*, 46 (2), 485–497,

doi: 10.1109/JSSC.2010.2091322. URL http://ieeexplore.ieee.org/stamp/stamp.jsp? arnumber=5661865.

17 Tanzawa, T., Takano, Y., Watanabe, K., and Atsumi, S. (2002) High-voltage transistor scaling circuit techniques for high-density negative-gate channel-erasing nor flash memories. *IEEE Journal of Solid-State Circuits*, 37 (10), 1318–1325, doi: 10.1109/JSSC.2002.803045.

18 Zheng, W.M., Lam, C.S., Sin, S.W., Lu, Y., Wong, M.C., Seng-Pan, U., and Martins, R. (2015) Capacitive floating level shifter: Modeling and design, in *TENCON 2015 - 2015 IEEE Region 10 Conference*, pp. 1–6, doi: 10.1109/TENCON.2015.7373013.

19 Ke, X. and Ma, D.B. (2018) A 3-to-40V V_{IN} 10-to-50MHz 12W isolated GaN driver with self-excited t_{dead} minimizer achieving 0.2 ns/0.3ns t_{dead}, 7.9% minimum duty ratio and 50 V/ns CMTI, in *2018 IEEE International Solid - State Circuits Conference - (ISSCC)*, pp. 386–388, doi: 10.1109/ISSCC.2018.8310346.

20 Rindfleisch, C. and Wicht, B. (2020) 11.3 A one-step 325V to 3.3-to-10V 0.5W resonant DC-DC converter with fully integrated power stage and 80.7 efficiency, in *2020 IEEE International Solid-State Circuits Conference - (ISSCC)*, pp. 194–196, doi: 10.1109/ISSCC19947.2020.9063150.

21 Rindfleisch, C. and Wicht, B. (2021) A resonant one-step 325 V to 3.3-10 V DC-DC converter with integrated power stage benefiting from high-voltage loss-reduction techniques. *IEEE Journal of Solid-State Circuits*, 56 (11), 3511–3520, doi: 10.1109/JSSC.2021.3098751.

22 Rindfleisch, C., Otten, J., and Wicht, B. (2022) A highly-integrated 20-300V 0.5W active-clamp flyback DCDC converter with 76.7% peak efficiency, in *2022 IEEE Custom Integrated Circuits Conference (CICC)*, pp. 1–2, doi: 10.1109/CICC53496.2022.9772834.

23 Lutz, D., Seidel, A., and Wicht, B. (2018) A 50V, 1.45ns, 4.1pJ high-speed low-power level shifter for high-voltage DCDC converters, in *ESSCIRC 2018 - IEEE 44th European Solid State Circuits Conference (ESSCIRC)*, pp. 126–129, doi: 10.1109/ESSCIRC.2018.8494292.

6

Protection and Sensing

This chapter focuses on protection and sensing circuits in the power stage. They often require voltage and current references, which are also covered. Power management has to deal with high voltages and large currents, and we need to protect the power stage and the connected loads from damage if the maximum ratings are exceeded. In addition, various conditions and quantities must be controlled during regular operation. Examples include zero-voltage crossing of the switching node and current sensing.

Figure 6.1 repeats the block diagram of the power stage similar to Fig. 5.2. Fundamental protection functions include:

- Overvoltage (in particular the drain–source voltage)
- Overcurrent (current sensing)
- Thermal protection
- Short circuits and open load (OL), mainly short-to-ground (SCG), short-to-battery (SCB)
- Under-voltage lock-out (UVLO), no gate driver operation if the gate supply is too low

6.1 Overvoltage Protection

6.1.1 High-Voltage Cascode

High-voltage transistors can be used to protect low-voltage transistors (LV). Figure 6.2 shows the implementation. The differential input transistor pair of the error amplifier in a linear regulator or a DC–DC converter may be an example. One input is tied to the resistive feedback divider, which is connected to the output. The low-voltage transistor must be protected if the output level exceeds its maximum ratings. Safety-critical designs require protection against failures associated with shorts to the converter's input and other high-voltages. Another scenario may use a low-voltage transistor as an enable switch, which turns on a regulator once V_{bat} is available after the high-voltage battery is connected. The same can happen during supply sequencing if the output voltage (high voltage) of the previous regulator is tied to the supply voltage pin (V_{bat}).

In Fig. 6.2, the gate of the low-voltage transistor (actually, the gate–source voltage V_{GS}) does not see any voltage higher than $V_{GS} = V_{bias} - V_{GSHV}$. It is ensured by the high-voltage transistor, which acts as a high-voltage cascode. A Zener diode may provide secondary protection, preventing the gate node from rising in case of leakage currents. Often V_{bias} can be an existing digital or analog supply rail such that the circuit of Fig. 6.2 can be implemented in an elegant way with minimum effort.

Design of Power Management Integrated Circuits, First Edition. Bernhard Wicht.
© 2024 John Wiley & Sons Ltd. Published 2024 by John Wiley & Sons Ltd.

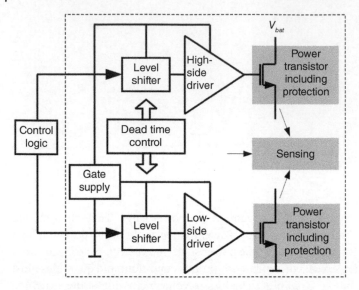

Figure 6.1 Power stage block diagram with protection.

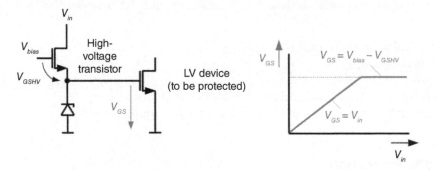

Figure 6.2 Overvoltage protection of a low-voltage transistor (LV) by a high-voltage cascode.

6.1.2 Active Zener Diode

The circuit in Fig. 6.3 is often called an active Zener diode as it combines a precise voltage level of a Zener diode with the current capability of a power transistor. It, therefore, forms a voltage clamp that can handle much larger currents than a regular Zener diode. The clamp level is set by the sum of the Zener voltage V_Z and the gate–source voltage of the power transistor at the target clamp current: $V_{clamp} = V_Z + V_{GS}$. The latter gives some variation of the clamp voltage. However, this simple circuit is sufficient for a wide range of applications.

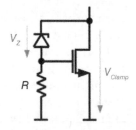

Figure 6.3 Active Zener diode.

6.2 Overvoltage Protection for Inductive Loads

Consider the low-side switch in Fig. 6.4a). The inductive load could be a solenoid, for instance. As long as the switch is active, a load current builds up. Because I_{load} is stored in the inductor, it will not instantaneously go to zero if the switch turns off and needs to flow somewhere. If the switch is off, the switching node can be considered as an infinite impedance that sinks I_{load}. Hence, V_{sw} will increase ideally to infinity. In reality, the switching node potential is limited by the parasitic resistance of the inductor and interconnects. Nevertheless, the voltage may exceed V_{bat} by far and rise above the maximum drain–source voltage of the low-side switch. A similar scenario applies to the high-side switch (not shown). It should be noted that switching nodes are limited, however, toward the negative side. The body diode of the switch prevents V_{sw} from falling below $-V_F \sim -0.7\,\mathrm{V}$.

A Zener diode provides a simple way of protecting against overvoltage, Fig. 6.4b). Its Zener voltage needs to be lower than the maximum V_{DS} ratings of the power switch. For example, a Zener diode with 5.8 V protects a 10 V transistor. The Zener diode provides a free-wheeling path for the inductor and ensures that the stored energy (I_{load}) can be dissipated safely. The current decreases at a rate determined by $\Delta I_{load}/\Delta t = V_L/L$ with $V_L = V_{bat} - V_Z < 0$. The higher the voltage across the inductor, the faster the energy is dissipated. As soon as the load current reaches zero, there is no risk of damage anymore. The Zener diode turns off, and V_{sw} approaches V_{bat}. The Zener diode can also be placed parallel to the coil. A regular PN diode may be used similarly to Fig. 2.7b). Due to the small value of V_F, the current ramp-down may be slow. As the major drawback of the solution in Fig. 6.4b), the Zener diode needs to have a high current rating identical to the power transistor because it needs to carry the full load current.

Figure 6.4c) shows a drain–gate clamp circuit where clamping diodes are placed between the drain and gate. They draw only microamperes of current during clamping. Hence, the concept is well suitable for integrated power stages. The energy in the coil is dissipated by the power transistor, which is rated for the maximum load current. The drain–gate clamp consists of a Zener diode (which can also be a stack of multiple Zener diodes) and a PN diode. Let us assume the power switch gets turned off after a significant load current has built up in the inductor. As for the unprotected switch in Fig. 6.4a), the switching node attempts to rise to high potential. Eventually, the drain clamp starts to conduct and lifts the gate voltage of the power transistor. This way, the transistor remains on as long as there is energy in the coil. The drain–source voltage (identical to V_{sw}) is clamped at a voltage equal to $V_F + V_Z + V_{GS}(I_{load})$, which is set to be greater than V_{bat}. This way, the diodes in the drain–gate clamp will turn off after the energy is dissipated and $V_{sw} = V_{bat}$. The gate driver pulls the gate of the power transistor to zero and turns the switch fully off. Why do we always need the PN diode in series to the Zener diode? At turn-on, the gate driver provides $V_{GS} = V_{drv}$ and V_{sw} is at ground potential. Without the PN diode, the Zener diode would get forward-biased and conduct a large current, which loads the gate driver.

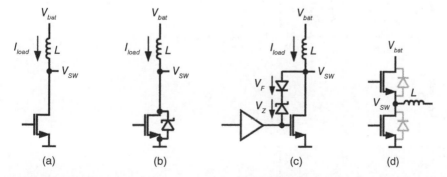

Figure 6.4 Overvoltage protection for inductive loads: a) low-side switch; b) drain–source clamp; c) drain–gate clamp; d) half-bridge.

The same concept can also be implemented to a high-side switch, where the voltage at the switching node may be pulled to negative values (theoretically to $-\infty$). Figure 6.5 shows an example [1]. A stack of integrated Zener diodes follows the approach of shorting collector and emitter (see Section 3.5.4 and Fig. 3.21). The collector (the third terminal) is connected to the next higher emitter potential according to the ratings in the given technology. Two PN diodes compensate for the temperature drift of the Zener diodes (Section 3.5.4) to achieve a stable clamping level over temperature. Adding up all voltages, V_{DS} is clamped at 47 V and would protect a 55 V-transistor, for instance. For $V_{bat} = 14$ V the switching node approaches -33 V. Even though this is an extreme value based on the experiments reported in [1], it nicely demonstrates the behavior. The current-source gate driver may operate from a charge pump. It can sink and source a current. We will discuss the role of MD shortly.

During the clamping, the diode stack works against the gate driver, which is usually very strong (see Chapter 5). A trade-off can be achieved using a segmented gate driver, which weakly starts the transition and goes into strong driving mode afterward. Some designs detect the clamping case, for instance, by sensing the forward voltage drop at the PN diode and bringing the gate driver into a tri-state mode during clamping.

An alternative method is shown in Fig. 6.5. The power transistor's gate potential will be pulled below ground during clamping. Any n-area at the gate driver output would lead to a forward-biased junction without a current limit. For this reason, the LDMOS transistor MD is placed in the driver's pull-down path. Only the body diode is used with a p-doped source (see Figs. 3.6 and 3.7) connected to the gate driver output. During clamping, the driver will be in a high-impedance state. In the settled off-state, MD will prevent that the source reaches zero. It is rather limited to the forward voltage of MD's body diode. With $V_F \sim 0.7$ V, this is still below the threshold voltage of typical power transistors.

Finally, we look at the half-bridge in Fig. 6.4d). Thanks to the body diodes of both transistors, the switching node is clamped in between $V_{bat} + V_F$ and $-V_F$. Hence, bridges are inherently protected

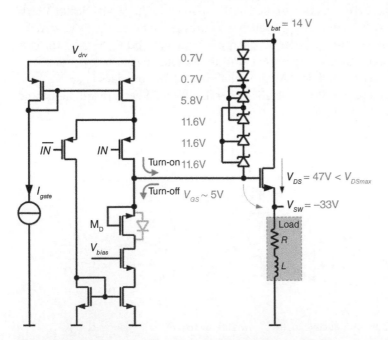

Figure 6.5 Drain–gate clamp example for a high-side power stage.

against overvoltage. However, as discussed in Section 2.5.4 (Fig. 2.10), the battery connection is never ideal, and the parasitic power loop inductance may still lead to excessive drain–source voltages. Careful system design is mandatory, and clamps at the battery level may be required.

6.3 Temperature Sensing and Thermal Protection

Most power management circuits inherently have some power dissipation and lead to an increase in the die temperature. Temperature sensing circuits can be used to adjust control parameters depending on the actual temperature and even to trigger the shutdown – thermal shutdown – of the power stage in case of over-temperature. The maximum junction temperature limit depends on the particular semiconductor technology. It is typically 150 °C with a trend toward 160 °C and above in more advanced process nodes. See Section 12.9 on thermal design related to IC floorplanning, packaging, and printed-circuit board (PCB) assembly.

Main contributors to on-chip power dissipation include integrated power switches and linear regulators, but also gate drivers and charge pumps are often significant. Chapters 2 and 5 explore various loss mechanisms and how they can be addressed. Even for an optimum design, the die will still have significant power dissipation, usually associated with excessive load currents. Therefore, current sensing and overcurrent protection (see Section 6.6) are primarily installed to prevent the junction temperature exceeds the upper limit.

As a secondary protection, a thermal shutdown function is often mandatory in addition to limiting the current. As it is intended as a secondary protection, the threshold is usually set above the maximum guaranteed junction temperature, i.e., toward 200 °C. Figure 6.6 shows a typical temperature-sensing circuit with a thermal shutdown function. The system will shut down all or some power stages as soon as the thermal shutdown signal *TSD* goes to the 1-level. Depending on system requirements, the power stage will be activated after some time or after the die temperature has dropped by about 10 °C. In both cases, a thermal hysteresis will be achieved. If the root cause that heats the die has stayed the same after several *TSD*-cycles, the application may stop the power stage activity completely and flag an error on the system level.

In the circuit of Fig. 6.6, the PMOS current mirror keeps the emitter currents of the bipolar devices Q1 and Q2 equal. For more precision, cascode current mirrors may be preferred at the expense

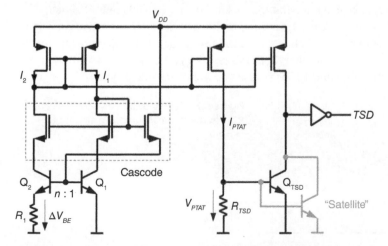

Figure 6.6 Thermal shutdown circuit based on an IPTAT generation.

of reduced headroom toward V_{DD}. Q2 has n-times more emitter area than the unit transistor Q1, i.e., Q2 consists of n parallel unit transistors with the corresponding emitter, base, and collector terminals connected. A ratio of $n = 8$ is often convenient for a well-matched layout because eight unit devices of Q2 can surround a single device Q1 in a three-by-three array. This way, Q1, and Q2 form a Brokaw loop where the difference in base-emitter voltage ΔV_{BE} leads to a voltage drop across resistor R_1. With

$$I_1 = I_S \exp \frac{V_{BE1}}{V_T} \text{ and } I_2 = nI_S \exp \frac{V_{BE2}}{V_T},$$

we get

$$\Delta V_{BE} = V_{BE1} - V_{BE2} = \frac{kT}{q} \ln n \propto T, \tag{6.1}$$

where the temperature voltage $V_T = \frac{kT}{q}$ is determined by Boltzmann's constant k, the elementary charge q, and the absolute temperature T. As Q2 consists of n parallel devices, the bipolar transistor's saturation current I_S is scaled by n. The cascode block has two purposes. (1) it ensures equal collector potential and better matching between the bipolar transistors. It (2) supplies both base currents from V_{DD} ensuring the symmetry in contrast to delivering the base current from one of the branches as part of I_1 or I_2. For more precision, Q1 may get $(n-1)$ dummy devices connected to its collector and base while the emitter is floating (similar to Fig. 6.9). Both collector areas get identical in size and show equal leakage currents (toward the substrate) of the reverse-biased collector–substrate junction.

The resistor R_1 converts ΔV_{BE} to a current-proportional-to-absolute-temperature I_{PTAT}, a well-known relationship in fundamental analog circuits:

$$I_{PTAT} = \frac{\Delta V_{BE}}{R_1} = \frac{1}{R_1} \frac{kT}{q} \ln n \propto T \tag{6.2}$$

The Brokaw loop and all devices in the two branches can be part of a global bandgap voltage reference circuit (see Section 6.4). The I_{PTAT} current is mirrored to the output and causes a voltage drop V_{PTAT} across R_{TSD}:

$$V_{PTAT} = I_{PTAT} R_{TSD} = \frac{R_{TSD}}{R_1} \frac{kT}{q} \ln n \propto T \tag{6.3}$$

So far, the circuit in Fig. 6.6 represents a temperature sensor with V_{PTAT} as its output signal. The current mirror ratio can be chosen as greater than unity and needs to be considered accordingly. Figure 6.7 depicts V_{PTAT} over temperature. While I_{PTAT} shows some nonlinearity versus temperature due to the temperature coefficient of resistor R_1, the voltage V_{PTAT} is determined by the resistor ratio R_{TSD}/R_1. Global variations will cancel if both resistors are of the same type (for instance,

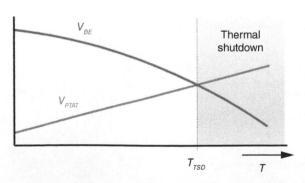

Figure 6.7 Temperature behavior of V_{BE} and V_{PTAT} in the thermal shutdown circuit of Fig. 6.6.

poly-silicon). However, as the die will see a nonuniform temperature distribution and the resistors are most likely not placed nearby, the temperature variation will cancel only to some extent.

The temperature sensing function can be further expanded to implement a thermal shutdown function. For this reason, one or multiple bipolar transistors Q_{TSD} can be added as shown in Fig. 6.6. If V_{PTAT} exceeds the base-emitter forward voltage of Q_{TSD}, a logic signal $TSD = 1$ is given (at temperature T_{TSD}). The base-emitter voltage shows a strong and nonlinear temperature dependence, as indicated in Fig. 6.7. The temperature coefficient has a negative sign with a typical value of $-2\,\frac{mV}{K}$ at room temperature ($T \approx 300\,K$). Consequently, V_{BE} drops from approximately $600\,mV$ to about $300\,mV$ near the thermal shutdown threshold toward $200\,°C$.

While Q1 and Q2 can be part of a global biasing block, Q_{TSD} should be placed near the power stage known as the primary heat source. As a benefit of the circuit in Fig. 6.6, multiple devices can be connected in parallel to Q_{TSD}. These *satellites* can be distributed across the die close to multiple blocks with significant power dissipation. The bipolar device with the highest local temperature will have the lowest forward voltage and trigger *TSD*. With multiple satellites, it is impossible to distinguish which causes thermal shutdown (they form a logical wired-OR). This information is usually optional if junction over-temperature occurs somewhere on the die. For more details on thermal design, refer to Section 12.9. Figure 12.21 provides an example of the temperature distribution on the die obtained from thermal simulation.

Example 6.1 *For a thermal shutdown temperature of $T_{TSD} = 190\,°C = 463\,K$, we assume a bipolar forward voltage $V_{BE} = 290\,mV$. Calculate R_1 and R_{TSD} for $n = 8$ such that the emitter currents of Q1 and Q2 do not exceed $10\,\mu A$ at room temperature.*

From Eqns. (6.1) and (6.3), we can calculate the resistor ratio:

$$\frac{R_{TSD}}{R_1} = \frac{V_{PTAT}(T_{TSD})}{\frac{kT}{q}\ln n} = \frac{190\,mV}{8.62 \cdot 10^{-5}\,\cancel{V/K} \cdot 463\,\cancel{K}}{\cancel{}}\,\ln 8 = 3.5 \tag{6.4}$$

At room temperature ($T = 300\,K$) Eqn. (6.1) yields $\Delta V_{BE} = 83\,mV$. In order to keep the emitter currents in Q1 and Q2 below $10\,\mu A$, we choose $R_1 = 10\,k\Omega$, which gives $I_1 = I_2 = 8.3\,\mu A$. From Eqn. (6.4), $R_{TSD} = 35\,k\Omega$. We use unit resistors of $5\,k\Omega$ and place two and seven of them in series to implement R_1 and R_{TSD}, respectively.

6.4 Bandgap Voltage and Current Reference

The IPTAT generation in Fig. 6.6 can be used to implement a bandgap voltage reference, as shown in Fig. 6.8a). This circuit is used in almost every IC for global biasing generation but also to generate threshold levels for protection and sensing circuits. We will use it, for instance, for the power-on reset in Section 6.9.

If we add a resistor R_2 below the two bipolar transistors Q1 and Q2, its voltage drop V_2 will also be proportional to temperature. The fundamental principle of the bandgap voltage reference is to add V_{BE} (negative temperature coefficient (TC)) and V_2 (positive TC). This way, the circuit can be designed such that the temperature drifts of both V_2 and V_{BE} cancel each other, see Fig. 6.8b).

The current through R_2 is the sum of I_1 and I_2 with $I_1 = I_2 = I_{PTAT}$ (see Eqn. (6.2)). Hence,

$$V_2 = 2I_{PTAT}R_2 = 2\frac{R_2}{R_1}\frac{kT}{q}\ln n \propto T. \tag{6.5}$$

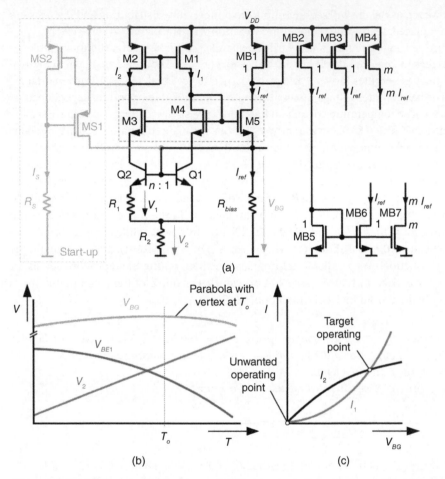

(a)

(b)   (c)

Figure 6.8 a) Bandgap voltage and current reference including start-up circuit; b) base-emitter voltage of transistor Q1, VPTAT voltage V_2, and the resulting bandgap reference voltage V_{BG} over temperature; c) operating points related to the start-up behavior.

Combined with the base-emitter voltage V_{BE} of Q1, a bandgap reference voltage V_{BG} will be obtained at the base node of Q1 and Q2:

$$V_{BG} = V_2 + V_{BE} = \frac{E_g}{q} \sim 1.2\,\text{V} \tag{6.6}$$

The name *bandgap* reference refers to the fact that a temperature coefficient of zero is achieved if V_{BG} is close to the silicon bandgap voltage, which is E_g/q with the bandgap energy E_g and the elementary charge q. Hence, the typical value of V_{BG} is 1.2 V. The temperature variation of the base-emitter voltage can be approximated by

$$\frac{\partial V_{BE}}{\partial T} = \frac{V_{BE} - E_g/q}{T}. \tag{6.7}$$

At room temperature, $T = 300\,\text{K}$, and with $E_g/q = 1.2\,\text{V}$, $V_{BE} = 0.6\,\text{V}$, we can estimate the temperature coefficient to be $-2\,\text{mV/K}$. Based on Eqn. (6.5), the temperature drift of V_2 is

$$\frac{\partial V_2}{\partial T} = 2\frac{R_2}{R_1}\frac{k}{q}\ln n = \frac{V_2}{T}. \tag{6.8}$$

The expressions in Eqns. (6.7) and (6.8) need to have opposite signs and identical absolute values to achieve zero temperature coefficient of V_{BG}:

$$-\frac{\partial V_{BE}}{\partial T} = \frac{\partial V_2}{\partial T} \tag{6.9}$$

This condition results in Eqn. (6.6) and confirms the approximation $V_{BG} = E_g/q \sim 1.2\,\text{V}$. Since the TC of V_{BE} depends on the actual temperature, zero TC can only be achieved at a specific temperature T_o, as illustrated in Fig. 6.8b). The resulting curve of V_{BG} versus temperature assumes a parabolic shape with its vertex at T_o. Equation (6.9) also results in a sizing equation for R_1 and R_2:

$$\frac{R_2}{R_1} = \frac{-\frac{\partial V_{BG}}{\partial T}}{2\frac{k}{q}\ln n} \tag{6.10}$$

Example 6.2 *Derive the value of R_2 for $R_1 = 10\,\text{k}\Omega$ and $n = 8$ (ratio between Q1 and Q2) if we aim for zero temperature drift at room temperature ($T = 300\,\text{K}$).*

From Eqn. (6.10),

$$R_2 = \frac{-\frac{\partial V_{BG}}{\partial T}}{2\frac{k}{q}\ln n} R_1 = \frac{2\,\text{mV/K}}{2\frac{8.62\cdot10^{-5}\,\text{V/K}}{q}\ln 8} 10\,\text{k}\Omega = 55.8\,\text{k}\Omega \tag{6.11}$$

6.4.1 Start-Up Circuit

During start-up, the collector currents I_1 and I_2 in the bandgap circuit of Fig. 6.8a) need to ramp up as a function of the base-emitter voltages of Q1 and Q2. Due to the unequal emitter areas (1 and n), the collector currents will show different behavior with respect to V_{BE}, as depicted in Fig. 6.8c). The current I_2 in the stronger device Q2 will ramp up more steeply in the beginning while I_1 takes over if V_{BG} gets larger. The intersection is the wanted operating point at $V_{BG} \sim 1.2\,\text{V}$. However, Fig. 6.8c) also shows that the origin at $I_1 = I_2 = 0$ represents a second but unwanted operating point (it would result in $V_{BG} = 0$). For this reason, all bandgap circuits require a start-up circuit that pulls the base voltage up such that the circuit finds its final operating point. One example of a start-up circuit is shown in Fig. 6.8a) to the left of the main circuit. If the supply voltage V_{DD} starts to ramp up from zero, R_S will keep the gate of MS1 at ground level. This way, MS1 conducts and provides the base currents of Q1 and Q2. Consequently, I_1, I_2 as well as V_{BG} ramp up to eventually assume the target operating point, as shown in Fig. 6.8c). MS2 and M2 form a current mirror. Once the circuit approaches the target operating point, the current I_2 of M2 gets mirrored to MS2. This way, the voltage drop across R_S turns off MS1 and deactivates the start-up circuit. One of the failure reasons for bandgap circuits is that the start-up circuit either does not kick-in properly or does not fully disable. Therefore, it is crucial to verify the start-up behavior over worst-case corners.

6.4.2 Reference Current Generation

The resistor R_{bias} between the base terminals of Q1 and Q2 and ground in Fig. 6.8a) allows to generate a reference current $I_{bias} = V_{BG}/R_{bias}$ with ideally zero TC. This reference current can be scaled up or down and distributed to any subcircuit across the die by multiple current mirrors. MB1–MB7 shows an example of how various source and sink currents can be generated. As V_{BG} is nearly constant, the accuracy of I_{bias} depends mainly on R_{bias}. Some designs use an off-chip resistor, such as a precision resistor of the E96 series, which comes with 0.1% tolerance and zero temperature coefficient. Integrated resistors always show considerable temperature coefficients of both positive and negative polarity. For this reason, R_{bias} mostly requires some trimming. For instance, a poly-silicon resistor may have a TC of $1 \times 10^{-3} \, 1/K = 0.1\%/K$. It corresponds to 10% variation over 100K in addition to an absolute process variation of $\pm 10\%$ and more. Another option would be to combine resistor types with opposite temperature coefficients.

Example 6.3 *Determine which external resistor out of the E96 series should be placed to generate a bias current $I_{bias} = 40 \, \mu A$. The nominal bandgap voltage is $V_{BG} = 1.22 \, V$.*

The ideal theoretical value of the bias resistor is

$$R_{bias} = \frac{V_{BG}}{I_{bias}} = \frac{1.22 \, V}{40 \, \mu A} = 30.5 \, k\Omega. \tag{6.12}$$

The resistor values of any standard E-series are defined by

$$R(m) = 10^{(m-1)/x} \tag{6.13}$$

with the index $m = (1 \ldots \infty)$ and the series identifier x, which is 96 in this example. For $m = 430$ we find $R = R_{bias} = 30.5 \, k\Omega$ as the closest value to the ideal resistance of Eqn. (6.12).

Using current mirrors like MB1–MB7 in Fig. 6.8a), the current $I_{bias} = 40 \, \mu A$ can be scaled and distributed to any subcircuits.

6.4.3 Accuracy

The typical accuracy of bandgap references is about 50 ppm/K, which corresponds to a relative change of $5 \cdot 10^{-5} \, 1/K = 5 \cdot 10^{-3}\%/K$. This corresponds to a relative change of 0.5% over a temperature range of 100 K. Assuming $V_{BG} = 1.2 \, V$, the absolute variation is as low as 6 mV. Such a high absolute precision over a wide temperature range can usually only be achieved by design measures to address excessive leakage at higher temperatures and to maintain high symmetry between the two branches of I_1 and I_2. In addition, trimming is required, as described further below. While the cascode devices M3–M5 Fig. 6.8a) are already essential for good accuracy, Fig. 6.9 show an extended bandgap circuit with various improvements that contribute to higher accuracy. In summary, these are the design measures for high accuracy of V_{BG} and, subsequently, of I_{bias}:

(1) M5 avoids asymmetry as it provides the base currents from V_{DD}.
(2) The cascode devices M3 and M4 keep the collector voltages of Q1 and Q2 identical and avoid asymmetry due to Early effect and collector-to-substrate leakage.
(3) In parallel to Q1 $(n-1)$ dummy devices (QD) are placed. This way, the collector areas in both branches are equal. Together with the balanced collector potential of item (2), this results in equal leakage of the reverse biased collector-to-substrate junctions. As this leakage current contributes equally to I_1 and I_2, circuit symmetry is ensured up to high temperatures (keep in mind that the leakage current doubles every 10 °C). In particular, this is crucial in bandgap references for temperatures above 100 °C.
(4) M1, M2, M6, and M7 form a cascode current mirror, which improves the matching between I_1 and I_2. As a drawback, the headroom toward V_{DD} decreases. Therefore, alternative techniques like a wide-swing current mirror can be utilized, especially in low-voltage designs.

6.4.4 Trimming

Over corners and because the above calculations are based on approximations (see Eqn. (6.7)), the bandgap voltage V_{BG} will not precisely reach its targeted absolute value nor the zero TC point at T_o

Figure 6.9 A bandgap voltage reference that can be trimmed and maintains accuracy at high temperatures.

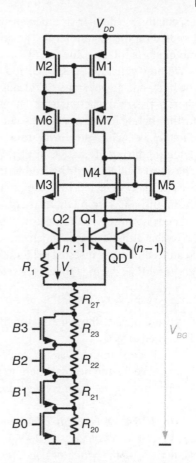

(see Fig. 6.8b)). Trimming allows to adjust the bandgap reference circuit such that these goals can be achieved. The temperature coefficient of the bandgap voltage V_{BG} correlates with the absolute value of V_{BG}. According to Eqn. (6.8), the temperature coefficient of V_2 is proportional to V_2 itself. Since V_{BG} is the sum of V_{BE} and V_2 (see Eqn. (6.6)), the TC of V_{BG} will be positive if V_{BG} comes out higher than the target value and, vice versa, lower V_{BG} corresponds to a negative overall TC. Therefore, trimming can be accomplished by making R_2 trimable because it determines the absolute value of V_2. A fraction of R_2 can be implemented in binary-weighted segments as shown in Fig. 6.9 for the case of 4-bit trimming. Four digital control bits Bx with $x = 0 \ldots 3$ turn on switches that bypass the binary-weighted segments R_{2x}. This way, the initial resistor R_{2T} can be increased in $2^4 - 1 = 15$ steps, which usually gives sufficient resolution. The W/L-ratios of the switch transistors can also scale in a binary-weighted fashion to ensure that each segment is fully shorted if not selected.

6.5 Short Circuits and Open Load

Short circuits and open load cases are common failure cases. The typical scenario is that a loose piece of wire falls into an electronics module. Or a piece of metal falls off and connects an arbitrary IC pin to a nearby potential. Therefore, short circuits to nearby supply rails and to ground are critical, defined as short-to-battery (SCB) and short-to-ground (SCG). Similarly, the connection between the power stage and the load may get disconnected, resulting in an open load (OL) failure.

Especially in safety-critical power management designs, these failure cases must be detected and handled on the system level. The goal is to prevent damage and severe injuries, for instance, if a motor starts rotating due to a SCG.

We now consider SCB, SCG, and OL for the low-side transistor. The same scenario applies to the high-side power stage. SCB leads to large drain currents for the low-side transistor because the short bypasses the actual load. Therefore, SCB is usually detected by current sensing, as discussed further below in this chapter (Section 6.6). The remaining SCG and OL cases are detected in the off-state, i.e., while the power transistor is turned off. As indicated in Fig. 6.10, the switching node potential during off-state is at V_{bat}. We first consider only the SCG case (not OL simultaneously). The short will connect the load to the ground, and V_{sw} falls below the threshold V_{ref1}. This way, SCG is easily detected by a comparator, which flags $SCG = 1$. A second comparator is needed to distinguish SCG from OL. Its trigger level is $V_{ref2} > V_{ref1}$. Hence, it gives $OL = 1$ during SCG.

Let us now remove the piece of wire that produced the short and cut the load connection instead. So we have the OL case, which causes a floating switching node. A biasing circuit consisting of I, R, and a diode D is inserted to identify OL. The current I pulls V_{sw} to a defined potential in between V_{ref1} and V_{ref2} during OL while it does not influence normal operation,

$$V_{sw,OL} = V_{DD} - V_F - R \cdot I. \tag{6.14}$$

For the OL case, the comparators flag $SCG = 0$ and $OL = 1$, which allows us to identify OL in difference to SCG.

The diode D in the biasing circuit isolates V_{DD} from the switching node and needs to block the V_{bat} voltage during normal operation. Also the comparator inputs cannot be directly connected to V_{sw} for large V_{bat}. A high-voltage cascode, according to Fig. 6.2, will be a suitable protection circuit.

Example 6.4 *A bandgap reference voltage is available for V_{ref1} and two times that value for V_{ref2}, i.e., $V_{ref1} = 1.2\,\text{V}$ and $V_{ref2} = 2 \cdot V_{ref1} = 2.4\,\text{V}$. Design I and R in the biasing circuit such that the OL value $V_{sw,OL}$ of the switching node is exactly in the middle between V_{ref1} and V_{ref2}. Furthermore, $V_{DD} = 3.3\,\text{V}$ and $V_F = 0.7\,\text{V}$.*

$V_{sw,OL}$ calculates to $1.8\,\text{V}$. We have to choose a reasonable value of R. For $R = 10\,\text{k}\Omega$ Eqn. (6.14) yields

$$I = \frac{V_{DD} - V_F - V_{sw,OL}}{R} = \frac{3.3\,\text{V} - 0.7\,\text{V} - 1.8\,\text{V}}{10\,\text{k}\Omega} = 80\,\mu\text{A}.$$

We can further increase the resistor to reduce the power consumption due to the biasing current. $R = 100\,\text{k}\Omega$ and $I = 8\,\mu\text{A}$ would be an alternative solution.

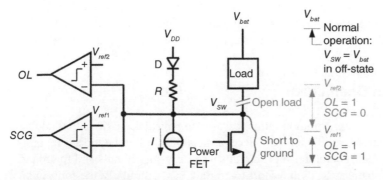

Figure 6.10 Short circuit to ground (SCG) and open load (OL) detection.

6.6 Current Sensing

6.6.1 Introduction

Current measurement is essential for acquiring, controlling, and optimizing the flow of energy. Not only on the IC level, compact and cost-effective current sensing circuits are a crucial element specifically for improving energy efficiency, power density, and solution size. Figure 6.11 shows typical locations for current sensing. It includes the total current drawn out of a battery (battery management systems (BMS)) and the current consumption of dedicated functions (circuit blocks). In the power stage, the drain currents of the low-side and the high-side transistor (branch currents), as well as the actual load current (phase current), must be measured. Due to the switching nature, the branch currents are not continuously available (i.e., not if the switch is turned off). Except for discontinuous operation, the phase current is usually available. However, the switching node shows steep common-mode transients and parasitic coupling, which makes current sensing more challenging.

We can distinguish between open-loop and closed-loop current sensing. Forghani-zadeh and Rincon-Mora [2] provides an overview of current sensing techniques for power management. Examples of open-loop sensing are overcurrent detection (like SCB in Section 6.5) and, in general, continuous current measurement. Closed-loop sensing is implemented in current limiting (Section 6.6.5) and current-mode control of DC–DC converters (Section 10.6). Current sensing is also applied to ensure the power transistor stays within its safe operating area, as described in Section 3.4.

The straightforward approach for current sensing would be to measure the drain–source voltage of the power switch. It is proportional to the drain current according to $V_{DS} = R_{DSon} \cdot I_D$ (see also Section 2.2). Besides the existing static losses of the power switch (see Eqn. (2.28)), this concept is nearly lossless. However, the precision is relatively poor as R_{DSon} varies over process, temperature, and voltage. In addition, the metallization resistance and the influence of the current distribution need to be considered. Nevertheless, several designs implement this approach as long as the relative accuracy is sufficient, for example, in DC–DC converters where an outer control loop compensates the current sensing error (see Section 10.6). For on-chip power stages, R_{DSon} sensing can be expanded to replica sensing with much better precision. Replica current sensing is investigated in Section 6.6.2.

Similar to R_{DSon} sensing, another fundamental approach is to measure the voltage drop across a sense resistor (shunt sensing), covered in Section 6.6.3. It requires inserting a discrete high-precision shunt resistor at additional cost and power dissipation. Operational or instrumentation amplifiers can be used for sensing. Their bandwidth, offset, common-mode rejection, etc., determine the achievable measurement accuracy and speed.

Figure 6.11 Typical current sensing locations.

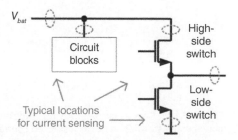

Most power management designs utilize a power inductor with a finite DC winding resistance (DCR). Hence, the voltage drop across the inductor measures its current. Section 6.6.4 is dedicated to this concept, known as DCR sensing.

Besides replica, shunt, and DCR sensing, plenty of concepts like Hall, fluxgate, anisotropic magneto resistive (AMR), and Rogowski coil current sensing exist. A comprehensive overview is given in [3]. These concepts are usually implemented as discrete current sensors and often measure the magnetic field associated with the current flow. Hence, galvanic isolation can be achieved. However, most concepts require extensive post-processing. Therefore, these concepts are usually unsuitable for co-integration on a power management IC.

6.6.2 Replica Sensing

Replica current sensing is a widely applied concept with good accuracy. It uses a fraction of the main power transistor to derive a replica current, which is a down-scaled copy of the actual load current [4–6]. The concept is illustrated in Fig. 6.12. In Section 3.2, we have seen that an LDMOS transistor consists of multiple fingers. We can design a sense transistor of one or a few fingers such that the W/L-ratios of the power switch MP and the sense transistor MS are ($M : 1$) where M can be as high as 1000 or more. Larger scaling factors M can also be achieved by stacking multiple sense transistors as shown in Fig. 6.12b). The transistor length adds and reduces the W/L-ratio of the entire sense transistor.

The sensing circuit in Fig. 6.13a) consists of a feedback loop formed by the amplifier, transistor MR, and resistor R_{out}. The loop aims to equalize the source voltages of MP and MS, i.e., $V_d = 0$. As both transistors have then identical gate, drain, and source voltages, their on-resistances and,

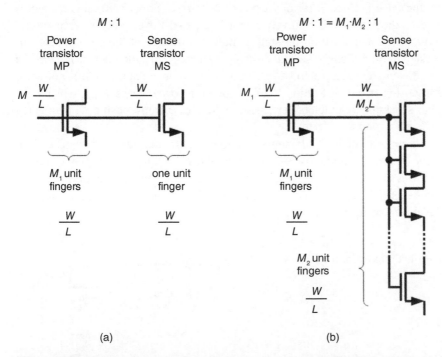

(a) (b)

Figure 6.12 Replica current sensing with the sense transistor MS implementation a) as one finger of the power transistor MP and b) as a stacked configuration of multiple fingers achieving large scaling factors *M*.

consequently, the drain currents scale by a factor M,

$$R_{MP} = \frac{R_{MS}}{M} \quad \text{and} \quad I_{in} = MI_{out}. \tag{6.15}$$

The larger the ratio M, the smaller the sense current I_{out}. Especially if the input current I_{in} is large, this may require implementing the replica fingers, as shown in Fig. 6.12b).

Figure 6.13b) shows the block diagram of the feedback loop. We apply small-signal analysis to calculate the closed-loop transfer function. Therefore, the signal variables are denoted by lowercase symbols. Before we start the analysis, a few considerations:

(1) We model the gain of the amplifier by a one-pole transfer function, which is sufficient in most cases: $A(j\omega) = \frac{A_o}{1+j\frac{\omega}{\omega_o}}$ (where A_o is the DC gain and ω_o the cut-off frequency)
(2) The feedback loop has only one high-impedance node, the output of the amplifier. Hence, the overall circuit can be considered a one-pole system, which is inherently stable.
(3) Transistor MR and R_{out} form a source follower. Small-signal analysis with the transconductance g_m of MR results in $i_{out} = g_m(v_R - i_{out}R_{out})$. The source follower's transfer behavior can be approximated by $\frac{i_{out}}{v_R} = \frac{g_m}{1+g_m R_{out}} \sim \frac{1}{R_{out}}$.

Now, all parts of the block diagram in Fig. 6.13b) are defined, and we can derive the closed-loop transfer behavior [6]. The voltage drops across MP and MS determine the feedback amplifier's differential input voltage v_d. Both R_{MP} and R_{MS} have a negative sign because MP and MS are tied to V_{bat}. Hence, the source voltages decrease if their drain currents go up. We can now put together two equations, one for the forward path from v_d to the output current I_{out} and another one for v_d as a function of I_{in} and I_{out}:

$$i_{out} = v_d A/R_{out} \tag{6.16}$$

$$v_d = i_{out}(-R_{MS}) - i_{in}(-R_{MP}) = i_{in}R_{MP} - i_{out}R_{MS} \tag{6.17}$$

Inserting Eqn. (6.17) in Eqn. (6.16) yields

$$\frac{i_{out}}{i_{in}} = \frac{A R_{MP}/R_{out}}{1 + A R_{MS}/R_{out}} \approx \frac{R_{MP}}{R_{MS}} = \frac{1}{M}. \tag{6.18}$$

In correspondence to Eqn. (6.15), we find the $M:1$ relation between the currents. In the denominator of Eqn. (6.18), the loop gain $T = A R_{MS}/R_{out}$ can be identified. This expression can also be found from the block diagram in Fig. 6.13b) if we multiply the expressions along the loop on the

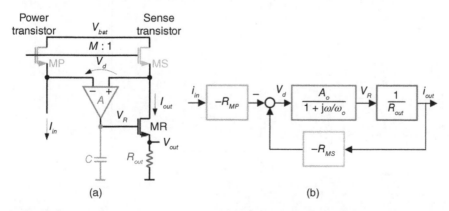

(a)　　　　　　　　　　(b)

Figure 6.13 Replica current sensing: a) schematic; b) block diagram.

right. The approximation in Eqn. (6.18) is based on the assumption that the loop gain is significantly greater than unity, which can be fulfilled in an actual design.

One of the major drawbacks of replica sensing is the limited bandwidth (speed). It is instructive to derive the gain-bandwidth product *GBW* of the loop gain. The DC gain is $T_o = A_o R_{MS}/R_{out}$. Thanks to the one-pole transfer function, the feedback amplifier's cut-off frequency is identical to ω_o. Hence,

$$GBW = \underbrace{A_o \cdot \omega_o}_{GBW(A)} \frac{R_{MS}}{R_{out}}. \tag{6.19}$$

The gain-bandwidth product of the amplifier $GBW(A) = A_o \cdot \omega_o$ is multiplied by the resistor ratio. In practical designs, R_{out} will be in the order of $10\,\text{k}\Omega$ to obtain a reasonable magnitude of V_{out}. At the same time, R_{MS} is at least a factor of 10 smaller. Consequently, the replica loop will always have a significantly lower bandwidth than the feedback amplifier. This observation is the reason why the speed of replica sensing circuits is limited. Nevertheless, it is still well-suitable for various applications, such as current-mode control of DC–DC converters at switching frequencies in the order of 5 MHz.

Coming back to the conceptual circuit in Fig. 6.13a), the overall accuracy depends on (1) the mismatch in the sense transistor ratio M, (2) the amplifier offset, and (3) on the absolute value of R_{out}, as well as (4) on the total loop gain. R_{out} usually needs trimming or has to be placed as a precise external resistor (e.g., E96 series).

Figure 6.14a) shows a widely used circuit implementation of replica sensing. A common-gate stage forms the differential feedback amplifier. The popularity of this circuit is probably due to its simplicity. However, there are also a few limitations. The drain current of MS always needs to be greater than I_{bias} (required by M2 and the current source underneath). This means for the input current,

$$I_{in} \geq MI_{bias}. \tag{6.20}$$

In addition, I_{bias} is subtracted from the current in MS, leading to an error. The real current $I_{out,r}$ in R_{out} is

$$I_{out,r} = \frac{I_{in}}{M} - I_{bias}. \tag{6.21}$$

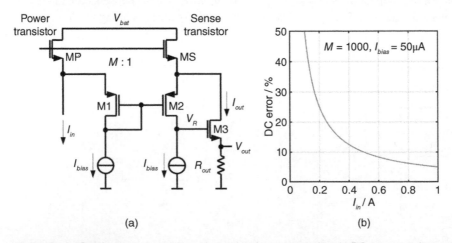

(a)

(b)

Figure 6.14 Replica current sensing: a) practical implementation; b) DC error over input current.

As defined in Eqn. (6.15), the ideal output current is $I_{out,i} = I_{in}/M$. Hence, we can determine the relative error:

$$err = \frac{I_{out,i} - I_{out,r}}{I_{out,i}} \cdot 100\% = M \frac{I_{bias}}{I_{in}} \cdot 100\% \tag{6.22}$$

The error increases toward lower values of I_{in}. Only for $I_{bias} \ll I_{in}/M$ the error disappears, which is difficult to achieve. Figure 6.14b) shows that the relative error reaches already 5% at $I_{in} = 1$ A and steeply rises to more than 50% if I_{in} approaches the lower boundary of MI_{bias} ($=50$ mA) (investigated in [7]).

By modifying the circuit as shown in Fig. 6.15), we can compensate for this error [6]. A current I_{comp} is taken from the input, with a negligible influence on I_{in}. It feeds into R_{out} to compensate for I_{bias}, which is subtracted from the current in MS. The output current calculates according to

$$I_{out} = \frac{I_{in}}{M} - I_{bias} + I_{comp}. \tag{6.23}$$

M3 has the same source–gate voltage like M1, hence, $I_{comp} = I_{bias}$. Under this condition, Eqn. (6.23) is identical to Eqn. (6.15) and zero error can be achieved. However, in Figure 6.15) the condition of Eqn. (6.20) holds.

For large supply voltages, the circuit in Fig. 6.15 requires various extensions like high-voltage cascodes to fulfill the voltage ratings. A fast high-side replica sensing circuit for 40 V is studied in [7].

Replica sensing can also be applied to a low-side power stage. Figure 6.16 shows a suitable circuit for an inductive load like in a DC–DC converter (Chapter 10). The operation of a half-bridge is described in Section 2.5.5 with the typical waveforms shown in Fig. 2.12. Accordingly, the load current in the inductor will commutate into the body diode of the low-side transistor as soon as the high-side turns off. After a dead time, the low-side transistor carries the load current. Hence, the switching node potential will be negative, i.e., $V_{sw} = -V_1 = -(I_{load}R_{DSon})$. The sensing transistor MS is tied with its drain to the switching node. The source of MS is connected to the input of the feedback amplifier, implemented as a common-gate stage. The source potential of M1 is at ground level, which is the set value for the feedback loop. The loop aims to control the sense voltage V_S at the source of MS to be equal to ground. The inset on the lower right of Fig. 6.16 shows the equivalent circuit of the feedback loop and reveals that it is identical to the loop of the initial replica sensing circuit in Fig. 6.13. Also, the closed-loop transfer function, as discussed in Eqns. (6.16)–(6.19), is the same. If the loop works properly, the drain–source voltages of the low-side power transistor MP and the sensing transistor MS will be equal, i.e., $V_1 = V_2$

Figure 6.15 Replica current sensing with I_{bias} compensation.

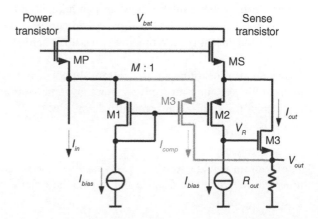

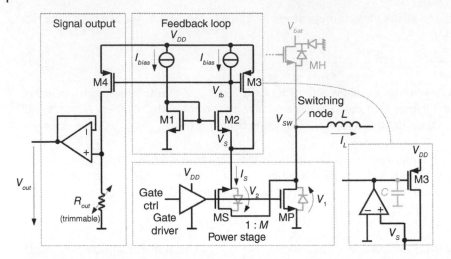

Figure 6.16 Replica current sensing at the low-side transistor for an inductive load (DC–DC converter).

with $V_1 = R_{MP}(I_L - I_s) \sim R_{MP}I_L$ and $V_2 = R_{MS}I_s$. We obtain the same relationship $R_{MP} = R_{MS}/M$ as in Eqn. (6.15). Also, the currents scale accordingly such that $I_s = I_L/M$ is a fraction of the load current. To get a ground-referred voltage V_{out} as a measure for I_L, we do not use a resistor R_{out} at the source of M3. Instead, we connect M4 in a current-mirror configuration with M3. Its current is a copy of I_s that is converted into V_{out} via R_{out}. Because R_{out} needs to be absolutely precise, it may need some trimming (e.g., with a binary weighted resistor array and non-volatile memory programming (EEPROM/Flash)). The circuit suffers from the fact that I_{bias} (M2) contributes to I_s and introduces an error similar to Eqn. (6.22). A compensation like in Fig. 6.15) can be added.

Because the high-side sensing circuit (Figs. 6.14 and 6.15) is connected to the battery voltage, it may need to be designed with higher voltage-rated transistors if V_{bat} is large. In contrast, the low-side circuit of Fig. 6.16 requires only low-voltage devices rated at V_{DD} (3.3 V for instance). Since low-voltage devices have lower parasitic capacitances, the gain bandwidth will be larger. We can also expect better matching. In conclusion, low-side sensing may be inherently faster and more precise.

6.6.3 Shunt Current Sensing

Discrete high-precision current sensing resistors are used for shunt sensing. They allow measuring any current flow using their voltage drop at high absolute accuracy with nearly zero temperature drift. Recently, a few integrated shunt-based concepts have been investigated [8]. These require compensation measures for the shunt resistor's spread, nonlinear temperature coefficient, and other errors.

Typical shunt locations are the battery supply or the output voltage of a DC–DC converter, as shown in Fig. 6.17a,b), respectively. For a given maximum current $I_{s,max}$ and a maximum voltage $V_{s,max}$ across the shunt its resistance is defined by

$$R_s = \frac{V_{s,max}}{I_{s,max}}. \tag{6.24}$$

The maximum voltage drop $V_{s,max}$ is typically in the range of 100 mV. As a drawback, the discrete shunt requires some board space and adds to the system cost. It also contributes static power loss

$$P_{loss} = I_s^2 R_s. \tag{6.25}$$

For $I_s = 1$ A and $R_s = 100$ mΩ the losses are already 100 mW. As shunts get larger for higher power ratings, parasitic inductive effects affect the achievable accuracy and maximum bandwidth.

If the shunt experiences a significant common-mode swing, an instrumentation amplifier must be used to sense the shunt voltage at any common-mode level. Nevertheless, the amplifier implementation becomes more straightforward if we place the shunt resistor such that one terminal connects to a fixed voltage. This way, common-mode variations will be small. Both setups in Fig. 6.17a,b) keep one shunt terminal stable enough to benefit from minimum common-mode variations. We also notice that the shunt is placed on the high side. This setup is usually preferred over low-side sensing because it does not cause a ground shift or ground disturbances. Another advantage of high-side current sensing is that it can detect a SCG failure of the load (see also Section 6.5).

Figure 6.17c) shows a typical sensing circuit that comprises a differential amplifier and resistor. Two matched resistors R_p support high common-mode levels. The common-mode voltage V_{cm} at the amplifier inputs is set by the resistor divider formed by resistors R_p and R_{c1},

$$V_{cm} = V_{high} \frac{R_{c1}}{R_p + R_{c1}}, \tag{6.26}$$

derived from the high-side rail voltage V_{high}. Assuming that V_{high} stays constant, the non-inverting input of the differential amplifier sees a constant bias voltage, which is identical to V_{cm}. This way, the negative feedback at the inverting amplifier input forms a negative gain stage. It comprises resistor R_{c2} and a second resistor R_p that connects to the shunt R_s. Since the negative shunt voltage drop $(-V_s)$ is the input to the negative gain stage, the overall (small-signal) gain is positive,

$$A_s = \frac{V_{out}}{V_s} = \frac{R_{c2}}{R_p}, \tag{6.27}$$

which corresponds to

$$V_{out} = R_s I_s \frac{R_{c2}}{R_p}. \tag{6.28}$$

For $I_s = 0$, V_{out} will be zero if we size R_{c1} and R_{c2} identical (both branches to the left and right of the shunt will be biased fully symmetrically). If the differential amplifier does not have a rail-to-rail output stage or if the input range of the subsequent stage requires some minimum level, R_{c2} should be smaller than R_{c1}. Alternatively, the bottom terminal of R_{c1} can connect to any reference voltage. It allows the output of the amplifier to be offset to some higher voltage with respect to ground.

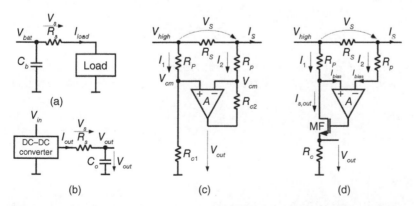

Figure 6.17 Shunt current sensing: typical shunt resistor locations at a) battery supply and b) a DC–DC converter output; c) sensing circuit with a differential amplifier and resistors; d) sensing circuit as used for replica sensing (see Fig. 6.13).

In conclusion, we can apply the following design guideline:

(1) Calculate the maximum shunt resistance according to Eqn. (6.24).
(2) Calculate the maximum losses using Eqn. (6.25) to ensure that it does not exceed the specified limit for the shunt.
(3) Calculate the ratio R_p/R_{c1} from Eqn. (6.26) such that V_{cm} stay within the input common-mode range of the differential amplifier.
(4) Select R_p, keeping the currents I_1 and I_2 at a reasonable level, typically in the order of milliamperes.
(5) Calculate R_{c1} from the result of step (3). R_{c2} is either set equal to R_{c1} or lower to comply with the output voltage range of the amplifier and the input range of the subsequent stage. Alternatively, R_{c1} connects to a reference voltage V_{ref}.
(6) The gain A_s can be determined from Eqn. (6.27). Check that it complies with the amplifier's output voltage range.
(7) The cut-off frequency of the current sensing circuit is defined by

$$f_o = \frac{GBW}{A_s} \tag{6.29}$$

with where GBW is the open-loop gain-bandwidth of the differential amplifier.

Example 6.5 *Follow the design guidelines and design a shunt sensing circuit according to Fig. 6.17c) for a maximum voltage drop across R_s of $V_{s,max} = 100$ mV and this specification: $V_{high} = 12$ V, $V_{cm} = 0.5$–4.5 V, $I_s = 0.05$–1 A, $GBW = 1$ MHz.*

(1) *Shunt resistance according to Eqn. (6.24):*

$$R_s = \frac{V_{s,max}}{I_{s,max}} = \frac{100\,\text{mV}}{1\,\text{A}} = 100\,\text{m}\Omega \tag{6.30}$$

(2) *Maximum losses (Eqn. (6.25)):*

$$P_{loss} = I_s^2\,R_s = (1\,\text{A})^2\,100\,\text{m}\Omega = 100\,\text{mW} \tag{6.31}$$

We can select a shunt resistor of 100 mΩ with a maximum power rating of 125 mW, which is available with 1% tolerance.

(3) *Now we calculate the ratio R_p/R_{c1} from Eqn. (6.26) such that V_{cm} stays within the input common-mode range of the differential amplifier. $V_{cm} = 4$ V appears as a suitable value assuming the amplifier operates from 5 V:*

$$\frac{V_{cm}}{V_{high}} = \frac{R_{c1}}{R_p + R_{c1}} \quad\longrightarrow\quad \frac{R_p}{R_{c1}} = \frac{V_{high}}{V_{cm}} - 1 = \frac{12\,\text{V}}{4\,\text{V}} - 1 = 2 \tag{6.32}$$

(4) *Select R_p, keeping the currents I_1 and I_2 at a reasonable level, typically in the order of milliamperes. The voltage drop across R_p is $(V_{high} - V_{cm}) = 8$ V. We decide for $R_p = 8$ kΩ, which results in branch currents of 1 mA.*

(5) *From steps (3) and (4):*

$$R_{c1} = R_p/2 = 4\,\text{k}\Omega \tag{6.33}$$

According to Eqn. (6.28), for $R_{c2} = R_{c1}$ the minimum given sense current $I_s = 0.05$ A corresponds to an output voltage of

$$V_{out} = R_s\,I_s\,\frac{R_{c2}}{R_p} = 100\,\text{m}\Omega \cdot 0.05\,\text{A}/2 = 2.5\,\text{mV}. \tag{6.34}$$

If the amplifier does not support output rail-to-rail operation, R_{c2} must be lowered.

(6) *The gain A_s can be determined from Eqn. (6.27):*

$$A_s = \frac{V_{out}}{V_s} = \frac{R_{c2}}{R_p} = 0.5 \tag{6.35}$$

Gain values lower than 1 are typical for this kind of sensing circuit.

(7) *Finally, we determine the cut-off frequency of the circuit according to Eqn. (6.29),*

$$f_o = \frac{GBW}{A_s} = \frac{1\,\text{MHz}}{0.5} = 2\,\text{MHz}. \tag{6.36}$$

The circuit in Fig. 6.17d) is similar to Fig. 6.17c). It also applies resistors R_p, which set the common mode and the current sensing gain concurrently. The circuit implements the feedback configuration used for replica sensing in Fig. 6.13 with the amplifier of Fig. 6.14. The common-mode is set to

$$V_{cm} = V_{high} - R_p \cdot I_{bias}. \tag{6.37}$$

Assuming zero differential input voltage at the feedback amplifier, we can put together the loop equation including R_s and both resistors R_p:

$$R_s I_s + R_p I_{bias} - R_p(I_{bias} + I_{s,out}) = 0 \tag{6.38}$$

Rearranging allows deriving the feedback current $I_{s,out}$, which converts to V_{out} via R_{out}. Hence,

$$V_{out} = R_c \cdot I_{s,out} = R_c \frac{R_s}{R_p} I_s = \frac{R_c}{R_p} V_s. \tag{6.39}$$

For $R_c = R_{c1}$, the gain $A_s = V_{out}/V_s$ is identical to the expression in Eqn. (6.27) found for Fig. 6.17c).

6.6.4 DCR Sensing

Inductor DC resistance current sensing, or DCR sensing, measures the voltage drop across the parasitic resistance of the inductor winding [9]. Compared to shunt sensing, no high-precision sense resistor is required, reducing component costs and power losses. There is also no additional voltage drop across a sense resistor, which is beneficial in low-voltage applications. In the ideal case, the voltage drop V_{DCR} across the inductor L is a measure of the inductor current I_L,

$$V_{DCR} = I_L R_{DCR}. \tag{6.40}$$

However, the inductor is modeled by a series connection of an ideal inductor L and its DC resistance R_{DCR}, as shown in Fig. 6.18a). Hence, the actual voltage drop across the inductor is expressed by

$$V_{DCR} = I_L \left(j\omega L + R_{DCR}\right) = I_L \cdot R_{DCR} \left(1 + j\omega \frac{L}{R_{DCR}}\right). \tag{6.41}$$

Consequently, Eqn. (6.40) is only valid up to the corner frequency $\omega_o = R_{DCR}/L$ at which the inductor impedance starts contributing to the total resistance. The setup in Fig. 6.18a) is usually restricted to near-DC frequencies.

Example 6.6 *Calculate the corner frequency ω_o for a 10-μH inductor with $R_{DCR} = 21.5\,\text{m}\Omega$.*

$$\omega_o = \frac{R_{DCR}}{L} = \frac{21.5\,\text{m}\Omega}{10\,\mu\text{H}} = 2.15 \times 10^3 \frac{1}{\text{s}} \tag{6.42}$$

It is indeed a low value, only suitable for sensing DC and slowly varying currents.

To make DCR sensing applicable for measuring fast current transients, the corner frequency ω_o can be compensated by inserting R_x and C_x as shown in Fig. 6.18b). We assume that the impedance

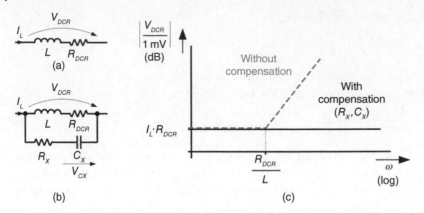

Figure 6.18 DCR current sensing: a) basic setup with the inductor modeled by its winding resistance R_{DCR} in series to the inductance L; b) compensated setup with R_x and C_x in which V_{cx} is a measure of the inductor current I_L; c) frequency behavior without and with compensation.

of the inductor itself is so low that the entire current I_L flows in the inductor, while the current I_x through R_x, C_x is negligible. R_x and C_x form a voltage divider, resulting in the voltage V_{cx} across C_x as indicated in Fig. 6.18b):

$$V_{cx} = V_{DCR} \frac{\frac{1}{j\omega C_x}}{R_x + \frac{1}{j\omega C_x}} = V_{DCR} \frac{1}{1 + j\omega R_x C_x} \tag{6.43}$$

Substituting V_{DCR} by Eqn. (6.41) yields

$$V_{cx} = I_L R_{DCR} \frac{1 + j\omega \frac{L}{R_{DCR}}}{1 + j\omega R_x C_x}. \tag{6.44}$$

Consequently, the voltage V_{cx} across the capacitor C_x represents the ideal DCR voltage according to Eqn. (6.40) if this condition is fulfilled:

$$R_x C_x = \frac{L}{R_{DCR}} \tag{6.45}$$

Figure 6.18c) shows the frequency behavior without and with compensation by R_x, C_x. In conclusion, DCR sensing measures the voltage V_{cx} as it represents the inductor current. Sense amplifiers used for shunt sensing (Section 6.6.3) can also be applied for DCR sensing with the typical circuits shown in Fig. 6.17.

Example 6.7 *Calculate the value of R_x for a 10-µH inductor with $R_{DCR} = 21.5\,m\Omega$ and $C_x = 220\,nF$.*
From Eqn. (6.45),

$$R_x = \frac{L}{C_x R_{DCR}} = \frac{10\,\mu H}{220\,nF \cdot 21.5\,m\Omega} = 2.11\,k\Omega \approx 2\,k\Omega. \tag{6.46}$$

As an advantage of DCR current sensing, process variations of the inductor winding resistance R_{DCR} are usually small. However, the strong temperature dependence is one of the significant drawbacks. Moreover, the sensing accuracy also depends on the variation of R_x, C_x.

Referred to the DCR value $R_{DCR,RT}$ at room temperature (27°C), the DC winding resistance at any temperature T is calculated from

$$R_{DCR}(T) = R_{DCR,RT} \left(1 + TC \left(T - 27°C\right)\right). \tag{6.47}$$

The linear temperature coefficient of the copper wire is $TC = 3930 \, \text{ppm/K} = 0.39\%/\text{K}$.

Example 6.8 *At room temperature the DCR value of a 10-μH inductor is given with $R_{DCR,RT} = 21.5 \, m\Omega$. How does the value change at $T = 60°C$? What is the relative error e_{rel} of the DCR current sensing referred to room temperature?*

Equation (6.47) results in

$$\begin{aligned} R_{DCR,60} &= R_{DCR,RT} \left(1 + TC \left(T - 27°C\right)\right) \\ &= 21.5 \, \text{m}\Omega \left(1 + 0.393 \times 10^{-2} \text{K}^{-1} \, 33\text{K}\right) = 24.3 \, \text{m}\Omega. \end{aligned} \tag{6.48}$$

This result corresponds to a relative error of

$$e_{rel} = \frac{|R_{DCR,RT} - R_{DCR,60}|}{R_{DCR,RT}} = \frac{|21.5 \, \text{m}\Omega - 24.3 \, \text{m}\Omega|}{21.5 \, \text{m}\Omega} \, 100\% = 13\%. \tag{6.49}$$

The error at 60°C is already significant. It will get worse at even higher temperatures.

In the sensing circuits of Fig. 6.17, the temperature influence can be compensated if the resistors have a suitable temperature coefficient. Referring to Eqn. (6.28), the term $\left(R_s \frac{R_{c2}}{R_p}\right)$ needs to have zero temperature drift. It can be accomplished if R_{c2} has a negative temperature coefficient or R_p has a positive coefficient.

6.6.5 Current Limit

For protection against overcurrent at the output, a current limit can be implemented. There are two options, constant current limiting and foldback current limiting. The equivalent circuit and the DC characteristics are shown in Fig. 6.19 [10]. The general scenario in Fig. 6.19a) indicates the current limit affects both the load (represented by the load resistance R_{load}) and the power stage of the supply (R_{sup}). The current limit I_{lim} is usually much higher than the nominal load current $I_{load,nom}$.

Constant Current Limit

One of the most obvious approaches to overload protection is to install a constant current limit at a given threshold. The load current I_{load} is not allowed to increase anymore once a critical value

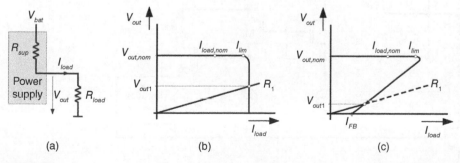

Figure 6.19 Current limiting: a) setup with power supply and load; b) constant current limit; c) foldback current limit.

I_{lim} is reached, Fig. 6.19b). The power supply output changes from a voltage to a current source. Consequently, the output voltage V_{out} depends on the load. Assuming a load resistance R_1, the output voltage reaches a value V_{out1}. The output current will not exceed the current limit even for a hard short circuit at the output (zero load resistance). A power supply with this kind of current limit still holds the output current to a specified value once the output load resistance decreases below a certain value. While this method protects the load from overcurrent, the internal power stage will dissipate excessive power. For $V_{out} = 0$, the full supply voltage V_{bat} will drop across R_{sup} as shown in Fig. 6.19a). The internal power dissipation reaches its maximum with the full current I_{lim}. This critical behavior is prevented by a foldback current limit, as covered further below.

Figure 6.20 shows a constant current limit example based on replica current sensing to provide a current limit in a linear voltage regulator (LDO, see Chapter 7). A sense transistor MS is placed in parallel to the main power transistor MP. A sense resistor R_s provides a voltage drop proportional to the load current I_{load}. As the input voltage can be large, two resistors R_p reduce the common-mode level at the input of the common-gate amplifier, similar to Fig. 6.17d). The feedback loop is identical to Fig. 6.14. As soon as the sense current $I_{s,out}$ exceeds I_{ref}, $V_{s,out}$ turns on M3, which, in turn, pulls down the gate of the main power transistor MP. As MP cannot deliver the load current, the output voltage of the LDO breaks down as part of the current limitation. Assuming equal source–gate voltages of M1 and M2, we can put together the loop equation including R_s and both resistors R_p:

$$R_s I_s + R_p I_{bias} - R_p(I_{bias} + I_{s,out}) = 0 \tag{6.50}$$

This approach is similar to Eqn. (6.38). We define the load current limit by $I_s = I_{lim}/M$ and $I_{s,out} = I_{ref}$ and insert these expression into Eqn. (6.50). This way,

$$I_{lim} = M I_{ref} \frac{R_p}{R_s}. \tag{6.51}$$

The error is due to the absolute precision of I_{ref}. Usually, the current limit can be simple, and perhaps only two bits of trimming may be sufficient for I_{ref}.

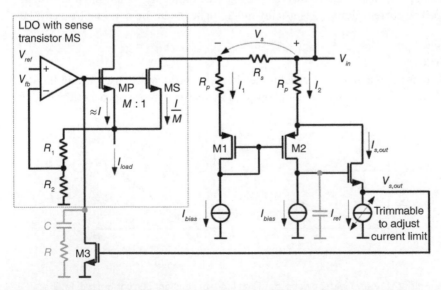

Figure 6.20 A linear voltage regulator with a current limit based on replica current sensing.

Foldback Current Limit

In this method, the output current decreases along with the decreasing output voltage as the load resistance gets smaller, Fig. 6.19c). Current foldback brings the load current to a much smaller value $I_{load} = I_{FB} \approx 0.2 - 0.3 I_{load,nom}$ under an abnormal load condition. In comparison to the constant current limiting, the power dissipation in R_{sup} (within the supply) is significantly reduced (note that $P_{diss} \propto I_{load}^2$). In conclusion, foldback protects both the load and the power stage in case of overcurrent.

The circuit in Fig. 6.21a) implements a foldback current limit. There are few publications on this circuit class; some early examples include [11, 12]. In Fig. 6.21a), the load current I_{load} is sensed by a shunt resistor R_s. As soon as I_{load} exceeds the maximum current limit I_{lim}, as defined in Fig. 6.19c), the feedback amplifier activates transistor MF. Consequently, MF pulls down the gate of the power device MP. As the driving capability of MP reduces, the output voltage will drop. Now the voltage divider consisting of R_1 and R_2 comes into play. With decreasing V_{out}, also the voltage V_1 across R_1 will drop. This way, the current limit value reduces gradually to implement foldback current limiting. The voltage across R_1 is

$$V_1 = \frac{V_{out} + I_{load} R_s}{R_1 + R_2} R_1 \approx V_{out} \frac{R_1}{R_2}. \tag{6.52}$$

The approximation neglects the voltage drop across the shunt resistor R_s and assumes that R_2 is much greater than R_1, which is usually fulfilled.

MF gets activated if V_p at the sense amplifier's non-inverting input exceeds V_{out}, which is applied to the inverting input. In that case, the load current I_{load} will be identical to the current limit I_{lim}. We can put together the loop equation exactly at $V_p = V_{out}$:

$$I_{lim} R_s - V_1 - V_{th} = 0 \tag{6.53}$$

With the approximation for V_1 from Eqn. (6.52) we get

$$I_{llm} = \frac{V_{th} + V_{out} \frac{R_1}{R_2}}{R_s}. \tag{6.54}$$

This expression confirms that the current limit I_{lim} is proportional to V_{out}, in line with the desired foldback behavior.

We can determine the maximum current limit by inserting the nominal output voltage $V_{out,nom}$ into Eqn. (6.54). For $V_{out} = 0$, also the voltage V_1 across R_1 will reach zero. This condition corresponds to the foldback current limit $I_{lim} = I_{FB}$. According Eqn. (6.54) the threshold V_{th} determines the foldback current limit:

$$I_{FB} = \frac{V_{th}}{R_s} \tag{6.55}$$

Figure 6.21b) shows a way to implement V_{th} with respect to V_{out}. As I_{th} is supplied from V_{bat}, the threshold will be available even if V_{out} approaches zero.

Even though Fig. 6.21 shows the foldback circuit for a linear regulator (LDO), the same concept can be applied to a switched-mode DC–DC converter. In that case, transistor MF needs to connect to the output of the error amplifier (see Fig. 10.5).

Based on the above equations, the foldback circuit can be designed according to this guideline:

(1) Determine the value of the shunt R_s depending on the voltage drop and the maximum power dissipation (see Section 6.6.3).
(2) Calculate the threshold V_{th} from Eqn. (6.55).

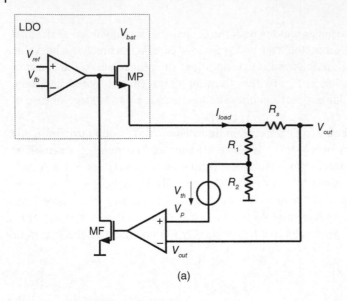

(a)

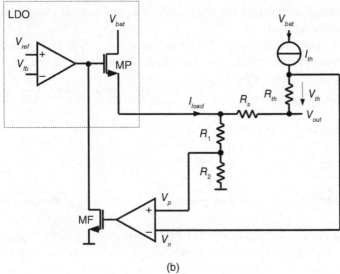

(b)

Figure 6.21 A linear voltage regulator (LDO) with foldback current limit: a) with a threshold V_{th} set by an ideal voltage source; b) V_{th} generated by R_{th} with respect to V_{out}. The threshold V_{th} sets the foldback current $I_{FB} = V_{th}/R_s$ (see Fig. 6.19c).

(3) Calculate the ratio R_1/R_2 from Eqn. (6.54).

(4) Determine the values of R_1 and R_2 for a given current in both resistors.

Example 6.9 *Design the foldback current limit circuit of Fig. 6.21b) for $I_{load,nom} = 100\,\text{mA}$, $I_{lim,max} = 2I_{load,nom}$, and $I_{FB} = 0.3I_{load,nom}$. The nominal output voltage is $V_{out,nom} = 3\,\text{V}$.*

For a typical maximum voltage drop $V_{s,max} = 100\,\text{mV}$ across the shunt resistor at the nominal load current $I_{load,nom} = 100$ mA the shunt resistor will be $R_s = 1\,\Omega$.

Knowing R_s, we can determine the threshold V_{th} from Eqn. (6.55) to achieve the given foldback limit $I_{FB} = 0.3 I_{load,nom} = 30\,\text{mA}$:

$$V_{th} = R_s \cdot I_{FB} = 1\,\Omega \cdot 30\,\text{mA} = 30\,\text{mV} \tag{6.56}$$

This threshold can be achieved according to Fig. 6.21b) by a current source $I_{th} = 10\,\mu\text{A}$ and a resistor $R_{th} = 3\,\text{k}\Omega$.

The resistor ratio R_1/R_2 follows from Eqn. (6.54):

$$\frac{R_1}{R_2} = \frac{R_s \cdot I_{lim,max} - V_{th}}{V_{out,nom}} = \frac{1\,\Omega \cdot 200\,\text{mA} - 30\,\text{mV}}{3\,\text{V}} = 56.7 \cdot 10^{-3} \tag{6.57}$$

If we limit the current through R_1 and R_2 to $I_R = 100\,\mu\text{A}$ the total resistance will be

$$R_1 + R_2 = \frac{V_{out,nom}}{I_R} = \frac{3\,\text{V}}{100\,\mu\text{A}} = 30\,\text{k}\Omega. \tag{6.58}$$

The absolute values of R_1 and R_2 follow from Eqns. (6.57) and (6.58):

$$R_2 = \frac{R_1 + R_2}{\frac{R_1}{R_2} + 1} = \frac{30\,\text{k}\Omega}{56.7 \cdot 10^{-3} + 1} = 28.4\,\text{k}\Omega \tag{6.59}$$

$$R_1 = \frac{R_1}{R_2} \cdot R_2 = 56.7 \cdot 10^{-3} \cdot 28.4\,\text{k}\Omega = 1.61\,\text{k}\Omega \tag{6.60}$$

6.7 Zero-Crossing Detection

The sensing of the zero-voltage crossing of the switching node is often required. Applications include zero-voltage switching and dead-time control as outlined in Section 2.9 and light-load detection in DC–DC converters, see Section 10.9.2. Figure 6.22 shows some typical circuits. For high battery voltages, the switching node potential needs to be scaled down into a low-voltage sensing domain. A simple solution is to place a voltage divider between the switching node and ground, Fig. 6.22a) [13]. Due to the fast transients (see also Table 5.3), a resistive and a capacitive divider are combined. Further below, we will investigate why a purely resistive divider is insufficient for fast V_{sw} transitions. The bias voltage V_{bias} can be set such that V'_{sw} stays consistently above ground level [13]. This detail is critical in the case of inductive switching when the switching node falls one diode voltage below ground. If the divider output voltage V'_{sw} falls below a reference level V_{ref}, the comparator indicates zero-crossing by setting signal *ZCD* to 1. Not only the zero level can be monitored by this circuit. Instead of the comparator, other circuits may be used to process the divider output further. For example, V'_{sw} may be stored by a sample-and-hold stage that consists of transistor MT (or even a transmission gate) and a storing capacitor C_H. This way, a cycle-by-cycle control of zero-voltage switching and dead time can be established. The drain-to-substrate junction, as well as the body diode of MP, would get forward biased if $V_{sw} < 0$ and would discharge C_H. It can be prevented by adequately setting $V_{bias} > 0$.

Let us consider a purely resistive divider to understand why an R-C divider is required to detect switching fast transitions. The resistors should not be too small to limit the power dissipation. For example, for a maximum resistor current of $100\,\mu\text{A}$ and a maximum switching node voltage of $V_{bat} = 40\,\text{V}$, the total resistance should be larger than $400\,\text{k}\Omega$. If the low-voltage domain is supplied

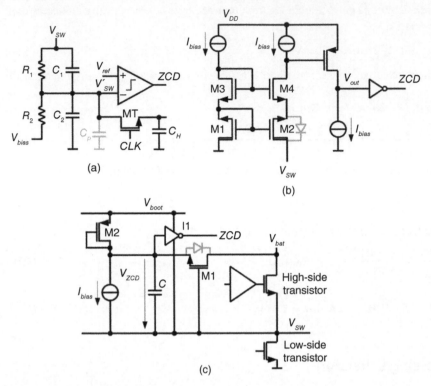

Figure 6.22 Zero-crossing detection circuits: a) R-C voltage divider at the switching node; b) common-gate sensing; c) body-diode-based detection.

from $V_{DD} = 5\,\text{V}$, we get $R_1 = 350\,\text{k}\Omega$ and $R_2 = 50\,\text{k}\Omega$. The resistive divider forms an R-C low-pass with the parasitic capacitance C_p at the comparator input or any other circuit connected to V'_{sw}. From Table 5.3 we can assume typical transition slopes of 50 V / ns. If we allow a cut-off frequency of the parasitic low-pass of $f_o = 1\,\text{GHz}$, the parasitic capacitance must not exceed $C_p = \frac{1}{2\pi R_1 f_o} = 0.45\,\text{fF}$. This requirement cannot be achieved in typical power technologies. The input capacitance of a minimum-size inverter is in the order of 10 fF (see Table 5.1). In consequence, the resistors would need to be orders of magnitude smaller. Excessive power dissipation is usually not acceptable. It will counteract any attempt to reduce power losses by zero-crossing detection. The original resistor values in the kΩ-range can be used if we combine them with a capacitive divider.

How about a purely capacitive divider? For high voltages, metal–metal capacitors will be the preferred capacitor type (see Section 4.1). C_1 can be sized to the minimum allowed capacitance of the given technology. Assuming $C_1 = 250\,\text{fF}$, C_2 calculates to 1.75 pF for the same parameters $V_{DD} = 5\,\text{V}$, $V_{bat} = 40\,\text{V}$ as above. A capacitor using only the third and fourth metal layers gives more distance toward the substrate, reducing the parasitic capacitance to ground. The values of C_1 and C_2 in this example are reasonable, but leakage currents will lead to an error at V'_{sw}. For this reason, a purely capacitive divider cannot be used.

In conclusion, an R-C voltage divider needs to be used. The resistive divider ensures proper biasing, while the capacitive divider dominates during fast transients and ensures sufficient speed. If the two-time constants in the R-C divider, $\tau_1 = R_1 C_1$ and $\tau_1 = R_2 C_2$ are equal, we achieve a constant flat run over frequency:

$$\frac{V'_{sw}}{V_{sw}} = \frac{R_2/(1+j\omega\tau_2)}{R_1/(1+j\omega\tau_1) + R_2/(1+j\omega\tau_2)} = \frac{R_2}{R_1 + R_2} \qquad (6.61)$$

In our example, $R_1 = 350\,\text{k}\Omega$, $R_2 = 50\,\text{k}\Omega$, $C_1 = 250\,\text{fF}$, and $1.75\,\text{pF}$ fulfill this condition. Equation (6.61) also reveals that the bias voltage V_{bias} at the bottom resistor R_2 (Fig. 6.22a) does not affect the small signal transfer function of the input divider. It should, however, not be too large and rather in the order of $<1\,\text{V}$ to avoid a DC biasing error.

The zero-crossing detection in Fig. 6.22b) is a comparator with a common-gate stage M3 and M4 protected by M1 and M2 against high-voltage at the switching node. M2 operates like a high-voltage cascode in Fig. 6.2. M1 provides its gate bias. For reasons of symmetry, M1 and M2 are identical transistor types. M1's drain is tied to the ground, but it is physically the source. M3 and M4 can be low-voltage devices, while M2 needs to block the full voltage swing of V_{sw}. As indicated in Fig. 6.22b), also the body diode remains reverse biased if V_{sw} is at high potential. High-voltage cascode blocking according to Fig. 6.2 can also be achieved if the gate connections on M1 and M2 are not diode-connected to M2's source but directly biased from V_{DD}. It gives even better symmetry, requires lower headroom, and can operate from lower V_{DD}-voltages.

Figure 6.22c) shows a circuit that is suitable to detect the zero-crossing of the drain–source voltage at the high-side transistor in a power stage (see Fig. 2.1) [14]. A typical application is to implement zero-voltage switching as described in Section 2.9. The sensing circuit is based on the body diode of a high-voltage transistor M1. The sensing circuit is implemented on the floating high-side supply domain (like the gate driver). In a typical application, it may be supplied by bootstrapping (V_{boot}), referred to the switching node. As long as V_{sw} is at low potential, the inverter I1 flags $ZCD = 0$ because its input voltage V_{ZCD} is at high level and close to V_{boot} (biased by the diode-connected M2 and I_{bias} with a buffer capacitor C). As soon as V_{sw} approaches V_{bat}, V_{ZCD} gets pulled to low-level and the inverter I1 signals $ZCD = 1$. Neglecting the voltage drop across the high-side switch, V_{sw} finally reaches V_{bat} and V_{ZCD} corresponds to the forward voltage $V_F \sim 0.7\,\text{V}$ of M1's body diode. Similarly, the concept can also be applied to the low-side transistor. In that case, the circuit refers to the ground, and the drain of M1 connects to the switching node.

6.8 Under-Voltage Lockout

In many applications, a supply level V_{in} (the battery supply V_{bat} or a sub-regulated rail like V_{DD}) has to be monitored to ensure operation within the permitted supply range. It is necessary to generate a reset signal whenever the supply voltage drops below specification because digital blocks can behave erratically. Similar requirements apply to analog circuits. In particular, the on-resistance of any power switch can only be guaranteed if the gate overdrive voltage $V_{GS} = V_{drv}$ is within the specification (see Eqn. (2.2)).

An under-voltage lockout (UVLO) function ensures that circuits can only operate if the supply is sufficiently high. This way, a safe start-up, as well as failure monitoring during operation, can be installed. UVLO works often in addition to power-on reset (POR, see Section 6.9). The situation when the supply drops below the permitted level is called brown-out. That is why the UVLO function is sometimes called brown-out reset (BOR).

The typical UVLO circuit in Fig. 6.23a) consists of a resistive divider with hysteresis and a comparator. The divider connects the input voltage V_{in} to be monitored. The down-scaled voltage V_{in}' is compared to a reference voltage (e.g., a bandgap voltage $V_{ref} \sim 1.2\,\text{V}$). M1 and R_3 generate a hysteresis to avoid bouncing and oscillations around the UVLO threshold. For $V_{in}' > V_{ref}$ $UVLO = 0$ and M1 is off. This defines the lower threshold level V_{LL} of V_{in} according to Fig. 6.23b),

$$V_{LL} = V_{ref}\frac{R_1 + R_2 + R_3}{R_2 + R_3}. \tag{6.62}$$

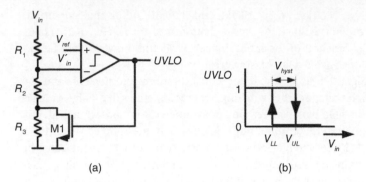

Figure 6.23 a) Under-voltage lockout circuit; b) transfer behavior with hysteresis.

For $V_{in} \leq V_{LL}$, the comparator output *UVLO* gets 1. M1 turns on and bypasses R3. This way, the resistive divider consists only of R_1 and R_2, which now defines the upper threshold level V_{UL},

$$V_{UL} = V_{ref} \frac{R_1 + R_2}{R_2}. \tag{6.63}$$

By subtracting both thresholds, we can calculate the hysteresis:

$$V_{hyst} = V_{UL} - V_{LL} = \frac{R_1 \cdot R_3}{R_2(R_2 + R_3)} \tag{6.64}$$

Example 6.10 *Determine the values of R_1, R_2, and R_3 to monitor a nominal supply voltage of $V_{in} = 3.3$ V. The specification defines $V_{LL} = 3.0$ V, $V_{hyst} = 100$ mV, $V_{ref} = 1.2$ V. During normal operation, the maximum current in the resistive divider should not exceed $I_{in} = 100$ μA.*

We easily find the total resistance to be $(R_1 + R_2 + R_3) = V_{in}/I_{in} = 33$ kΩ and the upper level $V_{UL} = V_{LL} + V_{hyst} = 3.1$ V. Equation (6.62) gives $R_2 + R_3 = 13.20$ kΩ. Subtracting it from the total resistance yields $R_1 = 19.80$ kΩ. By solving Eqn. (6.63) for R_2 and inserting R_1 we obtain $R_2 = R_1 V_{ref}/(V_{UL} - V_{ref}) = 12.5$ kΩ. Hence, $R_3 = 13.20$ kΩ $- 12.5$ kΩ $= 694.7$ Ω.

We want to build each resistor from a unit device R_o for a well-matching layout. We chose the unit resistor such that R_1, R_2, and R_3 are integer multiples of the unit device (or integer fractions for parallel configurations). To find a suitable unit device, we can increase the value of each resistor to stay within the I_{in} specification. As long as their ratios remain, it does not impact V_{LL}, V_{UL}, and the hysteresis. In this case, we decide for a unit resistor of $R_o = R_3/2 \approx 347$ Ω. With the chosen resistor values, we can recalculate all parameters. In this case, there is no noticeable deviation from the given values of V_{LL}, V_{UL}, and V_{hyst}. Also, $I_{in} = 100$ μA. The specification is fulfilled.

6.9 Power-on Reset

When the supply voltage of an IC ramps up at power-on, digital logic cells (registers, memory, and entire processors) need to be initialized by a dedicated reset pulse. This task is accomplished by a power-on reset (POR) circuit. POR is similar to and usually installed in addition to under-voltage lockout UVLO (Section 6.8). Figure 6.24a) shows the transient behavior of POR in comparison to UVLO. UVLO is more of a slowly reacting static level check that kicks in if V_{DD} drops typically by 10–20% after a few microseconds (t_{UVLO}). UVLO monitors if the digital supply voltage falls below its minimum guaranteed operating voltage. This effect is also known as brown-out (brown-out reset, see Section 6.8). In contrast, POR reacts much faster within a few hundred nanoseconds (t_{POR}) in case of a sudden drop of the supply voltage V_{DD} and at power-on. However, its threshold V_{POR} is

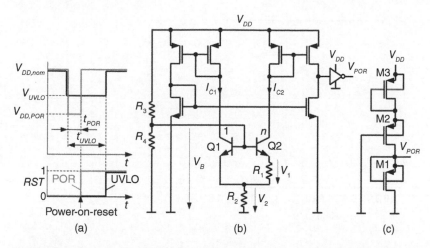

Figure 6.24 Power-on reset (POR): a) transient behavior in comparison to under-voltage lockout UVLO; b) bandgap-based POR circuit; c) POR circuit based on subthreshold operation.

much lower than the UVLO detection level V_{UVLO}. This way, short glitches will lead to a reset while at power-on the digital blocks get initialized by POR well before UVLO indicates that the minimum supply is reached and the IC starts operating. After UVLO or POR gives a reset signal as shown in Fig. 6.24a), $RST = 1$ is usually required to remain for ~ 1 ms until it goes back to zero. Sometimes, POR and UVLO are merged with a threshold set toward the lower end of the permitted supply range of the digital circuits.

The main difference to a UVLO circuit like in Fig. 6.23a) is that the POR circuit cannot rely on a reference voltage V_{ref} to be available. In particular, at power-on, also the reference will be at an undefined level. Hence the POR circuit requires an inherent threshold level. The circuit in Fig. 6.24b) uses a bandgap reference structure to provide the threshold (see Section 6.4 and Fig. 6.8). R_3 and R_4 form a voltage divider that provides the base voltage V_B of Q1 and Q2 as a fraction of V_{DD}. As explained for the IPTAT generation and bandgap voltage reference circuits in Sections 6.3 and 6.4, Q2 has n-times the emitter area of Q1. As long as V_B is lower than the bandgap voltage $V_{BG} \sim 1.2$ V, the collector current I_{C1} in Q1 will be lower than I_{C2} in Q2 (shown in Fig. 6.8c)). This way, the output signal V_{POR} remains at logical 0. The cross-over between I_{C1} and I_{C2} happens exactly at $V_B = V_{BG} \sim 1.2$ V and causes V_{POR} to toggle to 1 to provide a power-on reset pulse. The POR threshold is defined by

$$V_{DD,POR} = V_{BG} \left(\frac{R_3}{R_4} + 1 \right). \tag{6.65}$$

Figure 6.25 shows a transient simulation with the critical voltage and current waveforms. At very low values of V_{DD} (below 0.2 ms in Fig. 6.25), the POR signal V_{POR} experiences a glitch. This behavior is typical in POR circuits as all node voltages get initialized while V_{DD} ramps up. The glitch is usually not critical.

Example 6.11 *Determine the values of R_2 to R_4 for $n = 6$ and $R_1 = 7.5$ kΩ such that the POR threshold becomes $V_{DD,POR} = 2.4$ V. At normal supply $V_{DD} = 3.3$ V the maximum current in the voltage divider with R_3 and R_4 should not exceed $I_q = 50$ μA.*

We can calculate the ratio $\frac{R_2}{R_1}$ from Eqn. (6.10) and determine R_2:

$$R_2 = \frac{-\frac{\partial V_{BG}}{\partial T}}{2 \frac{k}{q} \ln n} R_1 = \frac{2 \, \text{mV/K}}{2 \frac{8.62 \cdot 10^{-5} \, \text{eV/K}}{q} \ln 6} 7.5 \, \text{k}\Omega = 48 \, \text{k}\Omega \tag{6.66}$$

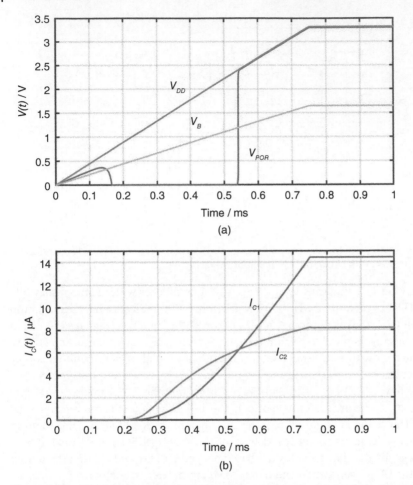

Figure 6.25 Simulated transient behavior of the power-on reset circuit for the parameters of Example 6.11: a) the base voltage of Q1 and Q2 ramps up proportionally to V_{DD}. At $V_B \sim 1.2$ V, the POR signal V_{POR} switches to 1. This corresponds to the cross-over point between both collector currents as shown in b).

The ratio R_3/R_4 is defined by Eqn. (6.65), which is easily found to be 1 in this case (assuming $V_{BG} \approx 1.2$ V $= 0.5\,V_{DD,POR}$).

In order to keep the current I_q through R_3 and R_4 below the limit of 50 μA total resistance needs to be $(R_3 + R_4) = V_{DD}/I_q = 66$ kΩ. Hence, both $R_3 = R_4 = 33$ kΩ.

The transient behavior of the POR circuit for the parameters of this example is shown in Fig. 6.25. When V_{DD} reaches the threshold $V_{DD,POR} = 2.4$ V, Fig. 6.25a) confirms that the base potential is at $V_B = V_{BG} \sim 1.2$ V. A rising transition at V_{POR} happens and causes a power-on reset as expected. The collector currents in Fig. 6.25b) confirm the cross-over behavior according to Fig. 6.8c).

Due to the bipolar junction transistors, the POR circuit will quickly respond and detect short supply drops. If no bipolar devices are available, the POR circuit can be implemented purely in Complementary Metal-Oxide-Semiconductor (CMOS) technology. One example is shown in Fig. 6.24c) [15]. It utilizes three stacked p-type (PMOS) transistors in subthreshold operation. This way, the circuit not only avoids the need for bipolar transistors but also draws minimal current from V_{DD}. This low standby power dissipation makes the circuit suitable as a minimum voltage detector in energy harvesting front-ends. The drain current of the PMOS devices is determined by

$$I_D = \frac{W}{L} I_{Do} e^{\frac{V_{SG}}{n \frac{kT}{q}}} \tag{6.67}$$

with the width-to-length ratio $\frac{W}{L}$ of the device, the process-dependent parameter I_{Do}, the parameter $n = 1 \ldots 4$, Boltzmann's constant k, and elementary charge q at temperature T. M1 has its source–gate shorted ($V_{SG} = 0$), which results in the drain current of M1,

$$I_{D1} = \frac{W_1}{L_1} I_{Do} e^0 = \frac{W_1}{L_1} I_{Do}. \tag{6.68}$$

The POR threshold $V_{DD,POR}$ corresponds to the sum of the source–gate voltages of M2 and M3, with their drain currents equal to I_{D1}. M2 and M3 are sized identically, hence,

$$V_{DD,POR} = V_{SG2} + V_{SG3} = 2V_{SG2} = 2n\frac{kT}{q} \ln\left(\frac{I_{D1}}{\frac{W_2}{L_2} I_{Do}}\right). \tag{6.69}$$

V_{SG2} can be derived from Eqn. (6.67). Equation (6.69) also indicates that M3 increases the POR threshold. Without M3, the level would be too low. Substituting I_{D1} in Eqn. (6.69) by the expression of Eqn. (6.68) yields the POR threshold,

$$V_{DD,POR} = 2n\frac{kT}{q} \ln \frac{\frac{W_1}{L_1}}{\frac{W_2}{L_2}}. \tag{6.70}$$

Example 6.12 *What is the POR threshold of the circuit in Fig. 6.24c) and the drain current in transistors M1–M3 at the POR threshold for a sizing of $\frac{W_1}{L_1} = 3000$ and $\frac{W_2}{L_2} = \frac{W_3}{L_3} = 10$? The subthreshold current is described by $n = 2$ and $I_{Do} = 10^{-12}$A.*

Using Eqn. (6.70), we obtain

$$V_{DD,POR} = 2n\frac{kT}{q} \ln \frac{\frac{W_1}{L_1}}{\frac{W_2}{L_2}} = 4 \frac{8.62 \cdot 10^{-5} \frac{eV}{K} \cdot 300K}{q} \ln\left(\frac{3000}{10}\right) = 590\,\text{mV}. \tag{6.71}$$

The drain current of M1–M3 is given by Eqn. (6.68):

$$I_D = \frac{W_1}{L_1} I_{Do} = 3000 \cdot 10^{-12}\text{A} = 3\,\text{nA} \tag{6.72}$$

The values of both $V_{DD,POR}$ as well as I_D are indeed relatively low.

References

1 Wendt, M., Thoma, L., Wicht, B., and Schmitt-Landsiedel, D. (2008) A configurable high-side/low-side driver with fast and equalized switching delay. *IEEE Journal of Solid-State Circuits*, 43 (7), 1617–1625, doi: 10.1109/JSSC.2008.923734.

2 Forghani-zadeh, H. and Rincon-Mora, G. (2002) Current-sensing techniques for DC-DC converters, in *The 2002 45th Midwest Symposium on Circuits and Systems, 2002. MWSCAS-2002.*, vol. 2, p. II, doi: 10.1109/MWSCAS.2002.1186927.

3 Xin, Z., Li, H., Liu, Q., and Loh, P.C. (2022) A review of megahertz current sensors for megahertz power converters. *IEEE Transactions on Power Electronics*, 37 (6), 6720–6738, doi: 10.1109/TPEL.2021.3136871.

4 Ma, D., Ki, W.H., Tsui, C.Y., and Mok, P. (2001) A 1.8 V single-inductor dual-output switching converter for power reduction techniques, in *2001 Symposium on VLSI Circuits. Digest of Technical Papers (IEEE Cat. No.01CH37185)*, pp. 137–140, doi: 10.1109/VLSIC.2001.934219.

5 Lee, C.F. and Mok, P. (2002) On-chip current sensing technique for CMOS monolithic switch-mode power converters, in *2002 IEEE International Symposium on Circuits and Systems (ISCAS)*, vol. 5, p. V, doi: 10.1109/ISCAS.2002.1010691.

6 Lam, H., Ki, W.H., and Ma, D. (2004) Loop gain analysis and development of high-speed high-accuracy current sensors for switching converters, in *2004 IEEE International Symposium on Circuits and Systems (ISCAS)*, vol. 5, p. V, doi: 10.1109/ISCAS.2004.1329936.

7 Renz, P., Lamprecht, P., Teufel, D., and Wicht, B. (2016) A 40 V current sensing circuit with fast on/off transition for high-voltage power management, in *2016 IEEE 59th International Midwest Symposium on Circuits and Systems (MWSCAS)*, pp. 1–4, doi: 10.1109/MWSCAS.2016.7870011.

8 Shalmany, S.H., Draxelmayr, D., and Makinwa, K.A.A. (2017) A ±36-A integrated current-sensing system with a 0.3% gain error and a 400- µA offset from −55^oC to +85^oC. *IEEE Journal of Solid-State Circuits*, 52 (4), 1034–1043, doi: 10.1109/JSSC.2016.2639535.

9 Dallago, E., Passoni, M., and Sassone, G. (2000) Lossless current sensing in low-voltage high-current DC/DC modular supplies. *IEEE Transactions on Industrial Electronics*, 47 (6), 1249–1252, doi: 10.1109/41.887952.

10 Mohan, N., Undeland, T.M., and Robbins, W.P. (2003) *Power Electronics. Converters, Applications and Design*, John Wiley & Sons, Inc., 3rd edn.

11 Xiao-jie, C. and Quan-yuan, F. (2005) A low-power high reliability CMOS current limit circuit, in *2005 Asia-Pacific Microwave Conference Proceedings*, vol. 5, p. 3, doi: 10.1109/APMC.2005.1607114.

12 Guo, J. and Leung, K.N. (2009) A fold-back current-limit circuit with load-insensitive quiescent current for CMOS low dropout regulator, in *2009 IEEE International Symposium on Circuits and Systems (ISCAS)*, pp. 2417–2420, doi: 10.1109/ISCAS.2009.5118288.

13 Wittmann, J., Barner, A., Rosahl, T., and Wicht, B. (2016) An 18 V input 10 MHz buck converter with 125 ps mixed-signal dead time control. *IEEE Journal of Solid-State Circuits*, 51 (7), 1705–1715, doi: 10.1109/JSSC.2016.2550498.

14 Xue, J. and Lee, H. (2016) A 2 MHz 12-100 V 90% efficiency self-balancing ZVS reconfigurable three-level DC-DC regulator with constant-frequency adaptive-on-time V^2control and nanosecond-scale ZVS turn-on delay. *IEEE Journal of Solid-State Circuits*, 51 (12), 2854–2866, doi: 10.1109/JSSC.2016.2606581.

15 Chen, P.H., Ishida, K., Ikeuchi, K., Zhang, X., Honda, K., Okuma, Y., Ryu, Y., Takamiya, M., and Sakurai, T. (2011) A 95 mV-startup step-up converter with Vth-tuned oscillator by fixed-charge programming and capacitor pass-on scheme, in *2011 IEEE International Solid-State Circuits Conference*, pp. 216–218, doi: 10.1109/ISSCC.2011.5746290.

7

Linear Voltage Regulators

Linear voltage regulators convert an input voltage into a lower output voltage. The basic principle is introduced Section 1.3.1. Invented in the 1970s, they are the most popular power management circuits. Multiple linear regulators can be found in almost all electronics; see the block diagrams in Fig. 1.7.

The famous series of the 78xx (part number with xx indicating the output voltage) regulators, designed in bipolar technology, is still used today. These regulators generate a fixed output voltage. The 7805, introduced in 1972, produces 5 V from any unregulated input voltage. Shortly after, the linear voltage regulator with an adjustable output voltage entered the market. It was described in an Electronic Design article by Robert Dobkin (National Semiconductor, later co-founder of Linear Technology) in 1977.

Fig. 2.2 shows how a low dropout between input and output voltage extends the battery time. The lower the dropout, the lower the input voltage that can be tolerated. Since dropout is essential, a linear regulator is often referred to as Low-Dropout Regulator or LDO in its widely used short form. This term is used to distinguish regulators with higher dropout voltage, which have been around in the early days. Today, LDO is used for any linear regulator. Typical dropout values of state-of-the-art regulators are in the order of 100 mV.

7.1 Fundamental Circuit and Control Concept

In the fundamental circuit in Fig. 7.1a) a p-type (PMOS) transistor MP is connected between the input and the output voltage of the LDO. MP is called the pass device. It can be a dedicated power transistor or, if the voltage ratings are fulfilled, a regular low-voltage transistor. We denote the input voltage by V_{in}, which can be a battery or sub-regulated supply voltage. MP forms a regulated resistance to achieve the source–drain voltage required for V_{out} to reach its specified set value. For this reason, V_{out} is divided into a feedback voltage V_{fb} by the resistive divider of R_1 and R_2,

$$V_{fb} = \frac{R_2}{R_1 + R_2} V_{out} = \alpha V_{out}. \tag{7.1}$$

We will use the symbol α for simplicity for the divider ratio. The output voltage is measured through the resistive divider to reduce the required input swing of the error amplifier. The user can set the output voltage by properly sizing the resistors. Especially for low output voltages, the resistive divider may not be needed.

The error amplifier determines the difference between V_{fb} and the set point V_{ref}. As this indicates the control error, this feedback amplifier is also called the error amplifier. The error is amplified

Design of Power Management Integrated Circuits, First Edition. Bernhard Wicht.
© 2024 John Wiley & Sons Ltd. Published 2024 by John Wiley & Sons Ltd.

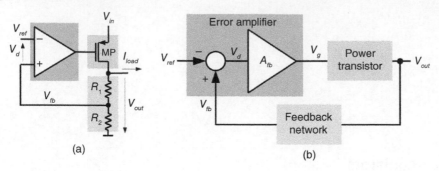

Figure 7.1 a) Fundamental circuit of a linear voltage regulator; b) block diagram.

by its differential gain A_{fb}. The error amplifier output controls the gate of the power transistor MP. In this way, a negative feedback loop is installed. We can check the correct behavior by verifying the signal flow around the loop. Let us assume that the LDO supplies a microcontroller that goes from standby to full operation, which leads to a positive load step in I_{load}. We can consider the LDO as a voltage source with a (low) source impedance. Therefore, V_{out} drops if I_{load} goes up. Consequently, V_{fb} drops, which feeds to the non-inverting input of the error amplifier. The output of the amplifier decreases and pulls down the gate of MP. With larger source–gate voltage MP delivers more drain current and counter-acts the increase in load current. In consequence, V_{out} gets stabilized and regulated as expected.

To further analyze the control loop, we can put together the block diagram of Fig. 7.1b). We can apply small-signal analysis to obtain the transfer behavior of the linear regulator. We assume the input voltage to be constant. Its small-signal value is zero; hence V_{in} connects to the ground in the small-signal equivalent circuit. This way, the (open-loop) output impedance at V_{out} consists of the feedback divider $R_1 + R_2$ in parallel to the output impedance r_{ds} of MP. It is convenient to determine which resistive path dominates to simplify the analysis. Using the channel-length modulation coefficient λ, we find that $r_{ds} = 1/(\lambda I_D)$. Because the drain current $I_D \sim I_{load}$ is large in a linear regulator, r_{ds} will be rather low. For $\lambda = 1/20\,\text{V}$ and $I_D = 20\,\text{mA}$ we can estimate $r_{ds} = 1\,\text{k}\Omega$ and even lower values if I_D gets larger. In contrast, the sum of R_1 and R_2 is rather in the order of 10–100 kΩ to keep their current consumption low. Hence, they can be neglected, and we only consider r_{ds}. Furthermore, we need the small-signal transconductance g_m of MP.

Two equations describe the small-signal behavior of the linear regulator:

$$v_{out} = i_d r_{ds} = v_{sg} g_m r_{ds} = (0 - v_g) g_m r_{ds} \tag{7.2}$$

$$v_g = \left(v_{fb} - v_{ref}\right) A_{fb} = \left(\alpha v_{out} - v_{ref}\right) A_{fb} \tag{7.3}$$

Inserting Eqn. (7.3) into Eqn. (7.2) results in

$$\frac{v_{out}}{v_{ref}} = \frac{g_m r_{ds} A_{fb}}{1 + \alpha g_m r_{ds} A_{fb}} \sim \frac{1}{\alpha}. \tag{7.4}$$

The term

$$T = \alpha g_m r_{ds} A_{fb} \tag{7.5}$$

in the denominator represents the loop gain of the LDO. For a well-operating feedback loop, the loop gain has to be much greater than unity. Hence, the transfer function simplifies as shown in Eqn. (7.4). Only the feedback network, i.e., the resistive divider, determines the output voltage as a function of V_{ref}:

$$V_{out} = \frac{V_{ref}}{\alpha} \tag{7.6}$$

Example 7.1 *Determine R_1 and R_2 for $V_{out} = 3.3\,V$ and $V_{ref} = 1.2\,V$. The current I_R through the resistors should not exceed $100\,\mu A$.*

$$R_1 + R_2 = \frac{V_{out}}{I_R} = \frac{3.3\,V}{100\,\mu A} = 33\,k\Omega$$

From Eqn. (7.6):

$$\alpha = \frac{V_{ref}}{V_{out}} = \frac{1.2\,V}{3.3\,V} = 0.363$$

Knowing $(R_1 + R_2)$ and α we can calculate R_2 from Eqn. (7.1):

$$R_2 = \alpha\left(R_1 + R_2\right) = 0.363 \cdot 33\,k\Omega = 12\,k\Omega$$

Finally,

$$R_1 = \left(R_1 + R_2\right) - R_2 = 33\,k\Omega - 12\,k\Omega = 21\,k\Omega.$$

What would be a good unit finger size for the resistors? Looking at the values of R_1 and R_2, $R = 3\,k\Omega$ would be a suitable unit resistor. Then, R_1 consists of seven fingers in series and R_2 of four devices. This way, good matching can be achieved, contributing to the accuracy of V_{out}.

Considering the power transistor as a resistive element between the LDO's input and output, using an n-type (NMOS) power transistor is an interesting alternative as it requires less layout area. Also, for very low input voltages, the PMOS type would no longer be suitable due to the gate-overdrive voltage lag. Figure 7.2a) shows the corresponding circuit along with the PMOS type in Fig. 7.2b) (repeated from Fig. 7.1a) for comparison). We notice a couple of differences between the two options. To ensure loop stability, the polarity of the error amplifier input changes. Also, the error amplifier is supplied from a charge pump (CP). The operation of charge pumps is covered in Chapter 8. With a charge pump, the gate of the power transistor MP can be driven to voltages above V_{in}. Why is this necessary? To ensure that MP remains turned on and delivers load current, the gate voltage must be higher than V_{out} by at least one threshold voltage V_{th} of MP. Without the

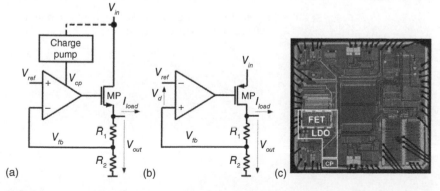

Figure 7.2 Fundamental LDO with a) PMOS and b) NMOS power transistor; c) die photo for an NMOS type. Source: Adapted from [1] Wittmann, J et al. (2013).

charge pump, the gate voltage would be limited to V_{in} and the dropout to $V_{in} - V_{out} = V_{th}$, which might be too large for many applications. Figure 7.2c) [1] illustrates the layout advantage of an NMOS type LDO. It shows the die photo of a multi-rail power supply containing an LDO and several switched-mode regulators. A PMOS power transistor would need 2–3 times more area than a NMOS, which exceeds the area of the NMOS transistor (FET) plus the charge pump (CP) in Fig. 7.1c). The transistor is designed for a dropout voltage of 200 mV at 100 mA load current.

7.2 Dropout Voltage

If the input voltage decreases below a certain limit, the control loop cannot keep the output constant anymore, and V_{out} will drop. This condition defines the dropout voltage as the minimum difference V_{do} between input and output at large load currents, usually at maximum I_{load}. Figure 7.3 shows the relationship between input and output voltage. It can be considered as the DC-version of the transient plot that is shown in Fig. 2.2 (Chapter 2) to illustrate the dropout due to the power transistor's on-resistance. Let us consider the PMOS-type LDO. For large values of V_{in}, V_{out} can be well controlled. If V_{in} goes down, the gate voltage of MP goes down as well because its on-resistance (proportional to $1/V_{SG}$) needs to be further reduced. This way, V_{out} stays constant at the present load current. If V_{in} decreases further, V_{out} will drop as well, as depicted in Fig. 7.3. At this point the regulator reaches its control limit and enters the dropout region. The diagram defines the two regions, the Control Region and the Dropout Region.

The dropout definition assumes the LDO is biased below the control limit. This is fulfilled if $V_{in} < V_{out}$. There is, however, no unique definition to that. Many LDO specifications define the dropout voltage $V_{do} = V_{in} - V_{out}$ at 95% of the nominal output voltage, $V_{in} = 0.95V_{out,nom}$, as indicated in Fig. 7.3. In the Dropout Region, the power transistor MP is fully turned-on and operates in its triode region. In contrast, MP operates predominately in the saturation region in the Control Region of the LDO.

Because the transistor operates in the triode region, we can derive the sizing equation for the W/L-ratio of the power transistor to meet the dropout specification. The drain current has to be greater than or at least equal to the maximum load current at which the dropout is specified. For the PMOS transistor,

$$I_D = \beta \left((V_S - V_G - V_{th})V_{do} - \frac{V_{do}^2}{2} \right) \geq I_{load,max}. \tag{7.7}$$

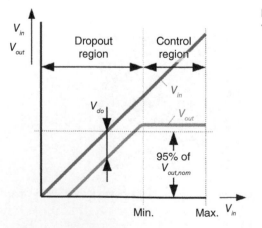

Figure 7.3 Relationship between input and output voltage of the LDO, dropout definition.

We can make the following substitutions,

$$\beta = \frac{W}{L}\mu_o C_{ox}, \quad V_S - V_G = V_{in,min} - 0, \quad \frac{V_{do}^2}{2} = 0, \tag{7.8}$$

and obtain

$$\frac{W}{L} \geq \frac{I_{load,max}}{\mu_o C_{ox}(V_{in,min} - V_{th})V_{do}}. \tag{7.9}$$

For the NMOS case, we assume that its gate is driven by the charge pump voltage, which is $V_{CP,min}$ in the worst case. Hence we derive the corresponding sizing equation:

$$\frac{W}{L} \geq \frac{I_{load,max}}{\mu_o C_{ox}(V_{CP,min} - V_{out} - V_{th})V_{do}} \tag{7.10}$$

The conditions in Eqns. (7.9) and (7.10) have to be fulfilled across all corner variations (temperature and process). As a rule of thumb, we calculate the nominal W/L-ratios from the sizing equations and use three times that value for an initial design. In the second step, the sizing will be finalized by a worst-case simulation.

Example 7.2 *A PMOS-type LDO requires $V_{do} = 150$ mV at $I_{load,max} = 100$ mA for these parameters: $V_{out,nom} = 1.8$ V, $V_{thp} = 0.6$ V, $\mu_o C_{ox} = 87 \, \mu A/V^2$. Calculate the nominal W/L-ratio of the power transistor, assuming the 95%-rule for the dropout. What is the corresponding width W if $L = 0.18 \, \mu m$.*
The input voltage at the dropout condition calculates to $V_{in,min} = 0.95V_{out,nom} = 1.71$ V.
Equation (7.9) yields the sizing:

$$\frac{W}{L} = \frac{I_{load,max}}{\mu_o C_{ox}(V_{in,min} - V_{th})V_{do}} = \frac{100 \, \text{mA}}{87 \, \mu A/V^2 (1.71 \, \text{V} - 0.6 \, \text{V})0.15 \, \text{V}} = 6903 \tag{7.11}$$

For $L = 0.18 \, \mu m$, we obtain $W = 1243 \, \mu m$.
As expected, this is a large transistor. It requires a considerable layout area like the design shown in Fig. 7.2c).

7.3 DC Parameters

Power efficiency is an important DC parameter, like in all power management circuits. In linear regulators, the current efficiency is another DC parameter. We are also interested in how changing load current and input voltage influence the regulated output voltage, defined as load and line regulation.

7.3.1 Power Efficiency

From the operation principle of the LDO, we know that excessive power is dissipated in the power transistor. Figure 7.1a) shows that the input current flows mainly to the output. But also, the bias currents of the error amplifier and all other circuits need to be delivered from the input. Also, the current through R_1 and R_2 contributes to the total input current. Summarizing all additional currents, we can define a quiescent current I_q. Hence, the input current is $I_{in} = I_{load} + I_q$. For the power efficiency, we obtain

$$\eta = \frac{P_{out}}{P_{in}} = \frac{V_{out}I_{load}}{V_{in}I_{in}} = \frac{V_{out}I_{load}}{V_{in}(I_{load} + I_q)} \approx \frac{V_{out}}{V_{in}}. \tag{7.12}$$

The efficiency is often expressed in percent by multiplying the result of Eqn. (7.12) by 100%. In normal operation, if the load current I_{load} is large, I_q is negligible, and the efficiency is only defined

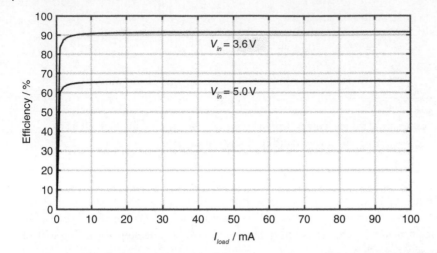

Figure 7.4 Efficiency of the linear voltage regulator for $V_{out} = 3.3$ V, $I_q = 100$ μA and two different input voltages V_{in}. The closer V_{in} is to V_{out}, the higher the efficiency. For large I_{load} the efficiency η can be approximated by V_{out}/V_{in}. The two cases result in $\eta = 66\%$ and $\eta = 91.6\%$, respectively.

by the output-to-input voltage ratio. It confirms poor efficiency for large voltage drops across the power transistor. If the voltage difference between input and output is small, linear regulators show good efficiency. That's why power management systems often use highly efficient switched-mode converters in series with an LDO. The DC–DC converter handles a large step-down ratio and provides an input voltage to the LDO, typically a few hundred millivolts above the target output voltage of the LDO. This way, the LDO is very power efficient as well. In addition, due to its linear control without any switching operation, the LDO provides a "clean" output voltage.

Figure 7.4 confirms that the efficiency stays nearly constant over a wide load current range. The quiescent current cannot be neglected at light load, and the efficiency degrades significantly. The growing demand for high efficiency, especially in battery-operated applications, has driven the trend toward LDOs with low I_q. It requires ultra-low power control loop designs, including low-power reference voltage generation. Typical values of I_q are in the range of 10–100 μA.

7.3.2 Current Efficiency

This parameter relates the quiescent current I_q to the load current, expressed by

$$\eta_I = \frac{I_{load}}{I_{in}} 100\% = \frac{I_{load}}{(I_{load} + I_q)} 100\%. \tag{7.13}$$

For large load currents, the current efficiency approaches 100%. In contrast, it drops significantly at light loads. For this reason, recent LDO designs aim to scale I_q with the load current. It can be achieved by sensing the load-dependent variation of the output voltage and adjusting the bias currents in the control loop.

Example 7.3 *Calculate the current efficiency for the parameters given in Fig. 7.4 for* $I_{load} = 100$ mA *and* $I_{load} = 1$ mA.

$I_{load} = 100$ mA: $\eta_I = \frac{I_{load}}{(I_{load} + I_q)} 100\% = \frac{100 \text{ mA}}{(100 \text{ mA} + 0.1 \text{ mA})} 100\% = 99.9\%$

$I_{load} = 1$ mA: $\eta_I = \frac{1 \text{ mA}}{1.1 \text{ mA}} 100\% = 90.9\%$

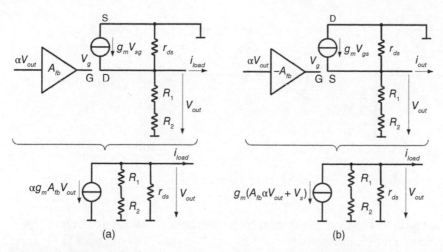

Figure 7.5 Small-signal equivalent circuits to derive the load regulation for an LDO with a) PMOS and b) NMOS power transistor.

7.3.3 Load Regulation

Since there is small but non-zero closed-loop output impedance, we can expect a variation of V_{out} as a function of I_{load}. This relationship is defined as load regulation,

$$\left|\frac{\Delta V_{out}}{\Delta I_{load}}\right| \approx \left.\left|\frac{v_{out}}{i_{load}}\right|\right|_{v_{in}=0} = \frac{1}{\frac{R_2}{R_1+R_2}g_m A_{fb}} = \frac{1}{\alpha g_m A_{fb}}. \tag{7.14}$$

Equation (7.14) is valid for NMOS and PMOS types. It can be derived by small-signal analysis (indicated by lower-case symbols) from the equivalent circuits shown in Fig. 7.5. Due to the large resistance, the current through $r_{ds} \parallel (R_1 + R_2)$ can be neglected in both cases. Under this assumption, we can readily derive expressions for i_{out} from the simplified circuits at the bottom of Fig. 7.5. Both expressions are nearly identical. For the PMOS case in Fig. 7.5a), we get

$$i_{out} = -\alpha g_m A_{fb} v_{out}, \tag{7.15}$$

while the NMOS type in Fig. 7.5b) results in

$$i_{out} = -g_m \left(1 + \alpha A_{fb}\right) v_{out} \sim -\alpha g_m A_{fb} v_{out}. \tag{7.16}$$

Finally, we get the load regulation according to Eqn. (7.14).

Example 7.4 *What is the load regulation for $g_m = 10\,\text{mA/V}$, $A_{fb} = 300$, $R_1 = 20\,\text{k}\Omega$, $R_2 = 10\,\text{k}\Omega$? We obtain the divider ratio $\alpha = \frac{R_1}{R_1+R_2} = 1/3$ from the resistor values. Hence, using Eqn. (7.14),*

$$\left|\frac{\Delta V_{out}}{\Delta I_{load}}\right| = \frac{1}{\alpha g_m A_{fb}} = \frac{1}{\frac{1}{3}10\,\text{mA/V}\,300} = \frac{1\,\text{mV}}{1\,\text{mA}}.$$

We keep in mind that this result for the load regulation is valid for both the PMOS and NMOS types of the LDO.

7.3.4 Line Regulation

The output voltage varies not only depending on the load current. It is also a function of the input voltage called line regulation. Figure 7.6 shows the small-signal equivalent circuits that allow to

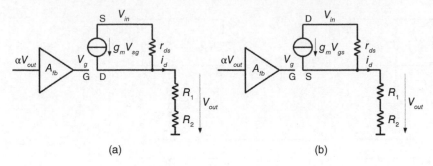

Figure 7.6 Small-signal equivalent circuits to derive the line regulation for an LDO with a) PMOS and b) NMOS power transistor.

derive an expression for the line regulation. Note that the line regulation assumes a constant output current, which means that the small-signal value i_{load} is zero. Therefore, for both types in Fig. 7.6, the drain current i_d flows through the feedback divider and results in

$$v_{out} = (R_1 + R_2)i_d. \tag{7.17}$$

For the PMOS type in Fig. 7.6a),

$$i_d = g_m v_{sg} + \frac{v_{in} - v_{out}}{r_{ds}} \approx g_m v_{sg}. \tag{7.18}$$

The approximation neglects the current through r_{ds}. With

$$v_{sg} = v_{in} - \alpha A_{fb} v_{out}, \tag{7.19}$$

inserting Eqn. (7.18) into Eqn. (7.17) yields the line regulation of the PMOS type LDO,

$$\left| \frac{\Delta V_{out}}{\Delta V_{in}} \right| \approx \left| \frac{v_{out}}{v_{in}} \right|_{i_{load}=0} = \frac{g_m(R_1 + R_2)}{1 + g_m(R_1 + R_2)\alpha A_{fb}} \approx \frac{1}{\alpha A_{fb}}. \tag{7.20}$$

The approximation is valid in practical designs with a loop gain $g_m(R_1 + R_2)\alpha A_{fb}$ greater than unity.

For the NMOS type in Fig. 7.6b) we obtain

$$i_d = g_m v_{gs} + \frac{v_{in} - v_{out}}{r_{ds}} \tag{7.21}$$

and

$$v_{gs} = v_g - v_s = -\alpha A_{fb} v_{out} - v_{out} = -(1 + \alpha A_{fb})v_{out}. \tag{7.22}$$

Inserting Eqns. (7.21) and (7.22) in Eqn. (7.17) yields

$$v_{out} = -(R_1 + R_2)\left(g_m(1 + \alpha A_{fb}) + \frac{1}{r_{ds}}\right)v_{out} + \frac{R_1 + R_2}{r_{ds}}v_{in}. \tag{7.23}$$

Neglecting the term $1/r_{ds}$ and rearranging results in the line regulation for the NMOS type LDO,

$$\left| \frac{\Delta V_{out}}{\Delta V_{in}} \right| \approx \left| \frac{v_{out}}{v_{in}} \right| = \frac{(R_1 + R_2)/r_{ds}}{1 + (R_1 + R_2)g_m(1 + \alpha A_{fb})} \approx \frac{1}{g_m r_{ds}\alpha A_{fb}}. \tag{7.24}$$

Again the approximation assumes that the loop gain in the denominator is greater than unity, which is usually fulfilled in practical designs.

Comparing Eqns. (7.20) and (7.24), we note that the line regulation for the NMOS type is lower by a factor $g_m r_{ds}$, which corresponds to the intrinsic gain of a single transistor. With typical values of the intrinsic gain toward 100, the LDO with an NMOS power transistor shows significantly better line regulation than a PMOS type. This observation is intuitive because the PMOS's source terminal is connected to V_{in}, directly influencing the drain current. In contrast, the drain of the NMOS at the input has only little impact on the channel current.

Example 7.5 *Calculate the line regulation for the NMOS-type LDO using the exact expression of Eqn. (7.24) and the approximation if $r_{ds} = 1\,k\Omega$. Keep the same parameters like in Example 7.4, i.e., $g_m = 10\,mA/V, A_{fb} = 300, R_1 = 20\,k\Omega, R_2 = 10\,k\Omega, \alpha = \frac{R_1}{R_1 + R_2} = 1/3$.*
With the full equation,

$$\left| \frac{\Delta V_{out}}{\Delta V_{in}} \right| = \frac{(R_1 + R_2)/r_{ds}}{1 + (R_1 + R_2)g_m(1 + \alpha A_{fb})}$$
$$= \frac{30\,k\Omega/1\,k\Omega}{1 + 30\,k\Omega\,10\,mA/V(1 + \frac{1}{3}300)} = 0.99\,mV/V. \tag{7.25}$$

The approximation yields

$$\left| \frac{\Delta V_{out}}{\Delta V_{in}} \right| \sim \frac{1}{g_m r_{ds} \alpha A_{fb}} = \frac{1}{10\,mA/V\,1\,k\Omega\,\frac{1}{3}300} = 1\,mV/V. \tag{7.26}$$

The results of Eqns. (7.25) and (7.26) are nearly identical. Hence, the approximation is valid very well for the given parameter set.

7.4 The Error Amplifier

Referring to the fundamental LDO circuit of Fig. 7.1, we have already discussed the design of the feedback divider (R_1 and R_2) as well as the sizing of the power transistor. The other essential block is the error amplifier with its input pins connected to the feedback voltage V_{fb} and the reference voltage V_{ref}. While the load and line regulation benefit from an error amplifier with significant gain A_{fb} (it appears in the denominator), the investigation of the dynamic stability further below will show that too much loop gain has a negative impact. Hence, the typical gain of the error amplifier is in the range of 40–60 dB. A symmetrical amplifier like in Fig. 7.7 or a folded-cascode stage are widely used. With the small-signal transconductance g_{ma} of the input differential pair, the output

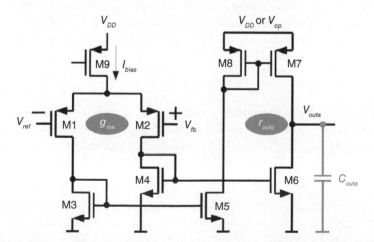

Figure 7.7 The symmetrical amplifier is used as a typical error amplifier.

resistance r_{outa} and the parasitic output capacitance C_{outa}, the AC-behavior of the error amplifier in Fig. 7.7 is

$$A_{fb} = \frac{g_{ma} r_{outa}}{1 + j\omega r_{outa} C_{outa}} \tag{7.27}$$

with

$$r_{outa} = r_{ds6} \parallel r_{ds7}. \tag{7.28}$$

The slew rate (SR) at the amplifier's output is important for the LDO's transient response. It is the rate at which the error amplifier output can change over time and refers to the capacitive charging and discharging of C_{outa}. The maximum output current is I_{bias} (assuming that the differential input voltage is so large that one device of the differential pair conducts the full bias current). Hence,

$$SR = I_{bias}/C_{outa}. \tag{7.29}$$

The capacitance C_{outa} can be reduced by inserting a buffer (see Fig. 7.11). With a given value of C_{outa}, the slew rate can only be improved by increasing the bias current I_{bias}. Some LDOs apply adaptive biasing depending on their load current; see Section 7.6.2.

The error amplifier is supplied by V_{DD}, usually derived from the LDO's input voltage V_{in} via a pre-regulator. Such a pre-regulator can be a shunt regulator as described in Section 7.12. As this regulator is not very precise, some designs use it only during start-up and supply the error amplifier (and all other LDO circuitry) from the LDO's output voltage if the voltage levels are suitable (see Fig. 7.23 in Section 7.12). As the NMOS-type LDO may be driven from a charge pump, the output stage of the error amplifier can be connected to the charge pump as indicated by V_{cp} in Fig. 7.7. With $V_{cp} > V_{DD}$, the transistors in the amplifier's output stage may need to have higher voltage ratings.

The error amplifier's power supply rejection (PSR) plays an important role in the PSR of the entire LDO, as discussed in Section 7.8. It can be expressed in general as

$$PSR = 20 \text{ dB log} \left| \frac{1}{A_p} \right| = 20 \text{ dB log} \left| \frac{v_{dd}}{v_{out}} \right| \tag{7.30}$$

with the power supply gain

$$A_p = \frac{v_{out}}{v_{dd}} = G_{mp} \left(r_{outa} \parallel C_{outa} \right). \tag{7.31}$$

The power supply transconductance G_{mp} translates the supply voltage ripple v_{dd} into a small-signal output current. That current flows into the amplifier output impedance $\left(r_{outa} \parallel C_{outa} \right)$ and causes a signal v_{out}. Based on the method in [2], the power supply transconductance for the amplifier in Fig. 7.7 can be approximated by

$$G_{mp} = \frac{1}{r_{ds5} \parallel r_{ds7}} + j\omega \left(C_{p5} + C_{outa} \right), \tag{7.32}$$

which assumes that the input differential pair has zero mismatch and all current mirrors have a 1-to-1 ratio. C_{p5} refers to the parasitic capacitance at the drain node of M5. Let us consider the DC case ($\omega = 0$). If we insert G_{mp} as well as r_{outa} from Eqn. (7.28) into Eqn. (7.31) we obtain

$$A_p = \frac{r_{ds6} \parallel r_{ds7}}{r_{ds5} \parallel r_{ds7}} \approx 1. \tag{7.33}$$

This result means that any supply ripple will couple without attenuation to the amplifier output. Even though this indicates poor PSR of the error amplifier itself, in some cases, it will benefit the PSR of the entire LDO in conjunction with a p-type power transistor (see Section 7.8).

The PSR looks much different if the error amplifier of Fig. 7.7 is implemented in a complementary way (i.e., flipped along the horizontal axis with PMOS and NMOS devices exchanged). The approximation of the power supply gain according to [2] is then

$$A_p = \left(\frac{1}{r_{ds6}} - \frac{1}{r_{ds5}} \right) r_{outa} \approx 0. \tag{7.34}$$

According to Eqn. (7.30), this result corresponds to infinite PSR, which means that the supply noise is fully suppressed and not present at the amplifier output. Also, this scenario will be discussed with respect to the LDO PSR in Section 7.8.

7.5 Frequency Behavior and Stability

7.5.1 PMOS Type

By adding the relevant capacitances to the LDO circuit of Fig. 7.1a), we obtain the AC equivalent circuit in Fig. 7.8. It allows us to derive the frequency behavior of the linear regulator. We notice two relevant capacitances at the output node and the gate of the power transistor. The bypass or buffer capacitor C_o stabilizes the output voltage V_{out} in case of fast load current transients. Typical values are in the microfarad range. Multi-level ceramic capacitors are the preferred capacitor type, while traditionally, electrolyte capacitors have been used. In any case, the capacitor has an equivalent series resistance R_{ESR}, which varies over a wide range of typically $10\,\text{m}\Omega$–$10\,\Omega$. The value of R_{ESR} is often not specified in detail by the capacitor vendors. Because the W/L-ratio of the power transistor is large, its gate capacitance dominates the parasitic capacitance C_{par}. This way, C_{par} forms the parasitic output capacitance C_{outa} of the error amplifier according to Fig. 7.7.

Based on Fig. 7.8, we can rewrite the loop gain of Eqn. (7.5):

$$T(s) = g_m z_{out}(s) A_{fb}(s) \tag{7.35}$$

$A_{fb}(s)$ represents the frequency behavior of the error amplifier given by Eqn. (7.27). $z_{out}(s)$ is the (open-loop) node impedance at the LDO output. Assuming $r_{ds} \ll (R_1 + R_2)$ as discussed earlier, we find

$$z_{out}(s) = r_{ds} \parallel \left(R_{ESR} + \frac{1}{j\omega C_o} \right) \approx r_{ds} \frac{1 + j\omega R_{ESR} C_o}{1 + j\omega r_{ds} C_o}. \tag{7.36}$$

The approximation is valid for $R_{ESR} \ll r_{ds}$, which is usually fulfilled in practical circuits. From Eqns. (7.27), (7.35) and (7.36), we can derive the DC loop gain T_o as well as two poles and one zero. The two poles of the loop gain correspond to the two relevant nodes in the LDO where C_o and

Figure 7.8 Equivalent circuit for the frequency behavior of the PMOS-type linear regulator.

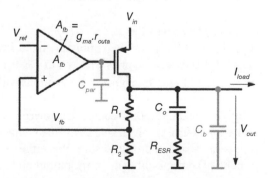

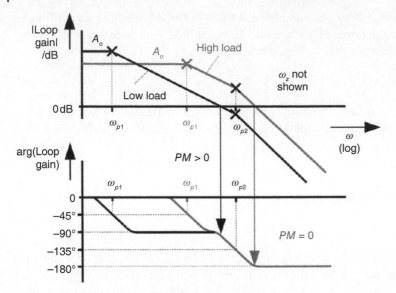

Figure 7.9 Bode plot of the LDO loop gain.

C_{par} are present – the output node of the LDO and the output node of the error amplifier:

$$T_o = g_m r_{ds} A_{fb}(0) = g_m r_{ds} g_{ma} r_{outa}$$
$$\omega_{p1} = \frac{1}{r_{ds} C_o}$$
$$\omega_{p2} = \frac{1}{r_{outa} C_{par}}$$
$$\omega_z = \frac{1}{R_{ESR} C_o}$$

(7.37)

The loop gain is a two-pole transfer function, which gives a phase shift of $2 \cdot 90°$. The Bode plot in Fig. 7.9 indicates that the loop will be marginally stable with zero degrees of phase margin *PM*. Figure 7.9 omits the zero ω_z as it usually appears at high frequencies far above the unity gain frequency of the loop gain. Since the power transistor forms a common-source gain stage, its parasitic drain–gate capacitance causes a Right Half-Plane Zero (RHPZ), related to the Miller effect. The RHPZ degrades stability by adding another $-90°$ of phase shift. However, the RHPZ typically occurs at high frequencies, above the unity-gain frequency of the closed regulator loop, and can be neglected. Nevertheless, the Miller effect on the parasitic drain–gate capacitance contributes to the value of the effective gate capacitance C_{gate}.

Frequency Compensation

Equation (7.37) reveals one important characteristic of any LDO. At least one pole is load dependent. In this case, ω_{p1} varies with the LDO's load current I_{load} because $r_{ds} \propto \frac{1}{I_{load}}$. This behavior is illustrated in Fig. 7.9. In modern applications with load currents ranging from µA to A, the position of the dominant pole moves over a wide range of frequencies.

In conclusion, the LDO control loop design challenge is ensuring its dynamic stability over the entire load current range. Due to the load dependency, it is not sure which is the dominant pole. A general design approach ensures that either ω_{p1} or ω_{p2} remains the dominant pole. The conventional LDO designs often make the output node dominant. They ensure the poles do not cross and stay at a distance with sufficient phase margin. In contrast, capacitor-less LDOs remove the output buffer capacitor C_o and make the power transistors gate node the dominant pole (see Section 7.10).

Let us assume that ω_{p1} is the dominant pole exactly as shown in Fig. 7.9. As shown in Fig. 7.9, the high-load case is most critical for stability because $\omega_{p1} \propto I_{load}$ moves up toward ω_{p2}. Four options for improving the loop stability are shown in Fig. 7.10 and discussed below. The goal is to achieve a sufficient phase margin of typically 60°.

1) *Lowering the DC loop gain T_o*: This can be achieved by reducing the gain A_{fb} of the feedback amplifier. However, several DC parameters of the LDO get worse for low A_{fb} such as load and line regulation (see Eqns. (7.14), (7.20), and (7.24)). Also, the overall precision of V_{out} suffers because the error by neglecting the "1+"-term in the denominator in Eqn. (7.4) gets larger.
2) *Reduce ω_{p1}*: Increasing C_o shifts ω_{p1} to lower frequencies, further apart from ω_{p2}. However, as ω_{p1} is the dominant pole, it determines the speed how fast the LDO can counteract a load step, for instance. Decreasing ω_{p1} makes the LDO slower.
3) *Compensation via ω_z by choice of R_{ESR}*: As R_{ESR} is not well predictable, this option is somewhat limited to a unique lab setup and unsuitable for a series product. Nevertheless, some designs

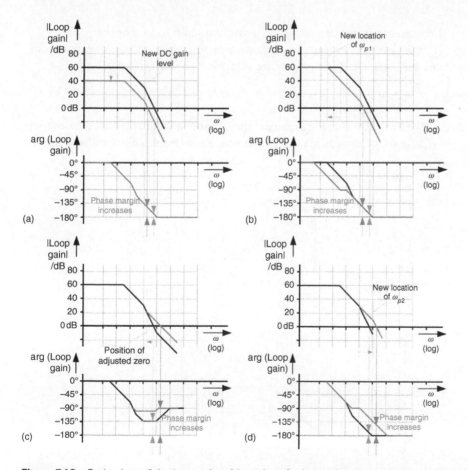

Figure 7.10 Bode plots of the loop gain with options for improving the dynamic stability of the LDO: a) lowering the DC loop gain; b) reducing ω_{p1}; c) canceling the non-dominant pole by the zero; d) increasing ω_{p2}.

place an additional resistor in series to R_{ESR}. As this does not track the load variation, this option is limited to LDOs with well-defined loads.

4) *Increase ω_{p2}*: Lowering the capacitance C_{par} shifts the second pole ω_{p2} to higher frequencies and increases the phase margin. However, C_{par} is dominated by the gate capacitance of the power transistor and cannot be changed for a given design as this is required to fulfill the dropout requirements. Therefore, a buffer stage is commonly inserted between the error amplifier and the power transistor, reducing the impedance at that node.

Inserting a Buffer

For option 4, Fig. 7.11a) shows how the control loop benefits from inserting a buffer [3]. One more pole occurs because the circuit gets one more node. However, both poles appear at higher frequencies than the original setup without buffer, as illustrated in Fig. 7.11b). The second pole ω_{p2} is determined by the parasitic capacitance C''_{par} that the buffer has to drive, which is still dominated by the large gate capacitance. We can assume $C''_{par} \sim C_{par}$. Because of the low output impedance $r_{outb} \ll r_{outa}$ of the buffer, the pole shifts to

$$\omega'_{p2} = \frac{1}{r_{outb} C''_{par}} \gg \omega_{p2}. \tag{7.38}$$

The third pole ω_{p3} is related to the output node of the error amplifier. Its output impedance r_{outa} remains unchanged, but the parasitic capacitance C'_{par} is much smaller than the gate capacitance C_{par} of the power transistor. Hence, the pole ω_{p3} occurs at much higher frequencies,

$$\omega_{p3} = \frac{1}{r_{outa} C'_{par}} > \omega_{p2}. \tag{7.39}$$

Inserting a buffer can be as simple as using a source follower as shown in Fig. 7.11c). Its output impedance $r_{outa} \sim 1/g_m$ is in the order of 10 kΩ. In comparison, the error amplifier (see Section 7.4)

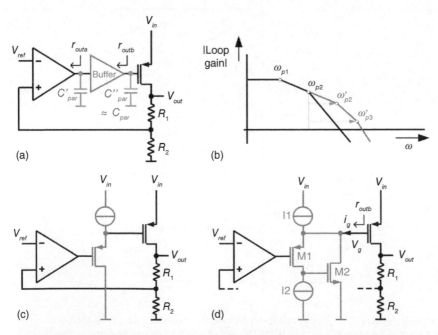

Figure 7.11 Inserting a buffer: a) concept indicating the influence on the node impedances; b) bode plot with the shifted pole locations; c) source follower; d) super-follower.

usually has a much larger output impedance of typically $r_{outa} = 1\,\text{M}\Omega$. We can estimate, that ω'_{p2} will occur at least one decade higher than ω_{p2}. The super-follower in Fig. 7.11d) achieves even lower output impedance r_{outb} [3]. M2 and the current source I2 add a local feedback loop to the source follower M1 resulting in

$$i_g = g_{m2}v_{gs2} + \frac{v_{gs2}}{r_2} \sim g_{m2}v_{gs2} \tag{7.40}$$

$$v_{gs2} = r_2 \left(g_{m1}v_g + \frac{v_g - v_{gs2}}{r_{ds1}} \right) \Rightarrow v_{gs2} \sim g_{m1}r_{ds1}v_g \tag{7.41}$$

with the source impedances r_1 and r_2 of I1 and I2. The approximations assume $r_2 \rightarrow \infty$. Inserting Eqn. (7.41) into Eqn. (7.40) yields

$$r_{outb} = \frac{v_g}{i_g} = \frac{1}{g_{m1}(g_{m2}r_{ds1})}. \tag{7.42}$$

The term $(g_{m2}r_{ds1})$ corresponds to the intrinsic gain of a single MOS transistor with a typical value in the order of 100 (down to 10–20 in advanced digital CMOS technologies). Compared to the basic source follower in Fig. 7.11c), the output impedance of the super-follower is reduced by this factor, which offers several design options. The drain current, as well as the W/L-ratio of M1, may be reduced. Since the dominant pole location ω_{p1} according to Eqn. (7.37) increases with the load current ($r_{ds} \propto \frac{1}{I_{load}}$) [3] proposes to adaptively increase also the current I2. This way, r_{outb} decreases for larger loads and shifts ω'_{p2} further away from ω_{p1} to maintain stability over the entire load range.

7.5.2 NMOS Type

Figure 7.12 shows the AC equivalent circuit of the NMOS-type LDO. As for the PMOS type LDO, we can calculate the loop gain and derive the DC gain as well as the pole-zero locations:

$$
\begin{aligned}
T_o &= \alpha A_{fb}(0) = \alpha g_{ma}r_{outa} \\
\omega_{p1} &= \frac{g_m}{C_o} \\
\omega_{p2} &= \frac{1}{r_{outa}C_{par}} \\
\omega_z &= \frac{1}{R_{ESR}C_o}
\end{aligned} \tag{7.43}
$$

We obtain identical expressions for ω_{p2} and ω_z as for the PMOS case in Eqn. (7.37). However, the DC gain T_o misses the intrinsic gain factor $g_m r_{ds}$ because the power transistor in Fig. 7.12 is configured

Figure 7.12 Equivalent circuit for the frequency behavior of the NMOS-type linear regulator.

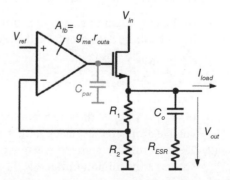

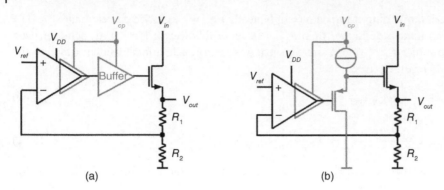

Figure 7.13 Adding a buffer stage in the NMOS-type LDO allows connecting the charge pump: a) concept; b) using a source-follower.

as a source follower. In contrast, it forms a common-source gain stage in the PMOS type LDO, contributing a gain of $g_m r_{ds}$. For the same reason, the pole is now determined by the transconductance g_m of the power transistor and not by its output resistance r_{ds}. Nevertheless, also g_m is load dependent ($g_m \propto I_{load}$). Therefore, the NMOS-type LDO shows similar load dependence as the PMOS type. In conclusion, the same loop stability options as for the PMOS-type LDO in Section 7.5.1 apply.

Also, the NMOS-type LDO benefits from inserting a buffer, as shown Fig. 7.13. It even provides a convenient option to connect the charge pump. The error amplifier's output stage connects to the charge pump as explained in Section 7.4 (see also Fig. 7.7).

7.6 Transient Behavior

In many applications, the load current changes quickly. A microcontroller may change from standby to full load leading to a steep increase of its supply current by ΔI_{load} with a rise time of a few nanoseconds. The LDO cannot respond instantaneously to a fast load transient because there is some inherent delay due to the finite speed of the control loop.

7.6.1 Voltage Under- and Over-Shoot

The major mechanisms determining the transient response are shown in Fig. 7.14. Due to the complex dependencies and influence of parasitics like the equivalent series resistance (ESR) of the output capacitor, approximations are used to estimate the voltage under and over-shoot and the corresponding settling times. We will discuss the transient behavior step-by-step based on Fig. 7.14:

(1) In the very beginning, only the output capacitor C_o supplies the excessive current ΔI_{load}. The output voltage decreases from its steady-state value $V_{out,nom}$ by a constant rate,

$$\frac{\Delta V_{out}}{\Delta t} = \frac{\Delta I_{load}}{C_o}. \tag{7.44}$$

The equivalent series resistance R_{ESR} of the output capacitor adds an initial instantaneous drop of $(R_{ESR} \cdot \Delta I_{load})$ at the very beginning. This drop may be negligible (see Example 7.6). The ESR voltage drop remains steady until the minimum value $V_{out,min}$ is reached.

(2) The control loop starts to react by gradually increasing the drain current of the power transistor. At the minimum value $V_{out,min}$ of the output voltage, the drain current equals the new

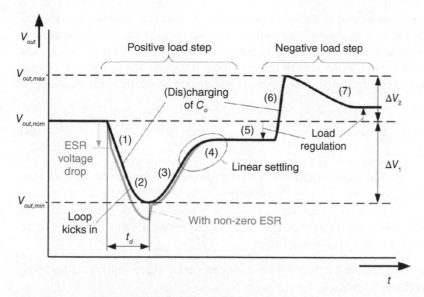

Figure 7.14 LDO transient behavior for a positive and a negative load step. The gray curve includes the equivalent series resistance (ESR) of the output capacitor C_o.

value of the load current. The output capacitor stops discharging. If we consider the ESR, the voltage drop goes to zero again. Figure 7.14 indicates that the ESR may significantly influence the voltage droop but is not as critical for the overall settling delay. For a worst-case estimation of the voltage droop $\Delta V_1 = (V_{out,nom} - V_{out,min})$, we can assume that the slope from Eqn. (7.44) goes on during the entire response delay t_d,

$$\Delta V_1 = \frac{\Delta I_{load}}{C_o} t_d. \tag{7.45}$$

Since the response time t_d of the LDO depends on various parameters, it is often based on simulation or prior knowledge. It can be estimated by summing up the closed-loop bandwidth (BW) of the LDO and the slew rate at the error amplifier output, $SR = I_{bias}/C_{outa}$. Inserting a buffer as shown in Fig. 7.11 improves the slew rate because C_{outa} is formed by the buffer input capacitance C'_{par}, which is much smaller than the gate capacitance C''_{par}. We can assume that the load current step ΔI_{load} requires the gate voltage to be adjusted by ΔV_{gate} ($\approx \Delta I_{load}/g_m$).

$$t_d = \frac{1}{BW} + C'_{par}\frac{\Delta V_{gate}}{I_{bias}} = \frac{1}{BW} + \frac{\Delta V_{gate}}{SR} \tag{7.46}$$

The closed-loop bandwidth BW can be extracted from a simulation. It can be considered to be inverse proportionally to C_o.

(3) The power transistor's current exceeds the load current, bringing up the output voltage again.
(4) Linear settling of the control loop moves V_{out} back to its final value. The transient behavior is determined by the phase margin of the LDO's loop gain. There may even be ringing if the system is under-damped.
(5) The final value of V_{out} is determined by the load regulation of the LDO. It is a DC parameter, introduced in Section 7.3. As we can consider the LDO output as a voltage source with finite source resistance, a positive load step results in a value of V_{out}, which is lower than the original steady-state value $V_{out,nom}$.

(6) A negative load step shows similar behavior, except that the standard LDO cannot actively pull down its output. In the very beginning, V_{out} moves up similar to (1) by a slope as defined in Eqn. (7.44). The response delay t_d for the negative load step is determined by Eqn. (7.46), except that the closed-loop bandwidth (BW) of the LDO is different because the initial load current (before the load step) is relatively high. Therefore, the bandwidth is usually larger, resulting in a shorter delay t_d.

$$\Delta V_2 = \frac{\Delta I_{load}}{C_o} \left(\frac{1}{BW} + \frac{\Delta V_{gate}}{SR} \right) \tag{7.47}$$

(7) The loop kicks in, and the power transistor is now controlled such that the load current discharges C_o until V_{out} reaches its final value determined by the load regulation. A smaller over-shoot ΔV_2 can be achieved by adding a pull-down power transistor controlled by the error amplifier and a buffer. The flipped voltage follower LDO as described in Section 7.11 (see Fig. 7.21), also provides a strong pull-down path.

As a general conclusion, the LDO will have a different response delay depending on the load step direction, resulting in different deviations during under and over-shoot.

Example 7.6 *Estimate the response delay as well as the voltage under- and over-shoot for a positive and a negative load step of $\Delta I_{load} = 100$ mA. Check if the LDO keeps the maximum output voltage deviation within $\pm 10\%$ for $V_{out,nom} = 3.3$ V for two cases: $R_{ESR} = 0$ and $R_{ESR} = 50$ mΩ. Parameters: $C_o = 10$ μF, $BW = 100$ kHz/200 kHz (low/high load), $C'_{par} = 70$ pF, $SR = 0.14$ V/μs, $\Delta V_{gate} = 2$ V.*
We first assume $R_{ESR} = 0$. From Eqn. (7.46):

$$t_{d1} = \frac{1}{BW} + \frac{\Delta V_{gate}}{SR} = \frac{1}{100\,\text{kHz}} + \frac{2\,\text{V}}{0.14\,\text{V/μs}} = 10\,\text{μs} + 14\,\text{μs} = 24\,\text{μs} \tag{7.48}$$

With the high load bandwidth of $BW = 200$ kHz, the response delay for the negative load step calculates to $t_{d2} = 19$ μs. Knowing the LDO response times $t_{d1,2}$, we can estimate the under- and over-shoot according to Eqns. (7.45) and (7.47),

$$\Delta V_1 = \frac{\Delta I_{load}}{C_o} t_{d1} = \frac{100\,\text{mA}}{10\,\text{μF}} 24\,\text{μs} = 240\,\text{mV}, \tag{7.49}$$

$$\Delta V_2 = \frac{\Delta I_{load}}{C_o} t_{d2} = \frac{100\,\text{mA}}{10\,\text{μF}} 19\,\text{μs} = 190\,\text{mV}. \tag{7.50}$$

With respect to $V_{out,nom} = 3.3$ V the $\pm 10\%$-limit corresponds to $\Delta V_1 = \Delta V_2 = 330$ mV. Hence, the LDO ensures this limit for both-step directions.
What if the output capacitor C_o has non-zero ESR? For $R_{ESR} = 50$ mΩ we need to add the voltage drop of $(R_{ESR} \cdot \Delta I_{load}) = 5$ mV across R_{ESR} to the under- and over-shoot values, resulting in $V_1 = 245$ mV and $V_2 = 195$ mV. The ESR influence is negligible in this case, and the deviation of V_{out} is still within the $\pm 10\%$ specification.

Inspecting the equations for the voltage under and over-shoot and for the response time, we observe that the output capacitor C_o plays a significant role in the transient response. While large C_o reduces the discharge rate at the beginning of a load step according to Eqn. (7.44) and the overall voltage droop V_1 as shown in Eqn. (7.45), it slows down the response because the loop bandwidth BW is inverse proportionally to C_o. Hence, the LDO will take longer to recover from a load step if C_o increases. Let us investigate this with another example.

Example 7.7 *Recalculate the values for t_d and V_1 from Example 7.6 if C_o increases from 10 to 22µF ($R_{ESR} = 0$).*

If we assume that the bandwidth scales indirectly proportional to C_o, the new closed-loop bandwidth will be $BW = 100\,\text{kHz} \cdot (10\,\mu\text{F}/22\,\mu\text{F}) = 45.5\,\text{kHz}$. Inserting into Eqn. (7.46) gives

$$t_d = 42\,\mu\text{s},$$

which allows us to calculate the voltage droop from Eqn. (7.45),

$$V_1 = 91\,\text{mV}.$$

This example confirms that larger C_o reduces the voltage under-shoot but slows down the settling time.

7.6.2 Fast Transient Techniques

In addition to the primary regulation loop, a second feedback loop can be added, enabling the LDO to respond faster to large-output transients. Figure 7.15 shows the block diagram of an LDO with a fast-transient loop. This loop bypasses the error amplifier partly or entirely and may even drive the power transistor directly. The fast path is a negative feedback loop with very high bandwidth because bandwidth determines the system's settling time. In contrast, the main loop determines the overall gain-bandwidth product and maintains a stable output voltage in the steady state. Inspecting the response delay in Eqn. (7.46), we notice that both the slew rate (*SR*) of the error amplifier and the overall loop bandwidth significantly influence the transient response.

7.6.3 Slew Rate Enhancement

According to Eqn. (7.29), more bias current I_{bias} improves the slew rate. However, since it contributes to the quiescent current of the entire LDO, larger I_{bias} will significantly reduce the current efficiency, as shown in Eqn. (7.13). Adaptive biasing is a common solution, as depicted in Fig. 7.16a). The bias current adaptively increases with the load current of the LDO. The circuit uses a sense transistor MS, which can be one finger of the main power transistor MP. This scenario is similar to the replica current sensing in Section 6.6 (Figs. 6.12 and 6.13) but without closed-loop feedback. M1 and M2 keep the source potential of MS and MP equal (i.e., the output voltage),

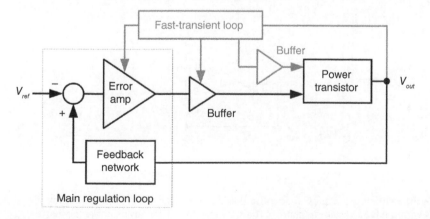

Figure 7.15 LDO with a secondary fast-transient loop to enhance the transient response.

while the bias current I_{bias1} of M1 is considered negligible compared to the load current. This way, the current in MS equals I_{load}/M. M3 mirrors this current to M4 and forms the adaptive bias current I_{bias3}, which adds to the initial constant bias I_{bias2}. The sum of I_{bias2} and I_{bias3} can be the bias current of the error amplifier. Another current mirror output branch can be added to adaptively supply the buffer reducing its output resistance (indicated by the dashed line in Fig. 7.16a)).

7.6.4 Loop Bandwidth

High-bandwidth loops are introduced for the capless LDO and the flipped voltage follower LDO in Sections 7.10 and 7.11, respectively. They inherently provide good transient characteristics. The Miller compensation in capless LDOs is based on a feedback capacitor, indicated by C_1 in Fig. 7.16b). Assuming a positive load step, the drop in V_{out} will couple almost instantaneously to the gate of the power device MP. It can be considered a fast-transient loop according to Fig. 7.15 that counteracts any load step by providing more or less current to the load before the main loop kicks in. Similarly, a bypass capacitor C_2 feeds the output variation directly to the error amplifier without dividing it by the ratio α set by R_1 and R_2 for the DC accuracy.

7.7 Noise in Linear Regulators

Every component of the LDO contributes noise, as shown in Fig. 7.17. Understanding the noise mechanisms in the LDO is essential to achieve a clean output voltage with minimum noise. The noise is usually a combination of thermal (white) noise and flicker noise ($1/f$-noise) calculated over the bandwidth of interest (the closed-loop bandwidth BW of the LDO). The standard approach for a noise analysis is to calculate the input-referred noise by dividing each noise source by the corresponding gain toward the input (open loop). In the LDO, the input usually is defined as the input of the error amplifier. The noise sources are uncorrelated and must be added by the root mean square. This way, the input-referred noise of the LDO is

$$V_{n,in}^2 = V_{nref}^2 + V_{na}^2 + \frac{V_{np}^2}{A_{fb}^2} + \left(\frac{R_1}{R_2}\right)^2 V_{nr1}^2 + V_{nr2}^2 \qquad (7.51)$$

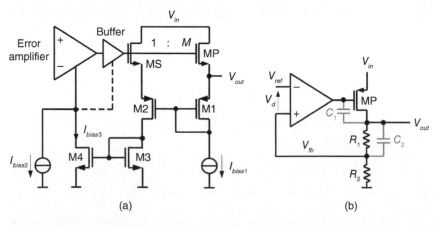

(a) (b)

Figure 7.16 Transient enhancement techniques: a) adaptive biasing: the bias current of the error amplifier and the buffer increases with the load current to improve the slew rate for a faster transient response. b) inserting coupling capacitors C_1 and C_2.

with the reference voltage noise V_{nref}^2, the input-referred noise of the error amplifier V_{na}^2, the noise of the power transistor V_{np}^2, and the noise contributed by both feedback resistors V_{nr1}^2 and V_{nr2}^2. Since the noise of the power device MP is divided by the gain of the error amplifier, its noise contribution is usually negligible. R_1 and R_2 contribute thermal noise according to

$$V_{nr}^2 = 4kTRBW \qquad (7.52)$$

with Boltzmann's constant k and temperature T. Lower resistor values lead to lower noise at the expense of area and current consumption.

Often the contributions from MP, R_1 and R_2 can be neglected and Eqn. (7.51) simplifies to

$$V_{n,in}^2 = V_{nref}^2 + V_{na}^2. \qquad (7.53)$$

Moreover, usually, the error amplifier can be optimized for low noise, and the voltage reference circuit is the primary noise source of the LDO.

An R-C low-pass filter can be inserted between the bandgap output and the error amplifier input as indicated in Fig. 7.17. The cut-off frequency of this filter f_o can be as low as possible to filter out most of the noise coming from the reference:

$$V_{nref,LP}^2 = \frac{V_{nref}^2}{1 + (f/f_o)^2} = \frac{V_{nref}^2}{1 + (2\pi R_{LP}C_{LP}f)^2} \qquad (7.54)$$

The bandwidth of the reference can be set to a few hertz. Assuming a constant spectral noise density v_n^2 (white noise and thermal noise), the root-mean-square (RMS) value of the reference noise is $V_{nref,LP}^2 = v_n^2 f_o$. If the RMS value V_{nref}^2 is known at full bandwidth (BW) the filtered noise can be derived from

$$V_{nref,LP}^2 = V_{nref}^2 \frac{f_o}{BW}. \qquad (7.55)$$

Example 7.8 *A capacitor $C_{LP} = 100$ nF is placed externally. What is the value of R_{LP} to achieve a bandwidth of $f_o = 1$ Hz?*

$$f_o = \frac{1}{2\pi R_{LP}C_{LP}} \rightarrow R_{LP} = \frac{1}{2\pi C_{LP}f_o} = \frac{1}{2\pi \cdot 100 \text{ nF/s}} = 1.59 \text{ M}\Omega \qquad (7.56)$$

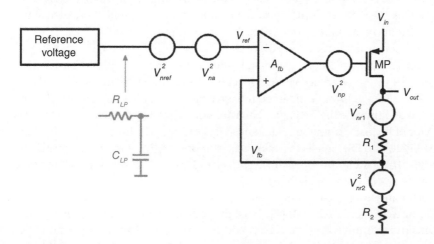

Figure 7.17 LDO with equivalent noise sources.

Finally, we obtain the output-referred noise by multiplying the input-referred noise by the closed-loop gain of the LDO. The closed-loop gain is simply the reciprocal $1/\alpha$ of the feedback resistor ratio as described by Eqn. (7.1). As this is only a factor, the input-referred noise is equally meaningful. However, the scaling by $1/\alpha = V_{out}/V_{ref}$ implies that the noise increases as V_{out} gets larger with respect to the reference voltage V_{ref}.

Example 7.9 *We assume that the LDO noise is fully described by Eqn. (7.53) and that the other noise sources are negligible. For a bandwidth BW = 100 kHz the noise is given by $V_{nref} = 500\,\mu V_{rms}$ and $V_{na} = 1.5\,\mu V_{rms}$. Determine the input referred LDO noise for the full bandwidth BW and the case that the reference noise is filtered with $f_o = 1$ Hz like in Example 7.8.*

At full bandwidth, the amplifier noise is much lower than the reference noise, hence,

$$V_{n,in} \sim V_{nref} = 500\,\mu V_{rms}. \tag{7.57}$$

According to Eqn. (7.55), at $f_o = 1$ Hz the reference noise reduces to

$$V_{nref,LP} = V_{nref}\sqrt{\frac{f_o}{BW}} = 500\,\mu V_{rms}\sqrt{\frac{1\,Hz}{100\,kHz}} = 1.58\,\mu V_{rms}. \tag{7.58}$$

The filtered reference noise $V_{nref,LP}$ gets small, and the error amplifier's noise contribution V_{na} cannot be neglected anymore. The total input referred noise at $f_o = 1$ Hz is now

$$V_{n,in} = \sqrt{V_{nref}^2 + V_{na}^2} = \sqrt{(1.58\,\mu V_{rms})^2 + (1.5\,\mu V_{rms})^2} = 2.18\,\mu V_{rms}. \tag{7.59}$$

7.8 Power Supply Rejection

The power supply rejection (PSR) is similar to the line regulation discussed in Section 7.3. However, while the line regulation is a DC parameter, PSR measures the voltage ripple at the LDO output with respect to a ripple at V_{in} over a wide frequency range. PSR is a measure of how well the LDO can suppress ripple at various frequencies injected from its input:

$$PSR = 20\,dB\ \log\left|\frac{1}{A_p}\right| = 20\,dB\ \log\left|\frac{v_{in}}{v_{out}}\right| \tag{7.60}$$

A_p is the power supply gain (see Eqn. (7.31)). In a small-signal approach, v_{in} and v_{out} represent the (ripple) amplitudes at the corresponding terminals. The expression in Eqn. (7.60) is sometimes called the power supply rejection ratio (PSRR). However, in amplifier theory, PSRR is related to the ratio of the signal gain (from signal input to output) to the power supply gain A_p.

Various superimposed effects determine the PSR of the LDO, mainly (1) the PSR capability of the LDO regulation loop and (2) the PSR of the error amplifier in conjunction with the power stage itself. In addition, the PSR of the reference voltage generator has a direct influence but will not be further discussed here. Item (1) can be obtained if we repeat the small-signal analysis for the line regulation (Fig. 7.6) and include the frequency behavior determined by parasitic capacitances and by the AC behavior of the error amplifier. Equations (7.20) and (7.24) indicate that high loop gain is crucial for good PSR. The PSR equals the open-loop gain until ω_{p1}. Therefore, a typical LDO can have as much as 80 dB of PSR at low frequencies (below 1 kHz) while it starts to roll off at the dominant pole frequency of a few tens of kilohertz.

The impact of the error amplifier's PSR (item (2) above) can be studied by comparing the p-type and the n-type LDO (for example, in Fig. 7.2) [4]. Since the source terminal of the PMOS power transistor is highly sensitive to input voltage noise, it would be beneficial to have the error amplifier also sensitive such that the gate of MP is tightly coupled to V_{in}. If both the source and the gate experience the same power supply noise, the overall PSR of the power transistor and the error amplifier will be maximized. As an example, the error amplifier of Fig. 7.7 can be used because it results in $PSR = 0$ (see Eqn. (7.33)). In contrast, the n-type power transistor with its drain at V_{in} would provide the best PSR if its gate has no coupling to V_{in}. In that case, the error amplifier needs to have good PSR, which is the case for the complementary version of the amplifier in Fig. 7.7 with PMOS differential pair, see Eqn. (7.34). As a rule of thumb, a p-type LDO should have an error amplifier with a low PSR, while an n-type LDO benefits from an error amplifier with a high PSR. This observation is in accordance with the findings in [4].

7.9 Soft-Start

If the input voltage is applied, the LDO will start up. Some LDOs even have a dedicated enable signal to initiate their start-up. Usually, the output buffer capacitor C_o is fully discharged (by the load) and represents a low-resistive load ($\sim 1/j\omega C_o$). In consequence, the feedback input of the error amplifier will be at zero level, corresponding to a maximum error voltage $|V_{ref} - V_{fb}|$. This way, the power transistor gets turned on with as much driving capability as possible providing maximum current to the load. Known as inrush current, this current charges C_o until V_{out} reaches the target value. The output voltage sees a steep transient dV_{out}/dt, which can be in the order of V/ns, Fig. 7.18. This steep transition and the inrush current are not desired in any application. The reasons include (1) component stress, in particular, at the power transistor and at C_o, which causes degradation and accelerated aging, and (2) parasitic coupling effects, which may lead to ringing at V_{out} and internal nodes.

These effects can be circumvented by adding a soft-start function to the LDO that reduces the transition slope to typically 1–10 V/ms. The circuit in Fig. 7.19a) provides a soft-start function in a very effective way by modifying the error amplifier (see also Fig. 7.7). The soft-start circuit consists of a capacitor C_S charged by a current source I_S. As long as the LDO is off, its inverted enable signal $\overline{EN}$ is 1. The pull-down transistor ME keeps the voltage V_S across C_S at zero. At start-up, $\overline{EN}$ changes to 0 and deactivates ME. With its gate initially at zero, transistor MS bypasses transistor M1 in the input differential stage. This way, the voltage ramp across C_S defines the reference voltage instead of V_{ref}. The slope of V_S can be set by I_S. The control loop will maintain V_{out} to follow V_S proportionally, i.e., $V_{out} = V_S/\alpha$ (see also Eqn. (7.1)):

$$\frac{dV_{out}}{dt} = \frac{I_S}{\alpha C_S} \tag{7.61}$$

For slopes in the order of V/ns, the current I_S will be in the nA-range. Larger currents can be used at the expense of larger capacitors C_S. The global biasing in power management ICs often operates with currents in the order of microamps. Figure 7.19b) shows a cascode current mirror that converts a μA current into a nA output current. The current mirror ratio in this example is 100:1 because the

Figure 7.18 V_{out}-transition of the LDO without and with soft-start.

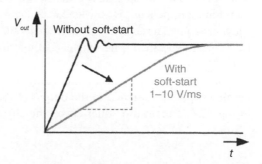

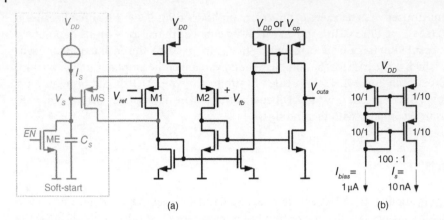

Figure 7.19 Soft-start: a) error amplifier with soft-start function; b) nA-current generation.

input transistors have a 10 times larger width while the output devices are 10 times longer. Such a design will not provide good matching. However, the soft-start requirements must be considered a protecting function, usually not specified with tight limits. It is usually acceptable to bring the slope at V_{out} down to a few volts per millisecond.

Example 7.10 *Which current I_S is required to achieve a soft-start slope of $dV_{out}/dt = 2\,V/ms$ if $C_S = 2\,pF$ and $\alpha = 2.5$?*

From Eqn. (7.61),

$$I_S = \frac{dV_{out}}{dt}\,\alpha\,C_S = 2\frac{V}{ms} \cdot 2.5 \cdot 2\,pF = 10\,nA. \tag{7.62}$$

The current mirror in Fig. 7.19b) could generate this low current from a 1 µA input.

7.10 Capacitor-Less LDO

This LDO class, also called capless LDO, does not have an external output buffer capacitor C_o. Removing the large output capacitor reduces the printed-circuit board (PCB) space and the overall cost. It makes the LDO suitable for highly integrated SoC designs. Most capless LDOs still comprise an internal (on-chip) output capacitor in addition to parasitic capacitance, which is mainly due to the power transistor. This way, C_o is in the range of a few hundred picofarads. With such small values of C_o, the pole ω_{p1} (see Eqn. (7.37)) at the LDO output becomes non-dominant. Instead, the dominant pole occurs at the error amplifier output. Consequently, the loop bandwidth increases to react much faster to any load transients, which takes over the buffering function instead of C_o. The non-dominant pole ω_{p1} is now load dependent. Therefore, a capless LDO is prone to instability at a low load when a minimum distance separates the two poles. The LDO has a required minimum load current specified. Alternatively, an internal current source can provide a minimum current from V_{out} to ground.

Capless LDOs usually install multi-stage error amplifiers to achieve high gain bandwidth. The dominant pole is assigned to the output of the first stage using Miller compensation as shown in Fig. 7.20. Due to the Miller effect, the effective capacitance C_1 at that node is approximately

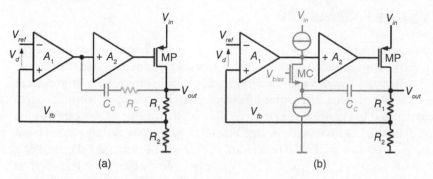

Figure 7.20 Capacitor-less LDO: a) multi-stage feedback with Miller compensation; b) inserting a cascode stage to eliminate the RHPZ.

formed by C_c multiplied by the voltage gain between the left and right connection of C_c. For $R_c = 0$, Fig. 7.20a) gives

$$C_1 = A_2 g_m r_{ds} C_c \text{ and} \tag{7.63}$$

$$\omega_{p1} = \frac{1}{r_{outa} A_2 g_m r_{ds} C_c}. \tag{7.64}$$

The expression for the dominant pole ω_{p1} includes the error amplifier output impedance r_{outa} as defined in Fig. 7.7 (Section 7.4). The power transistor accounts for g_m and r_{ds}.

As a major drawback of Miller compensation, a right half-plane zero (RHPZ) occurs. The denominator of the loop gain expression (i.e., the forward path transfer function v_{out}/v_d, see Fig. 7.20a)) contains the expression $(A_2 g_m - sC_c)$, which yields

$$\omega_z = \frac{A_2 g_m}{C_c} \tag{7.65}$$

for the RHPZ. A common technique to eliminate the RHPZ is to insert a series resistor R_c as indicated in Fig. 7.20a). This way,

$$\omega_z = \frac{A_2 g_m - 1/R_c}{C_c}. \tag{7.66}$$

The RHPZ cancels for $R_c = 1/(A_2 g_m)$. The zero can also be moved to the left half-plane to benefit from the positive phase shift. However, this technique has several disadvantages: (1) g_m varies depending on the load current, and (2) the compensation does not track process and temperature variations very well. Even if R_c is implemented with a transistor of the identical type as the power transistor, the load dependency remains. Figure 7.20b) shows a Miller compensation technique, also known as Ahuja compensation according to [5], that inserts a cascode stage to eliminate the RHPZ. Assuming a capless LDO in which its output capacitor C_o is lower than the resulting Miller-C, the dominant pole remains approximately identical to the expression in Eqn. (7.64). The non-dominant pole occurs at the output of the LDO:

$$\omega_{p2} = \frac{A_2 g_m}{C_c} \tag{7.67}$$

By inspecting Eqns. (7.64) and (7.67), we can conclude that both poles are widely spaced at high load while the minimum phase margin occurs at low load. This observation matches the conclusion above at the beginning of this section.

7.11 Flipped Voltage Follower LDO

The flipped voltage follower (FVF), as shown in Fig. 7.21 achieves low output impedance of the LDO such that it can respond quickly to changes in load current and maintains good steady-state load regulation [6–8]. MP is the power transistor in Fig. 7.21a). Transistor MF and the current source I_{bias} form the FVF. MF operates as a source follower, biased by a "flipped" current source I_{bias}. This difference compared to a conventional source follower (like in Fig. 7.13b)) explains the name "flipped voltage follower." The reference voltage V_{ref} at the gate of MF defines the set point of the LDO's output, $V_{out} = V_{ref} + V_{SG,MF}$. The generation of V_{ref} will be discussed shortly. For a positive load step, V_{out} will drop and, consequently, $V_{SG,MF}$. It pulls down the gate voltage V_{gate} of the power transistor to source a higher current to the load. The operation of this negative feedback loop counteracts the load step and stabilizes V_{out} at its target level $V_{ref} + V_{SG,MF}$. There is, however, a finite DC output resistance r_{out}. With the transconductance parameters g_m and g_{mf} of MP and MF, respectively, and the small-signal output resistances r_{dsb} and r_{dsf} of I_{bias} and MF, r_{out} can be derived as follows:

$$v_{gate} = \left(g_{mf}v_{out} + \frac{v_{out}}{r_{dsf} + r_{dsb}} \right) r_{dsb} \approx g_{mf}r_{sdb}v_{out} \tag{7.68}$$

$$i_{load} = -g_m v_{gate} - g_{mf}v_{out} - \frac{v_{out}}{r_{dsf} + r_{dsb}}$$

$$= -\left(g_m g_{mf}r_{sdb} + g_{mf} + \frac{1}{r_{dsf} + r_{dsb}} \right) v_{out} \approx -g_m g_{mf}r_{sdb}v_{out} \tag{7.69}$$

$$r_{out} = \frac{v_{out}}{-i_{load}} = \frac{1}{g_m g_{mf}r_{sdb}} \tag{7.70}$$

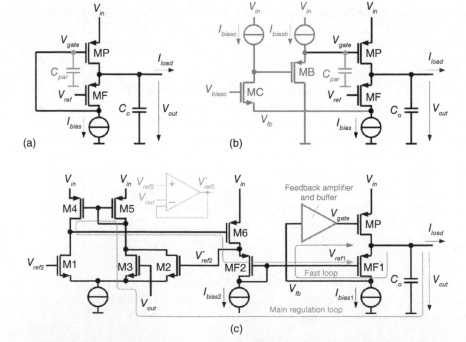

(a)   (b)   (c)

Figure 7.21 LDO with flipped voltage follower (FVF): a) concept; b) with increased loop gain by inserting a folded-cascode gain stage and a source-follower; c) multi-loop FVF LDO with reference generation.

The minus sign at i_{load} in Eqn. (7.70) is due to the definition of i_{load} (Fig. 7.21a)). The output resistance r_{out} is much lower than $1/g_{mf}$ if MF was used as a conventional source follower. The output resistance is reduced by the DC loop gain of the FVF LDO, which can be approximated by

$$T_o = g_m r_{dsb}. \tag{7.71}$$

Hence,

$$r_{out} = \frac{1}{T_o g_{mf}}. \tag{7.72}$$

The DC loop gain is in the order ~100, determined by the intrinsic gain and even higher due to the large W/L-ratio of MP (large g_m). The output resistance is inversely proportional to the loop gain T_o. We note that the loop gain and the output resistance vary with the load (because $g_m \propto I_{load}$). Hence, r_{out} reaches its minimum at maximum load.

The poles can be approximated by

$$\omega_{p1} = \frac{1}{r_{dsb} C_{par}}, \tag{7.73}$$

$$\omega_{p2} = \frac{g_{mf}}{C_o}. \tag{7.74}$$

Most FVF LDOs are designed as capless regulators with the dominant pole ω_{p1} at the gate of the power transistor. Nevertheless, also conventional output-pole dominant designs are possible. As stated for the capless LDO in Section 7.10, C_o is only a small on-chip capacitance in the order of 100 pF. It defines the non-dominant pole ω_{p2}. To ensure that the poles are sufficiently separated, a compensation capacitance can be placed in parallel to I_{bias} (which adds to C_{par} and shifts ω_{p1} to lower values). In first order, the poles ω_{p1} and ω_{p2} show no load dependence. However, at larger load currents, the equivalent load resistance has to be considered in parallel to r_{dsb} and $1/g_{mf}$, respectively. This way, stability is critical when the poles move together at high loads.

Capless LDOs require a control loop with high gain-bandwidth. The limited loop gain of the FVF LDO, expressed by Eqn. (7.71), can be improved by adding a folded cascode gain stage as shown in Fig. 7.21b) [8]. In addition, also a buffer stage is inserted that can strongly drive the large parasitic gate capacitance C_{par} of MP. Similar to the concept in Fig. 7.11, the pole associated with the gate of MP moves to higher frequencies. A third pole occurs at the output of the folded cascode stage, also at high frequencies. The concept of Fig. 7.21b) increases the gain-bandwidth and enables MP to be turned off at light load. The behavior is not possible for the configuration in Fig. 7.21a).

The concept of the FVF LDO considered so far does not lead to good DC accuracy and load regulation because any variation of $V_{SG,MF}$ is of significant influence. Figure 7.21c) shows a multi-loop approach of a FVF LDO, which comprises a dedicated control loop (main loop) that improves the DC accuracy [9]. The fast loop remains as in Fig. 7.21b) consisting of a FVF. This fast loop may contain the same buffer as shown in Fig. 7.21b) but any other buffer like a basic source follower according to Fig. 7.11 or the super-follower of Fig. 7.11d) will be suitable.

Transistors M1–M6 form a voltage follower that generates a buffered replica of the reference voltage V_{ref2}, the set value for V_{out}. This way, the gate voltage of MF1 and MF2 gets $V_{ref1} = V_{ref2} - V_{SG,MF2}$. One side of the input differential pair of the voltage follower is split into two transistors, M2 and M3 (see also the inset with the amplifier symbol in Fig. 7.21c)).

M2 forms an inner loop that incorporates the negative feedback for the voltage follower operation to provide V'_{ref2}. The outer main loop is formed by M3, which connects to the LDO's output voltage V_{out}. The outer loop feedback is stronger than the inner loop feedback by weighting the W/L-ratios of M2 and M3. Lu et al. [9] suggests M1 to M3 to scale in a ratio 4:1:3. In steady-state, V'_{ref2} equals V_{ref2}. However, during a transient, if V_{out} drops below the reference V_{ref2}, V'_{ref2} increases such that

V_{gate} pulls down and increases the driving strength of the power transistor MP as to restore the output voltage to V_{ref2}.

The main loop bandwidth is slower than the fast FVF loop, such that the slow loop achieves excellent DC output regulation, while improved transient response is enabled by the fast high-frequency FVF loop. For this reason, the FVF concept belongs to the transient enhancement techniques described in Section 7.6.2.

7.12 The Shunt Regulator

The supply voltage generation can often be simple for non-critical blocks. Therefore, no LDO with closed-loop control is required. If we keep the gate of an NMOS power transistor at a constant level, we obtain the shunt regulator as shown in Fig. 7.22. Due to the missing control loop, this circuit is no longer a linear *tegulator*. Note that the name is commonly used, but the conventional shunt regulator assumes the power transistor to be placed in parallel to the load. Nevertheless, we will stick to the term shunt regulator referring to Fig. 7.22.

The basic circuit in Fig. 7.22a) sets the gate voltage of MP by a Zener diode and, concurrently, we get $V_{out} \approx V_Z - V_{GS}$. The bias current I_{bias} can be in the order of a few microamperes. Without a precise regulation loop, V_{GS} will vary with the load current. If the W/L-ratio of MP is sufficiently high, the transconductance of MP will be large such that V_{out} stays within an acceptable range in the order of a few hundred millivolts. Besides its simple structure, as a key advantage, the shunt regulator is an open loop circuit, which counteracts load changes immediately. Due to the transconductance of MP, there is still negative feedback present.

In case of a positive load step, the drop in V_{out} may couple to the gate of MP via its gate–source capacitance. Therefore, a buffer capacitor C_B helps stabilize the reference level at the gate. The modified circuit in Fig. 7.22b) cancels V_{GS} in the first order such that V_{out} is approximately identical to V_Z. The additional biasing transistor MB should preferably be the same type as MP, but it will have a much smaller W/L-ratio. As the component set in typical technologies provides only a limited number of Zener voltages, the Zener diode may be complemented by MOS transistors in diode configuration or by PN-diodes to achieve the target output voltage. In particular, if V_{out} is below ~ 2 V, a diode-connected n-type transistor is a suitable replacement for the Zener diode.

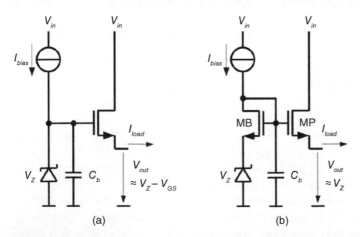

(a)   (b)

Figure 7.22   a) Basic shunt regulator circuit and b) modified version with $V_{out} \approx V_Z$.

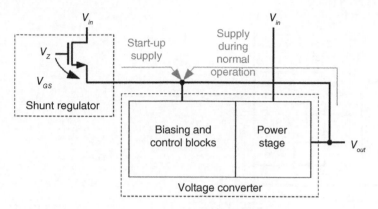

Figure 7.23 The shunt regulator as a start-up supply.

The shunt regulator can be applied as a coarse supply for non-critical analog and digital blocks. It can also act as a pre-regulator to an LDO, for instance, to design the LDO with devices of lower voltage ratings.

The Shunt Regulator as a Start-Up Supply

The shunt regulator makes an excellent initial bias for any voltage converter. It supplies all circuit parts required during start-up. Once the regulator works, its output voltage will take over the role of the shunt regulator and supply all blocks, including the ones in the start-up. This concept is shown in Fig. 7.23. In the case of a switched-mode converter, its output voltage is generated much more power efficiently. This way, the overall power efficiency will benefit, while the shunt regulator, with its low power efficiency, is usually acceptable as a coarse supply at start-up. According to Fig. 7.23, the implementation would require the final output voltage V_{out} to be slightly higher than the target output level of the shunt regulator. Hence, the power transistor of the shunt regulator will turn off after the main voltage converter is fully operational. As a rule of thumb, the difference between both output voltages should be a few hundred millivolts. It must be guaranteed over corners to avoid excessive leakage through the shunt regulator. If a power-on reset or power-good signal is available, the shunt regulator can be turned off after power-up by pulling its gate low.

7.13 Digital LDOs

A digital low-dropout regulator (DLDO) replaces the *analog* pass device MP of the fundamental LDO according to Fig. 7.1 by an array of transistors that get activated by a *digital* controller. The required resistance between V_{in} and V_{out} is achieved by controlling the number of parallel transistor elements. Figure 7.24 shows the block diagram of a DLDO. For comparison, the conventional analog LDO is called ALDO.

With the implementation of control in the digital domain, DLDOs have become popular as a power supply for digital loads in advanced CMOS technologies. A good overview can be found in [10–13].

Despite the inherently lower efficiency, DLDOs are attractive as a fully-integrated voltage regulator because they do not need large passive components (inductors, capacitors) in the power path.

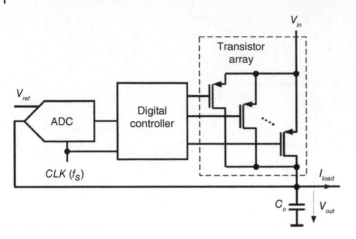

Figure 7.24 Block diagram of a digital linear regulator (DLDO).

For this reason, DLDOs benefit from technology scaling toward advanced CMOS process nodes and are suitable for low supply voltages (see Fig. 1.13). They can be made fully synthesizable and, thus, are compatible with the digital design flow and provide technology portability.

The key parameters are the transient response to dynamic load steps (droop), PSR, and output voltage ripple.

The block diagram of the DLDO in Fig. 7.24 consists of three main blocks:

(1) an analog-to-digital converter (ADC) used as a quantizer of the error voltage
(2) a digital controller
(3) a transistor array

In addition, an output buffer capacitor C_o is used like in the analog LDO. C_o must be large enough to instantly supply the current during fast load transients to keep the droop low.

There is also a trend toward hybridization combining DLDO and ALDO techniques to achieve small output ripple and fast transient response [11].

7.13.1 The Transistor Array

The transistors can be uniformly sized or binary-weighted. This way, the transistor array works as a digital-to-analog converter (DAC), converting the control word into resistance between V_{in} and V_{out}. The most common implementation uses digital PMOS transistors like in Fig. 7.24. The digital controller can directly drive them so that the source–gate voltage changes between 0 and the (digital) supply V_{in} without needing a level shifter. Unlike in the ALDO, the transistors always operate as switches in the triode region. Since they do not need to be biased in the saturation region, their W/L-ratio gets relatively small, thus, saving layout area. As a disadvantage of PMOS transistors, their conductance changes with V_{in} (the source connection), and the DLDO becomes sensitive to changes in V_{in}. This effect results in poor PSR (see Section 7.8). Often, the DLDO is used as a sub-regulator after a DC–DC converter (see Section 1.4 and Fig. 1.7), which has some output voltage ripple, and if high PSR is essential. Therefore, also NMOS transistors are used in DLDOs. With their drain connected to V_{in} and operating in the saturation region, they improve the PSR. However, the saturation region requires a larger W/L-ratio to achieve the same voltage drop. In addition, NMOS devices may require a charge pump exactly like the ALDO described in Section 7.1.

7.13.2 The Analog-to-Digital Converter

The ADC quantizes the error between the output voltage V_{out} and the reference voltage V_{ref}, which defines the set point of V_{out}. There is usually no resistive feedback divider like in the ALDO. Instead, V_{ref} can be adjustable (for instance, for dynamic voltage and frequency scaling (DVFS) as explained in Section 1.7.3). The ADC is commonly simply a clocked comparator with a binary output triggered by the sampling clock *CLK*. Fast transient techniques, like the dead zone concept described below, use multiple comparators with different thresholds. Due to inherent positive feedback, the clocked comparator can be high-speed with minimum propagation delay. High-performance DLDO designs use multi-bit ADCs.

7.13.3 The Digital Controller

Digital control converts the quantized error signal of the ADC into a digital output to selectively turn on or off the devices in the transistor array. The controller output has a bit width of N, which correlates to the resolution of the DLDO. It can be encoded in thermometer code (for uniformly sized pass devices), unsigned binary integer formats, or 2's complement. The controller's primary design challenge is finding a trade-off between performance (droop, PSR) and power consumption. Therefore, the controller is often some shift register directly connected to the transistor array. It acts as an integrator. In the digital domain, PID control (proportional–integral–derivative control) can easily be implemented. However, to ensure fast response, it has to run at high clock frequency f_s (*CLK*) (see Fig. 7.24) and comes at the expense of increased power consumption.

7.13.4 Transient Response

In the case of fast load transients, the digital feedback loop may not respond immediately, mainly due to the sampling delay in the quantizer. Hence, V_{out} will deviate from V_{ref}. We will review common control techniques in DLDO in this regard. The quantizer (ADC) can be triggered in a synchronous (time-driven) and an asynchronous (event-driven) way. Synchronous triggering operates at a fixed sampling frequency f_s, changing the controller output by only one bit per clock cycle. Therefore, it takes up to 2^N cycles for V_{out} to settle close to V_{ref}. The settling time is $2^N \cdot 1/f_s$. Therefore, only a high-frequency clock will ensure a fast response.

Alternatively, asynchronous control can provide an ultra-fast response without using the sampling clock. Instead, a local high-speed clock updates the output as long as the ADC (usually only a comparator) detects a deviation of V_{out}. Similarly, adaptive clocking can be applied as shown in Fig. 7.25a). Two comparators form a window comparator with thresholds V_{ref1}, V_{ref2}, below and above V_{ref}. The sampling frequency increases if the window comparator detects that V_{out} exceeds the window (below and above). The underlying P-control quickly returns V_{out} within the window. Once in the window, V_{out} is controlled using I-control (like synchronous triggering). An asynchronous operation can reduce the feedback latency. However, it requires more complex hardware, resulting in more extensive power and area consumption.

Advanced techniques for fast transient response of the DLDO also include feedforward and non-linear control. Feedforward control estimates the transient slope of V_{out} to measure the required load current. They immediately supply the projected amount of charge for V_{out} to recover from the droop. Non-linear control and non-linear search schemes aim to improve the dynamic load regulation performance. Longer computation times may cause latency and counteract the speed advantage.

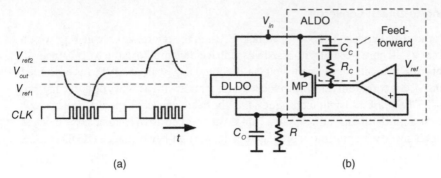

Figure 7.25 DLDO techniques: a) adaptive clocking; b) feedforward power supply noise cancelation.

Similar to the transient response techniques of ALDOs in Section 7.6.2, DLDO may implement multi-loops with different types of feedback depending on the error size. This way, the DLDO can improve dynamic response while reducing power and area overhead.

7.13.5 Power Supply Rejection

Another issue faced by the DLDO is PSR. Especially when power constraints limit the clock frequency and response time, the PSR of the DLDO is inherently worse than the PSR of the ALDO. This is because the coupling from V_{in} to V_{out} between sampling points is determined by a resistive divider formed by the total on-resistance R_{on} of the DLDO pass device and the equivalent load resistance $R = V_{out}/I_{load}$. The PSR can be approximated by

$$PSR = 20 \, \text{dB} \, \log \left(\frac{v_{in}}{v_{out}} \right) = 20 \, \text{dB} \, \log \left(\frac{R + R_{on}}{R} \right) \tag{7.75}$$

where v_{out} and v_{in} are small-signal voltages that represent the ripple (see also Eqn. (7.30)). The PSR ranges theoretically from infinity for $v_{out} = 0$ (no power supply noise at the output) to 0 dB if $v_{out} = v_{in}$ (supply noise couples fully into the output). According to Eqn. (7.75), high PSR can be achieved for $R_{on}/R \to \infty$, which implies that the PSR gets worse at low load (large R).

Example 7.11 *Estimate the PSR for $R_{on} = 1 \, \Omega$, $V_{out} = 1 \, \text{V}$, and $I_{load} = 100 \, \text{mA}$.*
For the given values of V_{out} and I_{load}, the equivalent load resistance is $R = 10 \, \Omega$. Inserting into Eqn. (7.75) yields

$$PSR = 20 \, \text{dB} \log \left(1 + \frac{R_{on}}{R} \right) = 20 \, \text{dB} \log \left(1 + \frac{1 \, \Omega}{10 \, \Omega} \right) = 0.83 \, \text{dB}. \tag{7.76}$$

This poor PSR means that almost the full input supply ripple will be present at V_{out}.

Several techniques for improving the PSR have been proposed [11, 12]. Figure 7.25b) shows an example of a hybrid technique where an ALDO operates in parallel to the DLDO [14]. It implements a feedforward technique, which cancels the supply noise. The series combination of C_c and R_c between the supply and the gate node of the ALDO power transistor MP suppresses the coupling of supply noise by modulating the conductance of MP. If the supply noise causes V_{in} the drop, MP pulls V_{out} up, while the resistive divider R, R_{on} attempts to reduce V_{out}.

7.13.6 Limit Cycle Oscillations

With the quantization introduced through the ADC and the transistor array, acting as a DAC, the DLDO has inherent steady-state limit cycle oscillations (LCO) of the output voltage at frequencies lower than the sampling frequency f_s. They are undesirable as both the frequency and magnitude of the oscillations and their electromagnetic interference (EMI) are difficult to predict. Only if the digital control output forces V_{out} steadily to the zero error bin of the ADC LCO can be eliminated. This requirement is often fulfilled if the DAC has a higher resolution than the ADC. However, more special cases lead to LCO.

7.13.7 Summary

In summary, digital LDOs provide several advantages, such as small die area, process scalability, and low supply voltage, which makes them an attractive power management solution in advanced CMOS technologies. On the downside, fast transient response, PSR, low output voltage ripple, and general accuracy require a power-speed trade-off.

References

1 Wittmann, J., Neidhardt, J., and Wicht, B. (2013) EMC optimized design of linear regulators including a charge pump. *IEEE Transactions on Power Electronics*, 28 (10), 4594–4602, doi: 10.1109/TPEL.2012.2232785.

2 Steyaert, M. and Sansen, W. (1990) Power supply rejection ratio in operational transconductance amplifiers. *IEEE Transactions on Circuits and Systems*, 37 (9), 1077–1084, doi: 10.1109/31.57596.

3 Lee, H. and Al-Shyoukh, M. (2010) Stability and transient response enhancement techniques for low-dropout regulators, in *2010 53rd IEEE International Midwest Symposium on Circuits and Systems*, IEEE, Seattle, WA, USA, pp. 580–583, doi: 10.1109/MWSCAS.2010.5548894.

4 Gupta, V., Rincon-Mora, G., and Raha, P. (2004) Analysis and design of monolithic, high PSR, linear regulators for SoC applications, in *IEEE International SOC Conference, 2004. Proceedings.*, pp. 311–315, doi: 10.1109/SOCC.2004.1362447.

5 Ahuja, B. (1983) An improved frequency compensation technique for CMOS operational amplifiers. *IEEE Journal of Solid-State Circuits*, 18 (6), 629–633, doi: 10.1109/JSSC.1983.1052012.

6 Carvajal, R., Ramirez-Angulo, J., Lopez-Martin, A., Torralba, A., Galan, J., Carlosena, A., and Chavero, F. (2005) The flipped voltage follower: a useful cell for low-voltage low-power circuit design. *IEEE Transactions on Circuits and Systems I: Regular Papers*, 52 (7), 1276–1291, doi: 10.1109/TCSI.2005.851387.

7 Surkanti, P.R., Garimella, A., Manda, M., and Furth, P.M. (2017) On the analysis of low output impedance characteristic of flipped voltage follower (FVF) and FVF LDOs, in *2017 IEEE 60th International Midwest Symposium on Circuits and Systems (MWSCAS)*, pp. 17–20, doi: 10.1109/MWSCAS.2017.8052849.

8 Surkanti, P.R., Garimella, A., and Furth, P.M. (2018) Flipped voltage follower based Low Dropout (LDO) voltage regulators: a tutorial overview, in *2018 31st International Conference on VLSI Design and 2018 17th International Conference on Embedded Systems (VLSID)*, pp. 232–237, doi: 10.1109/VLSID.2018.68.

9 Lu, Y., Ki, W.H., and Yue, C.P. (2014) 17.11 A 0.65ns-response-time 3.01ps FOM fully-integrated low-dropout regulator with full-spectrum power-supply-rejection for wideband communication systems, in *2014 IEEE International Solid-State Circuits Conference Digest of Technical Papers (ISSCC)*, pp. 306–307, doi: 10.1109/ISSCC.2014.6757446.

10 Wang, Z., Kim, S.J., Bowman, K., and Seok, M. (2022) Review, survey, and benchmark of recent digital LDO voltage regulators, in *2022 IEEE Custom Integrated Circuits Conference (CICC)*, pp. 01–08, doi: 10.1109/CICC53496.2022.9772734.

11 Huang, M., Lu, Y., and Martins, R.P. (2021) Review of analog-assisted-digital and digital-assisted-analog low dropout regulators. *IEEE Transactions on Circuits and Systems II: Express Briefs*, 68 (1), 24–29, doi: 10.1109/TCSII.2020.3040393.

12 Huang, M., Lu, Y., and Martins, R.P. (2020) A comparative study of digital low dropout regulators. *Journal of Semiconductors*, 41 (11), doi: 10.1088/1674-4926/41/11/111405.

13 Akram, M.A., Hwang, I.C., and Ha, S. (2020) Architectural advancement of digital low-dropout regulators. *IEEE Access*, 8, 137838–137855, doi: 10.1109/ACCESS.2020.3012467.

14 Liu, X., Krishnamurthy, H.K., Na, T., Weng, S., Ahmed, K.Z., Ravichandran, K., Tschanz, J., and De, V. (2019) 14.7 A modular hybrid LDO with fast load-transient response and programmable PSRR in 14nm CMOS featuring dynamic clamp tuning and time-constant compensation, in *2019 IEEE International Solid-State Circuits Conference – (ISSCC)*, pp. 234–236, doi: 10.1109/ISSCC.2019.8662343.

8

Charge Pumps

8.1 Introduction

Charge pumps are capacitive DC–DC converters. They are also called voltage multipliers because the output voltage is usually higher than the input voltage. It will be the main focus of this chapter. Nevertheless, some charge pumps generate negative output voltages or implement a step-down conversion. Since charge pumps are typically applied for bias generation, we distinguish them from switched-capacitor DC–DC converters covered in Chapter 9. Nevertheless, they are based on the same physical mechanism: charge redistribution within a capacitor network.

8.1.1 Fundamental Circuit and Operation

The conceptual charge pump circuit in Fig. 8.1a) consists of a pumping capacitor C_p and four switches. Two are controlled by a clock signal φ_1, complementary to a second clock φ_2. The resulting equivalent circuits in each clock phase are shown in Fig. 8.1b,c), respectively. Both clocks are non-overlapping signals as depicted in Fig. 8.1d) to avoid any cross-currents (charge loss from C_p) during clock transitions. For $\varphi_1 = 1$, the capacitor C_p is connected to the input and will be charged to V_{in}, referred to as the refresh phase. Hence, $V_{cp} = V_{in}$. During this clock phase, the complementary clock φ_2 is at 0, and the connected switches are off. Therefore, there is no connection to V_{out}. In the second clock phase, the switches at φ_1 turn off, while $\varphi_2 = 1$. This phase is called the charge redistribution phase. The top plate of C_p connects to the charge pump output and its bottom plate to the input. Kirchhoff's voltage law yields

$$V_{out} = V_{in} + V_{cp} = 2V_{in}. \tag{8.1}$$

The charge pump circuit forms a voltage doubler. However, this is only true during phase φ_2 ($\varphi_2 = 1$). To achieve a stable DC output voltage V_{out}, we need to add an output buffer capacitor C_o as shown in Fig. 8.1e). Both capacitors, C_p and C_o, can be implemented as on-chip capacitances or discrete external components. On-chip integration is limited by the required output power (load current I_{load}), as will be investigated below in Section 8.2. The circuit in Fig. 8.1e) is called the Dickson charge pump (or Dickson type charge pump) after the author who published an early example of an IC-level integrated charge pump back in 1976 [1]. Dickson implemented seven pumping stages with MOS transistors in a diode configuration.

Design of Power Management Integrated Circuits, First Edition. Bernhard Wicht.
© 2024 John Wiley & Sons Ltd. Published 2024 by John Wiley & Sons Ltd.

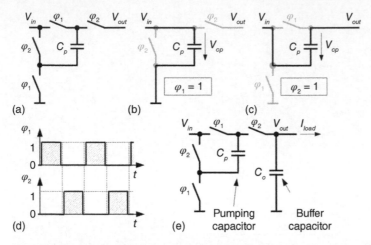

Figure 8.1 Voltage doubler charge pump: a) Concept and equivalent circuit for b) $\varphi_1 = 1$ and c) $\varphi_2 = 1$; d) timing diagram; e) final circuit with pumping and output buffer capacitor (Dickson charge pump).

8.1.2 Charge Pumps Applications

Now that we understand the essential operation of a charge pump, many applications will benefit if a higher or negative voltage can be generated from any given system voltage. Typical applications include low-voltage EEPROM and Flash memory programming at voltage levels of 10–20 V. In digital design, negative body biasing is a well-established method to reduce static leakage by increasing the threshold voltage. A suitable charge pump circuit generates the negative voltage. Finally, in power management, we can improve the available gate overdrive voltage to reduce the on-resistance of power switches (see Eqn. (2.2) in Chapter 2) and the associated static power losses. When combined with a bootstrap gate supply, a charge pump can enable the high-side power switch to remain turned on continuously (see Section 5.11).

8.1.3 General Characteristics

Assume that all capacitors in Fig. 8.1e) have been fully discharged before the clocks get activated at $t = 0$. Figure 8.2 shows the simulated transient waveforms. We can derive three general

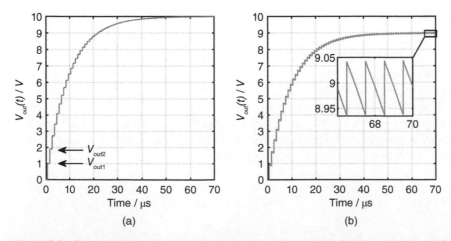

Figure 8.2 Start-up transients of the charge pump output voltage for $V_{in} = 5$ V, $C_p = 100$ pF, $C_o = 900$ pF, $f_{sw} = 1$ MHz: a) without load; b) for $I_{load} = 100$ μA.

characteristics of charge pumps from these curves. Let us first consider the case for zero load ($I_{load} = 0$) in Fig. 8.2a). We notice a step-wise ramp-up of the output voltage due to charge redistribution from phase φ_1 to φ_2 when C_p charges C_o. For this reason, the voltage step gets smaller the higher V_{out} is. In a final application, it may be crucial that the charge pump ramps up within a given time. We will analyze this charge pump behavior in Section 8.2.1.

With a non-zero load current I_{load}, Fig. 8.2b) reveals two more effects. The output voltage does not reach its ideal value of $2V_{in}$. A significant voltage drop ΔV_{out} occurs. We also observe a voltage ripple at V_{out} due to the switching nature of the charge pump.

8.2 Analysis of the Fundamental Charge Pump

We will now analyze the three effects identified in Section 8.1: Step-wise ramp-up, voltage droop, and output voltage ripple. It will lead to essential design guidelines for charge pumps.

8.2.1 Step-Wise Ramp-Up

The charge balance will help us to derive the ramp-up behavior in Fig. 8.2. The initial condition for $t \leq 0$ corresponds to $\varphi_1 = 1$ and $\varphi_2 = 0$ and $V_{cp} = V_{in}$. We assume that C_o is fully discharged. Hence,

$$\text{Initial condition:} \quad \varphi_1 = 1, \varphi_2 = 0 \rightarrow Q_o = 0, Q_p = C_p V_{in}, \tag{8.2}$$

with the respective charge Q_o and Q_p stored on C_o and Q_p. At $t = 0$, φ_2 switches to 1 and connects C_p to C_o:

$$\text{Clock cycle 1:} \quad \varphi_1 = 0, \varphi_2 = 1 \rightarrow Q_{o1} = C_o V_{out1}, Q_p = C_p(V_{out1} - V_{in}) \tag{8.3}$$

V_{out1} denotes the output voltage level in the first clock cycle. Since the original charge defined by Eqn. (8.2) does not change, the total charge in C_o and C_p in Eqn. (8.3) is equal the one in Eqn. (8.2). The charge balance yields:

$$C_o V_{out1} + C_p(V_{out1} - V_{in}) = C_p V_{in} \tag{8.4}$$

Rearranging allows to calculate the output voltage V_{out1} in the first clock cycle,

$$V_{out1} = 2V_{in} C_p/(C_p + C_o). \tag{8.5}$$

In the second half-period, the clocks change to $\varphi_1 = 1$, $\varphi_2 = 0$, and C_p gets refreshed from the input source V_{in}. Hence,

$$\text{Refresh 1:} \quad \varphi_1 = 1, \varphi_2 = 0 \rightarrow Q_{o1} = C_o V_{out1}, Q_p = C_p V_{in}. \tag{8.6}$$

The charge in the next clock cycle is defined similarly to Eqn. (8.3):

$$\text{Clock cycle 2:} \quad \varphi_1 = 0, \varphi_2 = 1 \rightarrow Q_{o2} = C_o V_{out2}, Q_p = C_p(V_{out2} - V_{in}) \tag{8.7}$$

Setting the charge values of Eqn. (8.6) and Eqn. (8.7) equal,

$$Q_{o1} + Q_p = C_o V_{out2} + C_p(V_{out2} - V_{in}), \tag{8.8}$$

yields the output voltage during the second clock phase,

$$V_{out2} = \frac{2V_{in} C_p + V_{out1} C_o}{C_p + C_o}. \tag{8.9}$$

From this expression, we can derive a recursive equation that describes the output voltage $V_{out,n}$ after the nth clock cycle,

$$V_{out,n} = \frac{2V_{in}C_p + V_{out,n-1}C_o}{C_p + C_o}. \tag{8.10}$$

Example 8.1 *For the given parameters, calculate the voltage levels in clock phases 1 and 2 of Fig. 8.2a).*

$$V_{out1} = 2V_{in}C_p/(C_p + C_o) = 2 \cdot 5\,\text{V} \cdot 100\,\text{pF}/(100\,\text{pF} + 900\,\text{pF}) = 1\,\text{V} \tag{8.11}$$

$$V_{out2} = \frac{2V_{in}C_p + V_{out1}C_o}{C_p + C_o} = \frac{2 \cdot 5\,\text{V} \cdot 100\,\text{pF} + 1\,\text{V} \cdot 900\,\text{pF}}{100\,\text{pF} + 900\,\text{pF}} = 1.8\,\text{V} \tag{8.12}$$

Both results match the values in Fig. 8.2a).

8.2.2 Voltage Droop for Nonzero Load

The root cause for the voltage dip in Fig. 8.2b) is the finite charge delivery capability of the charge pump. In steady-state, the charge delivered from V_{in} to V_{out} per cycle has to be equal to the charge drawn by the load,

$$\Delta Q_{out} = \Delta Q_{load}. \tag{8.13}$$

The load current results in $\Delta Q_{load} = I_{load}T = I_{load}/f_{sw}$. The voltage difference in V_{out} between clock phase n and $(n-1)$ defines ΔQ_{out}. Equation (8.13) can be rewritten accordingly,

$$C_o(V_{out,n} - V_{out,n-1}) = I_{load}/f_{sw}. \tag{8.14}$$

Substituting $V_{out,n}$ by Eqn. (8.10) yields

$$C_o\left(\frac{2V_{in}C_p + V_{out,n-1}C_o}{C_p + C_o} - V_{out,n-1}\right) = I_{load}/f_{sw}. \tag{8.15}$$

We can simplify this expression by defining $V_{out} = V_{out,n-1}$ assuming only small changes of V_{out} per clock cycle. In most practical cases, we can further assume $C_o \gg C_p$ (the buffer capacitor is much greater than the pump capacitor). Hence,

$$C_p(2V_{in} - V_{out}) = I_{load}/f_{sw}, \tag{8.16}$$

and

$$V_{out} = 2V_{in} - \frac{I_{load}}{f_{sw}C_p} = 2V_{in} - \Delta V_{out}, \tag{8.17}$$

which results in a simple expression for the output voltage droop,

$$\Delta V_{out} = \frac{I_{load}}{f_{sw}C_p}. \tag{8.18}$$

Interestingly, the droop does not depend on the output buffer capacitor C_o. Intuitively, we can explain this by recalling that the root cause is the finite charge delivery, determined by C_p and not by C_o.

Figure 8.3 Output resistance of the charge pump.

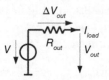

The charge pump can be modeled as a voltage source with a source resistance R_{out}, as shown in Fig. 8.3. Therefore, ΔV_{out} can be seen as the voltage drop across R_{out} due to the load current:

$$R_{out} = \frac{\Delta V_{out}}{I_{load}} = \frac{1}{f_{sw}C_p} \tag{8.19}$$

R_{out} is defined as the output resistance of the charge pump. The lower this resistance, the closer V_{out} gets to its ideal value defined by Eqn. (8.1) for the doubler charge pump of Fig. 8.1. Equation (8.19) identifies two design variables that define R_{out}, the clock frequency f_{sw} and the pumping capacitor C_p. Low R_{out} can be achieved if both values are large. However, f_{sw} and C_p have design restrictions. Capacitive charging losses, electro-magnetic interference, and parasitic coupling effects increase with the switching frequency f_{sw}. Larger C_p requires more layout area or may lead to a larger footprint in the case of discrete capacitors. The sizing equation for the pumping capacitor C_p can be derived from Eqn. (8.18):

$$C_p \geq \frac{I_{load}}{f_{sw}\Delta V_{out}} \tag{8.20}$$

Example 8.2 *Calculate the voltage droop ΔV_{out} in Fig. 8.2b). Which value of C_p will be required if the load current increases to 1 mA?*

From Eqn. (8.18),

$$\Delta V_{out} = \frac{I_{load}}{f_{sw}C_p} = \frac{100\,\mu A}{1\,MHz \cdot 100\,pF} = 1\,V. \tag{8.21}$$

The voltage droop is exactly 1 V as expected from the transient curve in Fig. 8.2b).

According to Eqn. (8.20), 10x larger load current (I_{load} = 1 mA) requires to increase C_p by the same factor. We get C_p = 1 nF, which can barely be integrated with standard CMOS at reasonable area consumption. Nevertheless, we can use a discrete external capacitor. As an option, we could increase the switching frequency also by 10× to f_{sw} = 10 MHz. This way, the charge pump can deliver I_{load} = 1 mA from the original capacitor C_p = 100 pF.

8.2.3 Output Voltage Ripple

While the droop is determined only by C_p, the output ripple depends on the output buffer capacitor C_o, assuming $C_o \gg C_p$, which is usually the case. For a maximum peak-to-peak ripple $\Delta V_{out,pkpk}$, C_o is given by this expression:

$$C_o \geq \frac{I_{load}}{f_{sw}\Delta V_{out,pkpk}} \tag{8.22}$$

The value of C_o is independent of C_p. However, both capacitors are determined by I_{load} and f_{sw}. If C_p has the same order of magnitude as C_o, its contribution has to be taken into account, as discussed for capacitive DC–DC converters in Section 9.6, refer to Eqn. (9.63) for details. During phase φ_2, C_p is connected to V_{out} and provides an additional buffer as studied in [2].

Example 8.3 *Calculate the voltage ripple $\Delta V_{out,pkpk}$ in Fig. 8.2b). How can we reduce the ripple at the given load current?*

Rearranging Eqn. (8.22), the ripple voltage is

$$\Delta V_{out,pkpk} = \frac{I_{load}}{f_{sw}C_o} = \frac{100\,\mu A}{1\,MHz \cdot 100\,pF} = 111\,mV. \tag{8.23}$$

The result fits very well with the graph in Fig. 8.2b). Increasing both the switching frequency and the buffer capacitance reduces the ripple magnitude.

8.3 Influence of Parasitics

The analysis of the charge pump in Section 8.2 shows how the output voltage drops in the presence of a non-zero load current as expressed by Eqn. (8.18). We can expect that parasitics also influence the output voltage. In particular, the parasitic capacitances and the finite on-resistance of the switches have to be considered as illustrated in Fig. 8.4.

8.3.1 Parasitic Capacitances

We first consider Fig. 8.4 for $R_{on} = 0$. The capacitance C_{par} in Fig. 8.4 accounts for the parasitic top-plate capacitance of C_p. The parasitic bottom-plate capacitance is not considered because it does not influence V_{out}. However, it can add significant power losses, as studied in Section 8.5.

To investigate the influence of the parasitic top-plate capacitance C_{par}, we can repeat the analysis presented in Section 8.2 including C_{par}. During refresh ($\varphi_1 = 1$), the charge on C_{par} is $Q_{par} = C_{par}V_{in}$. During charge redistribution when $\varphi_2 = 1$, C_{par} connects to the output, hence, $Q_{par} = C_{par}V_{out}$. Consequently, we can rewrite Eqn. (8.4),

$$C_oV_{out1} + C_p(V_{out1} - V_{in}) + C_{par}V_{out1} = C_pV_{in} + C_{par}V_{in}, \tag{8.24}$$

and we finally obtain an expression similar to Eqn. (8.17),

$$V_{out} = V_{in} + V_{in}\frac{C_p}{C_p + C_{par}} - \frac{I_{load}}{f_{sw}(C_p + C_{par})}, \tag{8.25}$$

or in a more general form,

$$V_{out} = 2V_{in} - V_{loss} - R_{out}I_{load}, \tag{8.26}$$

with

$$V_{loss} = V_{in}\frac{C_{par}}{C_p + C_{par}}, \tag{8.27}$$

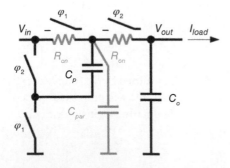

Figure 8.4 Fundamental charge pump with parasitic capacitance C_{par} and switch resistance R_{on}.

and a new expression for the equivalent output resistance

$$R_{out} = \frac{1}{f_{sw}(C_p + C_{par})}. \tag{8.28}$$

For $C_{par} = 0$, Equations (8.17) and (8.25) are equal. From Fig. 8.4, we can conclude that C_{par} forms a capacitive divider with the pumping capacitor C_p. Intuitively, this explains the second term on the right-hand part of Eqn. (8.25). It reveals that C_{par} increases the droop of V_{out} further in addition to the influence of the load current. Equation (8.26) introduces the voltage V_{loss} to account for the drop in V_{out}. Interestingly, C_{par} also reduces the droop of V_{out} as it appears in the denominator of the corresponding term to the right in Eqn. (8.25). C_{par} appears parallel to C_o during phase φ_2 and helps to buffer the output voltage. Consequently, the equivalent output resistance of Eqn. (8.19) reduces and can be expressed by Eqn. (8.28). Usually, the charge-sharing effect dominates such that any parasitic capacitance reduces V_{out}.

8.3.2 Finite On-Resistance

In a real charge pump design, transistors will implement the switches. The on-resistance R_{on}, as indicated in Fig. 8.4, corresponds to the R_{DSon} of the switches as described in Chapter 2, see Eqn. (2.2) in Section 2.2. If we consider the finite resistance, V_{out} reduces further. However, usually, the voltage drop corresponding to the capacitive term containing C_{par} in Eqn. (8.25) dominates because the switch resistance is small enough to ensure complete recharging of the pumping capacitor. This consideration relates to the definition of the slow and fast switching limit (SSL and FSL) as explained for capacitive DC–DC converters in Section 9.5.4 (see also Fig. 9.8). Charge pumps usually operate in SSL. Hence, for the design of charge pumps, the on-resistance of the switches can be neglected, and Eqn. (8.25) is sufficient.

8.4 Charge Pump Implementation

The switches in charge pumps can be implemented using diodes or transistors. While the diodes simplify the design (they do not need level shifters), the charge pump performance suffers from the diode forward drop. In any case, we need to carefully observe the layout to eliminate the influence of parasitic bipolar transistor structures.

8.4.1 Charge Pumps with Diodes

Figure 8.5 shows an implementation of the fundamental charge pump of Fig. 8.1e) with diodes as switches. The bottom plate of C_p is controlled by an oscillator that toggles between V_{in} and ground. This circuit is one of the most simple implementations because the diodes automatically control the charge transfer, synchronized to the oscillator's switching frequency f_{sw}. No dedicated clock

Figure 8.5 Charge pump of Fig. 8.1e) with the switches implemented by diodes.

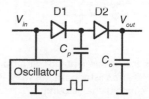

generation is required for φ_1 and φ_2. As one of the drawbacks, V_{out} is significantly reduced due to the forward voltage drop V_F of both diodes. Because of the diodes, the initial charge pump output voltage is $(V_{in} - 2V_F)$ (i.e., C_o is precharged to that level). Assuming the oscillator voltage to be at 0, the pump capacitor C_p is initially charged to $(V_{in} - V_F)$. Diode D1 conducts while D2 blocks similar to the corresponding switches of Fig. 8.1 during the refresh phase ($\varphi_1 = 1$). Likewise, if the oscillator switches to 1, D1 blocks and D2 gets forward biased such that C_p transfers charge to the output capacitor C_o. In steady-state, the maximum achievable output voltage is

$$V_{out} = 2(V_{in} - V_F). \tag{8.29}$$

The forward voltage V_F has to be taken into account two times because C_p charges to $(V_{in} - V_F)$ during refresh while another V_F is subtracted when C_p connects to V_{out}. For $V_{in} = 5\,\text{V}$ and $V_F = 0.6\,\text{V}$, V_{out} would be limited to 8.8 V. The finite output resistance and parasitic mechanisms as outlined in Sections 8.2 and 8.3 further reduce V_{out}. One additional loss mechanism is related to the diodes because integrated diodes are always associated with a parasitic bipolar junction transistor. The details are described in Section 3.5 along with mechanisms that reduce the parasitic bipolar junction transistor. The diode implementation of Fig. 3.19 with surrounding highly n-doped isolation gives the best results.

As long as the achievable value of V_{out} given by Eqn. (8.29) is sufficient, the diode-based charge pump circuit according to Fig. 8.5 is a favorable choice. If the diode drop is unacceptable, transistors must be used as switches.

8.4.2 Charge Pumps with Transistor Switches

It is instructive to consider how to replace the diodes of the Dickson charge pump in Fig. 8.5 by transistor switches. We will see that this is not straightforward and requires some effort. We will first emphasize diode D1. Figure 8.6a) shows a circuit in which D1 is replaced by a p-type (PMOS) transistor M1. This charge pump will work, but it has a significant drawback related to the body diode of M1 (see also Fig. 2.5 in Chapter 2). If the oscillator output voltage goes to 1 ($V_{ctrl} = V_{in}$),

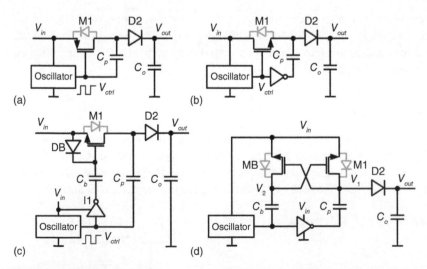

Figure 8.6 Charge pumps with transistor switches: Replacing diode D1 in Fig. 8.5 by a) a PMOS and b) by an NMOS transistor does not work in the first place; c) correct n-type transistor implementation; d) basic cross-coupled charge pump.

not only D2 but also the body diode of M1 will conduct. Some charge stored on C_p will flow back to V_{in} and will not be delivered to the output. This behavior is inefficient, leading to more V_{out} droop.

Alternatively, an n-type (NMOS) transistor can be used as a switch as shown in Fig. 8.6b). However, this configuration has even more drawbacks. In addition to the backflow of charge to V_{in} though M1's body diode, the pumping capacitor C_p cannot charge higher than $V_{in} - V_{th}$, limited by the threshold voltage V_{th} of M1. Because $V_{th} \approx V_F$, the limitation of V_{out} according to Eqn. (8.29) holds. The transistor must be placed in reverse polarity as shown in Fig. 8.6c) for an n-type switch to prevent the charge backflow via the body diode. The circuit utilizes local bootstrapping with an additional diode DB and a capacitor C_b. As explained in Section 5.11 for a general power stage, the operation is identical to active bootstrapping. If $V_{ctrl} = 1$, the inverter I1 pulls the bottom plate of C_b to ground and C_b charges to $(V_{in} - V_F)$. If V_{ctrl} changes to "0," inverter I1 connects the bottom plate of C_p to V_{in} such that the gate voltage of M1 reaches $(2V_{in} - V_F)$. M1 sees sufficient gate overdrive to turn on. Even though the source terminal of M1 is now connected to V_{in}, electrically, the drain acts as the source, and the channel current flows toward C_p. This way, C_p charges fully to V_{in}. In conclusion, this charge pump circuit solves both drawbacks of Fig. 8.6b) associated with the body diode and with the threshold voltage of M1.

The circuit in Fig. 8.6d) goes one step further and replaces the bootstrap diode DB with a transistor MB. MB and M1 form a symmetrical structure with complementary outputs V_1 and V_2. As an essential circuit, this cross-coupled charge pump is widely used. The steady-state output level reaches $V_{out} = 2V_{in} - V_F$. For example, $V_{out} = 9.4$ V for $V_{in} = 5$ V and $V_F = 0.6$ V. This is much better than $V_{out} = 8.8$ V as given by Eqn. (8.29) for the diode circuit in Fig. 8.5.

To get close to the ideal level $V_{out} = 2V_{in}$, also diode D2 needs to be replaced by a transistor. Developing the underlying idea of the cross-coupled charge pump in Fig. 8.6d) further leads to the circuit in Fig. 8.7, which was presented in 2001 [3] and, as a conceptual circuit, also in [4], but has been around in the industry long before that time. The two NMOS transistors M1A and M1B, and the two capacitors C_{p1}, C_{p2} correspond to M1, MB, C_p, C_b in Fig. 8.6d). The two PMOS devices, M2A and M2B, are used instead of D2. This way, all transistors in Fig. 8.7 form cross-coupled CMOS inverters IA, IB as indicated in the upper-right corner of the schematic. As the main benefit, this charge pump provides dual-phase operation with two pumping capacitors C_{p1} and C_{p2}. Compared to Fig. 8.6d), those capacitors provide inherent bootstrapping to drive all transistors properly but also transfer charge to the output. For $V_{ctrl} = 0$, C_{p1} pulls the input of inverter IA (M1A, M2A) to low level. Concurrently, the other inverter IB sees the inverted input signal via C_{p2}. Hence, NMOS M1B and PMOS M2A conduct. M1B recharges C_{p1} from V_{in} while M2A transfers charge from C_{p2} toward the output to recharge C_o and to contribute to the load current I_{load}. In the other clock phase, $V_{ctrl} = 1$, M1A and M2B turn on to refresh C_{p2} and to connect C_{p1} to the output. Assuming that the

Figure 8.7 A dual-phase charge pump.

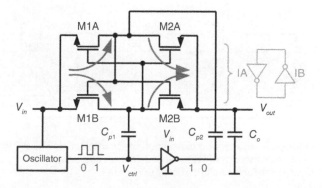

voltage at the pumping capacitors can fully settle within the associated phase of the clock period, the maximum achievable output voltage becomes

$$V_{out} = 2V_{in}, \tag{8.30}$$

which is much better than the diode-based charge pump (Eqn. (8.29)).

Due to the two-phase operation, the charge pump can deliver two times higher load current I_{load} compared to one-phase designs like all other circuits discussed in this chapter. Also, for identical I_{load}, the output voltage ripple reduces by 50%. Alternatively, for the same maximum load current, we can operate at half the switching frequency (benefit for switching losses and electromagnetic interference (EMI)) or reduce the size of C_{p1}, C_{p2} by a factor of two.

While the dual-phase operation described above works very well in steady-state, the start of the charge pump requires a closer look. If all capacitors are discharged, the transistors cannot turn on. So how does the charge pump start? The capacitors get charged via the body diodes of all transistors. Independent of the actual voltages at the inverter input and output nodes, the body diodes of both NMOS transistors M1A and M1B charge C_{p1} and C_{p2}, respectively, to $V_{in} - V_F$. This level is usually sufficient to drive the switches and start the pumping activity. The body diodes of the PMOS switches M2A and M2B precharge the output buffer capacitor C_o.

8.4.3 The Parasitic Bipolar Junction Transistor

At least one body diode may get forward-biased if the node voltage exceeds the regular input voltage due to the pumping operating of charge pump circuits. Therefore, reviewing the body diode's parasitic effect in any charge pump circuit is mandatory. We will do this exercise for the circuit in Fig. 8.7. Due to the symmetry, we consider only one inverter. Section 3.5 describes diode structures and their parasitic bipolar junction transistors. Figure 8.8a) shows the cross-section of one inverter of Fig. 8.7 with its NMOS and PMOS device (left and right). The challenge is to identify (1) the parasitic bipolar transistor structures and (2) the proper connection for the back gate and isolation, which are both depicted in Fig. 8.8b). The NMOS transistor structure is identical to the diode

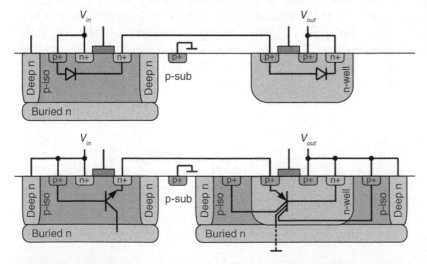

Figure 8.8 a) Layout cross-section of the inverters in Fig. 8.7 with the body diodes; b) parasitic bipolar transistors associated with the body diodes and biasing of the back gate, isolating deep n-well, and buried layers to reduce parasitic substrate currents.

shown in Fig. 3.19, associated with a parasitic NPN transistor, while the PMOS switch corresponds to Fig. 3.18 forming a parasitic PNP device.

Without the proper isolation, a significant fraction of the charge will not be delivered to the output and, instead, will flow into the substrate (as parasitic collector current) toward the ground. This effect has been one of the primary reasons for the total malfunction of the charge pump in actual designs. It simply fails to start up because the parasitic bipolar junction transistor dumps most current into the substrate, so the output capacitor cannot be charged.

8.5 Power Efficiency

The charge pump can be considered a DC–DC converter with a power conversion efficiency given by

$$\eta = \frac{P_{out}}{P_{in}} = \frac{\overline{V}_{out} I_{load}}{V_{in} \overline{I}_{in}}. \tag{8.31}$$

This expression considers the average values of the output voltage $\overline{V}_{out}$ and the input current $\overline{I}_{in}$. Due to the switching nature of the charge pump, there will be some output voltage ripple (see Eqn. (8.22) and also Example 8.3). The input current ripple is due to the inrush current recharging each pumping capacitor. While the charge pump output resistance R_{out} is essential if the minimum output voltage V_{out} is a key requirement (e.g., to ensure that the on-resistance of a power transistor driven by the charge pump does not exceed the upper limit), the efficiency is critical in battery-operated designs. It also provides the basis for thermal design in case of excessive losses. There are many sources for power loss $P_{loss} = P_{in} - P_{out}$ in charge pumps. The most important ones are illustrated in Fig. 8.9 for the diode-based implementation of the Dickson voltage doubler. Inserting the losses into Eqn. (8.31) yields

$$\eta = \frac{P_{out}}{P_{in}} = \frac{(2V_{in} - \Delta V_{out} - V_{loss1} - V_{loss2} - \cdots)I_{load}}{V_{in}(2I_{load} + I_{loss1} + I_{loss2} + I_{loss3} + \cdots)}. \tag{8.32}$$

Section 8.2 describes the loss contributions related to R_{out}, which lead to an output voltage drop of ΔV_{out} in the numerator according to Eqn. (8.18). The power dissipation in R_{out} also adds to the denominator's input power P_{in}. In addition, the output voltage reduces due to the two diode drops V_{loss1} and the charge sharing with the parasitic top plate capacitance of C_p (V_{loss2}). Input power will be dissipated by the control logic (the oscillator connected to the bottom plate of C_p), denoted by I_{loss1} in Fig. 8.9. These losses comprise mainly short circuit losses (see Eqn. (5.18) in Chapter 5).

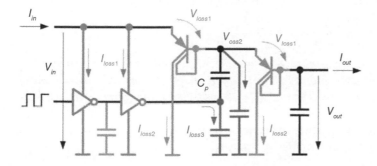

Figure 8.9 Major losses in a Dickson-type charge pump.

Substrate currents I_{loss2} due to parasitic bipolar structures as described in Section 8.4 also add to the input current. Further loss contributions are due to the parasitic bottom plate capacitance of C_p. While it does not contribute to R_{out}, it gets charged from V_{in} every time during the refresh phase ($\varphi_1 = 1$ as defined in Fig. 8.1). A current I_{loss3} flows. This charge is dissipated toward the ground for $\varphi_1 = 0$ without delivering any contribution to the load current I_{load} at the output of the charge pump.

For a practical analysis, expressing the charge pump efficiency in terms of output power P_{out} and power loss P_{loss} may be more convenient. We can rewrite Eqn. (8.32),

$$\eta = \frac{P_{out}}{P_{out} + P_{loss}} = \frac{\overline{V}_{out}I_{load}}{\overline{V}_{out}I_{load} + P_{loss}}, \tag{8.33}$$

with the average output voltage $\overline{V}_{out}$ according to Eqn. (8.26). The loss voltage V_{loss} in Eqn. (8.26) needs to be expanded to incorporate not only the charge sharing due to C_{par} but also the forward voltage drop of the diodes. As an advantage of Eqn. (8.33), the individual loss contributions can be calculated and added to the denominator,

$$P_{loss} = V_{in}(I_{loss1} + I_{loss2} + I_{loss3} + \cdots). \tag{8.34}$$

It is instructive to analyze various loss components and their influence on the charge pump efficiency.

Example 8.4 *Consider the diode-based circuit of Fig. 8.5 with its loss components shown in Fig. 8.9. The given parameters are $V_{in} = 4\,V, f_{sw} = 1\,MHz, I_{load} = 10\,\mu A$. We initially assume ideal diodes with zero forward voltage drop ($V_F = 0$).*

(a) *Calculate the value of C_p that ensures $V_{out} \geq 7\,V$. What is the output resistance R_{out} and the charge pump efficiency?*
Based on Eqn. (8.17) the output voltage droop is $\Delta V_{out} = 2V_{in} - V_{out} = 1\,V$. From Eqn. (8.18),

$$C_p = \frac{I_{load}}{f_{sw}\Delta V_{out}} = \frac{10\,\mu A}{1\,MHz \cdot 1\,V} = 10\,pF. \tag{8.35}$$

The output resistance is given by Eqn. (8.19):

$$R_{out} = \frac{\Delta V_{out}}{I_{load}} = \frac{1\,V}{10\,\mu A} = 100\,k\Omega \tag{8.36}$$

This yields a power loss of $P_{loss,o} = I_{load}^2 R_{out} = 10\,\mu W$ while the output power is $P_{out} = V_{out}I_{load} = 70\,\mu W$. Hence, the power efficiency can be calculated according to Eqn. (8.33):

$$\eta = \frac{P_{out}}{P_{out} + P_{loss}} = \frac{70\,\mu W}{70\,\mu W + 10\,\mu W} = 87.5\% \tag{8.37}$$

(b) *Calculate the power loss in the two identical inverters (Fig. 8.9) that drive the bottom plate of C_p. Consider only short circuit losses according to Eqn. (5.18) in Chapter 5. Take the same parameters like in Example 5.4, but $\beta_{inv} = 700\,\mu A/V^2$ and $t_{rf} = 500\,ps$. What is the efficiency if these losses get added to the losses in R_{out}?*
If we denote the short circuit loss energy of Example 5.4 by $E_{sc,o}$, the losses of one inverter can be readily calculated by scaling $E_{sc,o}$ by the different values of β_{inv} and t_{rf}:

$$E_{sc} = \frac{\beta_{inv}}{12}(V_{drv} - 2V_{th})^3 t_{rf} = \frac{700}{80} \cdot \frac{500}{135} \cdot E_{sc,o} = 32.4 \cdot 42\,fJ = 1.36\,pJ \tag{8.38}$$

The power loss for two inverters is

$$P_{loss1} = 2 E_{sc} f_{sw} = 2.72\,\mu\text{W}, \tag{8.39}$$

which adds to the loss $P_{loss,o}$ in the output resistance of item (a). The total power loss $P_{loss} = P_{loss,o} + P_{loss1} = 12.72\,\mu\text{W}$ reduces the charge pump efficiency to

$$\eta = \frac{P_{out}}{P_{out} + P_{loss}} = \frac{70\,\mu\text{W}}{70\,\mu\text{W} + 12.72\,\mu\text{W}} = 84.6\%. \tag{8.40}$$

(c) *Calculate the output voltage and the power loss if the diode forward drop is approximated by $V_F = 0.5\,\text{V}$. What is the total power efficiency if the losses get added to the losses of the previous items?*

The output voltage reduces by $2 V_F = 1\,\text{V}$ to $V_{out} = 6\,\text{V}$. Hence, the output power is now $P_{out} = V_{out} I_{load} = 60\,\mu\text{W}$, which reduces the efficiency already. However, in addition, there is also power loss associated with V_F: $P_{loss2} = 2 V_F I_{load} = 10\,\mu\text{W}$ increases the total loss to $P_{loss} = P_{loss,o} + P_{loss1} + P_{loss2} = 22.72\,\mu\text{W}$. The efficiency becomes

$$\eta = \frac{P_{out}}{P_{out} + P_{loss}} = \frac{60\,\mu\text{W}}{60\,\mu\text{W} + 22.72\,\mu\text{W}} = 72.5\%. \tag{8.41}$$

(d) *Assume that the pumping capacitor C_p has a parasitic bottom-plate capacitance $C_{bot} = 0.3\,C_p$ (but no top-plate capacitance, yet). Calculate the power loss and the charge efficiency.*

The parasitic capacitance $C_{bot} = 0.3\,C_p = 3\,\text{pF}$ is charged via the inverter from V_{in}. The power dissipation is

$$P_{loss3} = C_{bot} V_{in}^2 f_{sw} = 3\,\text{pF} \cdot (4\,\text{V})^2 \cdot 1\,\text{MHz} = 48\,\mu\text{W}, \tag{8.42}$$

the total power loss $P_{loss} = 70.7\,\mu\text{W}$ and the efficiency

$$\eta = \frac{P_{out}}{P_{out} + P_{loss}} = \frac{60\,\mu\text{W}}{60\,\mu\text{W} + 70.7\,\mu\text{W}} = 45.9\%. \tag{8.43}$$

(e) *What is the influence on the output voltage, power loss, and efficiency if C_p also has a parasitic top-plate capacitance $C_{top} = 0.18\,C_p$.*

Equation (8.26) describes the resulting output voltage if a parasitic top-plate capacitance $C_{par} = C_{top} = 0.18\,C_p = 1.8\,\text{pF}$ is present. However, the expression assumes ideal switches ($V_F = 0$). To incorporate the forward voltage V_F, we need to replace V_{in} in Eqn. (8.26) by ($V_{in} - V_F$). Hence,

$$V_{out} = (V_{in} - V_F) - (V_{in} - V_F)\frac{C_{top}}{C_p + C_{top}} - \frac{I_{load}}{f_{sw}(C_p + C_{top})} = 5.62\,\text{V}. \tag{8.44}$$

The output power is further reduced to $P_{out} = V_{out} I_{load} = 56.2\,\mu\text{W}$. The output resistance decreases as given by Eqn. (8.28),

$$R_{out} = \frac{1}{f_{sw}(C_p + C_{par})} = \frac{1}{1\,\text{MHz} \cdot 11.8\,\text{pF}} = 84.7\,\text{k}\Omega, \tag{8.45}$$

and results in $P_{loss4,o} = I_{load}^2 R_{out} = 8.47\,\mu\text{W}$. The charge delivered to C_{top} causes additional power dissipation $P_{loss4,c}$ from V_{in}. Since this charge is not provided directly from V_{in}, but via the pump capacitor C_p, half of the energy is lost. Hence, two times the charging loss is dissipated from V_{in}. The voltage swing at C_{top} is $V_c = V_{out} - V_{in} + 2 V_F = 2.62\,\text{V}$ and, therefore,

$$P_{loss4,c} = 2 C_{top} V_c^2 f_{sw} = 2 \cdot 1.8\,\text{pF} \cdot (2.62\,\text{V})^2 \cdot 1\,\text{MHz} = 24.7\,\mu\text{W}. \tag{8.46}$$

The power loss due to C_{top} is $P_{loss4} = P_{loss4,o} + P_{loss4,c} = 33.2\,\mu\text{W}$. We obtain the total losses $P_{loss} = 93.9\,\mu\text{W}$ by substituting the original losses $P_{loss,o}$ of item a) in the output resistance by

Table 8.1 Analysis of the charge pump efficiency in Example 8.4. Except for P_{loss} the values are cumulative.

	V_{out} (V)	P_{loss} (µW)	P_{out} (µW)	P_{in} (µW)	η (%)
(a) R_{out}	7	10	70	80	87.5
(b) Inverters	7	2.72	70	82.7	84.6
(c) Diodes	6	10	60	82.7	72.5
(d) C_{bot}	6	48	60	131	45.9
(e) C_{top}	5.62	33.2	56.2	150	37.4

P_{loss4}. It corresponds to a charge pump efficiency of

$$\eta = \frac{P_{out}}{P_{out} + P_{loss}} = \frac{56.2\,\mu W}{56.2\,\mu W + 93.9\,\mu W} = 37.4\%. \tag{8.47}$$

The entire loss analysis is summarized in Table 8.1

From Example 8.4 and the overview in Table 8.1 we notice that some loss components lead to excessive power dissipation P_{loss}, but they do not reduce V_{out} like the inverters in item (b) and C_{bot} in item (d). In addition, other loss contributions cause V_{out} to drop, e.g., the diodes in item (c) and C_{top} in item (e). In conclusion, the charge pump efficiency can drop significantly. As a rule of thumb, we can assume 50% for a typical charge pump power efficiency.

8.6 Cascading of Pumping Stages

The fundamental charge pump provides a maximum output voltage of $2\,V_{in}$, and it will be reduced by various losses, as studied in Section 8.5. Multiple charge pump stages can be cascaded if the output voltage level is insufficient. Figure 8.10 illustrates how the doubler circuit of Fig. 8.1e) is extended by one more stage comprising a second pumping capacitor C_{p2} that is driven by another inverter. All inverters are typically supplied from V_{in}. During $\varphi_1 = 1$, C_{p1} charges to V_{in} while C_{p2} delivers charge to the output (C_o). In the second half of the clock cycle ($\varphi_2 = 1$), the top plates of C_{p1} and C_{p2} get connected. Consequently, C_{p1} charges C_{p2}, while its bottom-plate is pulled to V_{in}. During this phase, the bottom plate of C_{p2} is at ground level. Therefore, C_{p2} can be charged up to $2\,V_{in}$. During $\varphi_1 = 1$ its bottom plate goes to V_{in}. This way, V_{out} can ideally reach $3\,V_{in}$.

In conclusion, the maximum node voltages increase from $V_1 = 2\,V_{in}$ toward $V_2 = V_3 = 3\,V_{in}$. Fortunately, these voltages occur if the bottom-plates of C_{p1} and C_{p2} are at V_{in}. Nevertheless, this

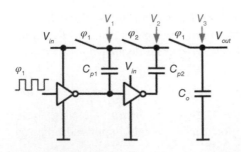

Figure 8.10 Cascading of charge pump stages. The node voltages scale up toward the output and require components with higher voltage ratings.

defines the maximum voltage ratings of the capacitors C_{p1}, C_{p2}, and C_o accordingly: $V_{max}(C_{p1}) = V_{in}$, $V_{max}(C_{p2}) = 2V_{in}$, $V_{max}(C_o) = 3V_{in}$

Similar considerations apply to the switches. The transistor between V_1 and V_2 sees the maximum drain-source voltage of $2V_{in}$, while the other switches need to withstand only V_{in}.

Even more stages can be cascaded. For N stages the output voltage as given in Eqn. (8.25) changes to

$$V_{out} = V_{in} + N\left(V_{in}\frac{C_p}{C_p + C_{par}} - \frac{I_{load}}{f_{sw}(C_p + C_{par})}\right). \tag{8.48}$$

For $N = 1$ Eqn. (8.48) corresponds to the doubler charge pump, while $N = 2$ applies for Fig. 8.10. Due to the losses, $N = 5$ is usually a reasonable maximum number of pumping stages. Any charge pump, including, for example, the dual-phase charge pump of Fig. 8.7, can be cascaded.

8.7 Other Charge Pump Configurations

Based on the principle of "pumping" charge, not only voltages greater than the input voltage can be generated. The step-down conversion is covered separately in Chapter 9. But charge pumps are also well-suitable to generate negative voltages. Figure 8.11a) shows the conceptual circuit. During the refresh phase, $\varphi_1 = 1$, C_p charges to V_{in}. However, the polarity is reversed in comparison to the doubler charge pump (see Fig. 8.1e)). Consequently, during $\varphi_2 = 1$ the top-plate potential of C_p pushes the output voltage toward $V_{out} = -V_{in}$ as shown in Fig. 8.11b). The graph also confirms step-wise charging. All effects, including the influence of parasitics, can be analyzed based on the charge balance as described in Sections 8.2 and 8.3. In case of non-zero load current, V_{out} will be limited to $V_{out} = -V_{in} + \Delta V_{out}$. There will also be some ripple with a sizing rule for C_o similar to Eqn. (8.22). While the maximum ratings of the capacitors and the switches are not critical in Fig. 8.11, care needs to be taken due to the negative voltage at the output and the intermediate switching node at the top plate of C_p. Any n-tank at those nodes will cause the PN-junction to

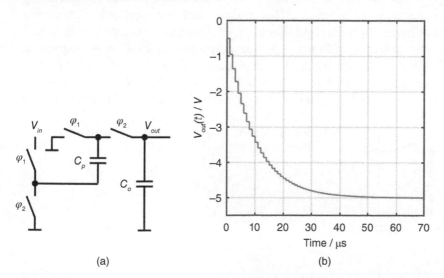

(a) (b)

Figure 8.11 a) An inverting charge pump; b) transient behavior at start-up with the ideal output voltage reaching $V_{out} = -V_{in}$ (same parameters as in Fig. 8.2).

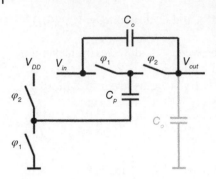

Figure 8.12 Charge pump with two input voltages (V_{in} and V_{DD}) and C_o tied to V_{in}. The ideal output voltage reaches $V_{out} = V_{in} + V_{DD}$.

get forward-biased. For this reason, the switch connected to V_{out} cannot be an n-type transistor. Otherwise, V_{out} would be clamped to $-V_F$, the forward voltage of the drain-bulk junction.

Especially for gate drive applications at higher supply voltages V_{bat}, the pumping capacitor needs to see only a swing equal to the desired gate overdrive voltage V_{drv} of the power transistor (see Chapter 5). For this purpose, the fundamental charge pump of Fig. 8.1e) can be modified as shown in Fig. 8.12. There are two input voltage levels, a high-voltage input $V_{in} = V_{bat}$ and low-voltage input $V_{drv} = V_{DD}$. Besides that, the operation is identical to Fig. 8.1e). The charge pump generates an ideal output voltage of $V_{out} = V_{in} + V_{DD}$. The value of V_{out} is well-suitable for a gate supply because it is higher than V_{in} by exactly one gate overdrive $V_{drv} = V_{DD}$. For example, in a 48 V-application with $V_{in} = 48$ V and $V_{drv} = V_{DD} = 5$ V, the charge pump provides $V_{out} = 53$ V. For such large voltages, Fig. 8.12 shows one more modification. The output buffer capacitor C_o can be tied to V_{in} instead of ground. This way, the voltage across C_o is only V_{DD}, and a capacitor type of much lower voltage ratings can be used. The maximum voltage of C_p remains V_{in}.

8.8 Current-Source Charge Pumps

Conventional charge pumps, like all circuits covered so far in the chapter, utilize hard-switching charge transfer from one capacitor to another. This results in inrush currents and high current spikes if the voltage difference between capacitors is large (at start-up and large load currents). Excessive EMI noise may be observed (Section 12.3) besides stress in the switches and capacitors. These effects can be eliminated by utilizing the current source that drives the bottom plate of C_p as shown in Fig. 8.13. During the refresh phase when $\varphi_1 = 1$ the current source I_A charges C_p slowly to V_{in}. For $\varphi_2 = 1$, the charge transfer to the output is limited by I_B. Compared to hard-switching

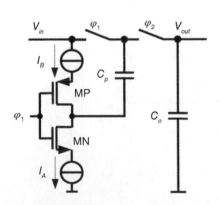

Figure 8.13 Current-source charge pump: Limited inrush current reduces stress and EMI.

designs, the current is much lower, and so is the charge transfer. Hence, the clock period $T = 1/f_{sw}$ must be large enough to ensure that the full charge can be delivered onto and from C_p. Basic current mirrors can implement the current sources I_A, I_B. The charge pump design in [5] uses $I_A = I_B = 50\,\mu A$ as opposed to current spikes of approximately 2.5 mA in the conventional hard-switching design. Operating at 6.38 MHz, the electromagnetic emission reduces by as much as $\sim 50\,dB\,\mu V$.

8.9 Charge Pumps Suitable as a Floating Gate Supply

Charge pumps are widely used in gate supply circuits. Chapter 5 shows that a buffer capacitor is installed in every gate driver that provides the gate charge and any other static or dynamic current (in the level shifter, etc.). Section 5.11 describes bootstrapping as a standard method to supply a n-type high-side power transistor. However, bootstrapping requires the switching node to return to zero periodically. It also prevents long on-times of the high-side transistor to ensure that the gate supply level remains within specification. These limitations can be solved using charge pumps instead of a bootstrap supply. The charge pump in Fig. 8.14a) [6] supports continuous on-times of the power transistor while $V_{sw} = V_{SSHS}$ is not required to return to ground. Hence, it is well-suitable for switched-capacitor and hybrid DC–DC converters (Chapters 9 and 11), especially if switching nodes alternate between non-zero voltages. This charge pump has been adapted from an input sampling switch in an analog-to-digital converter [7, 8]. The circuit operates from a single clock signal φ. During the recharge phase ($\varphi = 1$), shown in Fig. 8.14b), the pumping capacitor charges to the input voltage V_{in} while the buffer capacitor C_o supplies the gate driver and all attached circuitry in the power stage.

When the clock toggles to $\varphi = 0$, C_p transfers its charge to C_o, this way, C_o gets recharged frequently. It ensures that the power transistor can be kept turned on continuously. The body diodes are shown for reference, but the charge pump relies on switches. Hence, the output voltage is not reduced by any diode drops. Inverter I1 controls MP3 with respect to the floating high-side domain (V_{SSHS} and V_{DDHS}). This way, C_p connects in parallel to C_o without any voltage drop in between. Ideally, V_{out} reaches V_{in}. The implementation in [6] achieves a typical efficiency of 50%, see also [9].

There are a few critical voltage limitations in the circuit of Fig. 8.14: (1) For $\varphi = 0$, the source node of MP2 sees $V_{SSHS} + V(C_p) = V_{bat} + V_{in}$ with the high-voltage input V_{bat} of the power stage (the drain potential of the power transistor). Depending on the voltages V_{SSHS} and $V(C_p)$, a level shifter may be required for transistor MP2 (see Section 5.12). However, MN1 and MN3 can be directly connected to a ground-referred clock φ. (2) The other limitation stems from the maximum drain-source voltage at MP3. When $\varphi = 1$, the drain of MP3 is pulled to the ground by MN3 while its source potential reaches $V_{bat} + V_{in}$. (3) Also the drain-source voltage of MN2 needs to withstand V_{bat}.

If the power stage operates from higher voltages (V_{bat} in the order of 12 V), MP2, MP3, and I1 can be replaced by a diode. In addition, a diode can be inserted instead of MP1, which eases the high-voltage implementation at the expense of losses due to the diode drop. Consequently, this leads to suitable designs like the ones shown in Fig. 8.15.

The self-boosted charge pump [10, 11] shown in Fig. 8.15a) is similar to the fundamental diode-based charge pump in Fig. 8.5, except that the output buffer capacitor C_o is not referred to ground but to the floating high-side ground (the switching node). More specifically, this charge pump can be seen as a modification of Fig. 8.14, suitable for high switching-node voltages, with MP1, MP2, and MN3 replaced by diodes and a pull-up resistance R.

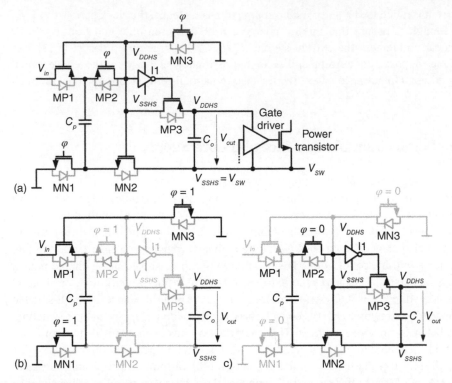

Figure 8.14 Charge pump to provide a floating gate supply V_{out} referred to V_{SSHS}: a) Full schematic with the power stage connected; equivalent circuit for b) $\varphi = 1$, and c) $\varphi = 0$.

MN2 disconnects the switching node from the bottom plate of C_p if $\varphi = 1$. The gate of MN2 is pulled to the ground by MN1. At the same time, C_p gets recharged from V_{in} through D1 and D3. During the other phase, $\varphi = 0$, MN1 turns off such that the resistor R can pull up the gate of MN2, which connects the bottom plates of both capacitors C_p and C_o. In consequence, C_p recharges C_o via D2. Its simple control by one clock signal φ and transistor MN1 and the self-boosting, enabled by resistor R, make the circuit attractive for many applications. However, the obvious drawback is static current consumption due to the resistor R, which is a function of V_{SW}. Also, the diode drop V_F significantly reduces the achievable output voltage, ideally $V_{out} = V_{in} - 3\,V_F$. Diodes D1 and D2, as well as transistors MN1 and MN2, need to be high-voltage devices.

The alternative charge pump in Fig. 8.15b) [11, 12], is also controlled by a single clock signal φ and a ground-referred CMOS inverter (MN1, MP1). The pumping capacitor C_p gets recharged out of the switching node ($V_{SSHS} = V_{sw}$) as long as $\varphi = 1$ (C_p's bottom-plate potential $V_b = 0$) and the switching node is at high potential (usually if the attached power switch is turned on). Like for Fig. 8.15a), this happens independent of the actual voltage level at the switching node. This way, power stages with nodes that do not return to zero (ground) can be supported. The charge transfer from C_p toward the buffer capacitor C_o is initiated by a short pulse $\varphi = 0$ while the switching node is at high potential as indicated in the timing diagram as part of Fig. 8.15b). The ideal output voltage reaches $V_{out} = V_{in} - 2\,V_F$, one diode drop better than Fig. 8.15a). In a practical design, transistors in the inverter must be strong enough to prevent its output voltage V_b getting pushed below ground or pulled above V_{in} due to coupling from C_p during transitions at the switching node. Otherwise, the body diodes of MN1 or MP1 would get forward conducting and may trigger parasitic bipolar transistor effects (see Section 3.3.1). Unlike Fig. 8.15a), only C_p needs to be a high-voltage component. Depending on the available technology, this may be the decision criteria

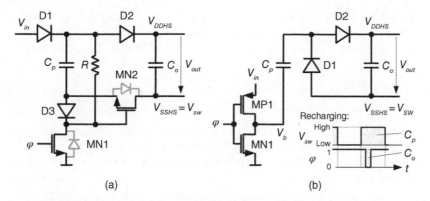

Figure 8.15 High-voltage floating gate supplies: a) Self-boosted charge pump and b) charge pump with push-pull diodes.

between both charge pumps in Fig. 8.15. Due to the static current and the voltage drop of $3V_F$, Fig. 8.15a) shows inherently lower efficiency, [11] states that Fig. 8.15b) reaches approximately 1.6 times higher power efficiency compared to the self-boosted charge pump for $V_{sw} = 20$ V.

8.10 Closed-loop Control

The charge pumps considered earlier in the chapter are all open-loop systems. Therefore, their output voltage varies over corners depending on parasitics and the load current. It is often sufficient as long as a minimum output level is reached (like providing the gate supply for a power transistor or the EEPROM programming voltage). Nevertheless, if required, adding a control loop can precisely control the output level.

Referring to Eqn. (8.17), we can take advantage of the fact that the output voltage drop ΔV_{out} is influenced by the charge pump's equivalent output resistance R_{out}. According to Eqn. (8.19) R_{out} is inversely proportional to C_p and f_{sw}. Influencing f_{sw} is usually more straightforward to be implemented. The hysteretic closed-loop control in Fig. 8.16a) utilizes clock gating to regulate V_{out} by influencing f_{sw}. The amount of charge delivered to the output can also be set by controlling the current source I_B shown Fig. 8.16b). This function can be accomplished by the current-source circuit of Fig. 8.13 (only the current source I_B is required). The linear control loop in Fig. 8.16b) is similar to the linear regulator (LDO) as described in Chapter 7 (see Fig. 7.1).

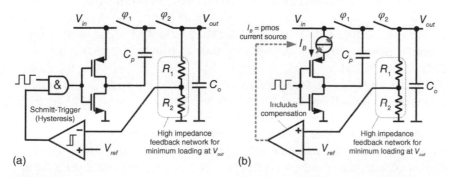

Figure 8.16 Charge pump with closed-loop control. Ways to control the charge delivery to the output include a) clock gating and b) linear control of the pumping capacitor's discharge current I_B.

If a fixed switching frequency is required (e.g., for EMI reasons, see also Section 12.3), the capacitance can be modulated as proposed in [13]. This technique also scales switching and bottom-plate losses with changes in load current and maintains a predictable fixed-frequency switching noise behavior.

References

1 Dickson, J. (1976) On-chip high-voltage generation in MNOS integrated circuits using an improved voltage multiplier technique. *IEEE Journal of Solid-State Circuits*, 11 (3), 374–378, doi: 10.1109/JSSC.1976.1050739.

2 Breussegem, T.V. and Steyaert, M. (2013) *CMOS Integrated Capacitive DC-DC Converters*, Springer Science+Business Media.

3 Pelliconi, R., Iezzi, D., Baroni, A., Pasotti, M., and Rolandi, P. (2001) Power efficient charge pump in deep submicron standard CMOS technology, in *Proceedings of the 27th European Solid-State Circuits Conference*, pp. 73–76.

4 Maksimovic, D. and Dhar, S. (1999) Switched-capacitor DC-DC converters for low-power on-chip applications, in *30th Annual IEEE Power Electronics Specialists Conference. Record. (Cat. No.99CH36321)*, vol. 1, pp. 54–59, doi: 10.1109/PESC.1999.788980.

5 Wittmann, J., Neidhardt, J., and Wicht, B. (2013) EMC optimized design of linear regulators including a charge pump. *IEEE Transactions on Power Electronics*, 28 (10), 4594–4602, doi: 10.1109/TPEL.2012.2232785.

6 Renz, P., Kaufmann, M., Lueders, M., and Wicht, B. (2019) 8.6 A fully integrated 85%-peak-efficiency hybrid multi ratio resonant DC-DC converter with 3.0-to-4.5 V input and 500 µA-to-120 mA load range, in *2019 IEEE International Solid- State Circuits Conference (ISSCC)*, IEEE, San Francisco, CA, USA, pp. 156–158, doi: 10.1109/ISSCC.2019.8662491.

7 Siragusa, E. and Galton, I. (2004) A digitally enhanced 1.8-v 15-bit 40-MSample/s CMOS pipelined ADC. *IEEE Journal of Solid-State Circuits*, 39 (12), 2126–2138, doi: 10.1109/JSSC.2004.836230.

8 Dessouky, M. and Kaiser, A. (1999) Input switch configuration suitable for rail-to-rail operation of switched-opamp circuits. *Electronics Letters*, 35 (1), 8, doi: 10.1049/el:19990028.

9 Renz, P., Kaufmann, M., Lueders, M., and Wicht, B. (2021) Switch stacking in power management ICs. *IEEE Journal of Emerging and Selected Topics in Power Electronics*, 9, 3735–3743, doi: 10.1109/JESTPE.2020.3012813.

10 Park, S. and Jahns, T. (2005) A self-boost charge pump topology for a gate drive high-side power supply. *IEEE Transactions on Power Electronics*, 20 (2), 300–307, doi: 10.1109/TPEL.2004.843013.

11 Lutz, D., Renz, P., and Wicht, B. (2018) An integrated 3-mW 120/230-V AC mains micropower supply. *IEEE Journal of Emerging and Selected Topics in Power Electronics*, 6 (2), 581–591, doi: 10.1109/JESTPE.2018.2798504.

12 Khoo, G., Carter, D., and McMahon, R. (2000) Analysis of a charge pump power supply with a floating voltage reference. *IEEE Transactions on Circuits and Systems I: Fundamental Theory and Applications*, 47 (10), 1494–1501, doi: 10.1109/81.886979.

13 Ramadass, Y.K., Fayed, A.A., and Chandrakasan, A.P. (2010) A fully-integrated switched-capacitor step-down DC-DC converter with digital capacitance modulation in 45 nm CMOS. *IEEE Journal of Solid-State Circuits*, 45 (12), 2557–2565, doi: 10.1109/JSSC.2010.2076550.

9

Capacitive DC–DC Converters

Capacitive DC–DC converters transform a DC input voltage into a DC output voltage using switches and capacitors. Alternatively, they are referred to as switched-capacitor converters or SC converters. Combining capacitors as energy-storing elements with switches enables voltage conversion without needing a resistive path that leads to excessive power losses like in a linear voltage regulator (Chapter 7). With the availability of high-density capacitances in state-of-the-art standard CMOS technologies, this class of converters became very popular for fully integrated point-of-load power supplies (power-supply-on-chip). For higher load currents, off-chip capacitors are used, which still result in compact solutions by co-integration (power-supply-in-package).

9.1 Introduction

The fundamental concept is identical to that of a charge pump covered in Chapter 8. Capacitive converters typically incorporate more complex topologies with more capacitors and switches supporting a wide operating range in terms of V_{in}, V_{out}, and I_{load}. Both conversion types, step-up and step-down, are possible. Besides offering lower losses, the step-up conversion is another advantage over linear voltage regulators, which can only provide an output voltage lower than the input level. The charge pump chapter covers the step-up conversion (Chapter 8). Therefore, the focus of this chapter will be on SC step-down converters. Figure 9.1a) shows the basic 2:1 SC converter. By combining more switches and capacitors, converters with several different conversion ratios can be implemented, as will be explored further below. The 2:1 converter in Fig. 9.1a) can be derived from the fundamental voltage doubler charge pump, as shown in Fig. 8.1 by swapping the input and output of the switch-capacitor network including C_o. There is a two-phase clock as defined in Fig. 8.1c). While C_1 is called the pumping capacitor in charge pumps, it is called the flying capacitor in SC converters.

The equivalent circuits in Fig. 9.1b,c) indicate that the converter changes between series and parallel configuration of the flying capacitor C_1 and the output capacitor C_o. Assuming the voltages across both capacitors are identical, we intuitively confirm the step-down voltage conversion ratio of 2:1 in the steady state.

We can draw one general conclusion by comparing the doubler charge pump and the 2:1 SC converter. Any step-up SC converter can be transformed into a step-down topology with the reciprocal conversion ratio by swapping the input and output connection of the SC network. This rule also applies vice versa such that any step-down topology can be converted into a step-up counterpart.

Design of Power Management Integrated Circuits, First Edition. Bernhard Wicht.
© 2024 John Wiley & Sons Ltd. Published 2024 by John Wiley & Sons Ltd.

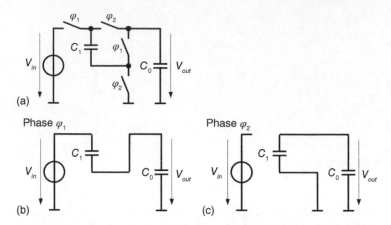

Figure 9.1 a) Basic 2:1 SC converter. The equivalent circuits during phases b) φ_1 and c) φ_2 reveal the alternating series–parallel configuration of C_o and C_1.

The DC input-to-output transfer behavior of SC converters can be modeled by the equivalent circuit, as shown in Fig. 9.2, also known as the equivalent impedance model. The model consists of an ideal DC transformer with the turns ratio 1: N_i as introduced in [1] and further developed in [2]. The transformer can be seen as an ideal voltage-controlled voltage source that provides $N_i V_{in}$ at its output. The model also includes the loss contributions due to the output resistance R_{out} and the extrinsic loss resistance R_p. $R_L = V_{out}/I_{load}$ represents a resistive load. R_{out} accounts for the output voltage droop. R_{out} is a function of the switching frequency, the capacitance of the flying capacitor(s), and the on-resistance of the switches. As will be investigated further below, R_{out} also depends on the particular SC converter topology. The resistance R_p parallel to the transformer's secondary side accounts for any extrinsic losses. These losses are mainly caused by the control of the switches, such as gate driver losses (see Section 2.7). Both R_{out} and R_p are influenced by various parasitics like the bottom and top-plate capacitances of the flying capacitors.

The ideal voltage conversion ratio is defined as the ratio between the voltages on both sides of the transformer in Fig. 9.2,

$$N_i = \frac{V_{int}}{V_{in}} = \frac{V_{out}|_{I_{load}=0}}{V_{in}}. \tag{9.1}$$

For inverting converter architectures, N_i can also have a negative sign. For zero load, the ideal conversion ratio N_i corresponds to the input-to-output voltage transfer function of the converter because the ideal output voltage V_{int} and the output voltage V_{out} are equal. Any load current causes

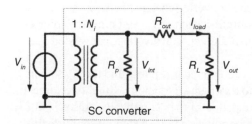

Figure 9.2 SC converter model with the ideal voltage conversion ratio N_i, the equivalent output resistance R_{out} and resistance R_p reflecting extrinsic losses. R_L models a resistive load.

a voltage drop across the equivalent output resistance R_{out} and reduces V_{out} to

$$V_{out} = N_i V_{in} - R_{out} I_{load}. \tag{9.2}$$

The model in Fig. 9.2 shows that R_{out} and the load resistance R_L form a voltage divider between $V_{int} = N_i V_{in}$ and V_{out}, resulting in the actual (real) voltage conversion ratio

$$N = \frac{V_{out}}{V_{in}} = \frac{R_L}{R_{out} + R_L} N_i \leq N_i. \tag{9.3}$$

In the case of non-zero load, the actual conversion ratio N is always lower than the ideal ratio N_i.

Neglecting the extrinsic losses ($R_p \rightarrow \infty$), we can estimate the conversion efficiency of the SC converter:

$$\eta = \frac{P_{out}}{P_{in}} = \frac{I_{load}^2 R_L}{I_{load}^2 R_{out} + I_{load}^2 R_L} = \frac{R_L}{R_{out} + R_L} = \frac{N}{N_i} = \frac{V_{out}}{N_i V_{in}} \tag{9.4}$$

Interestingly, the efficiency corresponds to the ratio N/N_i, which is identical to the V_{out}/V_{in} divided by N_i. For comparison, the efficiency of a linear regulator (LDO) is approximately given by the ratio V_{out}/V_{in}. Therefore, the SC converter usually has higher efficiency than the LDO.

Equation (9.4) reveals an important characteristic of SC converters. The lower the equivalent output resistance, the higher the efficiency. For $R_{out} = 0$ and $R_p \rightarrow \infty$ the efficiency approaches 100%. Therefore, SC converter designs aim to minimize R_{out} while keeping the extrinsic losses (gate driver, control block, etc.) low. Equation (9.4) also indicates that the efficiency does not depend on the load current in the first order (neglecting R_p).

Example 9.1 *Estimate the efficiency of a SC converter for $V_{in} = 5\,V$, $V_{out} = 1.8\,V$ if the ideal conversion ratio is $N_i = 1/2$?*

The given values of V_{in} and V_{out} correspond to an actual conversion ratio

$$N = \frac{V_{out}}{V_{in}} = \frac{1.8\,V}{5\,V} = 0.36. \tag{9.5}$$

Hence, we can calculate the efficiency utilizing Eqn. (9.4):

$$\eta = \frac{N}{N_i} = \frac{0.36}{0.5} = 72\% \tag{9.6}$$

The efficiency enhancement factor (*EEF*) as introduced in [3] allows for comparing the power conversion efficiency of any switched-mode step-down DC–DC converter η_{DCDC} to that of an ideal LDO η_{LDO} (see Chapter 7). The *EEF* relates the difference between both efficiencies to the efficiency of the DC–DC converter at the same output power and ideal conversion ratio:

$$EEF = \frac{\eta_{DCDC} - \eta_{LDO}}{\eta_{DCDC}} = 1 - \frac{\eta_{LDO}}{\eta_{DCDC}} \tag{9.7}$$

The efficiency of the LDO is given in Eqn. (7.12). The *EEF* value range is between zero (LDO and DC–DC converter have identical efficiency) and unity ($\eta_{DCDC} \gg \eta_{LDO}$).

Example 9.2 *Calculate the efficiency enhancement factor (EEF) for $V_{in} = 5\,V$, $V_{out} = 1.8\,V$ if the switched-capacitor converter has an efficiency of $\eta_{DCDC} = 72\%$ (see Example 7.1).*

The efficiency of the LDO is given by the ratio V_{out}/V_{in} (see Eqn. (7.12)), hence,

$$\eta_{LDO} = \frac{V_{out}}{V_{in}} = \frac{1.8\,V}{5\,V} = 36\%. \tag{9.8}$$

This results in the EEF

$$EEF = 1 - \frac{\eta_{LDO}}{\eta_{DCDC}} = 1 - \frac{36\%}{77\%} = 0.5. \qquad (9.9)$$

The model of Fig. 9.2 also determines the open-loop load regulation properties. If the target output voltage for a given input voltage is not exactly met by N_i, a voltage drop across R_{out} occurs. It will always be the case in applications with widely varying operating ranges of V_{in} and V_{out}. Adjusting R_{out} is the fundamental mechanism for controlling the output voltage of the SC converter. The availability of different SC converter topologies with various conversion ratios allows setting the ideal conversion ratio as close as possible to the actual ratio between V_{out} and V_{in}, hence, maintaining high conversion efficiency. The following section will study achievable conversion ratios for a given number of flying capacitors.

9.2 Realizable Ratios

Various SC converter topologies with specific conversion ratios can be implemented using multiple capacitors and switches. In 1995, the groundbreaking work on SC converters by Makowski and Maksimović [1] derived that all possible ideal conversion ratios N_i for a given number n of flying capacitors are determined by the term

$$N_i[n] = \frac{P}{Q}. \qquad (9.10)$$

This equation applies to two-phase designs, the most common class of SC converters. For multi-phase topologies, more ratios can be achieved at the expense of higher complexity (see Section 9.10). In Eqn. (9.10), P and Q are positive integers that are an element of an integer series $(1, F_{n+1})$ with F_{n+1} being the $(n + 1)$th element of the Fibonacci series F. With the zeroth and the first elements of the Fibonacci series F defined as 1, the successive elements are the sum of the preceding two elements. Hence, the zeroth to sixth elements are

$$F = 1, 1, 2, 3, 5, 8, 13, \ldots . \qquad (9.11)$$

Table 9.1 lists all realizable ratios for up to $n = 4$ flying capacitors assuming two-phase operation. For simplicity, only the step-down ratios are depicted. The step-up ratios can be obtained by taking the reciprocal value of all listed step-down values.

Table 9.1 reveals that the number of attainable ratios increases significantly with the number of flying capacitors. However, more than four capacitors are rarely used due to space and area constraints and growing complexity.

Table 9.1 Realizable ideal step-down voltage conversion ratios for n flying capacitors (in two-phase operation).

n	Realizable ratios
1	$\frac{1}{2}, 1$
2	$\frac{1}{3}, \frac{1}{2}, \frac{2}{3}, 1$
3	$\frac{1}{5}, \frac{1}{4}, \frac{1}{3}, \frac{2}{5}, \frac{1}{2}, \frac{3}{5}, \frac{2}{3}, \frac{3}{4}, \frac{4}{5}, 1$
4	$\frac{1}{8}, \frac{1}{7}, \frac{1}{6}, \frac{1}{5}, \frac{1}{4}, \frac{2}{7}, \frac{1}{3}, \frac{3}{8}, \frac{2}{5}, \frac{3}{7}, \frac{1}{2}, \frac{4}{7}, \frac{3}{5}, \frac{5}{8}, \frac{2}{3}, \frac{5}{7}, \frac{3}{4}, \frac{4}{5}, \frac{5}{6}, \frac{6}{7}, \frac{7}{8}, 1$

Table 9.2 All realizable ideal voltage conversion ratios N_i for $n = 3$ flying capacitors (in two-phase operation, shaded fields mark repetitive ratios).

$N_i = \frac{P}{Q}$	$Q = 1$	$Q = 2$	$Q = 3$	$Q = 4$	$Q = 5$
$P = 1$	$\frac{1}{1} = 1$	$\frac{1}{2}$	$\frac{1}{3}$	$\frac{1}{4}$	$\frac{1}{5}$
$P = 2$	$\frac{2}{1} = 2$	$\frac{2}{2} = 1$	$\frac{2}{3}$	$\frac{2}{4} = \frac{1}{2}$	$\frac{2}{5}$
$P = 3$	$\frac{3}{1} = 3$	$\frac{3}{2}$	$\frac{3}{3} = 1$	$\frac{3}{4}$	$\frac{3}{5}$
$P = 4$	$\frac{4}{1} = 4$	$\frac{4}{2} = 2$	$\frac{4}{3}$	$\frac{4}{4} = 1$	$\frac{5}{4}$
$P = 5$	$\frac{5}{1} = 5$	$\frac{5}{2}$	$\frac{5}{3}$	$\frac{5}{4}$	$\frac{5}{5} = 1$

For $n = 3$ flying capacitors, Table 9.2 shows how the ratios can be found based on Eqn. (9.3). In this case, the upper bound Fibonacci number is $F_4 = 5$, i.e., $1 \leq P \leq 5$ and $1 \leq Q \leq 5$. The shaded fields indicate repetitive ratios, which can be discarded. A total number of 19 ratios can be obtained, nine step-up ratios and nine step-down ratios, as well as $N_i = 1$.

SC converters with $n = 2$ flying capacitors provide four distinct conversion ratios. It includes $N_i = 1$, which is straightforward and will not be investigated further. One flying capacitor can achieve the $N_i = 1/2$ ratio, as shown in Fig. 9.1. With two capacitors, an interleaving topology can be implemented, or both capacitors are placed in parallel (as illustrated for multi-ratio converters in Figs. 9.21 and 9.22). SC converters with the remaining ratios, 1/3 and 2/3, are shown in Fig. 9.3. The behavior resembles the fundamental 2:1 converter in Fig. 9.1. The equivalent circuits in phases φ_1 and φ_2 reveal that all capacitors toggle between series and parallel configuration. Consequently, this kind of SC topology is called the series–parallel converter. All capacitors in Fig. 9.3a) get charged to the same voltage V_{out} during phase φ_2 (assuming ideal conversion and zero load). In phase φ_1, all capacitor voltages add up to V_{in}. Kirchhoff's laws are fulfilled in steady-state if all capacitors get charged to $V_{out} = V_{in}/3$. Intuitively, this confirms that the ideal conversion ratio is 3:1. It is worth noticing that it is independent of the actual capacitance values. The capacitances determine the charge transfer from phase to phase but not the steady-state conversion ratio.

The topology in Fig. 9.3b) provides the 3:2 conversion by connecting the flying capacitors C_1 and C_2 slightly different in comparison to the 3:1 topology. We can again verify the conversion ratio by inspection. Assuming $V_{out} = 2/3V_{in}$, during phase φ_1 the voltage across the parallel configuration of C_1 and C_2 will be $1/3V_{in}$. In phase φ_2, both capacitors appear in series and parallel to C_o. Hence, $V_{out} = 2/3V_{in}$ is fulfilled.

For a higher number of flying capacitors, the converter topologies of Fig. 9.3 can be expanded by adding more stages in the same pattern. We will explore this further in Section 9.3.

The question arises if other SC configurations achieve the same conversion ratio. The answer is yes, as confirmed in the following section.

9.3 Switched-Capacitor Topologies

The SC configuration for any conversion ratio as given by Eqn. (9.10) and Table 9.1 is not unique. Among many possible topologies, there are well-established configurations, namely the series–parallel, the Dickson, the ladder, and the Fibonacci topology. For 3:1 conversion, these topologies are shown in Fig. 9.4. We will discuss these topologies below, followed by a more comprehensive analysis of the ideal conversion ratio N_i and the equivalent output resistance R_{out} in Section 9.5.

9.3.1 Series–Parallel

Figure 9.4a) repeats the 3:1 series–parallel topology of Fig. 9.3a) with the switches numbered. The 3:1 topology extends the initial SC circuit of Fig. 9.1. It uses seven switches and two flying capacitors. Besides the number of components, it is interesting to compare the maximum voltages that occur across the switches and capacitors. The different converter topologies differ in device voltage ratings. Lower ratings will benefit the practical implementation. On the IC level, the lower the device ratings, the smaller the layout area. Higher device ratings typically also result in higher losses due to larger parasitic device capacitances and higher gate charges of the switches. It reduces the parallel resistance R_p in the model of Fig. 9.2. Transistors with higher voltage ratings usually also have a higher area-specific resistance (see Eqn. (3.3) in Section 3.2). By inspection, we find for Fig. 9.4a) that both S1 and S2 see a maximum voltage of $2V_{out}$ (S1 during φ_2 and S2 during φ_1, see also the series–parallel equivalent circuits in Fig. 9.3a). As an advantage of this topology, the maximum capacitor voltage of V_{out} is relatively low. Higher step-down ratios can be achieved by expanding the topology. For instance, the 3:1 topology converts into a 4:1 serial-parallel converter by adding one more capacitor and a switch φ_1 to the left (connected to V_{in}). An additional φ_2 switch needs to connect the right terminal of the new capacitor to the output node.

9.3.2 Dickson

This topology is based on the voltage multiplier proposed by Dickson in 1976 [4], discussed in the context of charge pumps in Chapter 8. The 3:1 step-down implementation in Fig. 9.4b) is the reciprocal of the three-stage Dickson charge pump of Fig. 8.10, which results from swapping the input and output connection of the SC network. Their internal pull-up and pull-down switches replace the inverters in Fig. 8.10. The pull-up switches are now tapped from V_{out}. Like the series–parallel converter, the Dickson topology uses two flying capacitors. Among the two capacitors, C_1 sees the maximum voltage of $2V_{out}$ during phase φ_1. The voltage at C_2 stays at V_{out} in both phases. Switch S2 needs to withstand $2V_{out}$, but all other switches need to block only V_{out}. Adding one more switch and capacitor between S1 and V_{in} increases the step-down ratio by one. This way, the Dickson topology

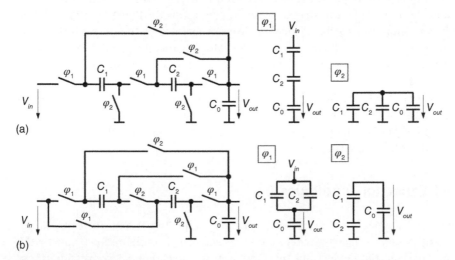

(a)

(b)

Figure 9.3 Series–parallel converters with two flying capacitors and equivalent circuits during each phase: a) 3:1 topology; b) 3:2 topology.

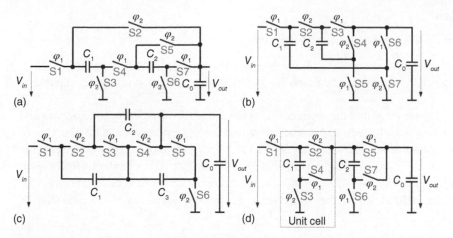

Figure 9.4 3:1 SC topologies: a) series–parallel; b) Dickson; c) ladder; d) Fibonacci.

can be expanded incrementally. In every step, the voltage ratings of the switches and capacitors increase by V_{out}. In any case, the maximum values occur at C_1 and S2.

9.3.3 Ladder

The ladder topology is one of the early voltage multipliers, first studied comprehensively by Lin and Chua in 1977 [5]. It implements a straightforward principle. The 3:1 topology in Fig. 9.4c) consists of two rows of flying capacitors, one row with C_1 and C_3 and the other with C_2. Each row can be seen as a fork that toggles alternately between V_{in} and V_{out} (upper row, i.e., C_2) or ground (lower row, i.e., C_1 and C_3). This way, the charge transfer from input-to-output is accomplished. As an advantage over other topologies, neither the switches nor the capacitors have to block maximum voltages higher than V_{out}. On the downside, the ladder converter needs one more capacitor than the other topologies. The step-down ratio can be extended by alternately adding another capacitor at the upper and lower rows. Finally, one more switch must be placed toward V_{in} on the left. Since each capacitor is charged to V_{out}, it can be used as a floating gate supply similar to a bootstrap capacitor (see Section 5.11). The lower potential (the right-hand terminal of each flying capacitor in Fig. 9.4c)) acts as the high-side ground for the floating gate driver. Only the ladder topology systematically offers this kind of benefit.

9.3.4 Fibonacci

Even though the Fibonacci series determines the number of attainable conversion ratios, the so-called Fibonacci topology in Fig. 9.4d) was introduced not before the early 1990s [1, 6]. The name stems from the fact that this topology can only provide step-down ratios of $1/F_{(n+1)}$ with the $(n + 1)$th element of the Fibonacci series F according to Eqn. (9.11). The next lower conversion ratio can be implemented by adding one more unit cell, indicated in Fig. 9.4d), between the right-hand connection of S5/S7 and V_{out}. This way, one more unit cell yields a 5:1 topology, and a second one forms the 8:1 converter. Note that complementary clock signals control the switches in neighboring unit cells. Section 9.10 explains how this topology can be utilized to provide a 4:1 conversion ratio by implementing a third switching phase (see Fig. 9.14). During φ_1, C_1 and S2 see the maximum voltage of $2V_{out}$. The same voltage needs to be blocked by S4 in phase φ_2 while C_1 remains at $2V_{out}$.

9.3.5 Conclusion

Overall, we can make a few general observations:

- For $N_i = 2$ all topologies are identical and correspond to the fundamental 2:1 stage of Fig. 9.1.
- Only some topologies can achieve the theoretically attainable conversion ratios investigated in Section 9.2.
- During start-up, the switches and the capacitors may see the entire input voltage. Clamping and pre-charging circuits must be installed to avoid violating the maximum voltage ratings. One or multiple shunt regulators (see Fig. 7.22) may be used to properly bias the individual nodes of the SC circuit at start-up. The implementation can be similar to Fig. 7.23 in the chapter on linear regulators. Also, high-resistive (power dissipation!) voltage dividers with resistors or diode-connected transistors can be helpful. Once the converter is settled, the pre-charge circuits may be disabled.

9.4 Gate Drive Techniques

Each power switch in the converter requires a dedicated gate control circuit comprising the level shifter, the gate driver, and the gate supply. The implementation of these circuits will have different levels of complexity depending on the actual topology and the voltage levels of the application. The gate drive concepts of Chapter 5 can be applied. Low-side transistors with their source terminal permanently connected to the ground can be driven directly, illustrated in Fig. 9.5a) for switch M1. All other switches are high-side switches. If their source node returns to ground, a bootstrap gate supply can be implemented as described in Section 5.11, shown for switch M2 in Fig. 9.5a). Not all switching nodes return to ground (zero) in each period. In this case, charge pump circuits can be used as introduced in Section 8.9. It applies, for instance, to M3 and M4 in Fig. 9.5a), where a charge pump with push-pull diodes (see Fig. 8.15b) is used.

Some SC converters utilize an auxiliary SC stage to generate proper gate supply levels. The voltage swing in many SC topologies can be approximated by integer multiples of V_{out} (discussed in Section 9.3). Therefore, the auxiliary supply stage needs to generate V_{out}, $2V_{out}$, $3V_{out}$, etc. It can be

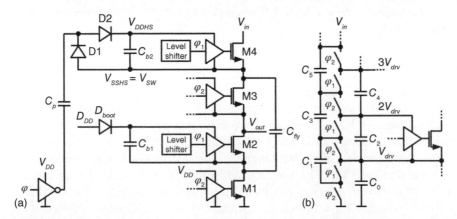

Figure 9.5 Gate drive techniques for SC converters: a) direct gate control at low-side transistor M1, bootstrapping applied to switch M2, charge pump of Fig. 8.15b) used for high-side switch M4; b) ladder SC converter as an auxiliary gate supply generator delivering multiples of the drive level V_{drv}.

accomplished by a ladder converter, as shown in Fig. 9.5b) (see also Fig. 9.4c) [7]. It represents a universal technique that provides gate drive signals regardless of the converter complexity. However, the maximum voltage ratings must be considered for widely varying V_{in}.

9.5 Charge Flow Analysis

This section introduces charge flow analysis as an effective method of analyzing complex SC converters. In contrast, even though always applicable, the charge balance analysis, as introduced in Section 8.2, has already become rather complicated for two flying capacitors. Charge flow analysis follows a similar idea to small-signal analysis of analog circuits because only the change of quantities is considered, not the absolute values. The method was developed by Seeman and Sanders in 2008 [2, 8] based on the fundamental work in [1].

In the first step, charge multiplier vectors are derived. Those vectors describe the topology based on the charge flow through all components. After the charge multiplier vectors are known, the performance of the SC topology can be analyzed and optimized in terms of ideal conversion ratio and equivalent output resistance. We will cover the charge flow analysis step by step below. We consider only two-phase converters. Section 9.10 will adopt the charge-flow analysis for multi-phase converters.

9.5.1 Charge Flow Vectors

To determine the conversion ratio and to model the behavior of a topology, the charge that flows between the components in the individual switching phases must be known. Charge flow vectors are defined, which contain the charge change q_j^i (the charge *flow*) of all components for each switching phase i. There are two different types of vectors. The first type accounts for the charge flow at the capacitors, while the other type considers the switches. It will be used to analyze the operation in the slow-switching limit (SSL) and fast-switching limit (FSL) as defined in Section 9.5.4.

Capacitor Charge Flow Vectors
This type of charge flow vector type has the form

$$\boldsymbol{q^i} = \begin{bmatrix} q_{out}^i & q_{c,1}^i & \cdots & q_{c,n}^i & q_{in}^i \end{bmatrix}.$$ (9.12)

The first vector element on the left of $\boldsymbol{q^i}$ is the charge that flows into the output. The last entry (right) is the charge the input voltage source V_{in} delivers. The flying capacitors are described by their average charge flow. The sign indicates whether the capacitor gets charged or discharged. If charge flows into the positive terminal of the capacitor, it charges the capacitor. The corresponding vector entry is positive. Vice versa, discharging results in a negative sign of charge flow (out of the capacitor). The input voltage source always delivers charge. Hence, its charge flow has always a positive sign. If the input source is disconnected, the charge flow may be zero. It is the case, for instance, if the left-most switch in the 2:1 converter of Fig. 9.1a) is turned off during phase $\varphi_2 (\varphi_1 = 0)$.

In the next step, the vectors according to Eqn. (9.12) are referred to as the total charge

$$q_{out} = q_{out}^1 + q_{out}^2 = \frac{I_{load}}{f_{sw}}$$ (9.13)

that is delivered to the output capacitor C_o per switching period $T = 1/f_{sw}$. The resulting vector $\boldsymbol{a^i}$ relates all charge flow quantities to the charge flowing into the output and thus to the load current

I_{load}:

$$\mathbf{a}^i = \begin{bmatrix} a^i_{out} & a^i_{c,1} & \cdots & a^i_{c,n} & a^i_{in} \end{bmatrix} = \begin{bmatrix} q^i_{out} & q^i_{c,1} & \cdots & q^i_{c,n} & q^i_{in} \end{bmatrix} / q_{out} \tag{9.14}$$

Each flying capacitor must have a net charge balance of zero over all phases. Otherwise, the capacitor voltage would increase or decrease with each clock cycle and grow to infinity in a positive or negative direction. Hence,

$$q^1_{c,j} + q^2_{c,j} = 0. \tag{9.15}$$

The charge flow vectors represent a linear equation system. Instead of solving these equations, the vector elements can typically be found by inspection. For example, consider the 2:1 converter of Fig. 9.1. If we define the charge that flows out of V_{in} during phase φ_1 to be q_x, as shown in Fig. 9.6, the non-normalized charge flow vectors during both phases are

$$\mathbf{q}^1 = \begin{bmatrix} q^1_{out} & q^1_{c,1} & q^1_{in} \end{bmatrix} = \begin{bmatrix} q_x & q_x & q_x \end{bmatrix}$$
$$\mathbf{q}^2 = \begin{bmatrix} q^2_{out} & q^2_{c,1} & q^2_{in} \end{bmatrix} = \begin{bmatrix} q_x & -q_x & 0 \end{bmatrix}. \tag{9.16}$$

Dividing these vectors by the total output charge flow $q_{out} = q^1_{out} + q^2_{out} = 2q_x$ according to Eqn. (9.13) yields the normalized charge flow vectors as defined by Eqn. (9.14):

$$\mathbf{a}^1 = \begin{bmatrix} a^1_{out} & a^1_{c,1} & a^1_{in} \end{bmatrix} = \begin{bmatrix} q_x & q_x & q_x \end{bmatrix} / 2q_x = \begin{bmatrix} \frac{1}{2} & \frac{1}{2} & \frac{1}{2} \end{bmatrix}$$
$$\mathbf{a}^2 = \begin{bmatrix} a^2_{out} & a^2_{c,1} & a^2_{in} \end{bmatrix} = \begin{bmatrix} q_x & -q_x & 0 \end{bmatrix} / 2q_x = \begin{bmatrix} \frac{1}{2} & -\frac{1}{2} & 0 \end{bmatrix} \tag{9.17}$$

Switch Charge Flow Vectors

The charge flows in the switches during phase i result in a vector of the form

$$\mathbf{a}^i_r = \begin{bmatrix} a^i_{r,1} & \cdots & a^i_{r,m} \end{bmatrix} = \begin{bmatrix} q^i_{r,1} & \cdots & q^i_{r,m} \end{bmatrix} / q_{out} \tag{9.18}$$

with q_{out} according to Eqn. (9.13). The elements $a^i_{r,j}$ represent the charge multipliers, defined as the charge flows through switch j during phase i. The values for switches that are off are zero. Again the charge quantities are normalized to the output charge q_{out} as defined in Eqn. (9.13). The index r refers to the switch resistance, contributing losses as part of R_{out} in FSL as analyzed in Section 9.5.4.

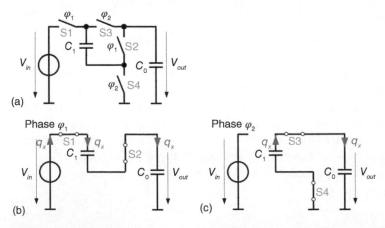

Figure 9.6 Charge flow analysis: a) the 2:1 SC converter of Fig. 9.1 with labeled switches convert into the equivalent circuits during phases b) φ_1, and c) φ_2 with the defined charge quantity q_x.

The basic 2:1 converter in Fig. 9.6 contains four switches. By defining the charge quantity q_x, as shown in Fig. 9.6b,c) the charge flow vectors of the switches can be found by inspection:

$$\boldsymbol{a_r^1} = \begin{bmatrix} a_{r,1}^1 & a_{r,2}^1 & a_{r,3}^1 & a_{r,4}^1 \end{bmatrix}$$

$$= \begin{bmatrix} q_x & q_x & 0 & 0 \end{bmatrix} / 2q_x = \begin{bmatrix} \frac{1}{2} & \frac{1}{2} & 0 & 0 \end{bmatrix}$$

$$\boldsymbol{a_r^2} = \begin{bmatrix} a_{r,1}^2 & a_{r,2}^2 & a_{r,3}^2 & a_{r,4}^2 \end{bmatrix}$$

$$= \begin{bmatrix} 0 & 0 & q_x & q_x \end{bmatrix} / 2q_x = \begin{bmatrix} 0 & 0 & \frac{1}{2} & \frac{1}{2} \end{bmatrix} \tag{9.19}$$

Each switch is turned on in phase 1 or 2 in a two-phase converter. Hence, a single vector can describe the charge flow [2],

$$\boldsymbol{a_r} = \begin{bmatrix} a_{r,1} & \cdots & a_{r,m} \end{bmatrix}. \tag{9.20}$$

The charge multiplier elements $a_{r,j}$ define the charge flow through each switch j during the phase in which the switch is on. For the switches that conduct during phase 1 ($\varphi_1 = 1$), the corresponding entries for $a_{r,j}$ can be calculated from the capacitor charge flows $a_{c,j}^1$. Likewise, the values for switches turned on during phase 2 can be determined from the coefficients $a_{c,j}^2$.

9.5.2 Charge Flow Vectors of Common Topologies

The first step when analyzing the charge flow is to draw the equivalent circuits during each phase, as shown in Fig. 9.7 for the four common topologies.

In the second step, we must mark the charge flow vectors for all capacitors and switches in each phase. Equation (9.15) provides a helpful rule stating that each phase's corresponding charge flow elements are opposite in sign but have equal absolute values. For this reason, it is convenient to define the charge quantity q_x of the serial-parallel converter first in phase φ_1 when all capacitors are in series. (Fig. 9.7a). For Dickson, it is better to start with phase φ_2 (Fig. 9.7b).

Out of the standard topologies, assigning the charge q_x in the Ladder converter in Fig. 9.7c) is most tricky. One suitable approach is to start with C_1 in phase φ_1. Since the polarity changes in phase φ_2, we also find the charge at C_2 by observing Kirchhoff's current law (KCL). Now we return to phase φ_1 and mark the charge at C_2 in the opposite direction. KCL gives the charge at C_3 during φ_1 as the sum of charge flow at C_1 and C_2, which is $2q_x$ in this case. This way, the charge flow at all three flying capacitors is also defined during φ_2. Finally, the output capacitor C_o sees a charge flow of q_x and $2q_x$ during phases φ_1 and φ_2, respectively.

For the Fibonacci topology, q_x is preferably defined in the series configuration during φ_2.

After all charge flows are determined, the total output charge according to Eqn. (9.13) can be calculated. Interestingly, all 3:1 topologies in Fig. 9.7 result in $q_{out} = 3q_x$. Now the charge flow vectors of the capacitors and switches as defined by Eqns. (9.14) and (9.18) can be derived. In those vectors, the charge elements get normalized to the total output charge q_{out}. Table 9.3 summarizes the charge flow vectors for each topology.

9.5.3 Ideal Conversion Ratio

Equation (9.13) defines the load current at the output of the SC converter to be $I_{load} = q_{out} f_{sw}$. Likewise, the average input current is $I_{in} = q_{in} f_{sw}$. The ideal conversion ratio assumes that the input power

$$P_{in} = V_{in} I_{in} = V_{in} q_{in} f_{sw} \tag{9.21}$$

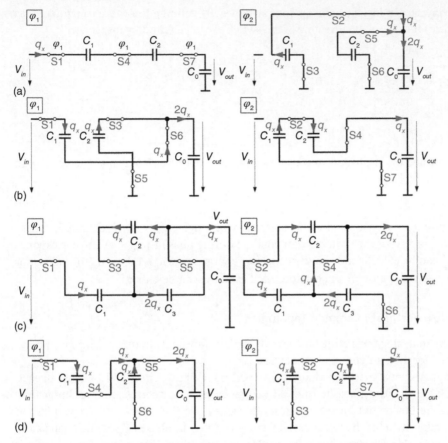

Figure 9.7 Charge flow analysis of common 3:1 topologies: a) series–parallel; b) Dickson; c) ladder; d) Fibonacci.

Table 9.3 Charge flow vectors of common 3:1 SC topologies (referred to Fig. 9.7).

Series–parallel	$\boldsymbol{a^1} = \begin{bmatrix} \frac{1}{3} & \frac{1}{3} & \frac{1}{3} & \frac{1}{3} \end{bmatrix}, \boldsymbol{a^2} = \begin{bmatrix} \frac{2}{3} & -\frac{1}{3} & -\frac{1}{3} & 0 \end{bmatrix}$
	$\boldsymbol{a_r} = \begin{bmatrix} \frac{1}{3} & \frac{1}{3} & \frac{1}{3} & \frac{1}{3} & \frac{1}{3} & \frac{1}{3} & \frac{1}{3} \end{bmatrix}$
Dickson	$\boldsymbol{a^1} = \begin{bmatrix} \frac{2}{3} & \frac{1}{3} & -\frac{1}{3} & \frac{1}{3} \end{bmatrix}, \boldsymbol{a^2} = \begin{bmatrix} \frac{1}{3} & -\frac{1}{3} & \frac{1}{3} & 0 \end{bmatrix}$
	$\boldsymbol{a_r} = \begin{bmatrix} \frac{1}{3} & \frac{1}{3} & \frac{1}{3} & \frac{1}{3} & \frac{1}{3} & \frac{1}{3} & \frac{1}{3} \end{bmatrix}$
Ladder	$\boldsymbol{a^1} = \begin{bmatrix} \frac{1}{3} & \frac{1}{3} & -\frac{1}{3} & \frac{2}{3} & \frac{1}{3} \end{bmatrix}, \boldsymbol{a^2} = \begin{bmatrix} \frac{2}{3} & -\frac{1}{3} & \frac{1}{3} & -\frac{2}{3} & 0 \end{bmatrix}$
	$\boldsymbol{a_r} = \begin{bmatrix} \frac{1}{3} & \frac{1}{3} & \frac{1}{3} & \frac{1}{3} & \frac{2}{3} & \frac{2}{3} \end{bmatrix}$
Fibonacci	$\boldsymbol{a^1} = \begin{bmatrix} \frac{2}{3} & \frac{1}{3} & -\frac{1}{3} & \frac{1}{3} \end{bmatrix}, \boldsymbol{a^2} = \begin{bmatrix} \frac{1}{3} & -\frac{1}{3} & \frac{1}{3} & 0 \end{bmatrix}$
	$\boldsymbol{a_r} = \begin{bmatrix} \frac{1}{3} & \frac{1}{3} & \frac{1}{3} & \frac{1}{3} & \frac{2}{3} & \frac{1}{3} & \frac{1}{3} \end{bmatrix}$

and the output power

$$P_{out} = V_{out}I_{load} = V_{out}q_{out}f_{sw} \tag{9.22}$$

are equal:

$$V_{in}q_{in}f_{sw} - V_{out}q_{out}f_{sw} \tag{9.23}$$

The assumption $P_{in} = P_{out}$ also means $R_{out} = 0$ and $R_p \to \infty$ (see Fig. 9.2). In consequence, the ideal conversion ratio as defined in Eqn. (9.1) can be calculated by the expression

$$N_i = \frac{V_{out}}{V_{in}} = \frac{q_{in}}{q_{out}} = \frac{a_{in}}{a_{out}} \tag{9.24}$$

with $a_{out} = a_{out}^1 + a_{out}^2$ (see Eqn. (9.13)) and $a_{in} = a_{in}^1 + a_{in}^2$. Equation (9.24) is a beneficial result as part of the charge flow analysis. The ideal conversion ratio can be determined from the input and output charge-flow vector elements.

Based on Eqn. (9.24) we confirm that the ideal conversion ratio of the 2:1 converter in Fig. 9.6 is

$$N_i = \frac{V_{out}}{V_{in}} = \frac{a_{out}}{a_{in}} = \frac{\frac{1}{2} + 0}{\frac{1}{2} + \frac{1}{2}} = \frac{1}{2}. \tag{9.25}$$

The charge flow vectors in Table 9.3 allow us to calculate the ideal conversion ratios for commonly used 3:1 topologies. As expected, all topologies result in $N_i = 1/3$.

9.5.4 Equivalent Output Resistance

Referring to the SC converter model in Fig. 9.2, the equivalent output resistance R_{out} accounts for charging losses in the flying capacitors:

$$E_{loss} = P_{loss}/f_{sw} = I_{load}^2 R_{out}/f_{sw} \tag{9.26}$$

Losses will be further investigated in Section 9.9. There are two asymptotic limits to the output resistance R_{out} concerning the switching frequency, the slow and fast-switching limits (SSL, FSL) [2, 8]. On a double-logarithmic scale, Fig. 9.8 shows the typical frequency characteristic of R_{out}. In the lower frequency range, the resistance decreases with frequency. This observation is intuitive, as more charges can be delivered in time. The SSL resistance assumes the switches and all interconnects have zero on-resistance. Only the flying capacitors determine R_{out}. At high switching frequencies, the FSL assumes that the period is so short that any charging/discharging process can be neglected, and the capacitors can be considered to stay at a constant voltage. They act effectively as constant voltage sources. Instead, the finite resistance of the switches and the interconnections dominate the output resistance. The equivalent output resistance in SSL and FSL operation will be analyzed below.

Figure 9.8 Equivalent output resistance R_{out} of an SC converter with its asymptotic limits, the slow switching limit (SSL), and the fast-switching limit (FSL). Both axes are drawn on a logarithmic scale.

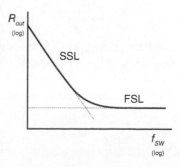

Slow-Switching Limit (SSL)

The capacitor charge flow vector according to Eqn. (9.14) is considered at lower switching frequencies. The change in charge at a given capacitor by an amount q requires the energy

$$\Delta E = \int_0^q V(Q)dQ = \frac{1}{2}\frac{q^2}{C}, \tag{9.27}$$

which corresponds to a change in the capacitor voltage by ΔV. The dissipated energy at the source is $E = q \cdot \Delta V$. The difference between the two energy quantities is the energy that is lost during the charging process:

$$E_{loss} = E - \Delta E = \frac{1}{2}\frac{q^2}{C} \tag{9.28}$$

Such losses occur in each switching phase in the flying capacitors C_j of the SC converter because their charge changes by q_j^i. The losses in the output capacitor can usually be neglected since C_o is much larger than the other capacitors. By summing up the losses from all charge flows of all n flying capacitors over all switching phases n_{ph}, the loss energy per clock cycle $T = 1/f_{sw}$ can be determined:

$$E_{loss} = \sum_{i=1}^{n_{ph}}\sum_{j=1}^{n} \frac{1}{2}\frac{\left(q_j^i\right)^2}{2\,C_j} \tag{9.29}$$

These losses are equal to the loss energy in R_{out} given by Eqn. (9.26). We assign $R_{SSL} = R_{out}$ to denote the SSL output resistance. Setting Eqns. (9.26) and (9.29) equal and solving for R_{out} yields

$$R_{SSL} = \sum_{i=1}^{n_{ph}}\sum_{j=1}^{n} \frac{f_{sw}\left(q_j^i\right)^2}{2\,C_j\,I_{load}^2}. \tag{9.30}$$

The load current I_{load} can be substituted by means of Eqn. (9.13):

$$R_{SSL} = \sum_{i=1}^{n_{ph}}\sum_{j=1}^{n} \frac{1}{2\,C_j f_{sw}}\left(\frac{q_j^i}{q_{out}}\right)^2 \tag{9.31}$$

The term q_j^i/q_{out} corresponds to the normalized charge flow elements a_j^i and can be substituted:

$$R_{SSL} = \sum_{i=1}^{n_{ph}}\sum_{j=1}^{n} \frac{\left(a_j^i\right)^2}{2\,C_j f_{sw}} \tag{9.32}$$

For SC converters with two switching phases ($n_{ph} = 2$), the absolute values of the charge flow in both phases are equal (see Eqn. (9.15)). Hence, for two-phase topologies Eqn. (9.32) simplifies to

$$R_{SSL} = \sum_{j=1}^{n} \frac{\left(a_j^1\right)^2}{C_j f_{sw}}. \tag{9.33}$$

According to Eqn. (9.15), the square of a_j^1 is identical in both phases, and either value can be used to calculate R_{SSL}. As a general observation, the equivalent output resistance decreases inversely proportional to the capacitance C_j and switching frequency f_{sw}.

For the fundamental 2:1 topology, we find

$$R_{SSL} = \frac{1}{4\,C_j f_{sw}} \tag{9.34}$$

with the charge flow elements of Eqn. (9.17).

Assuming $C_1 = C_2 = C$, the equivalent output resistance of the 3:1 series–parallel topology of Fig. 9.4a) becomes

$$R_{SSL} = \frac{1}{Cf_{sw}} \left(\left(\frac{1}{3}\right)^2 + \left(\frac{1}{3}\right)^2 \right) = \frac{1}{Cf_{sw}} \frac{2}{9} \approx \frac{0.222}{Cf_{sw}}. \tag{9.35}$$

Since the flying capacitors have identical absolute charge flows, Dickson and Fibonacci in Fig. 9.4b,d) have the same SSL output resistance as the series–parallel configuration. The ladder topology of Fig. 9.4c) contains three flying capacitors. Hence its output resistance can be expected to be larger. Assuming again equal capacitors $C_1 = C_2 = C_3 = C$, we get

$$R_{SSL} = \frac{1}{Cf_{sw}} \left(\left(\frac{1}{3}\right)^2 + \left(-\frac{1}{3}\right)^2 + \left(\frac{2}{3}\right)^2 \right) = \frac{1}{Cf_{sw}} \frac{2}{3} \approx \frac{0.667}{Cf_{sw}}, \tag{9.36}$$

which is three times higher than the result in Eqn. (9.35).

The charge flows a_j^i depend on the SC topology. The lower the charge flow components, the better the output resistance. The parameter

$$K_{SSL} = \sum_{j=1}^{n} |a_j| \tag{9.37}$$

can be defined as a metric that allows to compare SC converter topologies where n is the number of flying capacitors. Each element a_j is the sum of the charge flows over all phases for the particular capacitor C_j, i.e., $a_j = (a_j^1 + a_j^2)$ for two-phase conversion. For common topologies, K_{SSL} is evaluated further below in Section 9.7 (Table 9.4).

For multi-phase SC converters with more than two phases (investigated in Section 9.10) Eqn. (9.37) expands to

$$K_{SSL} = \sum_{i=1}^{n_{ph}} \sum_{j=1}^{n} \left| a_{c,j}^i \right|. \tag{9.38}$$

We will use Eqn. (9.38) in Example 9.12.

Intuitively, larger values of C_j and f_{sw} increase the load capability of the SC converter and, in turn, decrease the equivalent output resistance. However, eventually R_{out} cannot be further reduced because the FSL kicks in.

Fast-Switching Limit (FSL)
FSL is the asymptotic limit of the equivalent output resistance toward higher switching frequencies. Unlike for SSL, now the switch charge flow vectors as defined in Eqn. (9.18) are taken into account to calculate the equivalent output resistance $R_{FSL} = R_{out}$, dominated by the switches' non-zero on-resistance. If required, any other parasitic resistance in the circuit can also be considered, such as interconnections and the equivalent series resistance of the flying capacitors. The switching frequency f_{sw} is so high that the voltage changes at the capacitors are minimal and can be assumed as fixed voltage sources. The time constant between switch resistance and the (flying) capacitance is much larger than the period of the switching frequency.

Given a charge flow $q_{r,j}$ through a switch resistance R_j during a phase duration t_{on}, the current is

$$i_{r,j} = \frac{q_{r,j}}{t_{on}} = \frac{a_{r,j} q_{out}}{t_{on}} = \frac{a_{r,j} I_{load} T}{t_{on}} = \frac{a_{r,j} I_{load}}{D_i}. \tag{9.39}$$

This expression uses Eqn. (9.13) and considers the fact that the charge flow elements are normalized to the output charge, i.e., $a_{r,j} = q_{r,j}/q_{out}$ (see also Eqn. (9.14)). The switch current flows during phase i, which is a fraction of the clock period T, expressed by the duty cycle $D_i = t_{on,i}/T$.

With the switch being conducting for a time $t_{on,i}$, the average loss energy over one period T is

$$E_{loss,j} = \int_0^T R_j i_{r_j}^2 dt = \int_0^{t_{on,j}} R_j i_{r_j}^2 dt = t_{on,j} R_j i_{r_j}^2. \tag{9.40}$$

The switch current can be substituted by the expression from Eqn. (9.39):

$$E_{loss,j} = t_{on,j} R_i \left(\frac{a_{r_j} I_{load}}{D_j} \right)^2 = R_j \frac{\left(a_{r_j} I_{load} T \right)^2}{t_{on,i}} \tag{9.41}$$

Summing up the losses energies of all switches over all phases gives the total loss energy of the SC converter in FSL:

$$E_{loss} = \sum_{i=1}^{n_{ph}} \sum_{j=1}^{n_{sw}} E_{loss,j} = \sum_{i=1}^{n_{ph}} \sum_{j=1}^{n_{sw}} R_j \frac{\left(a_{r_j} I_{load} T \right)^2}{t_{on,j}} \tag{9.42}$$

The total losses also fulfill Eqn. (9.26) for $R_{out} = R_{FSL}$. Setting Eqns. (9.42) and (9.26) equal and solving for R_{FSL} gives

$$R_{FSL} = \sum_{i=1}^{n_{ph}} \sum_{j=1}^{n_{sw}} R_j \frac{\left(a_{r_j} \right)^2}{D_j}. \tag{9.43}$$

For the typical case of 50% duty cycle, $D_j = 0.5$, the expression simplifies to

$$R_{FSL} = 2 \sum_{i=1}^{n_{ph}} \sum_{j=1}^{n_{sw}} R_j \left(a_{r_j} \right)^2. \tag{9.44}$$

In two-phase operation and with the elements a_{r_j} of the switch charge flow vector as defined in Eqn. (9.20), Eqn. (9.44) turns into

$$R_{FSL} = 2 \sum_{j=1}^{n_{sw}} R_j \left(a_{r_j} \right)^2. \tag{9.45}$$

Based on this expression, we can use the charge flow vector given in Eqn. (9.19) to determine the FSL output resistance of the 2:1 stage:

$$R_{FSL} = 2R \tag{9.46}$$

This equation assumes all switches to have the same resistance R

In the same way, the switch charge flow vectors in Table 9.3 allow for calculating the output resistance of common 3:1 topologies. The FSL limit of the 3:1 series–parallel as well as the Dickson topology is

$$R_{FSL} = 2 \cdot R \cdot 7 \left(\frac{1}{3} \right)^2 = 2 \cdot R \cdot \frac{7}{9} \approx 1.55\,R. \tag{9.47}$$

The ladder topology has only six switches but charge flow elements with values of 2/3:

$$R_{FSL} = 2 \cdot R \left(4 \left(\frac{1}{3} \right)^2 + 2 \left(\frac{2}{3} \right)^2 \right) = 2 \cdot R \cdot \frac{4}{3} \approx 2.67\,R \tag{9.48}$$

The Fibonacci configuration again has seven switches like series–parallel and Dickson. In addition, one charge flow element has a value of 2/3. The equivalent output resistance is

$$R_{FSL} = 2 \cdot R \left(6 \left(\frac{1}{3} \right)^2 + \left(\frac{2}{3} \right)^2 \right) = 2 \cdot R \cdot \frac{10}{9} \approx 2.22\,R. \tag{9.49}$$

The calculations in Eqns. (9.47)–(9.49) assume equal switches. Section 9.8 will explain that the optimum SC converter design requires dedicated switch sizing.

Equation (9.44) indicates that the on-resistance of the individual switches defines the minimum achievable output resistance. The topology determines the charge flow components $a_{r,j}$. The parameter K_{FSL} is defined as the sum of the absolute values of the charge flows in each switch over all phases:

$$K_{FSL} = \sum_{j=1}^{n_{sw}} |a_{r,j}| \tag{9.50}$$

For two-phase topologies, $a_{r,j}$ are the elements of a single charge flow vector given by Eqn. (9.20). K_{FSL} will be utilized for the design of SC converters in Section 9.8.

Equation (9.50) is a simplified version in case of $n_{ph} = 2$ phases. The general expression, valid for multiple phases, is given by

$$K_{FSL} = \sum_{i=1}^{n_{ph}} \sum_{j=1}^{n_{sw}} \left| a_{r,j}^i \right|. \tag{9.51}$$

Total Output Resistance

Because the R_{out}-plot in Fig. 9.8 is drawn on a double-logarithmic scale, we can see that R_{FSL} is negligible in SSL operation and, vice versa, R_{SSL} is significantly lower than R_{FSL} at high switching frequencies. Based on this observation, [2] introduces this approximation of the total output resistance:

$$R_{out} = \sqrt{R_{SSL}^2 + R_{FSL}^2} \tag{9.52}$$

Seeman and Sanders [2] also proves that Eqn. (9.52) approximates R_{out} very well with the maximum error near the crossover region.

Full Derivation of R_{out}

As Eqn. (9.52) is an approximation, it is interesting to study the approach for the full analytical derivation of R_{out} [1, 8–10]. The total loss energy E_{loss} corresponds to the sum of E_i over all phases and all resistive components in the equivalent circuit of Fig. 9.9, indicated below by the term $\sum_i E_i$. E_{loss} is defined by Eqn. (9.26) as the energy dissipated in the equivalent output resistance R_{out} of the SC converter. Equating E_{loss} and $\sum_i E_i$, substituting I_{load} by $q_{out} f_{sw}$ (Eqn. (9.13)) and solving for R_{out} yields:

$$R_{out} = \frac{\sum_i E_i}{f_{sw} \cdot \left(\sum q_{out}^i \right)^2} \tag{9.53}$$

In each phase of the SC converter, the flying capacitors and switch resistances form an R-C circuit with lumped elements R_i and C_i, as shown in Fig. 9.9. The energy E_i dissipated during phase i is equal to

$$E_i = \int_0^{t_i} P_i(t)\mathrm{d}t = \int_0^{t_i} R_i \cdot I_i^2(t)\mathrm{d}t. \tag{9.54}$$

with the current $I_i(t)$ as defined in Fig. 9.9. Inserting Eqn. (9.54) into Eqn. (9.53) gives the full analytical solution for the equivalent output resistance:

$$R_{out} = \frac{\sum_i \left(\int_0^{t_i} R_i \cdot I_i^2(t)\mathrm{d}t \right)}{f_{sw} \cdot q_{out}^2 \cdot \left(\sum a_{out}^i \right)^2} \tag{9.55}$$

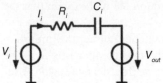

Figure 9.9   SC converter equivalent circuit in any phase *i*.

This expression is valid for any SC converter topology. Moreover, it applies to hybrid resonant SC converters covered in Chapter 11.

To determine R_{out} based on Eqn. (9.55), the current $I_i(t)$ needs to be known. For an SC converter, I_i is determined by a homogeneous first-order differential equation, which results in

$$I_i = A_i \cdot e^{-t/\tau} \tag{9.56}$$

with the R-C time constant $\tau = R_i C_i$. The factor A_i can be determined from the condition that the charge q_{out}^i delivered to the output of the converter in each phase i is equal to the integral of I_i over the duration t_i of phase i,

$$q_{out}^i = a_{out}^i \cdot q_{out} = \int_0^{t_i} I_i(t)\mathrm{d}t = A_i \cdot \tau \left(1 - e^{-t/\tau}\right). \tag{9.57}$$

Hence,

$$A_i = \frac{a_{out}^i \cdot q_{out}}{\tau \left(1 - e^{-t/\tau}\right)}. \tag{9.58}$$

Knowing A_i, Eqn. (9.56) describes the current I_i. The energy E_i dissipated during phase i is equal to

$$E_i = \int_0^{t_i} P_i(t)\mathrm{d}t = \int_0^{t_i} R_i \cdot I_i^2(t)\mathrm{d}t = A_i^2 \cdot R_i \cdot \frac{\tau}{2} \left(1 - e^{-2\,t_i/\tau}\right). \tag{9.59}$$

Inserting Eqn. (9.58) into Eqn. (9.59) gives a fraction with two exponential functions, which can be simplified as shown below:

$$\frac{1 - e^{-2x}}{(1 - e^{-x})^2} = \frac{(1 + e^{-x})(1 - e^{-x})}{(1 - e^{-x})^2}$$

$$= -\frac{e^{-2x} + 1}{e^{-2x} - 1} = -\coth\left(-\frac{x}{2}\right) = \coth\left(\frac{x}{2}\right) \tag{9.60}$$

This allows to express Eqn. (9.59) in an alternative form,

$$E_i = \frac{(q_{out} \cdot a_{out}^i)^2}{2C_i} \coth\left(\frac{t_i}{2R_i C_i}\right). \tag{9.61}$$

Inserting E_i into Eqn. (9.55):

$$R_{out} = \frac{\sum_i \left(\frac{(a_{out}^i)^2}{C_i} \coth\left(\frac{t_i}{2R_i C_i}\right)\right)}{2 f_{sw} \left(\sum a_{out}^i\right)^2} \tag{9.62}$$

The expressions for R_{SSL} and R_{FSL} in Eqns. (9.32) and (9.43), respectively, represent approximations of Eqn. (9.62) for $t_i \gg R_i C_i$ (SSL) and $t_i \ll R_i C_i$ (FSL). For small arguments, $\coth(x) \sim 1/x$ (FSL) while $\coth(\infty) = 1$ (SSL). Equation (9.52) approximates Eqn. (9.62) taking into account the contribution of both R_{SSL} and R_{FSL}.

9.6 Output Voltage Ripple

Due to the switching nature of SC converters, their output voltage V_{out} has an unavoidable (peak-to-peak) ripple $\Delta V_{out,pkpk}$ because the output capacitor is continuously discharged but recharged by discrete charge quantities. For a given load current I_{load} and switching frequency f_{sw}, the ripple is defined in [11] as

$$\Delta V_{out,pkpk} = \frac{I_{load}}{f_{sw} k_d C_o^*}, \tag{9.63}$$

where k_d is called the discharge fraction factor and C_o^* is the effective output buffer capacitance.

The discharge fraction factor k_d expresses how often per switching period charge is delivered to the output by the flying capacitor(s). In other words, C_o^* is discharged during the fraction T/k_d of the overall period $T = 1/f_{sw}$. In the case of a 2:1 converter, the factor is $k_d = 2$ because the charge transfer to the output happens twice within each period. This is in line with Eqn. (9.17), which reveals a charge flow of $a_{out}^1 = a_{out}^2 = \frac{1}{2}$ during both phases. The flying capacitor C_1 connects to the output in both phases φ_1 and φ_2, see Fig. 9.1.

For SC converters with large discrete capacitors, the output capacitor C_o can be sized much larger than the total capacitance C_{tot} of all flying capacitors ($C_o \gg C_{tot}$). We can approximate the effective output buffer capacitance C_o^* by C_o. However, in fully integrated capacitive converters, C_o can be similar in size to C_{tot}. Now, the effective output capacitance C_o^* consists of the actual output capacitor C_o and the amount of flying capacitance C_{tot}, which, depending on the switch position, is also connected to the output and contributes to the output buffering. This can be expressed by an implementation-dependent parameter k_t as introduced in [11]:

$$C_o^* = C_o + k_t C_{tot} \tag{9.64}$$

In the 2:1 cell of Fig. 9.1, for instance, the flying capacitor C_1 is connected in series to C_o during phase φ_1 and in parallel to C_o during phase φ_2. Hence, $C_{tot} = C_1$ contributes to C_o^* in both phases and $k_t = 1$. For more than one flying capacitor, k_t corresponds to the fraction of the total flying capacitance that is connected to V_{out},

$$k_t = \frac{C_t}{C_{tot}} = 0 \dots 1, \tag{9.65}$$

where C_t is the effective flying capacitance connected to the output node. Because $C_t \leq C_{tot}$, the range of k_t is between zero and 1. Parameter k_t is calculated for each phase, and the lower value is used for the ripple calculation [11] as this results in the worst-case ripple. These considerations are utilized in Section 9.8.3 for C_o sizing.

Example 9.3 *Determine the parameter k_t for the 3:1 Dickson and ladder topologies.*

For both topologies, Fig. 9.7 shows the equivalent circuits in each switching phase.

During φ_1, Dickson has both flying capacitors connected to V_{out} while their other terminal is kept at a DC level (V_{in} and ground). Hence, C_1 and C_2 are effectively connected in parallel, which gives $C_{t1} = C_1 + C_2$ as the effective flying capacitance connected to V_{out} during φ_1. In φ_2, both flying capacitors are switched in series. Hence, $C_{t2} = (C_1 \cdot C_2)/(C_1 + C_2)$. From Eqn. (9.65), we can calculate k_t in each phase,

$$k_{t1} = \frac{C_{t1}}{C_{tot}} = \frac{C_1 + C_2}{C_1 + C_2} = 1, \quad k_{t2} = \frac{C_{t2}}{C_{tot}} = \frac{C_1 \cdot C_2}{(C_1 + C_2)^2}. \tag{9.66}$$

The ladder topology has three flying capacitors. From Fig. 9.7c), we obtain $C_{t1} = C_1 + (C_2||C_3)$ and $C_{t2} = C_3 + (C_1||C_2)$. Hence,

$$k_{t1} = \frac{C_{t1}}{C_{tot}} = \frac{C_1 \cdot (C_2 + C_3)}{(C_1 + C_2 + C_3)^2}, \quad k_{t2} = \frac{C_{t2}}{C_{tot}} = \frac{C_3 \cdot (C_1 + C_2)}{(C_1 + C_2 + C_3)^2}. \tag{9.67}$$

We will utilizes these expressions when determining the size of C_o in Example 9.7.

9.7 Topology Selection

From the previous part of this chapter, several design criteria can be derived that characterize the behavior of SC converter topologies. They allow us to compare their performance and select the appropriate topology for a given design space. These are the most important design criteria:

- Realizable conversion ratio(s) (see Sections 9.2 and 9.5.2)
- Number of flying capacitors n
- Maximum capacitor voltage $V_{max,c}$
- Number of switches n_{sw}
- Maximum blocking voltage of the switches $V_{max,sw}$
- Ideal conversion ratio(s) (see Section 9.5.3)
- Achievable equivalent output resistance (see Section 9.5.4)
- Gate drive effort
- Control effort

It is instructive to investigate these items for the common topologies Series–Parallel, Dickson, Ladder, and Fibonacci (see Fig. 9.4). Their design criteria are discussed and derived in Sections 9.3 and 9.5. Key parameters are listed in Table 9.4. The equivalent circuits in two-phase operation of Fig. 9.7 allow us to obtain the charge flow vectors in Table 9.3. The circuits in Fig. 9.7 also reveal the voltages across the switches and capacitors during each phase as studied in Section 9.3.

After selecting a suitable topology, the design of an SC converter includes capacitor and switch sizing and optimization for minimum losses, as explored in Sections 9.8 and 9.9. This is also essential for more sophisticated converter designs like multi-phase and multi-ratio architectures, which are covered in separate sections. Also, control methods are addressed in a dedicated section further below.

9.8 Capacitor and Switch Sizing

The equivalent output resistance R_{out} of the SC converter depends on the sizes of the flying capacitors and the resistances of the switches. R_{out} must be minimized to keep the output voltage droop

Table 9.4 Parameters of common 3:1 SC topologies.

	n	$V_{max,c}$	n_{sw}	$V_{max,sw}$	K_{SSL}	K_{FSL}
Series–parallel	2	V_{out}	7	$2V_{out}$	$\frac{2}{3} = 0.67$	$\frac{7}{3} = 2.33$
Dickson	2	$2V_{out}$	7	$2V_{out}$	$\frac{2}{3} = 0.67$	$\frac{7}{3} = 2.33$
Ladder	3	V_{out}	6	V_{out}	$\frac{4}{3} = 1.33$	$\frac{8}{3} = 2.67$
Fibonacci	2	$2V_{out}$	7	$2V_{out}$	$\frac{2}{3} = 0.67$	$\frac{8}{3} = 2.76$

and the related power losses low. Seemann and Sanders [2] proposes a very effective but simple sizing method that we will introduce below.

Because of the switched-mode operation, the capacitive DC–DC converter will always show some output voltage ripple. The output capacitor C_o needs to be sufficiently large to fulfill the ripple specification.

9.8.1 Flying Capacitor Sizing

The sizing rule constrains the total capacitance to C_{tot}. The value of each flying capacitor can then be calculated proportionally to its charge flow in SSL, i.e.,

$$C_k = \frac{|a_{c,k}|}{\sum_{j=1}^{n}|a_{c,k}|}C_{tot} = \frac{|a_{c,k}|}{K_{SSL}}C_{tot} \tag{9.68}$$

with K_{SSL} according to Eqn. (9.37). Equation (9.68) is valid for two-phase operation (see Eqn. (9.75) below for SC converters with more than two phases).

The larger the charge flow element a_k, the larger the capacitor value C_k. Equation (9.68) also means that the SSL output impedance fulfills the condition

$$R_{SSL} = \frac{K_{SSL}^2}{C_{tot}f_{sw}}. \tag{9.69}$$

We can verify this condition for the 3:1 series–parallel topology of Fig. 9.4a). From Table 9.3, we know that the charge multipliers for the two capacitors are equal. According to Eqn. (9.68), also the capacitors will have the same value. Hence, we define $a_{c,1} = a_{c,2} = a_c$ and $C_1 = C_2 = C$. Equating Eqns. (9.69) and (9.33) and substituting K_{SSL} by Eqn. (9.37) gives

$$\frac{(|a_{c,1}| + |a_{c,2}|)^2}{C_{tot}f_{sw}} = \frac{a_{c,1}^2}{C_1 f_{sw}} + \frac{a_{c,2}^2}{C_2 f_{sw}}$$

$$\frac{4\,(a_c)^2}{C_{tot}f_{sw}} = 2\,\frac{(a_c)^2}{C f_{sw}}, \tag{9.70}$$

which is fulfilled exactly for $C_{tot} = C_1 + C_2 = 2C$. Since the capacitor charge flows are identical (Table 9.3), the 3:1 Dickson and Fibonacci topologies also result in equally sized flying capacitors.

Example 9.4 *Calculate the size of the flying capacitors for a capacitor budget of $C_{tot} = 8\,\text{nF}$ for the 3:1 Dickson and ladder topologies. What is the equivalent output resistance at $f_{sw} = 10\,\text{MHz}$?*

Since the charge flow in both flying capacitors of the Dickson topology is equal, the capacitors are identical in size, and we obtain $C_1 = C_2 = 4\,\text{nF}$. At $f_{sw} = 10\,\text{MHz}$ Eqn. (9.33) results in

$$R_{SSL} = \sum_{j=1}^{n=2}\frac{\left(a_j^1\right)^2}{C_j f_{sw}} = 2\frac{\left(\frac{1}{3}\right)^2}{4\,\text{nF}\,10\,\text{MHz}} = 5.6\,\Omega. \tag{9.71}$$

For the ladder topology, the charge flow vector during phase φ_1 is

$$\mathbf{a^1} = \begin{bmatrix} \frac{1}{3} & \frac{1}{3} & -\frac{1}{3} & \frac{2}{3} & \frac{1}{3} \end{bmatrix}. \tag{9.72}$$

From Eqn. (9.37):

$$C_1 = C_2 = \frac{|a_{c,1}|}{\sum_{j=1}^{n=3}|a_{c,1}|}C_{tot} = \frac{\frac{1}{3}}{\frac{1}{3} + \frac{1}{3} + \frac{2}{3}}8\,\text{nF} = \frac{1}{4}\cdot 8\,\text{nF} = 2\,\text{nF}$$

$$C_3 = \frac{|a_{c,3}|}{\sum_{j=1}^{n=3}|a_{c,1}|}C_{tot} = \frac{\frac{2}{3}}{\frac{1}{3} + \frac{1}{3} + \frac{2}{3}}8\,\text{nF} = \frac{1}{2}\cdot 8\,\text{nF} = 4\,\text{nF} \tag{9.73}$$

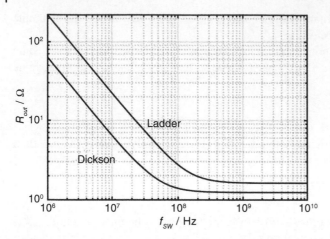

Figure 9.10 Equivalent output resistance for the Dickson and ladder topologies according to Examples (9.4)–(9.6).

Knowing the values of C_1, C_2 and C_3 we can calculate the output impedance at $f_{sw} = 10\,\text{MHz}$ by means of Eqn. (9.33):

$$R_{SSL} = \sum_{j=1}^{n} \frac{(a_j^1)^2}{C_j f_{sw}} = 2\frac{\left(\frac{1}{3}\right)^2}{2\,\text{nF}\,10\,\text{MHz}} + \frac{\left(\frac{2}{3}\right)^2}{4\,\text{nF}\,10\,\text{MHz}} = 22.2\,\Omega \tag{9.74}$$

Figure 9.10 shows the equivalent output resistance of both topologies over frequency. The graph includes the FSL resistance, covered below in Examples 9.5 and 9.6.

The SSL output resistance of the ladder topology is nearly four times larger. However, since C_1 in the Dickson converter needs to have higher voltage ratings (Table 9.4), which requires a larger area, its output resistance may be worse if the area is constrained instead of C_{tot}.

For SC converters with more than two phases, Eqn. (9.68) modifies to

$$C_k = \frac{\sum_{i=1}^{n_{ph}} |a_{c,k}^{(i)}|}{\sum_{i=1}^{n_{ph}} \sum_{j=1}^{n} \left| a_{c,k}^{(i)} \right|} C_{tot} = \frac{\sum_{i=1}^{n_{ph}} |a_{c,k}^{(i)}|}{K_{SSL}} C_{tot} \tag{9.75}$$

with the multi-phase representation of K_{SSL} according to Eqn. (9.38). Multi-phase converters are covered in Section 9.10, where Example 9.12 refers to Eqn. (9.75).

9.8.2 Switch Sizing

Like the capacitor sizing approach, [2] defines a total conductance G_{tot} as the sum of all switch conductances. Each conductance $G_{r,m}$ corresponds to the reciprocal of the switch on-resistance R_{DSon}. Since R_{DSon}, defined by Eqn. (2.2), is inversely proportional to the W/L-ratio of the transistor switch, the switch conductance is proportional to its channel width. Hence, using a total width W_{tot} instead of the conductance G_{tot} is convenient. W_{tot} is the sum of all transistor widths. Consequently, the width of each switch scales relative to its charge flow in FSL (two-phase operation):

$$W_{r,m} = \frac{|a_{r,m}|}{\sum_{j=1}^{n_{sw}} \left| a_{rj} \right|} W_{tot} = \frac{|a_{r,m}|}{K_{FSL}} W_{tot} \tag{9.76}$$

The parameter K_{FSL} is defined in Eqn. (9.50). As for the capacitor sizing, also the conductance of the switch needs to be proportional to its charge flow element $a_{r,m}$ (see Eqn. (9.20)). If all switches are sized according to Eqn. (9.76), the equivalent FSL output impedance can be expressed by

$$R_{FSL} - \frac{2K_{FSL}^2}{G_{tot}}. \tag{9.77}$$

Example 9.5 *Calculate the size and the on-resistance of the switches in the 3:1 Dickson topology for a total transistor width $W_{tot} = 8$ mm. We keep the transistor channel length at $L = L_{min} = 0.3\,\mu m$. The other parameters are $V_{th} = 0.6$ V, $V_{GS} = 5$ V, and $\mu_o C_{ox} = 75\,\mu A/V^2$. Determine R_{FSL} and verify if Eqn. (9.77) is fulfilled.*

We first recall the switch charge flow vector of the Dickson topology from Table 9.3,

$$\boldsymbol{a_r} = \begin{bmatrix} \frac{1}{3} & \frac{1}{3} & \frac{1}{3} & \frac{1}{3} & \frac{1}{3} & \frac{1}{3} & \frac{1}{3} \end{bmatrix}.$$

Equation (9.76) indicates that all switches get equal width W because all charge flow elements have the same value $a_r = 1/3$:

$$W = \frac{|a_r|}{\sum_{j=1}^{n_{sw}} |a_{r,j}|} W_{tot} = \frac{\frac{1}{3}}{7 \cdot \frac{1}{3}} W_{tot} = \frac{8\,mm}{7} = 1.14\,mm \tag{9.78}$$

The on-resistance of each switch can be calculated from Eqn. (2.2) (Chapter 2):

$$R_{DSon} = \frac{1}{\frac{W}{L}\mu_o C_{ox}\left(V_{GS} - V_{th}\right)} = \frac{1}{\frac{1.14\,mm}{0.3\,\mu m}75\,\mu A/V^2 4.4\,V} = 795\,m\Omega \tag{9.79}$$

We can now calculate R_{FSL} from Eqn. (9.44),

$$R_{FSL} = 2\sum_{i=1}^{n_{ph}} \sum_{j=1}^{n_{sw}} R_j \left(a_{r,j}\right)^2 = 2 \cdot 7 \cdot \left(\frac{1}{3}\right)^2 R_{DSon} = 1.56 \cdot 795\,m\Omega = 1.24\,\Omega. \tag{9.80}$$

To verify Eqn. (9.77), we can determine R_{FSL} from that equation. In the first step, we derive the total conductance as the sum of all switch conductances,

$$G_{tot} = 7\frac{1}{R_{DSon}} = 8.8\,A/V. \tag{9.81}$$

We also remember that K_{FSL} is the sum of the absolute values of all switch charge flows (see Eqn. (9.50)). Hence $K_{FSL} = \frac{7}{3} = 2.33$, which is also listed in Table 9.4.

Now we can calculate R_{FSL} from Eqn. (9.77):

$$R_{FSL} = \frac{2K_{FSL}^2}{G_{tot}} = \frac{2\,(2.33)^2}{8.8\,A/V} = 1.24\,\Omega \tag{9.82}$$

This value is identical to the result of Eqn. (9.80), which confirms Eqn. (9.77).

Figure 9.10 shows how the equivalent output resistance settles at $R_{FSL} = 1.24\,\Omega$ as the switching frequency goes up. For comparison, the curve for the ladder topology is drawn as well. Its FSL resistance is covered in the next example below.

Example 9.6 *Repeat Example 9.5 for the 3:1 ladder topology.*

The charge flow vector (Table 9.3)

$$\boldsymbol{a_r} = \begin{bmatrix} \frac{1}{3} & \frac{1}{3} & \frac{1}{3} & \frac{1}{3} & \frac{2}{3} & \frac{2}{3} \end{bmatrix}$$

indicates that there are two groups of switches with different sizing. Switches S1–S4 have a width of

$$W = \frac{|a_r|}{\sum_{j=1}^{n_{sw}} |a_{r,j}|} W_{tot} = \frac{\frac{1}{3} W_{tot}}{4 \cdot \frac{1}{3} + 2 \cdot \frac{2}{3}} = \frac{8\,\text{mm}}{8} = 1.0\,\text{mm}, \tag{9.83}$$

and a resistance of $R_{DSon} = 909\,\text{m}\Omega$. Similarly, the values for S5 and S6 are $W = 2.0\,\text{mm}$ and $R_{DSon} = 455\,\text{m}\Omega$.

The equivalent resistance according to Eqn. (9.44) gives

$$R_{FSL} = 2 \sum_{i=1}^{n_{ph}} \sum_{j=1}^{n_{sw}} R_j \left(a_{r,j}\right)^2 = 2 \left(4 \cdot \left(\frac{1}{3}\right)^2 909\,\text{m}\Omega + 2 \cdot \left(\frac{2}{3}\right)^2 455\,\text{m}\Omega \right)$$

$$= 2 \left(\frac{4}{9} 909\,\text{m}\Omega + \frac{8}{9} 455\,\text{m}\Omega \right) = 1.62\,\Omega. \tag{9.84}$$

With the total width being identical to Example 9.5, also the total conductance remains unchanged ($G_{tot} = 8.8\,\text{A/V}$). The same value R_{FSL} can be obtained from Eqn. (9.77). Figure 9.10 confirms that the ladder results in higher R_{FSL} than Dickson. Since the switches in the ladder require lower voltage ratings compared to Dickson, it can achieve lower R_{DSon} for a given layout area. Therefore, an optimized design may result in a smaller R_{FSL} difference between both topologies.

For multi-phase operation, Eqn. (9.76) changes to

$$W_{r,m} = \frac{\sum_{i=1}^{n_{ph}} |a_{r,j}^{(i)}|}{\sum_{i=1}^{n_{ph}} \sum_{j=1}^{n_{sw}} \left|a_{r,j}^{(i)}\right|} W_{tot} = \frac{\sum_{i=1}^{n_{ph}} |a_{r,m}^{(i)}|}{K_{FSL}} W_{tot} \tag{9.85}$$

with K_{FSL} as defined in Eqn. (9.38) for multi-phase SC converters.

9.8.3 Output Capacitor Sizing

The sizing of the output bypass capacitor C_o is directly related to the output voltage ripple $\Delta V_{out,pkpk}$ as discussed in Section 9.6 and expressed in Eqn. (9.63) with the effective output capacitance C_o^* given in Eqn. (9.64). Solving Eqn. (9.63) for C_o yields a sizing equation for the output capacitor,

$$C_o = \frac{I_{load}}{f_{sw} k_d \Delta V_{out,pkpk}} - k_t C_{tot}, \tag{9.86}$$

where C_{tot} is the total capacitance of all flying capacitors, and k_d, k_t are topology-dependent parameters defined in Section 9.6.

Example 9.7 *Take the flying capacitor values derived in Example 9.4 and calculate the value of C_o for the 3:1 Dickson and ladder topologies to meet the ripple specification $\Delta V_{out,pkpk} = 50\,\text{mV}$ at $f_{sw} = 10\,\text{MHz}$ and $I_{load} = 5\,\text{mA}$.*

Let us first discuss the discharge fraction factor k_d as introduced in Section 9.6. Because Dickson and Ladder have the flying capacitors connected to V_{out} in each phase, we always have $k_d = 2$.

The other topology-dependent parameter, k_t, is already derived in Example 9.3 for both 3:1 topologies. With the capacitor values from Example 9.4, $C_1 = C_2 = 4\,\text{nF}$ and $C_{tot} = 8\,\text{nF}$, we get for the Dickson topology

$$C_{t1} = C_1 + C_2 = 4\,\text{nF} + 4\,\text{nF} = 8\,\text{nF}, \quad k_{t1} = \frac{C_{t1}}{C_{tot}} = 1 \tag{9.87}$$

and

$$C_{t2} = \frac{C_1 \cdot C_2}{(C_1 + C_2)} = 2\,\text{nF}, \quad k_{t2} = \frac{C_{t2}}{C_{tot}} = \frac{2\,\text{nF}}{8\,\text{nF}} = 0.25. \tag{9.88}$$

According to Section 9.6, we take the lower value $k_t = 0.25$ for the sizing of C_o. Now, Eqn. (9.86) allows for calculating the output capacitance,

$$C_o = \frac{I_{load}}{f_{sw}k_d\Delta V_{out,pkpk}} - k_t C_{tot} = \frac{5\,\text{mA}}{10\,\text{MHz} \cdot 2 \cdot 50\,\text{mV}} - 0.25 \cdot 8\,\text{nF} = 3\,\text{nF}. \tag{9.89}$$

For the ladder topology, Example 9.4 calculates $C_1 = C_2 = 2\,\text{nF}, C_3 = 4\,\text{nF}$ for the same total capacitance of $C_{tot} = 8\,\text{nF}$. Based on Example 9.3, we can calculate the parameters C_t and k_t in the two phases:

$$C_{t1} = \frac{C_1 \cdot (C_2 + C_3)}{C_1 + C_2 + C_3} = 1.5\,\text{nF}, \quad k_{t1} = \frac{C_{t1}}{C_{tot}} = \frac{3.33\,\text{nF}}{8\,\text{nF}} = 0.188. \tag{9.90}$$

$$C_{t2} = \frac{C_3 \cdot (C_1 + C_2)}{C_1 + C_2 + C_3} = 2\,\text{nF}, \quad k_{t2} = \frac{C_{t2}}{C_{tot}} = \frac{5\,\text{nF}}{8\,\text{nF}} = 0.25. \tag{9.91}$$

We use $k_t = 0.188$ and calculate $C_o = 3.5\,\text{nF}$ similar to Eqn. (9.89).

Due to the lower value of k_t and C_t, the ladder topology needs a slightly larger output capacitor. We also notice that C_o and the flying capacitance are nearly the same size. For lower ripple, C_o has to be increased to much higher values than C_1 and C_2.

9.9 Loss Analysis and Efficiency

In the SC converter equivalent circuit of Fig. 9.2, losses are associated with the two equivalent resistances R_{out} and R_p. In addition to intrinsic losses due to the output resistance R_{out}, there are extrinsic loss contributions which are accounted for by R_p. We can distinguish various extrinsic losses, including losses related to the switch transistors (see Section 2.7), to the capacitors (see Section 4.1) and losses due to DC bias currents in any required circuit [12]. We will discuss the most critical losses below.

9.9.1 Intrinsic Losses

The power loss due to R_{out}, referred to as intrinsic loss, is determined by

$$P_{Rout} = I_{load}^2 R_{out}. \tag{9.92}$$

(see also Eqn. (9.26)). R_{out} reduces with frequency as determined by R_{FSL} according to Eqn. (9.33). Hence, the losses decrease inversely proportional to f_{sw} (see also Fig. 9.12).

Example 9.8 *Calculate the intrinsic power loss P_{Rout} of the 3:1 Dickson topology at $I_{load} = 100\,\text{mA}$ for the switch sizing according to Examples 9.4 and 9.5.*

With $R_{SSL} = 5.6\,\Omega$ from Eqn. (9.71) and $R_{FSL} = 1.24\,\Omega$ from Eqn. (9.80) the total output resistance follows from Eqn. (9.52):

$$R_{out} = \sqrt{R_{SSL}^2 + R_{FSL}^2} = \sqrt{(5.6\,\Omega)^2 + (1.24\,\Omega)^2} = 6.45\,\Omega \tag{9.93}$$

The intrinsic losses due to that output resistance are

$$P_{Rout} = I_{load}^2 R_{out} = (100\,\text{mA})^2\, 6.45\,\Omega = 64.5\,\text{mW}. \tag{9.94}$$

9.9.2 Switch Control Losses

These losses consist mainly of the gate charge losses as expressed by Eqn. (2.43) in Chapter 2, repeated here for convenience,

$$P_{gate} = Q_{gate}V_{drv}f_{sw} = C_{gate}V_{drv}^2 f_{sw}. \tag{9.95}$$

The P_{gate} contributions of all switches need to be summed up to obtain the total gate charge losses in the converter.

In fully integrated SC converters, the flying capacitors typically dominate the chip size, while the area consumption of the switches is usually negligible. Hence, the on-resistance of the switches can often be further reduced because the larger transistor width and its loss contribution can be tolerated.

The gate charge losses scale up if a dedicated charge pump is used for the gate supply. Including the charge pump efficiency η_{CP}, the losses become

$$P_{gate,CP} = \frac{P_{gate}}{\eta_{CP}}. \tag{9.96}$$

Example 9.9 *Calculate the gate charge losses of the 3:1 Dickson topology at $f_{sw} = 10\,\mathrm{MHz}$ for the switch sizing of Example 9.5, i.e., all switches have an on-resistance of $R_{DSon} = 795\,\mathrm{m\Omega}$. We do not consider the losses related to a charge pump or bootstrap gate supply, which might be required. All parameters are the same as in Example 9.5: $L = L_{min} = 0.3\,\mu m$, $V_{th} = 0.6\,\mathrm{V}$, $V_{GS} = 5\,\mathrm{V}$, $\mu_o = 500 \times 10^{-4}\,\mathrm{m^2/(Vs)}$, $C_{ox} = 1.5 \times 10^{-3}\,\mathrm{F/m^2}$.*

The on-resistance of the switch transistor corresponds to its active area $(W \cdot L)$. The transistor width W can be determined from Eqn. (9.79) or we directly take the value from Example 9.5 that is given in Eqn. (9.78): $W = 1.14\,\mathrm{mm}$.

With $L = L_{min}$ the gate capacitance is

$$C_{gate} = W \cdot L \cdot C_{ox} = 1.14\,\mathrm{mm} \cdot 0.3\,\mu m \cdot 1.5 \times 10^{-3}\,\mathrm{F/m^2} = 515\,\mathrm{fF}. \tag{9.97}$$

Inserting C_{gate} into Eqn. (9.95) results in the gate charge loss of a single switch,

$$P_{gate1} = C_{gate}V_{drv}^2 f_{sw} = 515\,\mathrm{fF} \cdot (5\,\mathrm{V})^2 \cdot 10\,\mathrm{MHz} = 129\,\mu W. \tag{9.98}$$

We need to multiply this value by the number of switches n_{sw}, which is seven in this case. Hence,

$$P_{gate} = n_{sw}P_{gate1} = 7 \cdot 129\,\mu W = 901\,\mu W. \tag{9.99}$$

At $f_{sw} = 10\,\mathrm{MHz}$, the gate charge losses are still low. It indicates we could increase the transistor width to reduce the R_{FSL} and related losses. Figure 9.12 shows the gate charge losses over frequency along with more loss components.

9.9.3 Parasitic Capacitor Bottom-Plate Losses

Integrated capacitors are always associated with parasitic capacitances from each plate to substrate (ground) (see Section 4.1). Figure 9.11a) illustrates the scenario for the 2:1 SC converter. Both the parasitic bottom-plate and top-plate capacitances, C_{tp}, C_{bp}, respectively, get charged from V_{in} during phase φ_1, Fig. 9.11b). In phase φ_2, C_{tp} delivers its charge toward the output. This way, the top-plate capacitance delivers a part of the load current and recharges C_o. Consequently, C_{tp} contributes to the output power. Due to the small size of C_{tp}, its charge redistribution loss is usually negligible. In contrast, C_{bp} is short-circuited to ground. Its charge is lost. Therefore, the bottom-plate capacitance

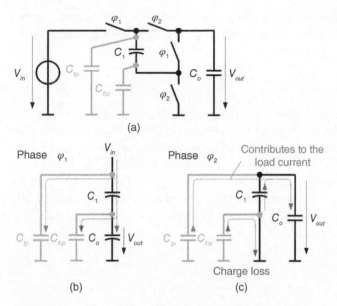

(a)

(b)   (c)

Phase φ_1 V_{in} Phase φ_2 Contributes to the load current

Charge loss

Figure 9.11 a) 2:1 SC converter with parasitic top-plate and bottom-plate capacitances C_{tp}, C_{bp} of the flying capacitor C_1; b) charge flow in phase φ_1, and c) in phase φ_2. C_{tp} contributes to the load current while the charge of C_{bp} is lost to ground.

C_{bp} is the main loss contributor. As described in Section 4.1, C_{bp} can be expressed as a fraction of the actual capacitor, i.e., $C_{bp} = \gamma C_1$. The factor γ depends on the capacitor type with typical values in the range of ~10% for MOS capacitors down to a few percent for MIM capacitors (see Section 4.1 and Table 4.1).

We can calculate to bottom-plate losses from

$$P_{bp} = \sum_{j=1}^{n} \Delta V_j^2 \, C_{bp,j} f_{sw} \tag{9.100}$$

with the voltage swing ΔV_j at the corresponding bottom-plate capacitor $C_{bp,j}$. The voltage swing can be found from the SC circuit by inspection, typically expressed in multiples of the output voltage V_{out}.

Example 9.10 *Determine the bottom-plate losses of the 3:1 Dickson topology at $f_{sw} = 50\,\text{MHz}$ for the capacitor sizing of Example 9.4, i.e., for $C_{tot} = 8\,\text{nF}$ and for $\gamma = 0.03\,(3\%)$. The output voltage is $V_{out} = 1.8\,\text{V}$.*

The Dickson topology as shown in Fig. 9.4b) with its two-phase equivalent circuit in Fig. 9.7b) contains two identical flying capacitors C_1, C_2 with a total capacitance of $C_{tot} = 8\,\text{nF}$. By inspection, we find that both capacitors see the same bottom-plate swing $\Delta V_1 = \Delta V_2 = V_{out}$. Hence,

$$P_{bp} = V_{out}^2 \, \gamma \, C_{tot} f_{sw} = (1.8\,\text{V})^2 \cdot 0.03 \cdot 8\,\text{nF} \cdot 50\,\text{MHz} = 38.9\,\text{mW}. \tag{9.101}$$

We can identify this value in Fig. 9.12, which shows P_{bp} over frequency for the parameters of this example.

Section 4.1 describes loss reduction techniques for parasitic bottom-plate capacitances. Since $C_{bp} > C_{tp}$, the flying capacitor C_1 in Fig. 9.11 can be flipped such that only the lower capacitance of C_{tp} contributes to the losses. Also, high-resistively biased wells can be placed underneath the flying capacitor C_1. Charge recycling techniques of the parasitic bottom-plate capacitance during the dead time between the phases φ_1 and φ_2 can be applied [13, 14]. It can be very effective if the input voltages and switching frequencies are low.

9.9.4 Static Losses

Any bias current contributes to the static current consumption I_{DC}. This current is delivered from V_{in} or an internal supply V_{DD}. V_{out} can be used as the internal supply V_{DD} in the SC converter depending on the voltage levels. Alternatively, V_{DD} can be generated by a separate regulator block like a simple SC converter stage or an LDO. The regulator has some power loss, expressed by its efficiency η_{sup}. For this reason, the static current I_{DD} drawn from V_{DD} will be different from the current I_{DC} that flows from V_{in}. Also, a shunt regulator may be used as an internal supply, which is usually required anyway to provide V_{DD} during start-up (similar to the concept in Fig. 7.23). The static power dissipation can be determined by

$$P_{DC} = V_{in} I_{DC} = \frac{V_{DD} I_{DD}}{\eta_{sup}}. \tag{9.102}$$

9.9.5 Loss Minimization

Figure 9.12 confirms that the intrinsic losses decrease with frequency while the extrinsic losses increase. Therefore, there is a loss minimum at the crossover region between these two types of losses. We can make more observations from Fig. 9.12:

(1) The total layout area of an SC power stage consists of capacitors and switches. A trade-off between R_{SSL} and R_{FSL} can be found by allocating more space to either part. Larger switch width reduces R_{FSL}. It leaves less area for the capacitors, and R_{SSL} increases. Also, the crossover between R_{SSL} and R_{FSL} will move to higher frequencies.

(2) In the lower frequency range where R_{SSL} dominates R_{out}, changing the value of the flying capacitors allows to trade-off P_{Rout} against P_{bp} at higher frequencies (note that P_{bp} increases with frequency).

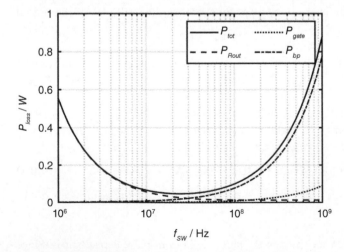

Figure 9.12 Total power loss and loss components in a 3:1 Dickson SC converter for the capacitor sizing of Example 9.4 ($C_{tot} = 8\,\text{nF}$, $\gamma = 0.03$ (3%), $V_{out} = 1.8\,\text{V}$).

(3) In the upper frequency range where $R_{out} \approx R_{FSL}$, increasing the width of the switch transistor results in lower R_{out}, while concurrently the gate charge losses P_{gate} go up and vice versa.

Based on this observation, several approaches have been proposed for loss minimization [11, 12]. However, they become relatively complex, and optimizing the design based on simulation may be more straightforward. As a starting point, we can operate the SC converter at the crossover frequency f_c at which $R_{SSL} = R_{FSL}$. Equating Eqns. (9.69) and (9.77) results in

$$f_c = \left(\frac{K_{SSL}}{K_{FSL}} \right)^2 \frac{G_{tot}}{2 C_{tot}}. \tag{9.103}$$

Example 9.11 *Calculate the crossover frequencies f_c for equivalent output resistance curves of the Dickson and ladder topology in Fig. 9.10.*

We can obtain the values for K_{SSL} and K_{FSL} from Table 9.4 (see also Eqns. (9.37) and (9.50)). The values of C_{tot} and G_{tot} are given or calculated in Examples 9.4–9.6, and they are identical for both topologies.

Dickson:

$$f_c = \left(\frac{K_{SSL}}{K_{FSL}} \right)^2 \frac{G_{tot}}{2 \cdot C_{tot}} = \left(\frac{0.67}{2.33} \right)^2 \frac{8.8\,\text{A/V}}{2 \cdot 8\,\text{nF}} = 45.5\,\text{MHz} \tag{9.104}$$

Ladder:

$$f_c = \left(\frac{K_{SSL}}{K_{FSL}} \right)^2 \frac{G_{tot}}{2 C_{tot}} = \left(\frac{1.33}{2.67} \right)^2 \frac{8.8\,\text{A/V}}{2 \cdot 8\,\text{nF}} = 136\,\text{MHz} \tag{9.105}$$

Figure 9.12 confirms that the minimum of the total losses of the Dickson topology occurs at a switching frequency close to f_c. If we know the dominant extrinsic loss component, there is an alternative way to find the optimum frequency. If the bottom-plate losses dominate like in Fig. 9.12, we can equate Eqns. (9.69) and (9.101) and solve for f_{sw}. In any case, starting from the rough value of f_{sw}, the design can be further optimized based on the observations in items (1)–(3) above. Note that the losses in Fig. 9.12 are plotted for a load current of 100 mA. The minimum will vary if the load current changes.

9.9.6 Total Losses

Taking the losses considered above into account, we get

$$P_{tot} = P_{Rout} + P_{gate} + P_{bp} + P_{DC}. \tag{9.106}$$

Additional loss components may be added, such as charging losses of isolation wells at any switching node. P_{tot} shows strong frequency dependence, because P_{Rout} reduces with frequency while P_{gate} and P_{bp} go up. Figure 9.12 shows the total losses for loss contributions considered in Examples 9.8–9.10.

9.9.7 Efficiency

The efficiency impact of the power losses in the equivalent output resistance has been covered by Eqn. (9.4) in the introduction to SC converters. Knowing the total losses P_{tot} the power efficiency can be refined,

$$\eta = \frac{P_{out}}{P_{in}} \cdot 100\% = \frac{P_{out}}{P_{out} + P_{tot}} \cdot 100\% \tag{9.107}$$

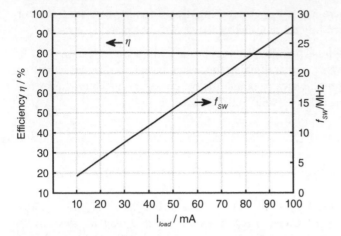

Figure 9.13 Power efficiency versus load current of the 3:1 Dickson SC converter at $V_{in} = 6\,V$, $V_{out} = 1.8\,V$ for the parameters of Examples 9.4, 9.5, 9.9, and 9.10.

with P_{tot} according to Eqn. (9.106), perhaps including additional losses. Figure 9.13 shows the efficiency for the losses of Fig. 9.12 over load current I_{load} at $V_{in} = 6\,V$. For each value of I_{load}, the control loop adjusts R_{out} according to Eqn. (9.2) by changing the switching frequency f_{sw}. Figure 9.13 also depicts that f_{sw} varies between 2.8 and 30.1 MHz. Over the entire load range, the converter operates in SSL because f_{sw} remains below the crossover frequency f_c as defined by Eqn. (9.103) (see also Eqn. (9.104) in Example 9.11). The efficiency stays almost flat at about 80%. Higher efficiency could be achieved by lowering the bottom-plate losses (lower factor γ). It can be accomplished by choosing different capacitor types, selecting another IC technology, and implementing the loss reduction techniques described in Section 4.1 for integrated capacitors. Lowering the bottom-plate losses, in turn, would also allow increasing the size of the flying capacitors to bring down R_{SSL} and f_c. This way, the SC converter could operate at a lower frequency with reduced contributions from P_{gate} and P_b.

9.10 Multi-phase SC Converters

The study of realizable conversion ratios in Section 9.2 indicates that a dedicated set of conversion ratios can be achieved for a given number of flying capacitors as expressed by Eqn. (9.10). For instance, for $n = 2$ flying capacitors, the ratios are $\frac{1}{3}$, $\frac{1}{2}$, $\frac{2}{3}$ (in addition to unity) as listed in Table 9.1. However, this only applies to two-phase operation. By adding more phases, more conversion ratios can be achieved. This way, the ideal conversion ratio may get closer to the actual ratio (defined by the given operating point of V_{in} and V_{out}). Consequently, the voltage drop across R_{out} will be lower, resulting in fewer losses and higher power conversion efficiency.

Typically, the converter is expanded by just one phase because this represents a good trade-off concerning effort and added complexity. How to introduce a third phase? A good starting point is to utilize one of the common topologies (see Section 9.3, Fig. 9.4) and to modify the clocking scheme such that the switches are activated differently in each phase. Often, additional switches are added to support the required capacitor configurations per phase. But remember that every switch increases the fast-switching limit output resistance R_{FSL}. While an intuitive approach can result in a suitable design, utilizing algorithms to synthesize multi-phase topologies as suggested in [15] may be worthwhile.

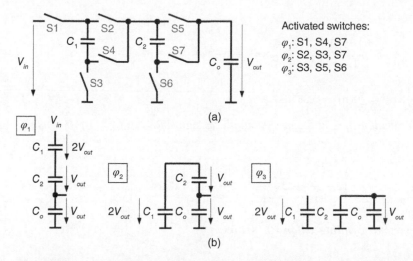

Figure 9.14 A three-phase 4:1 SC converter based on a) the Fibonacci topology (see Fig. 9.4d)) with switches S1–S7 activated per phase as shown on the right; b) equivalent circuits in each phase φ_1 to φ_3.

We stick with the intuitive approach. As an example, we study how to develop a three-phase SC converter based on the 3:1 Fibonacci topology (introduced in Fig. 9.4d), repeated in Fig. 9.14a)). The goal is to achieve a conversion ratio of 4:1 without any additional flying capacitors. In this case, we do not even need to add switches. We only need to activate the switches differently as indicated in Fig. 9.14a) (top right). The corresponding equivalent circuits per phase are shown in Fig. 9.14b). We can verify the achievable ideal conversion ratio by analyzing the voltages across each capacitor in quantities of V_{out}. Starting with the configuration that has a flying capacitor placed in parallel to C_{out} is recommended. In Fig. 9.14 this applies to C_2 during phase φ_3. Hence, the voltage at C_2 equals V_{out}. Generally, Kichhoff's Voltage Law (KVL) must be maintained in each phase. Therefore, phase φ_2 reveals that the voltage across C_1 is $2V_{out}$. Finally, in phase φ_1 KVL results in $V_{in} = 4V_{out}$, which perfectly confirms the ideal conversion ratio $N_i = V_{out}/V_{in} = \frac{1}{4}$. The reader can also prove the ideal conversion ratio via the charge flow analysis (Section 9.5).

Example 9.12 *Design the SC converter of Fig. 9.14 for a capacitor budget of $C_{tot} = 8\,nF$ for the flying capacitors and a total width of the switches of $W_{tot} = 8\,mm$. We keep the transistor channel length at $L = L_{min} = 0.3\,\mu m$. The other parameters are: $V_{th} = 0.6\,V$, $V_{GS} = 5\,V$, $\mu_o C_{ox} = 75\,\mu A/V^2$. Calculate both the equivalent output resistance R_{SSL} at $f_{sw} = 10\,MHz$ and R_{FSL}?*

We first determine the capacitor charge flow vectors in each of the three phases from the equivalent circuits in Fig. 9.14b):

$$a^1 = \begin{bmatrix} \frac{1}{4} & \frac{1}{4} & \frac{1}{4} & \frac{1}{4} \end{bmatrix}$$

$$a^2 = \begin{bmatrix} \frac{1}{4} & -\frac{1}{4} & \frac{1}{4} & 0 \end{bmatrix}$$

$$a^3 = \begin{bmatrix} \frac{1}{2} & 0 & -\frac{1}{2} & 0 \end{bmatrix} \tag{9.108}$$

Knowing the charge flows, the parameter K_{SSL} can be calculated from the multi-phase expression given in Eqn. (9.38),

$$K_{SSL} = \sum_{i=1}^{n_{ph}} \sum_{j=1}^{n} \left| a_{c,j}^{(i)} \right| = \frac{1}{4} + \frac{1}{4} + \left| -\frac{1}{4} \right| + \frac{1}{4} + 0 + \left| -\frac{1}{2} \right| = \frac{6}{4} = 1.5. \tag{9.109}$$

From Eqn. (9.75),

$$C_1 = \frac{\sum_{i=1}^{n_{ph}} |a_{c,1}^{(i)}|}{K_{SSL}} C_{tot} = \frac{\frac{1}{4} + |-\frac{1}{4}| + 0}{1.5} 8\,\text{nF} = \frac{1}{3} 8\,\text{nF} = 2.67\,\text{nF},$$

$$C_2 = \frac{\frac{1}{4} + \frac{1}{4} + |-\frac{1}{2}|}{1.5} 8\,\text{nF} = \frac{2}{3} 8\,\text{nF} = 5.33\,\text{nF}. \tag{9.110}$$

With C_1 and C_2 we can calculate the equivalent SSL output resistance R_{SSL} at $f_{sw} = 10\,\text{MHz}$ using Eqn. (9.32):

$$R_{SSL} = \sum_{i=1}^{n_{ph}} \sum_{j=1}^{n} \frac{(a_j^i)^2}{2 C_j f_{sw}} = \frac{1}{2 f_{sw}} \left(2 \frac{\left(\frac{1}{4}\right)^2}{C_1} + 2 \frac{\left(\frac{1}{4}\right)^2}{C_2} + 2 \frac{\left(\frac{1}{2}\right)^2}{C_2} \right) = 11.7\,\Omega \tag{9.111}$$

The switch charge flow vectors according to Eqn. (9.18) are

$$\boldsymbol{a_r^1} = \begin{bmatrix} \frac{1}{4} & 0 & 0 & \frac{1}{4} & 0 & 0 & \frac{1}{4} \end{bmatrix}$$

$$\boldsymbol{a_r^2} = \begin{bmatrix} 0 & \frac{1}{4} & \frac{1}{4} & 0 & 0 & 0 & \frac{1}{4} \end{bmatrix}$$

$$\boldsymbol{a_r^3} = \begin{bmatrix} 0 & 0 & 0 & 0 & \frac{1}{2} & \frac{1}{2} & 0 \end{bmatrix}. \tag{9.112}$$

As a next step, the K_{FSL}-parameter follows from the multi-phase representation in Eqn. (9.51):

$$K_{FSL} = \sum_{i=1}^{n_{ph}} \sum_{j=1}^{n_{sw}} \left| a_{r,j}^{(i)} \right| = 2.5 \tag{9.113}$$

The width of each switch transistor is given by Eqn. (9.76). Switches S1–S4 see a total charge flow of 1/4 over all phases. Hence, they have an equal size of

$$W_{r,m} = \frac{\sum_{i=1}^{n_{ph}} |a_{r,m}^{(i)}|}{K_{FSL}} W_{tot} = \frac{\frac{1}{4}}{2.5} 8\,\text{mm} = 0.8\,\text{mm}. \tag{9.114}$$

Switches S5–S7 will also be identical since their total charge flow is 1/2:

$$W_{r,m} = \frac{\frac{1}{2}}{2.5} 8\,\text{mm} = 1.6\,\text{mm} \tag{9.115}$$

We can now determine the corresponding switch resistance like in Eqn. (9.79) in Example 9.5. For S1–S4:

$$R_{DSon1} = \frac{1}{\frac{W}{L} \mu_o C_{ox} \left(V_{GS} - V_{th} \right)} = \frac{1}{\frac{0.8\,\text{mm}}{0.3\,\mu\text{m}} 75\,\mu\text{A/V}^2 (5\,\text{V} - 0.6\,\text{V})} = 1.14\,\Omega \tag{9.116}$$

For S5–S7:

$$R_{DSon2} = \frac{1}{\frac{1.6\,\text{mm}}{0.3\,\mu\text{m}} 75\,\mu\text{A/V}^2 (5\,\text{V} - 0.6\,\text{V})} = 568\,\text{m}\Omega \tag{9.117}$$

Like in Example 9.5 we can determine the total conductance G_{tot} similar to Eqn. (9.81):

$$G_{tot} = 4 \frac{1}{R_{DSon1}} + 3 \frac{1}{R_{DSon2}} = 8.8\,\text{A/V} \tag{9.118}$$

R_{FSL} can be calculated from Eqn. (9.77):

$$R_{FSL} = \frac{2 K_{FSL}^2}{G_{tot}} = \frac{2 (2.5)^2}{8.8\,\text{A/V}} = 1.42\,\Omega \tag{9.119}$$

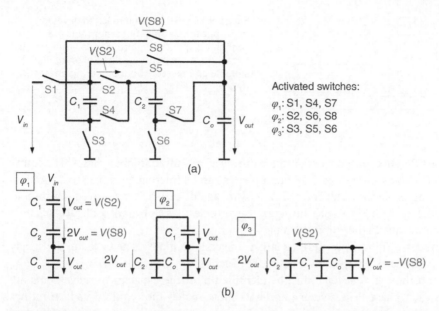

Figure 9.15 A three-phase 4:1 SC converter based on a) the Fibonacci topology with modified switch S5 and one additional switch S8; b) equivalent circuits in each phase φ_1, φ_2 and φ_3.

It is instructive to explore an alternative implementation of a multi-phase 4:1 converter based on the Fibonacci topology shown in Fig. 9.15a). Compared to the circuit in Fig. 9.14a), switch S5 taps from the top-plate of C_1 and switch S8 is added. The equivalent circuits in Fig. 9.15b) are similar to Fig. 9.14b), except that now C_2 sees the highest voltage of $2V_{out}$. This SC converter design is less favorable since the circuit will have the same multi-phase conversion ratio but needs one more switch. Moreover, it points to another effect, which frequently happens in multi-phase and multi-ratio (see Section 9.11) SC converters. Voltages with positive and negative polarity may occur across the switches in different configurations. In Fig. 9.15, this applies to switches S2 and S8. In phase φ_3 both switch voltages are $(-V_{out})$. A back-to-back switch configuration is needed to prevent charge backflow through the body diodes of S2 and S8. Section 2.11 outlines that this either doubles the switch resistance (higher R_{FSL}) or we need to increase the width and the length of both switches in the back-to-back configuration (4× larger layout area and gate charge). Alternatively, the back-gate bias voltage of S2 and S8 can be controlled to always connect to the source voltage (lower potential of the switch terminals) or any lower potential or ground.

Design Procedure of Multi-phase SC Converters

From the above considerations, we can generalize the development procedure for multi-phase SC converters:

1. Identify the target conversion ratio N_i and the number of phases to be implemented for the given input voltage, output voltage, and number of flying capacitors.
2. Arrange the flying capacitors in parallel and series connection between V_{in} and V_{out} as illustrated in Fig. 9.16a). This step defines phase φ_1 in which all capacitors are recharged from V_{in}. Mark the capacitor voltages in multiples of V_{out} (assuming the ideal steady state). KVL needs to be fulfilled.
3. Rearrange the capacitors in different parallel and series connections with respect to V_{out} during phase φ_2. The capacitor array is disconnected from V_{in}. The capacitor voltages remain as identified under step 2. Mark the voltages and ensure that KVL is fulfilled.

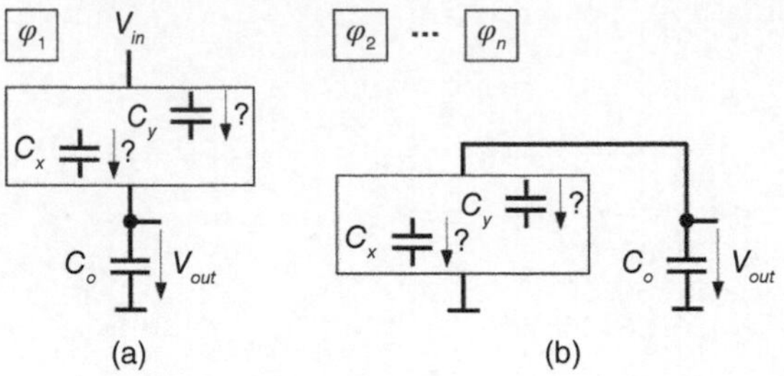

Figure 9.16 General approach to develop multi-phase SC converters: capacitors are configured a) to ensure proper recharging of all capacitors during phase φ_1 and b) to fulfill suitable charge redistribution during phases φ_2 to φ_n.

(a)　　　　　　　　　(b)

4. Repeat 3. with a different capacitor configuration during φ_3 and all additional phases. The equivalent circuits found for phases φ_2 and all other phases can be interchanged, i.e., their order is not relevant as long as the converter enters φ_1 afterward again. A capacitor can even stay disconnected if needed, like C_1 in Fig. 9.15b) during φ_3. Nevertheless, as a primary goal, all capacitors should aim to contribute to the output current.

5. Steps 2–4 are processed iteratively until the target conversion ratio N_i is achieved in the steady state. N_i can be verified by charge flow analysis (see Section 9.5.3, Eqn. (9.24)).

6. The switches and their connections must be identified after the capacitor arrangements are known during each phase. The switch transistors must match the required voltage ratings defined in multiples of V_{out} during each phase. The goal is to use the minimum number of switches with voltage ratings as low as possible.

The procedure can be verified for the multi-phase SC topologies in Figs. 9.14 and 9.15.

9.11 Multi-ratio SC Converters

SC converters demonstrate high power conversion efficiency only near the ideal voltage conversion ratio $N_i = V_{out}/V_{in}$. Any deviation of V_{out} from this ideal ratio increases the voltage drop in the equivalent output resistance R_{out} according to Fig. 9.2 and the corresponding power loss P_{Rout} as defined in Eqn. (9.92). Based on Eqn. (9.2). P_{Rout} can be expressed by

$$P_{Rout} = (N_i V_{in} - V_{out})I_{load}, \tag{9.120}$$

in which $(N_i V_{in} - V_{out})$ describes the voltage drop across R_{out}. Considering only the intrinsic losses P_{Rout}, Fig. 9.17 plots the efficiency as a function of V_{in} according to Eqn. (9.4) for various ideal conversion ratios. This graph represents the upper efficiency boundary that can theoretically be achieved. According to Eqn. (9.4), this efficiency value corresponds to the ratio of the actual and the ideal conversion ratio, N and N_i, respectively. Note that the actual efficiency will be lower due to the extrinsic losses described in Section 9.9.

Figure 9.17 shows that high efficiency can only be maintained over a wide input voltage range if the SC converter changes its (ideal) conversion ratio. It can be accomplished by multi-ratio SC converters. This converter class achieves many different conversion ratios N_i by reconfiguring the topology in various series–parallel combinations of the flying capacitors. Some of the early examples include [16] (based on Dickson) and [17], both published around 2010. The number of realizable ratios increases with the number of flying capacitors, as shown in Section 9.2. Note that the efficiency of an LDO according to Eqn. (7.12) will be identical to the curve for $n = 1$. It will continue to decrease if V_{in} increases beyond 3.6 V while the SC converter will switch to the next lower ratio (1/2 in this case). The significant increase in efficiency confirms the benefit of switched-mode SC conversion over linear regulation, particularly for SC converters with multiple ratios. We will investigate the most popular approaches for multi-ratio SC converters below.

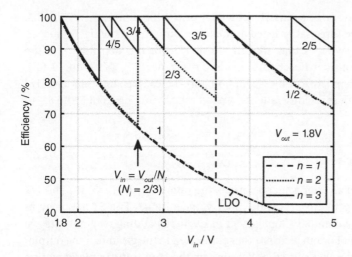

Figure 9.17 Theoretical power efficiency of an SC converter with $n = 1, 2, 3$ flying capacitors as a function of the input voltage V_{in} at $V_{out} = 1.8\,V$. The ratio transition happens at $V_{in} = V_{out}/N$ with the next lower conversion ratio N_i (shown for $V_{in} = 1.8\,V/(2/3) = 2.7\,V$).

9.11.1 Multi-ratio Implementation of Common SC Topologies

Figure 9.3 is an example of how two flying capacitors can provide two conversion ratios, in that case, 1/3 and 2/3. In addition, according to Table 9.1, two flying capacitors also provide the ideal ratio 1/2, which can be achieved by the fundamental 2:1 topology of Fig. 9.1a). A straightforward approach for the design of multi-ratio SC converters is to merge this 2:1 topology and the two configurations in Fig. 9.3. Figure 9.18 shows the resulting SC circuit as implemented in [18–20], for instance. Interestingly, for $N_i = 1/2$, switch S4 remains off, and the converter forms two separate 2:1 series–parallel stages that operate in parallel. This way, the effective flying capacitor corresponds to the sum of C_1 and C_2, minimizing the SSL output resistance and increasing the load current capability. Andersen et al. [19] uses only the 2/3 and 1/2 ratios but 16 interleaved slices (see Section 9.12) in parallel with a 1 nF deep trench cap each. It reduces the input current and output voltage ripples such that the output decoupling capacitor C_o can be omitted.

As a general characteristic of multi-ratio SC converters, additional switches are required compared to single-conversion-ratio topologies. The switches are only used in some ratio settings.

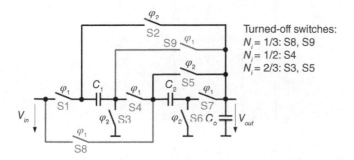

Figure 9.18 A multi-ratio SC converter with two flying capacitors supports three ideal conversion ratios $N_i = 1/3, 1/2, 2/3$. The converter is obtained by merging the single-ratio converters of Fig. 9.3 with the with the fundamental 2:1 series–parallel topology of Fig. 9.1.

They remain deactivated if the converter operates in other ratios. In the converter of Fig. 9.3, several switches are turned off in the three ratios listed at the top right of the drawing. Moreover, there may also be reconfiguration switches that are fully turned on or off, depending on the ratio. In any case, they need to be considered as part of the FSL resistance R_{FSL} (see Eqn. (9.44)). The multi-ratio approach can be applied to other common topologies. Ng and Sanders [16] implement the Dickson topology with seven different conversion ratios, ranging from 5:1 to 8:1 in half-integer steps.

9.11.2 Folding Dickson

Figure 9.19 shows a multi-ratio SC converter, which supports the ratios 1/5, 1/4, 1/3, 1/2 and 1/1 [21]. This way, the converter can operate over a wide input voltage range. The 5:1 topology in Fig. 9.19a) can be derived from the standard 3:1 topology of Fig. 9.4b) by adding two more stages. There are four flying capacitors. Their bottom plate is either connected to the ground or the output. While the original Dickson converter is also called the Dickson star, the multi-ratio implementation is called a folding Dickson converter. Folding is achieved by virtually merging the top and bottom plates of multiple flying capacitors. It creates one or more lumped flying capacitors, enabling different conversion ratios. In the 4:1 configuration of Fig. 9.19b), C_2 and C_3 are merged. Their top plates are permanently connected, while the bottom plates are controlled concurrently by the same signals (switches S8–S11). The conversion ratio in Fig. 9.19c) is $N_i = 1/3$. With S2 and S4 being turned on all the time, C_1 and C_2 as well as C_3 and C_4 merge. Therefore, the circuit is identical to the 3:1 Dickson topology of Fig. 9.4b). Finally, the 2:1 topology in Fig. 9.19d) has all flying capacitors placed in parallel. S2–S4 connect their top plates, and the bottom plates are controlled synchronously by switch pairs S6/S7 to S12/S13. If V_{in} is close to V_{out}, the converter will show the highest efficiency when the ratio is set to unity. $N_i = 1/1$ can be achieved from Fig. 9.19d) if the switches S6, S8, S10, S12 toward ground turn on constantly and, at the same time, S7, S9, S11, S13 remain turned off.

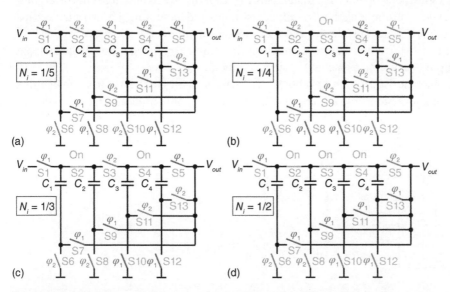

Figure 9.19 The folding Dickson topology in different conversion ratios: a) $N_i = 1/5$, b) $N_i = 1/4$, c) $N_i = 1/3$, and d) $N_i = 1/2$. The ratios are achieved by keeping some switches permanently activated (On) and assigning different control signals φ_1, φ_2 to each switch.

As an advantage of the folding Dickson converter, the switching transistors at the bottom plate of the capacitors only have to block V_{out} while the transistors connected to the top plate have to be rated for $2V_{out}$. Thus, low-voltage transistors can be used depending on the output voltage level. It allows for relatively simple control of the switching transistors by an extended bootstrap circuit [21]. The limitations of the folding Dickson include the fact that the number of conversion ratios is identical to the number of flying capacitors. In contrast, Table 9.1 (Section 9.2) shows that, in theory, with $n = 4$ flying capacitors, 22 step-down ratios can be realized. Also, Dickson does not provide a conversion ratio between 1/2 and 1. Nevertheless, the folding is attractive for high step-down ratios.

9.11.3 SAR SC Converters

Increasing the ratios using standard topologies increases the number of components, the converter complexity, and related losses. This drawback is addressed by the concept of successive approximation (register) SC converters (SAR SC converters) as proposed in [22, 23]. SAR SC converters cascade multiple 2:1 SC stages using pairs of configuration switches and a single switch at the end, as shown in Fig. 9.20. The converter provides a fine-grain output voltage with a resolution of

$$\Delta V_{out} = \frac{V_{in}}{2^{n_{st}}} \tag{9.121}$$

depending on the number of stages n_{st}. Each 2:1 stage is connected between two input levels V_{high}, V_{low}. It provides the average voltage between V_{high} and V_{low} at its output, i.e., $(V_{high} + V_{low})/2$. The high-voltage input of the subsequent stage is connected to V_{high} or the previous stage's output. The low-voltage input of the next stage is connected to either V_{low} or the previous stage's output. For a given binary configuration code B, the output voltage can be calculated from

$$V_{out} = (B + 1)\frac{V_{in}}{2^{n_{st}}}. \tag{9.122}$$

In the four-stage example of Fig. 9.20, for configuration code $B = 1001$, $V_{in} = 2\,\text{V}$ is converted to $V_{out} = 1.250\,\text{V}$. According to Eqn. (9.121) the voltage resolution is $125\,\text{mV}$ in this case. Consequently, decreasing the code to $B = 1000$ would result in $V_{out} = 1.125\,\text{V}$ by toggling the final switch. As an advantage of the SAR architecture, minimal change in configuration is required in case of conversion ratio adjustment.

Bang et al. [22] implements a 7-bit SAR SC converter with a cascade of a 4:1 converter followed by five 2:1 stages. All 2:1 stages are sized identically to handle the varying levels of current flow depending on configuration. The effective number of ratios is 117 (in theory $2^{n_{st}} = 127$) with a resolution of $31.25\,\text{mV}$ at $V_{in} = 4\,\text{V}$. Each stage is two-phase interleaved. Capacitive level shifters control all switches. MIM and MOS capacitors (see Section 4.1) are used for on-chip flying and decoupling capacitors. The total capacitance is $2.24\,\text{nF}$ (in $180\,\text{nm}$ CMOS).

Figure 9.20 reveals some drawbacks of the SAR SC converter. Since the capacitor network is connected by a large number of switches in each phase, the FSL resistance increases. Moreover,

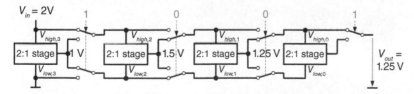

Figure 9.20 A 4-bit successive approximation SC converter (SAR SC converter) generating $V_{out} = 1.250\,\text{V}$ from $V_{in} = 2\,\text{V}$ for the configuration bits set to $B = 1001$.

for various code settings, including the depicted case $B = 1001$, the flying capacitor in the last 2:1 stage is not utilized. The slow-switching resistance increases as the capacitor does not contribute to delivering charge to the output. Fewer switches and better capacitance utilization are advantages of the RSC converter described below.

9.11.4 Recursive SC Converters

This class of multi-ratio SC converters [24–26] develops the idea of the SAR converter further. Recursive SC (RSC) converters keep the fundamental 2:1 stage as a unit cell shown in Fig. 9.21a) (see also Fig. 9.1). As one of the key advantages, RSC converters utilize 100% of the flying capacitance in the charge transfer across all conversion ratios. It is achieved by reconfiguring the 2:1 cells as a combination of cascade and parallel connections as illustrated in Fig. 9.21b–d). Moreover, switches and capacitors are sized according to the rules introduced in Section 9.8. Consequently, 2:1 SC stages with the highest current dynamically allocate the most capacitance and switch resources.

Starting with the fundamental 2:1 stage of Fig. 9.21a), there are two ways to implement cascading. Either all cells are configured in series, including the buffer capacitor (C_o in Fig. 9.21a)) at the output of each cell, as illustrated in Fig. 9.22a) [26]. Alternatively, [24] proposes a symmetric RSC, where each cell consists of two oppositely phased 2:1 stages, as shown in Fig. 9.22b) (also discussed in [23]). This interleaved operation eliminates the need for a buffer capacitor between cascaded stages (only the flying capacitor is needed). Keeping the buffer capacitor is more beneficial for higher values of V_{in} as it reduces the bottom-plate losses of the second time-interleaved stage [26]. Moreover, each switch may require a level shifter and dedicated gate supply. Their losses will counteract the benefits of the multi-phase operation. The cascading options shown in Fig. 9.22 can be applied to any other SC configuration.

Salem and Mercier [24] implement a 4-bit RSC converter ($n_{st} = 4$), which provides $(2^{n_{st}} - 1) = 15$ step-down ratios:

$$N_i = \left[\frac{1}{16}; \frac{1}{8}; \frac{3}{16}; \frac{1}{4}; \frac{5}{16}; \frac{3}{8}; \frac{7}{16}; \frac{1}{2}; \frac{9}{16}; \frac{5}{8}; \frac{11}{16}; \frac{3}{4}; \frac{13}{16}; \frac{7}{8}; \frac{15}{16} \right] \tag{9.123}$$

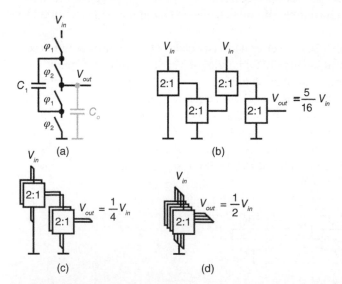

Figure 9.21 Recursive switched-capacitor converter: by reconfiguring multiple a) elementary 2:1 stages in cascade and parallel, various conversion ratios can be achieved such as b) 5/16; c) 1/4, and d) 1/2.

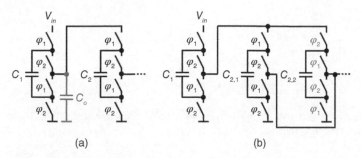

Figure 9.22 Cascading options for 2:1 SC stages: a) with an intermediate buffer capacitor C_o; b) time-interleaved operation without C_o.

For high resolutions, i.e., in $x/16$-ratios, all stages are cascaded. The example in Fig. 9.21b) achieves $N_i = 5/16$. The output of the second stage is $V_{out2} = 1/4\,V_{in} = 4/16\,V_{in}$. The third SC cell is connected between this voltage and V_{in} and adds half of the difference $(V_{in} - V_{out2})$ to V_{out2}, i.e., its output level is $V_{out3} = 4/16 + (16/16 - 4/16)/2 = 10/16$. The last stage divides this level by two and provides $V_{out4} = V_{out} = 5/16\,V_{in}$. If the last stage would be connected between V_{out3} and V_{in}, we would obtain $V_{out} = 13/16\,V_{in}$.

Ratios $x/8$, $x/4$, and $1/2$ do not require $1/16$ resolution. In these cases, cells are partially placed parallel to maximize capacitor utilization. Figure 9.21c,d) shows the cases $N_i = 1/4$ and $N_i = 1/2$. We will determine the optimum configuration and capacitor allocation below.

Cascading n_{st} 2:1 cells, we can find these expressions for the equivalent output resistances in SSL and FSL (see also Eqns. (9.124) and (9.125)) [25]:

$$R_{SSL} = \sum_{j=1}^{n_{st}} \left(\frac{1}{2^{n_{st}-j+1}} \right)^2 \frac{1}{C_j f_{sw}} \tag{9.124}$$

$$R_{FSL} = \sum_{j=1}^{n_{st}} \sum_{k=1}^{4} \frac{1}{2} \left(\frac{1}{2^{n_{st}-j}} \right)^2 R_{j,k} \tag{9.125}$$

The summation over k accounts for the four switches per 2:1 cell. Each switch resistance $R_{j,k}$ equals the two parallel switches in a symmetric RSC with two time-interleaved 2:1 stages. Applying the capacitor and switch sizing rules of Section 9.8, the capacitance and switch conductance have to match their relative charge flow,

$$C_j = \left(\frac{2^{j-1}}{2^{n_{st}} - 1}\, C_{tot} \right), \tag{9.126}$$

$$G_j = \frac{1}{4} \left(\frac{2^{j-1}}{2^{n_{st}} - 1}\, G_{tot} \right). \tag{9.127}$$

For $n_{st} = 4$ stages, Eqn. (9.126) results in

$$C_1 = \frac{1}{15}\, C_{tot}, \quad C_2 = \frac{2}{15}\, C_{tot}, \quad C_3 = \frac{4}{15}\, C_{tot}, \quad C_4 = \frac{8}{15}\, C_{tot}. \tag{9.128}$$

The output current and, accordingly, the charge flow doubles from stage to stage. Hence, the capacitance needs to scale similarly, with the highest value assigned to the last stage. To maintain 100% capacitance utilization, the authors in [24] split C_3 and C_4 into sub-cells as depicted in Fig. 9.23. Figure 9.24 shows the corresponding micrograph of the design presented in [24]. For the $x/16$-ratios, no segmentation is needed, and C_1 to C_4 are utilized in each stage, as shown in Fig. 9.25a). Figure 9.25b) shows an example for $x/8$-ratios, which always require three conversion stages. In these cases, the sub-cells of C_4 must be parallel to C_1, C_2, and C_3.

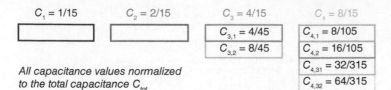

All capacitance values normalized
to the total capacitance C_{tot}

Figure 9.23 Capacitor sizing: to allow for recursive reconfiguration with 100% capacitor utilization, C_3 and C4 are split in two and four sub-cells, respectively.

For minimum R_{out}-losses (Eqn. (9.92)), the sub-cells of C_4 need to scale proportionally to the charge flow, exactly like C_1 to C_3 in Eqn. (9.128). Therefore, the total value of $C_4 = 8/15\,C_{tot} = 56/105\,C_{tot}$ is split into $C_{4,1} = 8/105\,C_{tot}$, $C_{4,2} = 16/105\,C_{tot}$ and $C_{4,3} = 32/105\,C_{tot}$.

In order to optimally support $x/4$-ratios with two conversion steps, $C_{4,3} = 32/105\,C_{tot} = 96/315\,C_{tot}$ is split one more time into two parts, $C_{4,31} = 32/315\,C_{tot}$ and $C_{4,31} = 64/315\,C_{tot}$. For the same reason, to optimally support $x/4$-conversion ratios, C_3 is split into two parts, as shown in Fig. 9.23. Each sub-cell, $C_{3,1}$ and $C_{3,2}$ can be placed in parallel to C_1 and C_2, respectively. Figure 9.25c) illustrates how the sub-cells of C_3 and C_4 are allocated for $N_i = 1/4$. Finally, for ratio 1/2, all flying capacitors are placed in parallel, as shown in Fig. 9.25d).

Now, we investigate how all capacitor sub-cells get connected efficiently. Figure 9.26a) shows the block diagram of an RSC converter with 3-bit resolution. Stage 3 comprises two segments, $C_{3,1}$ and $C_{3,2}$. A set of four reconfiguration switches $R_{1,2}$ connects stages 1 and 2 as introduced in [24]. Similarly, three pairs of switches $R_{2,3}$ connect stage 2 and the segments of stage 3. As an example, Fig. 9.26b,c) realize the configurations of Fig. 9.25b,c). For simplification, the allocation of C_4, as reported in [24], is not shown. Nevertheless, the RSC converter can be expanded by following the same scheme and adding more groups of reconfiguration switches. The reconfiguration switches are either controlled by φ_1/φ_2 or turned off. This way, the stages can be cascaded or connected in parallel. The 8:5-ratio in Fig. 9.26b) operates the reconfiguration switches $R_{1,2}$ instead of switches $S2_1$ and $S3_1$ (defined in Fig. 9.26a)). The output voltage of stage 1 is available at the

Figure 9.24 Die photo of a multi-ratio SC converter. Source: Loai G. Salem and Patick P. Mercier [24]/from IEEE. The chip contains four flying capacitors with 3 nF total capacitance. C_3 and C_4 consist of multiple segments, as shown in Fig. 9.23.

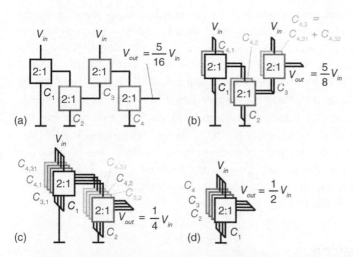

Figure 9.25 Capacitor allocation for a 4-bit RSC converter proportional to their charge flow: a) cascading C_1 to C_4 leads to $N_i = 5/16$; b) $x/8$ ratios require three conversion steps with sub-cells of C_4 assigned in parallel to C_1, C_2, and C_3; c) $x/4$ ratios consist of two stages with the sub-cells of C_3 and C_4 added to C_1 and C_2; d) all cells placed in parallel for $N_i = 1/2$.

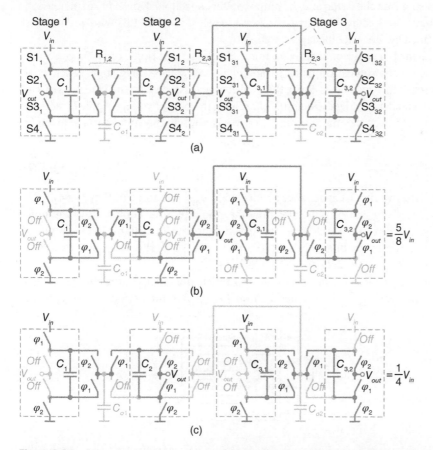

Figure 9.26 Recursive implementation: a) block diagram with 3-bit resolution (C_1 to C_3) including reconfiguration switches $R_{1,2}$, $R_{2,3}$; b) realization of ratio $N_i = 5/8$, as shown in Fig. 9.25b) by cascading C_1 and C_2 and connecting their output toward V_{in} by a parallel cell of $C_{3,1}$ and $C_{3,2}$; c) ratio $N_i = 1/4$ of Fig. 9.25c) with cascaded cells of C_1 and C_2 operating in parallel to the cascade of $C_{3,1}$ and $C_{3,2}$.

center node of $R_{1,2}$, which further connects to stage 2 via the corresponding reconfiguration switch of $R_{1,2}$, controlledr by φ_1. The other reconfiguration switch and $S1_2$ are permanently turned off. Moreover, also switches $S2_2$ and $S3_2$ are off. Instead, their function is replaced by the two reconfiguration switches $R_{2,3}$ at the output of the second stage. Consequently, stage 2 provides $\frac{1}{2} \cdot \frac{1}{2} V_{in} = \frac{1}{4} V_{in}$ and feeds it to the two segments of stage 3. Both segments operate in parallel in between $\frac{1}{4} V_{in}$ and V_{in}. As described above for Fig. 9.25b), the overall output voltage is $V_{out} = \frac{1}{4} V_{in} + (V_{in} - \frac{1}{4} V_{in})/2 = \frac{5}{8} V_{in}$. In Fig. 9.26c), stages 1 and 2 again form a cascade, but the output of stage 2 is directly connected to the output of the SC converter, resulting in $V_{out} = \frac{1}{4} V_{in}$. The capacitor segments in stage 3 are also utilized as an advantage of the RSC concept. They are cascaded like stages 1 and 2 with the output of the $C_{3,2}$-segment connected to V_{out}. This way, each RSC reconfiguration ensures both full capacitor utilization and optimum scaling in relation to the charge flow according to Eqns. (9.126) and (9.128).

9.12 Multi-phase Interleaving

Figure 9.27a) illustrates the output voltage ripple as described in Section 9.6 for single-phase operation. As indicated by Eqn. (9.63), a lower ripple can be achieved by increasing the switching frequency f_{sw} or increasing both the flying and the output buffer capacitors. However, an increase in the switching frequency is limited by the increase in losses (see Section 9.9). Likewise, an increase in capacitance is confined by the available layout area.

Multi-phase interleaving is a common method to minimize the output ripple of an SC converter [27, 28]. It reduces the output voltage ripple by paralleling n_{st} SC stages. Each stage is activated phase-shifted to the other converter stages. Since the total charge dissipated by the load per switching cycle is delivered in smaller equally divided sub-quantities, the output voltage ripple is reduced by a factor n_{st}, also referred to as the interleaving factor. We can refine the voltage ripple given in Eqn. (9.63) by

$$V_{out,pkpk} = \frac{I_{load}}{f_{sw} n_{st} k_d C_o^*},$$ (9.129)

where C_o^* denotes the effective flying capacitance connected to V_{out} as defined in Eqn. (9.64).

For the 2:1 SC converter, Fig. 9.27b) illustrates how the ripple reduces compared to the non-interleaved single-stage SC converter. Since each interleaved stage transfers charge to the output in each of the two phases, the optimum phase shift is $180°/n_{st}$. Figure 9.27c) shows the block diagram for $n_{st} = 4$ with the clock (CLK) shifted by $0°$, $45°$, $90°$ and $135°$.

Equation (9.129) reflects the fact that, in a multi-phase interleaving topology, the charge is delivered multiple times during each period, once per fraction $T/(n_{st} k_d)$ of the period, where k_d is the

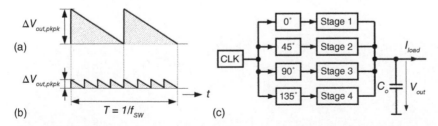

Figure 9.27 Multiphase interleaving: a) output voltage ripple $\Delta V_{out,pkpk}$ for a single-stage non-interleaved SC converter (2:1 SC stage); b) ripple; and c) block diagram in case of four-phase time-interleaving.

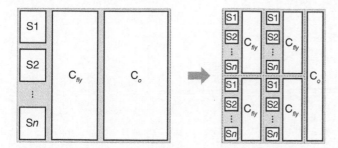

Figure 9.28 Scaling of the flying capacitors and switches for 4-phase interleaving. Source: Adapted from [11]. The non-interleaved SC converter (left) is replaced by four stages, with the capacitors and switches scaled down by a factor of four (right). The layout area remains nearly constant, but the output capacitor C_o gets four times smaller.

discharge fraction factor. For the basic 2:1 SC stage, $k_d = 2$. For a given ripple specification, the output capacitor size can be reduced by the interleaving factor n_{st}. At the same time, the size of the switches and the flying capacitors in each time-interleaved stage scale down by factor n_{st}. As illustrated in Fig. 9.28 [11], multiphase interleaving has nearly no area penalty, except for the additional control and wiring overhead. As the main benefit, the output capacitor can be scaled down by n_{st}. However, especially for higher input voltages (>5 V), the area and efficiency impact of the level shifters and gate supply for the switches in the additional phases may become significant. In that case, the cascading approach of Fig. 9.22a) may be a better choice.

9.13 Control Methods

SC converters need a control loop to provide a precise DC output voltage and a fast transient response. Recalling Eqn. (9.2), repeated below for convenience,

$$V_{out} = N_i V_{in} - R_{out} I_{load}, \tag{9.130}$$

two suitable control quantities exist: the ideal conversion ratio N_i and the equivalent output resistance R_{out}. For a narrow range of V_{in} and V_{out}, single-ratio topologies with constant N_i will be sufficient. Setting the conversion ratio in multi-ratio SC converters (Section 9.11) allows to achieve an ideal conversion ratio N_i that gets as close as possible to the actual ratio given mainly by the operating range of V_{in} and V_{out}. It forms an outer control loop, but a limited set of discrete conversion ratios determines the granularity. In addition, the ratios are spread nonlinearly (see Sections 9.2, 9.11 and also Fig. 9.17).

Therefore, precise control of V_{out} requires to adjust the equivalent output resistance R_{out}. The SSL and FSL expressions of R_{out} according to Eqns. (9.32) and (9.43) indicate

$$R_{SSL} \propto \frac{1}{C_{tot} f_{sw}} \quad \text{and} \quad R_{FSL} \propto \frac{R_{DSon}}{D}. \tag{9.131}$$

In SSL, the equivalent output resistance can be influenced by controlling the switching frequency f_{sw} and the amount of flying capacitance C_{tot}. When operating in FSL, the on-resistance R_{DSon} of the switch transistor and the duty cycle D of the particular switch allow to control R_{out}. We will discuss the various control options [11, 28, 29].

Switching frequency modulation, also called pulse-frequency modulation (PFM), is the most straightforward and most common control method in SC converters. There are many implementation examples. For instance, [30] uses a charge pump integrator that controls a voltage-controlled

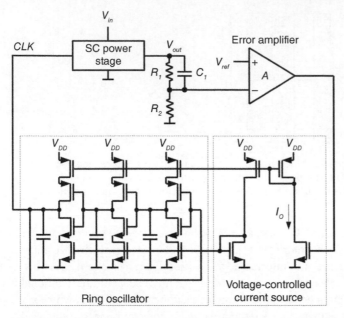

Figure 9.29 SC converter with a ring-oscillator-based control loop.

oscillator (VCO). A ring oscillator is used in [31]. That paper details deriving the transfer functions and applying the control theory.

Figure 9.29 shows the concept presented in [28]. The current-starved ring oscillator forms a VCO. The error amplifier determines any deviation of the SC converter's output voltage. V_{out} is sensed by a resistive divider to reduce the required input swing of the error amplifier. In this case, the error amplifier determines the bias current I_o that feeds the ring oscillator, which consists of three stages. If V_{out} drops, the inverting input of the error amplifier follows and increases the differential input of the error amplifier. The amplifier's output voltage rises. Consequently, I_o increases, leading to faster switching at each VCO stage. Higher f_{sw} counteracts the drop in the SC converter's output voltage V_{out}. This way, negative feedback is established. Various oscillator types can be used as a VCO, for instance, a relaxation oscillator with a variable current source (see also Section 11.4 and Fig. 11.6). In multi-ratio converters, a ring oscillator will be beneficial as each stage provides a shifted clock, which can directly be connected to the different phases (see Fig. 9.27c)).

As a general advantage of PFM control, the switching frequency scales proportionally to the load current, i.e., to the output power. This results in high light-load efficiency since the frequency-dependent losses (Section 9.9) scale down. On the other hand, the varying switching frequency leads to an unpredictable spectrum of generated noise (see electromagnetic interference in Section 12.3).

C_1 in parallel to R_1 in the resistive divider implements lead compensation of the control loop [11]. A stable control loop is obtained by positioning the zero at the error amplifier's 3 dB frequency and by shifting all higher-order poles beyond the switching frequency. Similar to the discussion for the error amplifier supply in an LDO (Section 7.4) the V_{DD}-supply for the control loop circuit blocks can be either derived from V_{in} or V_{out} in combination with a shunt regulator or some pre-bias during start-up (see Fig. 7.23).

Capacitance modulation [32] has a similar effect on R_{out} like PFM but without modifying f_{sw}. It requires an array of capacitors activated in parallel with increasing output power. While this positively impacts bottom-plate losses, it leads to reduced capacitance utilization since parts of C_{tot} do not contribute to the power conversion. Due to constant frequency operation, the light load efficiency will be lower than PFM because the switching losses do not reduce with the load.

In FSL, the modulation of the power transistor on-resistance R_{DSon} is an attractive control method known as switch conductance modulation (SwCR). See also Section 11.6.3 on SwCR in hybrid converters. It can be accomplished by switch segmentation or by varying the gate–source voltage of the power transistor. With the increasing integration capabilities of advanced CMOS technologies, switch segmentation can be implemented with fine granularity. Note that the deactivated segments still share common drain or source areas with the main switch. Therefore, the total parasitic capacitance of all switches will be present at the particular switching nodes with a negative impact on the power efficiency, especially for multi-megahertz operation. The gate–source voltage can vary the on-resistance with continuous resolution [16], but the I–V characteristic of the MOS transistor makes the on-resistance a highly nonlinear 1/x-function of V_{GS} (Eqn. (2.2)). The control loop may get overly sensitive and unstable if V_{GS} gets close to the V_{th}.

Another control option in FSL operation is the variation of the duty cycle D for all switches during each phase. It is accomplished by pulse-width modulation (PMW) in contrast to PFM in SSL. PWM is the standard control scheme in inductive DC–DC converters (see Fig. 10.1b) in Section 10.1). Operating in FSL, the switching frequencies and the frequency-dependent losses will be very high. Therefore, the efficiency at light load will be low. Nevertheless, the concept can be implemented well and allows continuous control of V_{out}.

References

1 Makowski, M. and Maksimovic, D. (1995) Performance limits of switched-capacitor DC-DC converters, in *Proceedings of PESC'95 - Power Electronics Specialist Conference*, vol. 2, pp. 1215–1221, doi: 10.1109/PESC.1995.474969.

2 Seeman, M.D. and Sanders, S.R. (2008) Analysis and optimization of switched-capacitor DC-DC converters. *IEEE Transactions on Power Electronics*, 23 (2), 841–851, doi: 10.1109/TPEL.2007.915182.

3 Wens, M. and Steyaert, M. (2011) *Design and Implementation of Fully-Integrated Inductive DC-DC Converters in Standard CMOS*, Springer, Dordrecht, Heidelberg, London, New York.

4 Dickson, J. (1976) On-chip high-voltage generation in MNOS integrated circuits using an improved voltage multiplier technique. *IEEE Journal of Solid-State Circuits*, 11 (3), 374–378, doi: 10.1109/JSSC.1976.1050739.

5 Lin, P. and Chua, L. (1977) Topological generation and analysis of voltage multiplier circuits. *IEEE Transactions on Circuits and Systems*, 24 (10), 517–530, doi: 10.1109/TCS.1977.1084273.

6 Harada, I., Ueno, F., Inoue, T., and Oota, I. (1991) Characteristics analysis of Fibonacci type SC transformer. *Transactions on IEICE Japan*, E75-A (6), 655–662.

7 Meyvaert, H., Piqu, V., Karadi, R., Bergveld, H.J., and Steyaert, M. (2015) 20.1 A light-load-efficient 11/1 switched-capacitor DC-DC converter with 94.7% efficiency while delivering 100 mW at 3.3V, in *2015 IEEE International Solid-State Circuits Conference - (ISSCC) Digest of Technical Papers*, pp. 1–3, doi: 10.1109/ISSCC.2015.7063074.

8 Seeman, M.D. (2009) *A Design Methodology for Switched-Capacitor DC-DC Converters*, Ph.D. thesis, EECS Department, University of California at Berkeley. URL http://www.eecs.berkeley.edu/Pubs/TechRpts/2009/EECS-2009-78.html.

9 Kimball, J. and Krein, P. (2005) Analysis and design of switched capacitor converters, in *20th Annual IEEE Applied Power Electronics Conference and Exposition, 2005. APEC 2005*, vol. 3, pp. 1473–1477, doi: 10.1109/APEC.2005.1453227.

10 Evzelman, M. and Ben-Yaakov, S. (2013) Average-current-based conduction losses model of switched capacitor converters. *IEEE Transactions on Power Electronics*, 28 (7), 3341–3352, doi: 10.1109/TPEL.2012.2226060.

11 Breussegem, T.V. and Steyaert, M. (2013) *CMOS Integrated Capacitive DC-DC Converters*, Springer Science+Business Media.

12 Le, H.P., Sanders, S.R., and Alon, E. (2011) Design techniques for fully integrated switched-capacitor DC-DC converters. *IEEE Journal of Solid-State Circuits*, 46 (9), 2120–2131, doi: 10.1109/JSSC.2011.2159054.

13 Tong, T., Zhang, X., Kim, W., Brooks, D., and Wei, G.Y. (2013) A fully integrated battery-connected switched-capacitor 4:1 voltage regulator with 70% peak efficiency using bottom-plate charge recycling, in *Proceedings of the IEEE 2013 Custom Integrated Circuits Conference*, pp. 1–4, doi: 10.1109/CICC.2013.6658485.

14 Butzen, N. and Steyaert, M.S.J. (2016) Scalable parasitic charge redistribution: design of high-efficiency fully integrated switched-capacitor DC-DC converters. *IEEE Journal of Solid-State Circuits*, 51 (12), 2843–2853, doi: 10.1109/JSSC.2016.2608349.

15 Karadi, R. (2020) Methodology and algorithm for synthesis of multi-phase switched-capacitor power converter topologies, in *2020 IEEE 21st Workshop on Control and Modeling for Power Electronics (COMPEL)*, pp. 1–8, doi: 10.1109/COMPEL49091.2020.9265838.

16 Ng, V. and Sanders, S. (2012) A 92 switched-capacitor DC-DC converter, in *2012 IEEE International Solid-State Circuits Conference*, pp. 282–284, doi: 10.1109/ISSCC.2012.6177016.

17 Su, L. and Ma, D. (2010) Monolithic reconfigurable SC power converter with adaptive gain control and on-chip capacitor sizing, in *2010 IEEE Energy Conversion Congress and Exposition*, pp. 2713–2717, doi: 10.1109/ECCE.2010.5618052.

18 Le, H.P., Seeman, M., Sanders, S.R., Sathe, V., Naffziger, S., and Alon, E. (2010) A 32nm fully integrated reconfigurable switched-capacitor DC-DC converter delivering 0.55W/mm² at 81% efficiency, in *2010 IEEE International Solid-State Circuits Conference (ISSCC)*, pp. 210–211, doi: 10.1109/ISSCC.2010.5433981.

19 Andersen, T.M., Krismer, F., Kolar, J.W., Toifl, T., Menolfi, C., Kull, L., Morf, T., Kossel, M., Brli, M., Buchmann, P., and Francese, P.A. (2014) 4.7 A sub-ns response on-chip switched-capacitor DC-DC voltage regulator delivering 3.7 W/mm^2 at 90% efficiency using deep-trench capacitors in 32 nm SOI CMOS, in *2014 IEEE International Solid-State Circuits Conference Digest of Technical Papers (ISSCC)*, pp. 90–91, doi: 10.1109/ISSCC.2014.6757351.

20 Renz, P., Kaufmann, M., Lueders, M., and Wicht, B. (2019) 8.6 A fully integrated 85%-peak-efficiency hybrid multi ratio resonant DC-DC converter with 3.0-to-4.5 V input and 500 µA-to-120mA load range, in *2019 IEEE International Solid- State Circuits Conference (ISSCC)*, IEEE, San Francisco, CA, USA, pp. 156–158, doi: 10.1109/ISSCC.2019.8662491.

21 Sarafianos, A. and Steyaert, M. (2015) Fully integrated wide input voltage range capacitive DC-DC converters: the folding Dickson converter. *IEEE Journal of Solid-State Circuits*, 50 (7), 1560–1570, doi: 10.1109/JSSC.2015.2410800.

22 Bang, S., Wang, A., Giridhar, B., Blaauw, D., and Sylvester, D. (2013) A fully integrated successive-approximation switched-capacitor DC-DC converter with 31 mV output voltage resolution, in *2013 IEEE International Solid-State Circuits Conference Digest of Technical Papers*, pp. 370–371, doi: 10.1109/ISSCC.2013.6487774.

23 Bang, S., Blaauw, D., and Sylvester, D. (2016) A successive-approximation switched-capacitor DC-DC converter with resolution of $v_{in}/2^N$ for a wide range of input and output voltages. *IEEE Journal of Solid-State Circuits*, 51 (2), 543–556, doi: 10.1109/JSSC.2015.2501985.

24 Salem, L.G. and Mercier, P.P. (2014) 4.6 An 85 integrated 15-ratio recursive switched-capacitor DC-DC converter with 0.1-to-2.2 V output voltage range, in *2014 IEEE International Solid-State Circuits Conference Digest of Technical Papers (ISSCC)*, pp. 88–89, doi: 10.1109/ISSCC.2014.6757350.

25 Salem, L.G. and Mercier, P.P. (2014) A recursive switched-capacitor DC-DC converter achieving $2^N - 1$ ratios with high efficiency over a wide output voltage range. *IEEE Journal of Solid-State Circuits*, 49 (12), 2773–2787, doi: 10.1109/JSSC.2014.2353791.

26 Lutz, D., Renz, P., and Wicht, B. (2016) 12.4 A 10mW fully integrated 2-to-13V-input buck-boost SC converter with 81.5% peak efficiency, in *2016 IEEE International Solid-State Circuits Conference (ISSCC)*, pp. 224–225, doi: 10.1109/ISSCC.2016.7417988.

27 Ma, D. and Luo, F. (2008) Robust multiple-phase switched-capacitor DC-DC power converter with digital interleaving regulation scheme. *IEEE Transactions on Very Large Scale Integration (VLSI) Systems*, 16 (6), 611–619, doi: 10.1109/TVLSI.2008.2000245.

28 Van Breussegem, T. and Steyaert, M. (2009) A 82% efficiency 0.5% ripple 16-phase fully integrated capacitive voltage doubler, in *2009 Symposium on VLSI Circuits*, pp. 198–199.

29 Villar-Piqué, G., Bergveld, H.J., and Alarcón, E. (2013) Survey and benchmark of fully integrated switching power converters: switched-capacitor versus inductive approach. *IEEE Transactions on Power Electronics*, 28 (9), 4156–4167, doi: 10.1109/TPEL.2013.2242094.

30 Le, H.P., Crossley, J., Sanders, S.R., and Alon, E. (2013) A sub-ns response fully integrated battery-connected switched-capacitor voltage regulator delivering 0.19W/mm² at 73% efficiency, in *2013 IEEE International Solid-State Circuits Conference Digest of Technical Papers*, pp. 372–373, doi: 10.1109/ISSCC.2013.6487775.

31 Souvignet, T., Allard, B., and Trochut, S. (2016) A fully integrated switched-capacitor regulator with frequency modulation control in 28-nm FDSOI. *IEEE Transactions on Power Electronics*, 31 (7), 4984–4994, doi: 10.1109/TPEL.2015.2478850.

32 Ramadass, Y.K., Fayed, A.A., and Chandrakasan, A.P. (2010) A fully-integrated switched-capacitor step-down DC-DC converter with digital capacitance modulation in 45 nm CMOS. *IEEE Journal of Solid-State Circuits*, 45 (12), 2557–2565, doi: 10.1109/JSSC.2010.2076550.

10

Inductive DC–DC Converters

Inductive DC–DC converters utilize an inductor for power conversion. As there are switches involved, they are also called switched-mode DC–DC converters or, more generally, switched-mode power supplies (SMPS). Even though it would also apply to the switched-capacitor DC–DC converters in Chapter 9, the term SMPS usually refers to inductor-based DC–DC converters, which is probably due to their longer history. It is also common to say DC–DC converters.

Inductive DC–DC converters have received significant attention because of their high efficiency compared to linear voltage regulators (LDO, Chapter 7). Due to their switching nature, the output voltage is not as "clean" as the voltage provided by an LDO. Unlike switched-capacitor DC–DC converters, which are only efficient close to their ideal conversion ratio, inductor-based converters can ideally resolve any conversion ratio. The key control concept is pulse-width modulation (PWM), but various other methods are also applied. For instance, pulse-frequency modulation (PFM) boosts efficiency at light load.

Thanks to the inductor, these converters can not only provide down-conversion (buck converter), but they can also step up the input voltage to higher values than V_{in} (boost converter). This function is not possible with a linear regulator (LDO). This chapter will explore the fundamental concepts and characteristics, starting with the buck converter. In the second step, we consider the boost and other converter types, including transformer-isolated designs. We will cover a broad range of power management IC design-related topics while, for some fundamentals, it will be helpful to refer to dedicated textbooks on power electronics like [1–3].

10.1 The Fundamental Buck Converter

Figure 10.1a) shows the fundamental concept of a switched-mode step-down DC–DC converter, also called buck converter. We will use this setup to derive the operating principle and the design requirements of DC–DC converters. The switches S1 and S2 implement PWM as the key concept of switched-mode conversion. As a general convention, for $PWM = 1$, the high-side switch S1 is closed. Consequently, we bring energy into the system (from V_{in}). S2 is controlled complementary to S1, which results in a pulsing switching node voltage V_{sw}, as indicated in Fig. 10.1b). There must be a dead time between the activation of S1 and S2 to avoid cross-conduction; see Section 2.8.

The passive components L and C fulfill two tasks: (1) They form a low-pass filter, which generates the DC–DC converter's output voltage as the average of the switching node voltage V_{sw}, and (2) they buffer the energy to supply the load while S1 is off. Also, the input voltage needs to be buffered by a capacitor (not shown in Fig. 10.1). In the ideal case (zero on-resistance), the switches' static on- and off-states are not associated with any losses. In an actual design, there will be finite

Design of Power Management Integrated Circuits, First Edition. Bernhard Wicht.
© 2024 John Wiley & Sons Ltd. Published 2024 by John Wiley & Sons Ltd.

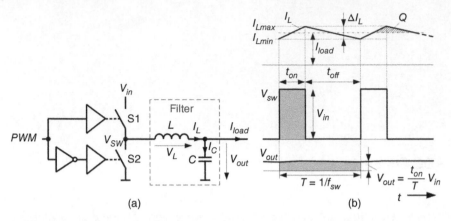

Figure 10.1 a) Basic switched-mode DC–DC converter with b) waveforms of the inductor current I_L and typical node voltages V_{sw}, V_{out}.

static losses, various switching losses, and losses due to the passive components. However, as a significant difference to the linear regulator (Chapter 7), the power path has no major resistive element. Therefore, inductive DC–DC converters can still reach power efficiencies much higher than 90%.

10.1.1 Inductor Current

The inductor voltage is defined by

$$v_L(t) = L\frac{di_L}{dt} = L\frac{\Delta i_L}{\Delta t}. \tag{10.1}$$

As the inductor voltage will change in polarity over one switching cycle, it will cause the inductor current to ramp up and down, as illustrated in Fig. 10.1b). Accordingly,

$$PWM = 1\,(t_{on}): \frac{\Delta i_L}{\Delta t} = \frac{V_{in} - V_{out}}{L}$$

$$PWM = 0\,(t_{off}): \frac{\Delta i_L}{\Delta t} = \frac{-V_{out}}{L}. \tag{10.2}$$

The average inductor current equals the load current I_{load}. For $I_L(t) > I_{load}$, a total charge Q will be delivered to recharge the output bypass or buffer capacitor C. We will use this observation for sizing the capacitor C in Sections 10.1.7 and 10.1.8.

10.1.2 On-/Off-Times

Replacing Δt by t_{on} and t_{off}, respectively, we can derive the on-/ off-times directly from Eqn. (10.2):

$$t_{on} = \Delta I_L \frac{L}{V_{in} - V_{out}} \tag{10.3}$$

$$t_{off} = \Delta I_L \frac{L}{V_{out}} \tag{10.4}$$

Note that, for t_{off}, we use the absolute value of the current ripple ΔI_L. Hence, the minus sign in front of V_{out} in Eqn. (10.2) disappears.

10.1.3 Volt-Second Balance

In steady-state, the net change in inductor voltage needs to be zero,

$$0 = \int_0^T v_L(t)\mathrm{d}t. \tag{10.5}$$

Assuming that S1 conducts during t_{on} and S2 during $t_{off} = T - t_{on}$:

$$0 = \int_0^{t_{on}} \left(V_{in} - V_{out} \right) \mathrm{d}t + \int_{t_{on}}^T \left(-V_{out} \right) \mathrm{d}t \tag{10.6}$$

If V_{in} and V_{out} do not change within one period T, we can express Eqn. (10.6) in a simple way:

$$\left(V_{in} - V_{out} \right) t_{on} = V_{out} t_{off} \tag{10.7}$$

Equation (10.7) defines the volt-seconds balance of the fundamental buck converter. Different expressions can be obtained for other types of DC–DC converters. However, the condition of Eqn. (10.5) must always be fulfilled in any inductive switched-mode DC–DC converter for stable steady-state operation.

10.1.4 Voltage Conversion Ratio

From Eqn. (10.7), we can derive the converter's voltage conversion ratio,

$$\frac{V_{out}}{V_{in}} = \frac{t_{on}}{T} = D. \tag{10.8}$$

The conversion ratio equals the duty cycle D, the ratio between the on-time t_{on} of switch S1 and the total period time T. Equation (10.8) applies to the basic buck converter according to Fig. 10.1 as long as the inductor current remains non-zero. This operation type is called continuous conduction mode (CCM) in difference to discontinuous conduction mode (DCM), which will be investigated further below.

Equation (10.8) indicates that the output voltage can be controlled by varying the pulse width at the switching node, as illustrated in Fig. 10.1b). To keep the volt-second balance (i.e., the energy balance), the area under each curve, V_{sw} and V_{out} over one period T has to be identical. Reducing the pulse width at V_{sw} will lower V_{out} and vice versa. It points to PWM as a fundamental control concept in DC–DC converters.

10.1.5 Current Ripple

The inductor current ramps up and down and leads to a well-defined inductor current ripple $\Delta I_L = I_{Lmax} - I_{Lmin}$. In order to determine the current ripple we calculate $T = t_{on} + t_{off}$ with t_{on} and t_{off} from Eqns. (10.3) and (10.4), respectively. Solving for ΔI_L yields

$$\Delta I_L = \frac{V_{out}}{L f_{sw}} \left(1 - \frac{V_{out}}{V_{in}} \right). \tag{10.9}$$

We immediately notice that the current ripple can be reduced by enlarging the inductor size and faster switching (larger f_{sw}). Note that the current ripple does not depend on the actual load current I_{load}. Nevertheless, the current ripple can be expressed relative to the (average) load current I_{load}:

$$\Delta I_{L,rel} = \frac{\Delta I_L}{I_{load}} \cdot 100\% \tag{10.10}$$

For typical designs, the rule-of-thumb is 40% current ripple (i.e., $\pm 20\%$ with respect to I_{load}).

10.1.6 Inductor Sizing

For a given current ripple target, Eqn. (10.9) provides the sizing equation of the inductor L:

$$L = \frac{V_{out}}{\Delta I_L f_{sw}} \left(1 - \frac{V_{out}}{V_{in}}\right) \qquad (10.11)$$

The switching frequency is the only way to reduce the inductor size for a given current ripple and fixed voltages. Higher f_{sw} (shorter storing times T) results in lower inductor values.

Example 10.1 *Calculate the inductance L, the on-time t_{on} and the off-time t_{off} of a DC–DC converter that operates at $f_{sw} = 1\,\text{MHz}$ for $V_{in} = 3.3\,\text{V}$, $V_{out} = 1.2\,\text{V}$, $I_{load} = 1\,\text{A}$ and $\Delta I_L = 0.4 \cdot I_{load} = 0.4\,\text{A}$.*
We insert all parameters into Eqn. (10.11),

$$L = \frac{V_{out}}{\Delta I_L f_{sw}} \left(1 - \frac{V_{out}}{V_{in}}\right) = \frac{1.2\,\text{V}}{0.4\,\text{A} \cdot 1\,\text{MHz}} \left(1 - \frac{1.2\,\text{V}}{3.3\,\text{V}}\right) = 1.909\,\mu\text{H} \sim 2\,\mu\text{H}. \qquad (10.12)$$

We calculate t_{on} and t_{off} according to Eqns. (10.3) and (10.4),

$$t_{on} = \Delta I_L \frac{L}{V_{in} - V_{out}} = 0.4\,\text{A} \frac{1.9\,\mu\text{H}}{3.3\,\text{V} - 1.2\,\text{V}} = 364\,\text{ns}, \qquad (10.13)$$

$$t_{off} = \Delta I_L \frac{L}{V_{out}} = 0.4\,\text{A} \frac{1.9\,\mu\text{H}}{1.2\,\text{V}} = 636\,\text{ns}. \qquad (10.14)$$

10.1.7 Output Voltage Ripple

The basic converter in Fig. 10.1 shows that the capacitor current is $I_C = I_L - I_{load}$. Over one period, we can assume the load current to be constant. Hence, I_C has a triangular shape with a positive sign as long as $I_L(t) > I_{load}$. Because of symmetry, this condition holds exactly for half the switching period time $T/2$ between t_1 and t_2, as shown in Fig. 10.2. During this time, capacitor C gets recharged by a total charge Q, also indicated in Fig. 10.1b). As Q corresponds to the area enclosed by the triangular current, we can calculate the size of the triangle,

$$Q = \int_0^{T/2} I_C(t)\mathrm{d}t = \frac{1}{2}\frac{\Delta I}{2}\frac{T}{2} = \frac{\Delta I\, T}{8}. \qquad (10.15)$$

Since the voltage across the capacitor is proportional to the integral of the capacitor current, the voltage ripple will presume a square function. The rising transition of I_C during the on-time t_{on} of the power stage corresponds to a square function open upwards with a minimum at t_1 at the

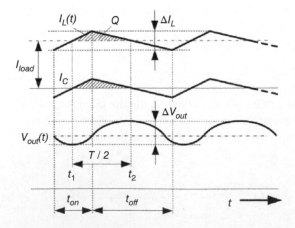

Figure 10.2 Waveforms of the inductor and capacitor currents and the output voltage V_{out} including current and voltage ripple.

zero-crossing of I_C. During the off-time t_{off}, the output voltage $V_{out}(t)$ follows a downward open quadratic function. Its maximum occurs at t_2 when I_C changes to negative values. Please note that the ripple is not symmetrical on the horizontal axis. The exact calculation shows that the minimum deviation is proportional to t_{on} while the peak value scales with t_{off}. We are usually interested in the total peak-to-peak voltage ripple ΔV_{out} in practical designs. We can derive it from Eqn. (10.15),

$$\Delta V_{out} = \frac{Q}{C} = \frac{\Delta I\, T}{8\,C}, \tag{10.16}$$

with ΔI according to Eqn. (10.9). This expression defines the output voltage ripple if the DC–DC converter operates in continuous conduction mode (CCM) where the inductor current never becomes zero. The voltage ripple is indirectly proportional to C. Lower current ripple ΔI leads to smaller voltage ripple. Therefore, utilizing Eqn. (10.9), ΔV_{out} also benefits from larger L. We can conclude that low ripple specifications require a trade-off regarding the size of passive components C and L.

The output ripple may be further affected if the capacitor has significant equivalent series resistance (ESR) and inductance components (ESL). The voltages across each parasitic component need to be added. The ESR voltage will be a triangular function proportional to I_C. The ESL voltage represents the change of the capacitor current. It is, therefore, a square function with its positive and negative values defined by the slope of the rising and falling transition of I_C during t_{on} and t_{off}, respectively.

10.1.8 Capacitor Sizing

For a given ripple specification of ΔV_{out} and ΔI, Eqn. (10.16) allows to calculate the required minimum value of the output capacitor:

$$C = \frac{\Delta I}{8 f_{sw}\, \Delta V_{out}} \tag{10.17}$$

Example 10.2 *What is the required output bypass capacitance C of the DC–DC converter of Example 10.1 if the ripple is specified to be* $\Delta V_{out} = 10\,\text{mV}$?
The current ripple in Example 10.1 is $\Delta I = 0.4\,\text{A}$ *and* $f_{sw} = 1\,\text{MHz}$. *Inserting into Eqn. (10.17):*

$$C = \frac{\Delta I}{8 f_{sw}\, \Delta V_{out}} = \frac{0.4\,\text{A}}{8 \cdot 1\,\text{MHz} \cdot 10\,\text{mV}} = 5.0\,\mu\text{F}. \tag{10.18}$$

A suitable discrete output capacitor would be $C = 6.8\,\mu\text{F}$. *If we can tolerate to slightly increase the ripple, we can select* $C = 4.7\,\mu\text{F}$.

10.1.9 Switches

Both switches are implemented by transistors, as shown in Fig. 10.3a). The entire switching scheme with the waveforms shown in Fig. 10.1b) are similar to half-bridge switching, as discussed in the power stage chapter in Section 2.5.5 (Fig. 2.12). Parasitics will always cause some ringing at the switching node, which requires the switches to be rated at a higher level than V_{in}. For integrated power stages, a typical design margin is 20% of V_{in} while discrete power stages may be selected for V_{DSmax} of up to $3\,V_{in}$.

For integrated power stages, the sizing and the available drive voltage V_{drv} determine the losses. Choosing the proper W/L-ratio allows for loss optimization as a trade-off between static and dynamic losses, as outlined in Section 2.7. The half-bridge configuration also requires dead-time control, as described in Section 2.8. The converter in Fig. 10.3a) comprises a low-side

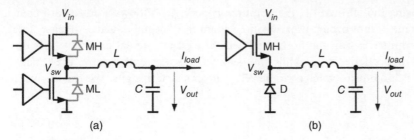

Figure 10.3 a) DC–DC converter with synchronous rectification: ML is controlled complementary to MH with a dead-time in between; b) as synchronous rectification: The low-side switch ML is replaced by a diode D. The inductor current commutates into diode D if the high-side switch MH turns off.

transistor and control circuits that must turn the switch ML on and off synchronized to the overall timing scheme.

We also notice that the full load current commutates into the body diode of the low-side switch during the dead time and pulls V_{sw} one forward voltage below ground. Once the low-side switch turns on, V_{sw} will still be at negative values but only by a few hundred millivolts because the low-resistive switch will short the body diode. As the body diode is associated with parasitic bipolar junction transistor structures (see Section 3.3.1), some DC–DC converter designs utilize a discrete Schottky diode in parallel to the low-side switch. Due to its low forward voltage of ~ 0.3 V, it prevents the body diode from conducting and eliminates any parasitic bipolar effects.

10.1.10 Asynchronous Rectification

Many DC–DC converters use a diode instead of the low-side switch, as shown in Fig. 10.3b). The diode is often placed externally and can even be a Schottky device, as described in the section on switches above. Besides saving the entire low-side switch with driver and gate control, it also eliminates the need for dead-time control (Section 2.8). After the high-side switch MH turns off, the load current will immediately commutate into the (low-side) diode. The diode will start to conduct the inductor current asynchronously. Therefore, this is called a DC–DC converter with asynchronous rectification or asynchronous DC–DC converter. In contrast, the configuration in Fig. 10.3a) is referred to as a converter with synchronous rectification or synchronous converter. Asynchronous DC–DC converters are easier to implement due to lower design effort. However, there are higher power losses because the full load current flows through diode D during the entire time t_{off} when the high-side switch is off. The losses are equal to

$$P_{dio} = \frac{t_{off}}{T} I_{load} V_F = (1 - D) I_{load} V_F. \tag{10.19}$$

These losses increase for low output voltages because the duty cycle D gets low. If losses are not critical, the inductive DC–DC converter with asynchronous rectification is usually preferred over a synchronous implementation.

10.2 Losses and Power Conversion Efficiency

Power conversion efficiency η is usually the key figure of merit of a DC–DC converter,

$$\eta = \frac{P_{out}}{P_{loss} + P_{out}}, \tag{10.20}$$

with the power losses described for the power stage in Section 2.7. Even though there are plenty of loss components, the general shape of the efficiency as a function of the load current becomes evident if we consider the conduction loss P_{cond}, the switching loss P_{sw}, and the gate charge loss P_{gate} (given in Eqns. (2.49), (2.51), and (2.53)), repeated here for convenience,

$$P_{cond} = I_{load}^2 R_{DSon}$$
$$P_{sw} = 2V_{in}I_{load}t_{tr}f_{sw}$$
$$P_{gate} = 2Q_{gate}V_{drv}f_{sw}. \qquad (10.21)$$

Note that these losses apply to the entire half-bridge. We further assume that the on-resistance of both switches is identical (R_{DSon}).

Based on Eqns. (10.20) and (10.21), Fig. 10.4 shows the losses and the efficiency as a function of the load current I_{load}. We use a log scale, which better shows the dependencies at light load. Figure 10.4a) shows that the gate charge loss remains constant and dominates at the lower end of the load current range. The conduction and switching losses take off at medium load and eventually exceed the gate charge loss. Note that the switching loss shows a linear increase while the conduction loss rises to the power of two. The latter is why the conduction loss dominates at high load and causes the efficiency to drop. The efficiency curve in Fig. 10.4b) has a typical shape with maximum efficiency at medium load. In this example, the efficiency stays above 90% over a range of about 50–500 mA. The peak efficiency is 92.6%.

The efficiency can be optimized by the switch sizing as investigated in Section 2.7 (see Fig. 2.18). This way, a trade-off between R_{DSon} and Q_{gate} can be found. The larger the gate current, the faster the voltage and current transition during the switching event (see Eqns. (2.21) and (2.22) in Section 2.5). This way, the switching node transition rise-fall time t_{rf} gets reduced, resulting in lower switching loss. Finally, we need to define an appropriate switching frequency. The lower f_{sw}

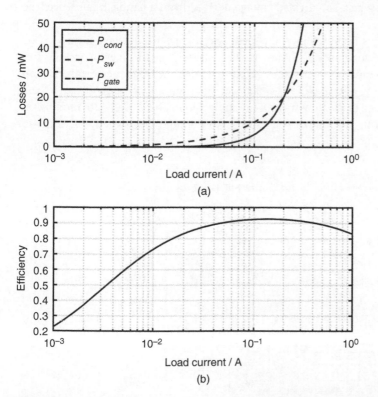

Figure 10.4 a) Losses in a DC–DC converter and b) power conversion efficiency over load current for these parameters: $f_{sw} = 10$ MHz, $V_{in} = 5$ V, $V_{out} = 3$ V, $R_{DSon} = 500$ mΩ, $Q_{gate} = 100$ pC, $V_{drv} = 5$ V, $t_{rf} = 1$ ns.

is, the higher the efficiency. However, the size of L and C and the ripple (current and voltage) will increase. With the trend to higher system voltages outlined in Section 1.5, V_{in} will cause higher baseline switching losses. Soft-switching techniques like resonant conversion and dead-time control, described in Section 2.9, help reduce switching losses while maintaining a sufficiently high switching frequency. A detailed analysis of losses and loss reduction techniques for High-Vin DC–DC converters can be found in [4].

10.3 Closing the Loop

To achieve a precise output voltage under all conditions, we need to add a control loop to the basic power stage of the DC–DC converter, shown in Fig. 10.5 for the buck converter of Fig. 10.1. The converter's output voltage V_{out} is divided down by a resistive divider. The controller aims to get V_{fb} equal to the set value V_{ref}. The resistors R_{fb1} and R_{fb2} are often placed externally such that V_{out} can be set by the user according to

$$V_{fb} = \frac{R_{fb2}}{R_{fb1} + R_{fb2}} V_{out} = \alpha V_{out}. \tag{10.22}$$

For the target $V_{fb} = V_{ref}$, the output voltage is defined as a function of the reference voltage:

$$V_{out} = \frac{V_{ref}}{\alpha} \tag{10.23}$$

As the minimum achievable value of V_{out} is equal to V_{ref}, this reference voltage can be set to values below 1V (e.g., 0.8V). It may be a sub-divided voltage derived from a bandgap reference (see Section 6.4).

The error amplifier determines by how much the feedback voltage V_{fb} deviates from V_{ref}. Its output voltage V_c is a measure for that *error*. The error amplifier needs to have a specific frequency behavior to ensure the control loop is stable. Because it achieves frequency compensation (by a passive R-C network Z_o), the overall block is called a compensator. Its output voltage V_c, the compensated error voltage, is used in the final stage of the control loop, the pulse-width modulator. That block converts the error voltage V_c into an on-time t_{on} (of the high-side switch) necessary to

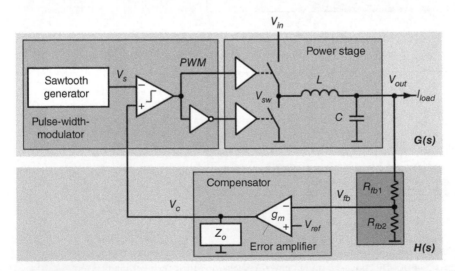

Figure 10.5 The fundamental step-down DC–DC converter with closed-loop control.

achieve the target $V_{fb} = \alpha V_{out} = V_{ref}$. The output quantity of the pulse-width modulator is the duty cycle $D = t_{on}/T$. Figure 10.5 shows a sawtooth generator as one of the widely used circuits for PWM generation. We will cover the details in Section 10.5.

There are various control methods. Some are constant-frequency, of which voltage-mode control (VMC) and CMC are widely used representations. On the other hand, hysteretic control and constant on-time or off-time control are popular variable-frequency control methods of DC–DC converters. Those, as well as VMC and CMC, will be investigated in the sections below.

10.4 Hysteretic Control

Hysteretic control is the most straightforward way of implementing a control loop as the controller is only a comparator, as shown in Fig. 10.6. Assuming a hysteresis V_{hyst} the upper trip point of the comparator is defined as $V_{tpu} = V_{ref} + V_{hyst}/2$ and, vice versa, the lower trip point will be $V_{tpl} = V_{ref} - V_{hyst}/2$. Assume V_{fb} to be greater than the upper trip point V_{tpu} such that the high-side switch turns off. The power inductor L delivers the load current, and V_{out} is buffered by C. If the load current continues to flow, it will gradually discharge C and cause V_{out} to drop. Consequently, V_{fb} drops, which is connected to the inverting input of the comparator. As soon as V_{fb} falls below V_{tpl}, the high-side switch will turn on again to bring another amount of energy into the system. The inductor current builds up again and, in turn, recharges C. It continues until V_{out} has reached its peak value V_{tpu}, which causes the comparator to toggle and turn off the high-side switch again. The entire sequence repeats continuously. In conclusion, this control method implements a ripple-controlled operation referring to the ripple of V_{out}. The hysteresis determines the output voltage ripple. The hysteresis should be small to achieve a low ripple. However, this makes the converter prone to false switching. Typical hysteresis values are in the order of 20–100 mV. A comparator example is given in the section on VMC below (Section 10.5, see Fig. 10.13).

V_{out} ramps up by energy transfer from V_{in} and ramps down by the load current I_{load}. The loop will always be stable and does not require any frequency compensation. Also, the comparator will react instantaneously, resulting in a fast load and line transient response. With only the comparator in the control part, hysteretic control achieves a low quiescent current. However, there are significant drawbacks to hysteretic control. There is no fixed frequency (the converter does not contain an oscillator). f_{sw} depends on many parameters, above all, on the hysteresis, the load current, and the values of the passives L, C. Consequently, the switching frequency is unpredictable, which many applications prefer to avoid. Reasons include EMI, usually suppressed by dedicated EMI filters optimized for a well-known frequency range (see EMI in Section 12.3). In particular, safety-critical applications like automotive do not accept unpredictable switching operations.

Figure 10.6 The DC–DC converter with hysteretic control.

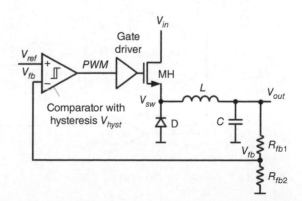

Moreover, hysteretic converters appear to be noise-sensitive, for instance, in the case of capacitive substrate coupling, leading to false switching and jitter, especially for small hysteresis values. For these reasons, other control methods, particularly VMC and CMC, are preferred. We will investigate them in great detail in Sections 10.5 and 10.6.

10.5 Voltage-Mode Control (VMC)

The DC–DC converter in Fig. 10.7 comprises a linear control loop consisting of an error amplifier and a pulse-width modulator. The controller senses the converter's output voltage through the feedback divider R_{fb1} and R_{fb2}. That is why this control method is referred to as VMC in comparison to CMC, which senses the inductor current (in addition to V_{out}) as covered further below in Section 10.6. VMC is a widely used and easily implemented control scheme.

10.5.1 Direct Duty Control

The pulse-width modulator consists of a comparator and a sawtooth generator that starts to ramp up from its zero level (its initial voltage) at the beginning of every switching cycle, illustrated in Fig. 10.8a). The maximum sawtooth amplitude is $V_{s,max}$. The output voltage V_c of the error amplifier (the error voltage) can be considered constant during one switching cycle. Its time constant is high because the frequency compensation limits its bandwidth in addition to the delay introduced by L and C in the power path. Every new switching period starts with a rising edge of *PWM* (the high-side switch turns on) because the comparator's input signals fulfill the condition $V_s < V_c$. When the ramp V_s reaches V_c, the PWM signal goes to "0" (the high-side switch turns off). We can apply the theorem on intersecting lines to express the duty cycle as given in Eqn. (10.8) by V_c and $V_{s,max}$:

$$D = \frac{t_{on}}{T} = \frac{V_c}{V_{s,max}} \tag{10.24}$$

From this equation, we can derive the transfer function of the pulse-width modulator. Its output quantity is the PWM signal, which, in fact, is the duty cycle D. The input signal is the error voltage V_c. Hence,

$$\frac{D}{V_c} = \frac{1}{V_{s,max}}. \tag{10.25}$$

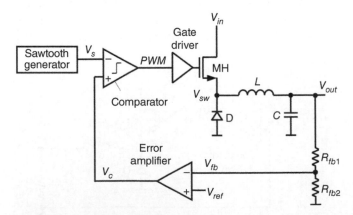

Figure 10.7 Voltage-mode control.

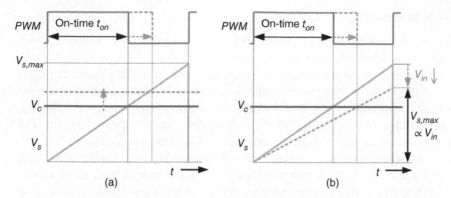

Figure 10.8   Pulse-width modulation in voltage-mode control: a) direct duty control with the dashed line indicating the response to a load current increase; b) voltage feedforward with the sawtooth magnitude being a function of V_{in}. The dashed line indicates the response to a decreasing input voltage V_{in}.

In conclusion, the transfer function of the pulse-width modulator is determined only by the magnitude of the sawtooth voltage $V_{s,max}$.

Figure 10.8a) illustrates the control behavior of the DC–DC converter in VMC. Assume a load step, i.e., an instantaneous increase of the load current I_{load}. Due to the finite output impedance, V_{out} will drop slightly. Because of the feedback divider, V_{fb} will drop, which is connected to the inverting input of the error amplifier. Consequently, a falling output voltage V_{out} will cause V_c to go up, as indicated by the dashed line in Fig. 10.8a). Now, the crossover point between the sawtooth voltage V_s and V_c will happen at a later point in time. The on-time t_{on} increases, and the duty cycle expands, keeping the high-side switch turned on for longer. Intuitively, more energy will be brought into the system to deliver the increased load current. As shown in Fig. 10.1b), more inductor current will build up to counteract the initial load step, confirming the proper operation of a negative feedback loop.

10.5.2   Voltage Feedforward

The major drawback of *direct duty control* in Section 10.5.1 is its poor line regulation. Any change of V_{in} varies the inductor current (see Eqn. (10.2)), but the duty cycle will be adjusted after the delay introduced by the L–C filter. A faster line response can be achieved by making the sawtooth magnitude in Fig. 10.7 dependent on the input voltage V_{in},

$$V_{s,max} = \frac{V_{in}}{K},\tag{10.26}$$

with a constant scaling factor K. It introduces *voltage feedforward*.

The transient curves in Figure 10.8b) explain the behavior for a drop of V_{in}. As the sawtooth magnitude decreases proportionally to V_{in}, the crossover point between V_s and V_c is reached later, which makes sense because lower V_{in} reduces the slope of the inductor current. Hence, the high-side switch needs to turn on for a longer time to deliver the same amount of output current.

In conclusion, VMC with voltage feedforward provides excellent line regulation because the duty cycle D changes instantaneously and precisely. However, the plant still has an $L - C$ time constant corresponding to a double-pole in the transfer function, making frequency compensation more challenging. It requires to understand the dynamic stability of the VMC DC–DC converter. For this reason, we need to look at the typical building blocks of the converter, and we will derive the transfer functions.

10.5.3 The Sawtooth Generator

The straightforward approach for generating a sawtooth ramp would be to charge a capacitor by a constant current, Fig. 10.9a). However, since the ramp will start from ground level, it may not fit into the common-mode range of V_c at the output of the error amplifier. It may not fit either to the input common-mode range of the comparator (even though the PMOS differential pair would support that, see the subsection below and Fig. 10.13). Shifting up the starting level by connecting the bottom plate of the capacitor to a reference voltage will be limited due to the finite source impedance of that voltage source. Therefore, an active integrator with the capacitor placed in feedback configuration is often a better choice, as shown in Fig. 10.9b). The reference voltage V_{ref} defines the starting level of the ramp. As it connects to the non-inverting input V_{ref}, it does not see a load, and its source impedance is not critical. Figure 10.9b) shows an extension by a simple circuit that generates an input-voltage dependent current $I_{in} = (V_{in} - V_{GS})/R_{in} \sim V_{in}/R_{in}$ (assuming $V_{in} \gg V_{GS}$) to achieve voltage feedforward to enhance the line transient response. I_{in} can be superimposed to a DC bias current I_{bias}.

Still, the capacitor cannot instantaneously be discharged at the end of the switching cycle. Figure 10.10a) indicates how the finite settling time leads to an undefined range during a time t_{fall} that limits the minimum duty cycle. According to Eqn. (10.8), lowering the duty cycle reduces the achievable voltage conversion ratio. The settling time t_{fall} is typically in the order of 10 ns. For long period times $T \gg t_{fall}$ this effect will be negligible, but for fast switching DC–DC converters

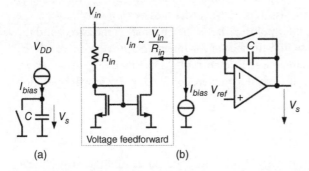

Figure 10.9 Sawtooth generation using a) a capacitor, b) an active integrator.

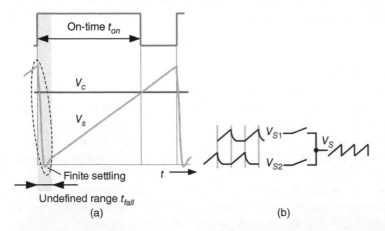

Figure 10.10 Sawtooth generation: a) impact of finite settling; b) time-interleaved operation.

operating at $f_{sw} \gg 1$ MHz this will impact the duty cycle. In that case, two sawtooth generators can be used in a time-interleaved fashion as illustrated in Fig. 10.10b). Their signals can be combined in the analog domain but also in the digital logic by superimposing the PWM signals from both integrator branches [5].

Example 10.3 *What is the maximum switching frequency f_{sw} if the DC–DC converter needs to support duty cycles of at least 10% ($D \geq 0.1$) and the settling time of the ramp generator is given by $t_{fall} = 10$ ns?*
From Eqn. (10.8):

$$D = \frac{t_{on}}{T} = t_{on} f_{sw} \rightarrow f_{sw} = \frac{D}{t_{on}} \tag{10.27}$$

For $t_{on} = t_{fall}$:

$$f_{sw} = \frac{D}{t_{fall}} = \frac{0.1}{10 \text{ ns}} = 10 \text{ MHz} \tag{10.28}$$

If the switching frequency reaches tens of MHz, not only the settling time but also the propagation delays of the comparator and the logic strongly impact the PWM duty cycle control. Figure 10.11 shows a sawtooth generation circuit compensating for comparator and logic delays, presented in [6]. The initial circuit uses a current source that charges a capacitor C as in Fig. 10.9a). Two gm-stages are added to form feedback loops, which exactly regulate the peak and valley trip point of the sawtooth voltage V_s. The target trip points are V_L and V_H, respectively. Consider the lower branch. Due to over-discharging, the valley point V_{valley} is initially lower than the target V_L. During the falling transition of the ramp voltage V_s signal S_L is at 1, which activates S_{ls} to store V_s onto capacitor C_{ls}. S_{ls} turns to 0 at the lower trip point V_{valley}. Subsequently, S_{hold} goes to 1 to sample V_{valley} onto C_h. The connected gm-stage compares $V_{hold} = V_{valley}$ to the set point value V_L.

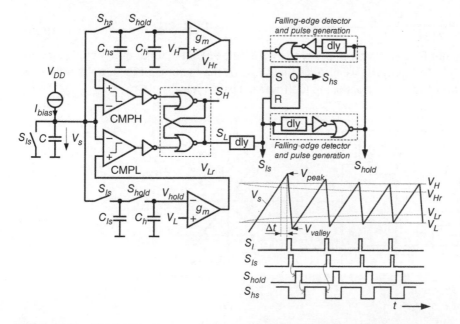

Figure 10.11 Sawtooth generator with comparator and logic delay compensation based on [6].

The output voltage V_{Lr} increases so that the comparator CMPL will be triggered slightly earlier. After multiple cycles, the regulated lower threshold V_{Lr} settles at its final value while the valley voltage of V_s is maintained at the target value V_L. The same mechanism applies to the upper branch that controls the peak threshold V_H. Due to the additional feedback loops, the sawtooth signal is no longer affected by the comparator and logic delay. It also tracks any delay variations over temperature, aging, etc. For the same reason, low-speed comparators can be used, and the overall feedback loop does not need to be fast. In [6] the comparator and the gm-stage consume only 16 µA and 1 µA, respectively.

10.5.4 The Error Amplifier

Symmetrical amplifiers like in Fig. 7.7, repeated in Fig. 10.12 for convenience, and cascode OTAs (operational transconductance amplifiers) are widely used. The same design guidelines as outlined for the LDO in Chapter 7 apply. The higher the gain, the lower the DC error, but the higher the risk for the loop to become unstable. The amplifier can also be an operational amplifier (opamp). If the error amplifier is implemented as a discrete component, opamps are usually preferred since they are more common than OTAs and readily available.

In contrast, the OTA is easier to implement in integrated DC–DC converters because, unlike an opamp, it does not require a power amplifier output stage. Besides the lower design effort, integrated DC–DC converters prefer OTA stages over opamps because frequency compensation can be implemented by placing a passive R-C network directly at its output. However, the resistive feedback divider appears as a damping factor in the controller transfer function $H(s)$. In contrast, in opamp designs, virtual ground at the error amplifier's input eliminates the influence of the feedback divider. For details, refer to Section 10.8.1.

The transfer function of an OTA is the transconductance $G_m = I_{out}/V_{ind}$ with the differential input voltage V_{ind}. For most differential stages like the basic configuration in Fig. 10.12 G_m is identical or proportional to the small-signal transconductance g_m of the transistors in the input differential pair,

$$G_m = g_m \propto \sqrt{\beta I_{bias}}. \tag{10.29}$$

From a design point-of-view, G_m can be set by the W/L-ratio (β) of the input differential pair and the bias current I_{bias}. In Fig. 10.12, we also see that the OTA may saturate if the absolute value of V_{ind} exceeds the overdrive voltage of the input differential pair (typically 200 mV).

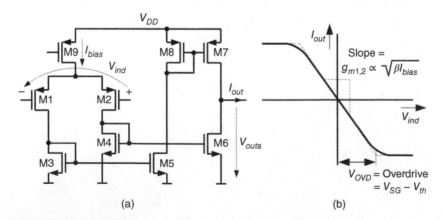

(a)   (b)

Figure 10.12 a) Basic error amplifier, and b) its transfer function.

10.5.5 The Comparator

Figure 10.13 shows a typical comparator circuit. It was used already in one of the earlier integrated DC–DC converters [7]. It utilizes a differential stage with positive feedback to achieve low propagation delay and some hysteresis. M3 and M5, as well as M4 and M6, are current mirrors with a mirror ratio M. Assume that V_{ind} is so large that M2 is fully turned off and M1 is fully turned on, carrying the full bias current I_{bias} provided by M11. M5 wants to sink $I_5 = MI_3$ with $I_3 = I_1$. As V_{ind} decreases, the upper trip point V_{tpu} will occur when $I_2 = I_5 = MI_1$ at

$$V_{tpu} = V_{SG2} - V_{SG1} = \sqrt{\frac{2I_2}{\beta_2}} + V_{th} - \sqrt{\frac{2I_1}{\beta_1}} - V_{th} = \sqrt{\frac{2}{\beta}}\left(\sqrt{I_2} - \sqrt{I_1}\right). \tag{10.30}$$

Due to symmetry, we can assume $\beta_1 = \beta_2 = \beta$. In order to determine V_{tpu} based on Eqn. (10.30) we need to substitute I_1 and I_2 considering

$$I_{bias} = I_1 + I_2 = I_1 + MI_1 \quad \rightarrow I_1 = \frac{I_{bias}}{1+M}, \tag{10.31}$$

and

$$I_2 = I_{bias} - I_1. \tag{10.32}$$

The lower trip point has the same absolute value as V_{tpu} but is inverted because of the symmetrical design of the comparator:

$$V_{tpl} = -V_{tpu} \tag{10.33}$$

Finally, the hysteresis is defined as

$$v_{hyst} = V_{tpu} - V_{tpl} = 2V_{tpu}. \tag{10.34}$$

Example 10.4 *Calculate the trip points and the hysteresis for $M = 3$. The bias current and transconductance are $I_{bias} = 10\,\mu A$ and $\beta_1 = \beta_2 = \beta = 200\,\mu A/V^2$, respectively.*
Equation (10.31) yields

$$I_1 = \frac{I_{bias}}{1+M} = \frac{10\,\mu A}{4} = 2.5\,\mu A \tag{10.35}$$

which can be inserted into Eqns. (10.30) and (10.34) to obtain

$$V_{tpu} = \sqrt{\frac{2}{\beta}}\left(\sqrt{I_{bias} - I_1} - \sqrt{I_1}\right) = 0.116\,V \tag{10.36}$$

and $V_{hyst} = 0.232\,V$.

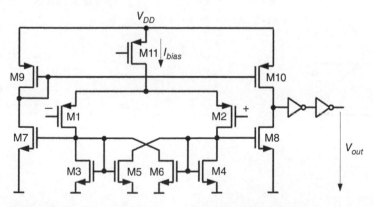

Figure 10.13 Comparator with hysteresis and positive feedback.

10.5.6 Closed-loop Transfer Function

The block diagram of the DC–DC converter in Fig. 10.5 consists of two parts: the control-to-output (plant) transfer function $G(s)$ and the feedback transfer function $H(s)$. Alternatively, $G(s)$ and $H(s)$ are called plant and controller transfer functions, respectively. The loop gain or, more precisely, the open-loop gain is easily defined by multiplying both expressions $G(s)$ and $H(s)$,

$$T = \frac{V_{fb}}{V_{ref}} = G(s)H(s). \tag{10.37}$$

Knowing the loop gain, we can determine the closed-loop transfer function,

$$\frac{V_{out}}{V_{ref}} = \frac{1}{\alpha} \frac{G(s)H(s)}{1 + G(s)H(s)} \approx \frac{1}{\alpha}. \tag{10.38}$$

For $G(s)H(s) \gg 1$, which has to be fulfilled for a properly designed control loop, the closed-loop transfer function is approximately given by the ratio of the feedback divider as provided by Eqns. (10.22) and (10.23).

10.5.7 Control-to-Output Transfer Function

$G(s)$ consists of the pulse-width modulator and the power stage. By multiplying both transfer functions we can derive the overall control-to-output transfer function,

$$G(s) = \frac{V_{out}}{V_c} = \frac{D}{V_c} \cdot \frac{V_{out}}{D} = \frac{1}{V_{s,max}} V_{in} \frac{1}{1 + s\frac{L}{R} + s^2 LC}. \tag{10.39}$$

The pulse-width modulator's transfer function is described by $D/V_c = 1/V_{s,max}$ according to Eqn. (10.25). For the power stage, we refer to Eqn. (10.8) and substitute the expression $\frac{V_{out}}{D}$ simply by the converter's input voltage V_{in}. The L–C low-pass filter results in a double-pole expression. $R = V_{out}/I_{load}$ accounts for the equivalent load resistance. The Bode plot of $G(s)$ in Fig. 10.14 indicates the characteristic behavior. The DC gain is given by $V_{in}/V_{s,max}$ where $V_{s,max}$ is the amplitude of the sawtooth voltage. Peaking occurs at the natural frequency $1/(2\pi\sqrt{LC})$ of the L–C filter's double-pole. Beyond that frequency, the gain decreases by -40dB per decade. The double-pole causes a phase shift of $-180°$. Hence, practically, there is no phase margin, and frequency compensation by the controller is mandatory; see Section 10.8.

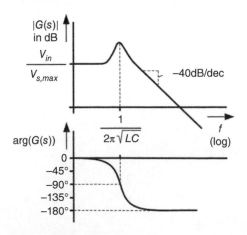

Figure 10.14 Bode plot of the control-to-output transfer function $G(s)$ of the buck converter in voltage-mode control with the DC gain $V_{in}/V_{s,max}$ and the double-pole of the L-C filter in the power path at $f = 1/(2\pi\sqrt{LC})$.

The output buffer capacitor C often has an equivalent series resistance R_{ESR}, which adds a zero in the expression of $G(s)$. This zero may be negligible if it occurs at high frequencies. If we want to take it into account, we can refine Eqn. (10.39):

$$G(s) = \frac{1}{V_{s,max}} V_{in} \frac{1 + sR_{ESR}C}{1 + s\frac{L}{R} + s^2LC} \qquad (10.40)$$

For converters with voltage feedforward, we can further substitute $V_{s,max}$ by Eqn. (10.26) and obtain an input-voltage independent expression of the control-to-output transfer function,

$$G(s) = K \frac{1 + sR_{ESR}C}{1 + s\frac{L}{R} + s^2LC}. \qquad (10.41)$$

10.5.8 Line-to-Output Transfer Function

While the control-to-output transfer function $G(s)$ is of primary importance, we also want to consider the response to line variations (V_{in}). It will tell us how V_{out} varies if V_{in} changes. Hence, the line-to-output transfer function is the ratio between V_{out} and V_{in}. For the fundamental buck converter, this is simply the duty cycle D given by Eqn. (10.8). Like for the control-to-output transfer function $G(s)$, we need to multiply the duty cycle by the double-pole frequency behavior of the L-C filter,

$$\frac{V_{out}}{V_{in}} = \frac{D}{1 + s\frac{L}{R} + s^2LC}. \qquad (10.42)$$

The control-to-output and the line-to-output transfer functions show the same double-pole behavior, leading to a phase shift of $2 \cdot (-90°)$. Now, the challenge is on the controller to achieve a stable closed-loop behavior with sufficient phase margin. A controller with integrating behavior adds another pole at DC. Thus, at least two zeros need to be inserted, i.e., a type III compensator will be required (see the section on compensator design below).

10.6 Current-Mode Control (CMC)

For VMC, Fig. 10.8 shows that PWM is achieved by comparing a ramp signal V_s from a sawtooth generator with the error signal V_c at the output of the compensator. The key idea of CMC is to sense the inductor current, which ramps up as long as the high-side switch is turned on ($PWM = 1$), and to use a proportional sense voltage $V_{sense} \propto I_L$ instead of V_s. Figure 10.15 shows the resulting block diagram of CMC. We replace the ramp generator of VMC with a current sensing stage, which generates a voltage V_{sense}. This voltage is proportional to the inductor current I_L. Hence, its transfer function is described by a resistance $R_{sense} = V_{sense}/I_L$. From Fig. 10.15, we also note that the current sensing path forms a second feedback loop, called the current or inner loop. The original control loop known from VMC remains unchanged, called voltage or outer loop. The block diagram of Fig. 10.15 reflects quite some complexity. However, we have covered all significant blocks before. In particular, this includes the power stage and devices in Chapters 2 and 3, the gate driver in Chapter 4, and the current sensing circuits in Section 6.6. Besides shunt sensing, shown as an example in Fig. 10.15, DCR-sensing and high-side replica sensing are typically implemented. The pulse-width modulator becomes simply a comparator. Figure 10.16 illustrates the most crucial

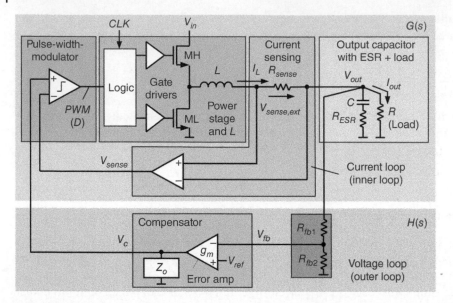

Figure 10.15 Current-mode control consists of two control loops, an inner current loop, and an outer voltage loop.

signal waveforms. Each switching period starts with a clock pulse *CLK* that sets the *PWM* signal to 1. The high-side power switch MH turns on. Consequently, there will be a positive voltage drop across the inductor, causing the inductor current I_L to increase. Once the current sensing output signal V_{sense} exceeds the error signal V_c, the comparator toggles and resets *PWM* to 0. Hence, MH turns off and, after some dead time (see Section 2.8), the low-side switch ML turns on. With a negative voltage across the inductor, its current I_L and V_{sense} will ramp down. An oscillator (not shown), running at f_{sw}, ensures fixed-frequency operation and sets the period time to $T = 1/f_{sw}$. CMC provides excellent line transient response similar to voltage feedforward in VMC because the inner loop directly senses the inductor current, which changes immediately with any variation of V_{in}. As an additional advantage, CMC can inherently limit the load current I_{load} by limiting the error voltage V_c. Note that V_c is inversely proportional to V_{out}. Large load current will cause V_{out} to

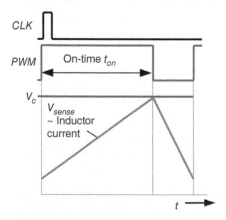

Figure 10.16 Current-mode control: Pulse-width modulation is achieved by comparing the error voltage V_c to a ramp voltage V_{sense}, proportional to the inductor current I_L.

drop. If V_c cannot go higher by clamping it in the error amplifier, the crossover point between V_c and V_{sense} will be reached earlier. This way, the average value of V_{sense} and, consequently, the load current is limited.

10.6.1 Transfer Function of the Current Loop

Despite the complexity, CMC brings several advantages, like improved transient response and options for limiting the load current. However, the major benefit and the main reason to go for CMC instead of VMC is that the entire DC–DC converter becomes a one-pole system with much easier loop stabilization. To identify why CMC is a one-pole system, we need to derive the transfer function of the current loop, the inner feedback loop. For this reason, we can redraw the block diagram of Fig. 10.15 and describe the sub-blocks by their transfer functions, as shown in Fig. 10.17.

The inner loop gets V_c as its input signal while the inductor current I_L is the output signal. We put together two equations to derive the transfer function I_L/V_c of the inner loop. The forward path is described by

$$I_L = (V_c - V_{sense})G_1 G_2. \tag{10.43}$$

We do not need to know the expressions for G_1 and G_2 as these will cancel later. The second equation describes the current sensing output,

$$V_{sense} = R_{sense} I_L. \tag{10.44}$$

We now substitute V_{sense} in Eqn. (10.43) by Eqn. (10.44) and obtain

$$\frac{I_L}{V_c} = \frac{G_1 G_2}{1 + G_1 G_2 R_{sense}} \approx \frac{1}{R_{sense}}. \tag{10.45}$$

This result is remarkable: The transfer function of the inner feedback loop simplifies to the reciprocal of the equivalent sense resistance. Of course, R_{sense} is the actual transfer function of the current sensing circuit (see also Section 6.6). It can be treated as a simple resistor if the current sensing has sufficient bandwidth, which is usually fulfilled because it needs to measure the inductor current

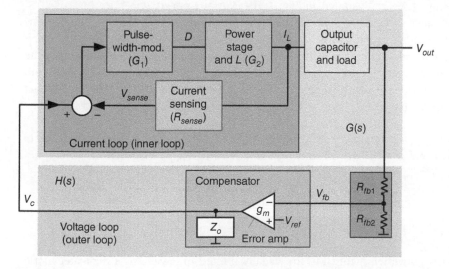

Figure 10.17 Current-mode control: Block diagram.

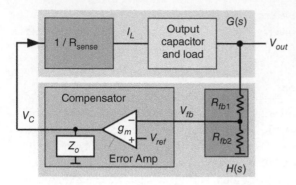

Figure 10.18 Current-mode control: Simplified block diagram with the inner loop represented by the current sensing equivalent resistance R_{sense}.

within the time frame given by the switching period. Likewise, the approximation in Eqn. (10.45) is typically fulfilled in practical designs ($G_1 G_2 R_{sense} \gg 1$). In conclusion, we obtain the block diagram of Fig. 10.18.

10.6.2 Control-to-Output Transfer Function

The forward path in Fig. 10.18 consists only of two blocks. For this reason, the control-to-output transfer function $G(s)$ is determined by R_{sense}, the output buffer capacitor C with its equivalent series resistance R_{ESR} and the load resistance $R = V_{out}/I_{load}$:

$$G(s) = \frac{V_{out}}{V_c} = \frac{R}{R_{sense}} \frac{1 + sR_{ESR}C}{1 + sRC} \tag{10.46}$$

It is a first-order system with only one pole. Hence, stabilization and frequency compensation will be much easier. The compensator design will be less complex (type II compensation, Section 10.8.3) than VMC (needs type III compensation).

10.6.3 The Initial Spike

The current sensing signal V_{sense} tends to be noisy. In particular, there will be a spike right at the beginning of the cycle as illustrated in Fig. 10.19. Various effects generate this so-called leading-edge spike: Steep switching transients, parasitic capacitances, and trace inductances, as well as instantaneous diode blocking, including reverse recovery (see Fig. 2.6). As a common practice, the digital control implements blanking right after the high-side switch turns on at the beginning of each cycle to prevent false triggering by this initial spike. Typical blanking times are $t_{blank} = 100\,\text{ns}$. This blanking limits the duty cycle range of the DC–DC converter. We can repeat Example 10.3 and estimate the maximum switching frequency of CMC compared to VMC.

Example 10.5 *What is the maximum switching frequency f_{sw} if the DC–DC converter needs to support duty cycles of at least 10% ($D \geq 0.1$) and the blanking time is $t_{blank} = 100\,\text{ns}$?*
From Eqn. (10.28) for $t_{on} = t_{blank}$:

$$f_{sw} = \frac{D}{t_{blank}} = \frac{0.1}{100\,\text{ns}} = 1\,\text{MHz} \tag{10.47}$$

Compared to Example 10.3, the switching frequency is 10 times lower.

In general, DC–DC converters with CMC are usually limited to lower single-digit megahertz values (<5 MHz). Nevertheless, these operating frequencies fit well to the limited bandwidth of

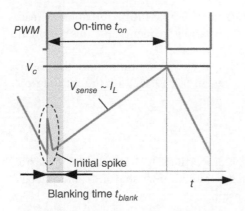

Figure 10.19 Current-mode control: Blanking of the initial spike.

current sensing circuits (see Section 6.6). Depending on the duty-cycle range, changing from peak-current to valley-current mode may allow to go to higher switching frequencies.

10.6.4 Subharmonic Oscillations

Assume a perturbation I_p of the inductor current, for instance, in case of a load current change. It will translate into a variation $V_{sense,p} = V'_{sense} - V_{sense}$ at the output of the current sensing stage. As long as the duty cycle D is below 50%, the perturbation will decay after a few switching cycles, as shown in Fig. 10.20a), i.e., $|I_p(T)| \leq |I_p(0)|$. Note that the crossover point between V_{sense} and V_c happens below 50% of the period, corresponding to $D \leq 50\%$. The switching point is reached earlier by

$$\Delta t = \frac{I_p(0)}{m_1}, \tag{10.48}$$

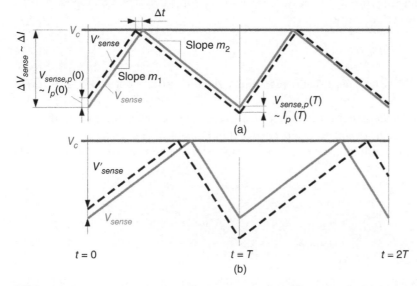

Figure 10.20 Subharmonic oscillations in case of an initial perturbation $I_p(0)$ of the inductor current for a duty cycle below and above 50%: a) $D < 50\%$: The perturbation decays after a few cycles; b) $D \geq 50\%$: Subharmonic oscillations occur. The waveforms are in the voltage domain where the current sensing signal $V_{sense,p}(0)$ represents the perturbation $I_p(0)$.

compared to the ideal behavior. For $D > 50\%$ Fig. 10.20b) indicates that the perturbation increases over time. Oscillations will occur at half of the converter switching frequency. For this reason, the phenomenon is called subharmonic oscillations.

Let us first investigate why this happens above 50% duty cycle. We first introduce the positive and the negative slopes m_1 and m_2, respectively, as indicated in Fig. 10.20. With the current ripple ΔI (see Fig. 10.1) Eqns. (10.2) and (10.7) yield

$$\frac{t_{on}}{t_{off}} = \frac{D}{1 - D} = -\frac{m_2}{m_1}. \tag{10.49}$$

Figure 10.20 shows that the absolute magnitude of V_{sense} is proportional to the total inductor current ripple ΔI as defined in Eqn. (10.9). ΔI can be calculated from both the rising and falling slope of the inductor current. Hence, we can derive this relationship:

$$\Delta I = I_p(0) + m_1(DT - \Delta t) = -m_2((1 - D)T + \Delta t) + I_p(T) \tag{10.50}$$

By substituting $(1 - D)$ by means of Eqn. (10.49) the duty-cycle related terms cancel and Eqn. (10.50) simplifies to

$$I_p(T) = I_p(0) - (m_1 - m_2)\Delta t. \tag{10.51}$$

With the expression for Δt from Eqn. (10.48) and by inserting Eqn. (10.49) we obtain

$$I_p(T) = I_p(0)\frac{m_2}{m_1} = -I_p(0)\frac{t_{on}}{t_{off}}. \tag{10.52}$$

This relationship can be generalized to express the perturbation after n cycles:

$$I_p(nT) = I_p(0)\left(-\frac{t_{on}}{t_{off}}\right)^n \tag{10.53}$$

In order to ensure that the perturbation decays we require $|I_p(T)| \leq |I_p(0)|$, which is only fulfilled if the ratio t_{on}/t_{off} is lower than unity,

$$\frac{t_{on}}{t_{off}} = \frac{D}{1 - D} < 1 \rightarrow D < 0.5. \tag{10.54}$$

Equation (10.54) confirms any perturbations in the inductor current decay for duty cycles below 50%

Are subharmonic oscillations harmful? Indeed, DC–DC converters will still operate properly. However, applications usually want to avoid having this kind of effect. The standard way of suppressing subharmonic oscillations is to add a negative slope m to the error voltage V_c. The

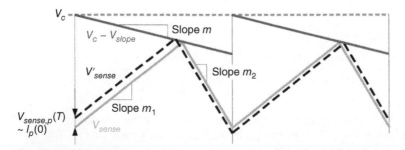

Figure 10.21 Slope compensation.

impact of such a slope compensation is shown in Fig. 10.21. Even for $D > 50\%$, any perturbation will disappear after a few cycles.

In order to find out how much slope m is required we adapt Eqn. (10.48) by taking the slope m into account:

$$\Delta t = \frac{I_p(0)}{m_1 - m} \tag{10.55}$$

Inserting the expression for Δt into Eqn. (10.50), we obtain an expression similar to Eqn. (10.51),

$$I_p(T) = I_p(0) \left(\frac{m_2 - m}{m_1 - m} \right). \tag{10.56}$$

We can bring this equation in a more general form to express the perturbation after n cycles:

$$I_p(nT) = I_p(0) \left(\frac{m_2 - m}{m_1 - m} \right)^n \tag{10.57}$$

From this equation, we can find the condition under which the perturbation decays:

$$\left| \frac{m_2 - m}{m_1 - m} \right| < 1 \rightarrow \left| \frac{1 - \frac{m}{m_2}}{\frac{m_1}{m_2} - \frac{m}{m_2}} \right| < 1 \tag{10.58}$$

To ensure this condition over the full duty cycle range up to 100% (which is the worst case), we set $\frac{m_1}{m_2} = -\frac{t_{off}}{t_{on}}$, as defined by Eqn. (10.49), to zero. Rearranging Eqn. (10.58) yields

$$m < \frac{m_2}{2}. \tag{10.59}$$

Note that, to obtain Eqn. (10.59), we need to multiply with m_2, which is a negative value and, hence, the relation sign changes. If we chose m to be half of m_2 or lower, subharmonic oscillations will be suppressed.

The measurements in Fig. 10.22 [8] show the inductor current after a perturbation is triggered by a step of the error voltage V_c. With slope compensation, Fig. 10.22a), the current ripple does not show any instabilities. Without slope compensation, Fig. 10.22b), subharmonic oscillations can be observed. The period time is twice as long as the switching period. Oscillations happen at $f_{sw}/2$, which is why they are called *subharmonic* oscillations. Figure 10.22b) also shows that these oscillations are disturbing but the converter still operates correctly.

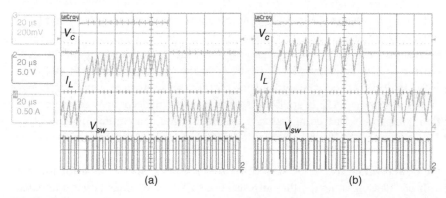

(a)   (b)

Figure 10.22 Measured transient behavior a) with and b) without slope compensation. Source: Adapted from [8].

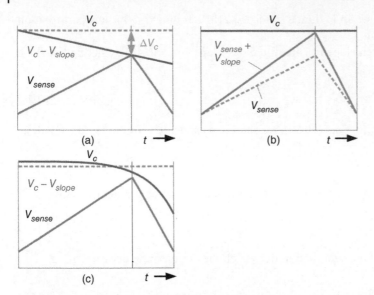

Figure 10.23 Options for slope compensation: a) V_{slope} is subtracted from the error voltage V_c; b) V_{slope} added to the current sensing signal V_{sense}; c) quadratic slope compensation.

To suppress subharmonic oscillations, Fig. 10.21 and Eqn. (10.59) suggest that we need to implement circuits for slope compensation. There are two options, as shown in Fig. 10.23. We can either subtract a slope signal V_{slope} from the error voltage V_c at the compensator output, Fig. 10.23a). Or we add V_{slope} to the ramp signal V_{sense}, Fig. 10.23b). In both cases, we can observe that slope compensation reduces the maximum value of V_{sense} by an amount ΔV_c in comparison to the case without slope compensation ($V_{sense,max} = V_c$). The behavior approaches VMC, which comes along with a weaker transient response. Moreover, since V_{sense} is proportional to the inductor current, the slope compensation reduces the load current capability of the DC–DC converter. For this reason, more sophisticated concepts like quadratic slope compensation have been proposed [8–10]. As illustrated in Fig. 10.23c), quadratic slope compensation will allow for higher peak values of $V_{sense} \propto I_L$ over a wide range of the duty cycle.

Figure 10.24 shows circuits for both options of linear slope compensation. Since subtraction and adding can easily be implemented in the current domain, both circuits use voltage-to-current conversion.

To generate the difference according to Fig. 10.23a), V-to-I converter blocks convert V_c and V_{slope} to corresponding currents I_c and I_{slope}, respectively. A typical V-to-I stage is shown on the right of Fig. 10.24. In that circuit, a feedback loop forces both inputs of the differential amplifier to have equal voltages. Hence, the output current will be $I = V/R$. In the slope compensation circuit in Fig. 10.24a), I_{slope} is fed into a PMOS current mirror such that we can subtract it from I_c as shown. Now, replica sensing (see Section 6.6.2) is applied to generate the signal voltages connecting to the PWM generation comparator. The comparator is identical to the comparator in the overall block diagram of Fig. 10.5, except for the polarity, which reflects that the entire sensing is applied to the high side. The example circuit of Fig. 10.13 could be used as a comparator. Transistor MP is the high-side power switch of the DC–DC converter, while the sense transistor MS has $1/M$-times the width of MP. Consequently, the on-resistance of MS is M-times larger compared to MP. This way, the non-inverting input of the comparator sees a voltage that is proportional to the slope-compensated error voltage $MR_{DSon}(I_c - I_{slope})$. The slope signal V_{slope} can be generated

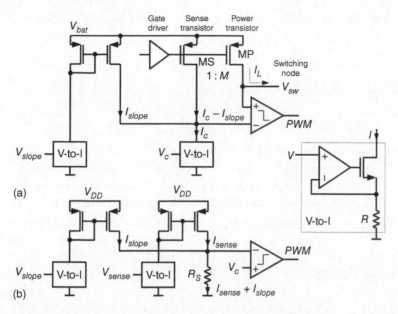

Figure 10.24 Slope compensation circuits: a) Concept according to Fig. 10.23a). The slope is subtracted from the error signal in the current domain and converted back into the voltage domain using a sense (replica) transistor MS. b) The slope and the current sensing signal are converted into the current domain and added. Resistor R_s converts the sum to a voltage fed to the comparator.

simply by connecting a current source to a capacitor. Also, an active integrator may be used similar to the ramp generator in VMC; see Fig. 10.9. The inverting input of the comparator is connected to the current sensing signal, which is $R_{DSon}I_L$ with the inductor current I_L. The high-side slope compensation in Fig. 10.24a) does not show any protection circuits and level shifters, which will be mandatory if V_{bat} gets large. The voltage ratings will be significantly lower if the same concept is implemented in the low-side domain. We only need to protect the comparator input from the switching node voltage, for instance, by the circuit of Fig. 6.2. However, this will only allow for sensing the current in the low-side switch, which is not sufficient in the case of a buck converter but well-suitable for a boost converter (see Section 10.10).

Figure 10.24b) implements slope compensation according to Fig. 10.23b) where the slope signal is added to the inductor current signal. The sum $(I_{sense} + I_{slope})$ is converted into a voltage via R_s and fed to the comparator, which compares to the error voltage V_c [7]. In DC–DC converters where the low-side current can be used for PWM generation, for instance, in boost converters, R_s can be replaced by a low-side switch. To be more precise, DC–DC converters may use an external current sense resistor (shunt) between the low-side transistor's source terminal and ground. In that case, the I_{sense} generation is replaced by the shunt and is not needed anymore. Only the slope signal I_{slope} has to be sourced into the switching node.

Example 10.6 *We want to implement slope compensation according to Fig. 10.24b) for the DC–DC converter of Example 10.1. The inductor current sensing has an equivalent sensing resistance of $R_{sense} = 1\,\Omega$ (i.e., $V_{sense} = I_L \cdot R_{sense}$, see Fig. 10.24b)).*

The negative slope of the inductor current during the off-time is given by

$$-\frac{\Delta I_L}{t_{off}} = -\frac{V_{out}}{L} = -\frac{1.2\,\text{V}}{1.9\,\mu\text{H}} = -631.6\,\text{mA/}\mu\text{s}. \tag{10.60}$$

The slope in Eqn. (10.60) corresponds to m_2. By applying the current sensing resistance, we define m_2 in the voltage domain,

$$m2 = -\frac{\Delta I_L}{t_{off}} \cdot R_{sense} = -631.6 \text{ mA/µs} \cdot 1\,\Omega = -631.6 \text{ mV/µs}. \tag{10.61}$$

Equation (10.59) defines the maximum compensation slope,

$$m < \frac{m_2}{2} = \frac{-631.6 \text{ mV/µs}}{2} = -315.8 \text{ mV/µs}. \tag{10.62}$$

Including some safety margin, we decide for $m = m_2/1.6 = -197.4 \text{ mV/µs}$.

As the next step, we must provide the ramp voltage V_s that connects to the V-I converter, as shown in Fig. 10.24b). We generate the slope m of V_s by a capacitor that gets charged by a current I_c. We decide for a reasonable capacitor value $C = 5$ pF and calculate the required charging current:

$$I_c = C \cdot |m| = 5 \text{ pF} \cdot 197.4 \text{ mV/µs} = 0.987 \text{ µA} \sim 1 \text{ µA}. \tag{10.63}$$

10.7 Constant On-Time Control

This control scheme, called COT control, keeps the advantages of constant-frequency CMC but does not require slope compensation. Even though it can be classified as a variable-frequency control method, it maintains constant frequency over a wide operating range by implementing voltage feed-forward. It provides excellent load transient response because the current loop can respond within one cycle by changing the length of the period. This way, COT delivers the speed advantage of hysteretic control with nearly constant frequency operation like VMC and CMC. It also benefits light-load efficiency because COT control reduces the switching frequency when entering discontinuous mode (DCM). As shown by the waveforms in Fig. 10.25a), the switching cycle starts if the current sensing signal V_{sense} falls below the compensated error voltage V_c. The high-side switch is turned on by a fixed on-time ($PWM = 1$). In conclusion, the comparator sets one trip point of the PWM generation, and a single shot on-time generator generates the other one. Figure 10.25b) shows an example of the PWM generation circuit for constant on-time control. The overall block diagram of the DC–DC converter is similar to constant-frequency CMC in Fig. 10.15. The outer loop is identical, with the same design requirements for the compensator (type II, see Section 10.8.3). The inner current loop consists of some inductor current sensing. However, the comparator senses the valley current, as shown in Fig. 10.25a). Therefore, replica sensing at the low-side switch can be implemented instead of sensing the current at the inductor (e.g., by a shunt like in Fig. 6.17). Hence, a design point of view, COT brings several benefits. See Section 6.6 for details on current sensing.

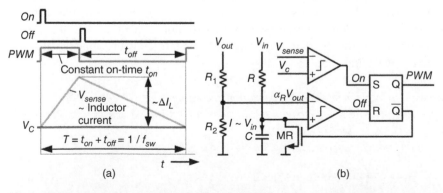

(a) (b)

Figure 10.25 Constant on-time current-mode control: a) circuit implementation; b) typical waveforms.

Recalling Eqn. (10.8), the on-time is defined by

$$t_{on} = \frac{1}{f_{sw}} \frac{V_{out}}{V_{in}}.$$ (10.64)

The on-time will only be constant for a given input and output voltage. However, the on-time generator in Fig. 10.25b) implements a feed-forward technique that keeps the switching frequency nearly independent of V_{in} and V_{out}. Voltage feed-forward is achieved by making the comparator trip point V_{trip} in the on-time generator output-voltage dependent,

$$V_{trip} = \frac{R_2}{R_1 + R_2} V_{out} = \alpha_R V_{out}.$$ (10.65)

For $V_{in} \gg V_{trip}$, we can assume that the capacitor C is charged by a current $I_c = V_{in}/R$. Then, the capacitor voltage becomes

$$V(t) = \frac{I_c}{C} t = \frac{V_{in}}{RC} t.$$ (10.66)

At the trip point, V_{trip} and $V(t_{on})$, given by Eqns. (10.65) and (10.66), are equal. Solving for $t = t_{on}$:

$$t_{on} = \alpha_R RC \frac{V_{out}}{V_{in}}$$ (10.67)

Comparing Eqns. (10.64) and (10.67) indicates that the switching frequency will be constant and defined by

$$f_{sw} = \frac{1}{\alpha_R RC}.$$ (10.68)

We can size R and C and the scaling ratio α_R to meet the target switching frequency. We also realize that the frequency depends on absolute values of R and C. In addition, second-order effects like offset voltage in the trip-point comparator cause some variation in frequency. Therefore, either R or C may need to be trimmed.

Figure 10.26a) shows the compensated error signal V_c (more precisely, a current domain equivalent level $\propto V_c$) and the inductor current for two different input voltages. The graph confirms that the period time T and the switching frequency f_{sw} do not change. This effect is due to voltage

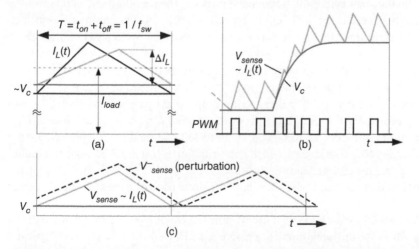

Figure 10.26 Constant on-time control: a) I_L and V_c for two different values of V_{in}; b) transient response for a positive load step with temporarily increased PWM frequency; c) COT does not need slope compensation, perturbations decay within one switching cycle.

feed-forward, as explained above. However, the current ripple changes. To supply the same load current I_{load}, the outer loop will adjust V_c accordingly.

Example 10.7 *We investigate the behavior of Fig. 10.26a) for a COT DC–DC converter at two different input voltages, $V_{in1} = 5.0\,V$ and $V_{in1} = 8.0\,V$. All other parameters remain constant in both cases, $V_{out} = 3.0\,V$, $V_{trip} = 0.75\,V$, $I_{load} = 200\,mA$ and $L = 10\,\mu H$. The target switching frequency is $f_{sw} = 2.5\,MHz$.*

From V_{out} and V_{trip} we can determine $\alpha_R = V_{trip}/V_{out} = 0.25$. Knowing α_R we can determine the R-C time constant from Eqn. (10.68),

$$RC = \frac{1}{\alpha_R f_{sw}} = \frac{1}{0.25 \cdot 2.5\,\text{MHz}} = 1.6\,\mu s. \tag{10.69}$$

We are flexible in choosing R and C and decide for $R = 80\,k\Omega$ and $C = 20\,pF$, which are convenient to be integrated.

For case 1 ($V_{in1} = 5.0\,V$) we calculate the on-time from Eqn. (10.67) by substituting $(\alpha_R RC)$ by $1/f_{sw}$ according to Eqn. (10.68),

$$t_{on1} = \frac{1}{f_{sw}} \frac{V_{out}}{V_{in1}} = \frac{1}{2.5\,\text{MHz}} \frac{3.0\,\text{V}}{5.0\,\text{V}} = 240\,\text{ns}. \tag{10.70}$$

By rearranging Eqn. (10.3), we can determine the current ripple for case 1,

$$\Delta I_{L1} = t_{on1} \frac{V_{in1} - V_{out}}{L} = 240\,\text{ns} \frac{5.0\,\text{V} - 3.0\,\text{V}}{10\,\mu\text{H}} = 48\,\text{mA}. \tag{10.71}$$

Knowing ΔI_{L1}, the off-time is defined by Eqn. (10.4),

$$t_{off1} = \Delta I_{L1} \frac{L}{V_{out}} = 48\,\text{mA} \frac{10\,\mu\text{H}}{3.0\,\text{V}} = 160\,\text{ns}. \tag{10.72}$$

We repeat the procedure for case 2 and obtain $t_{on2} = 150\,ns$, $\Delta I_{L2} = 75\,mA$, and $t_{off2} = 250\,ns$.

The results for both cases align with the inductor current waveforms in Fig. 10.26a). We also confirm that the duty cycle changes, but the switching frequency remains constant.

During a positive load step, Fig. 10.26b), the frequency rapidly increases because the off-time reduces. This way, the output voltage regulation happens faster than if the frequency is fixed. After the duration of the transient event, the frequency returns to the pseudo-fixed value given by Eqn. (10.68). Because of this mechanism, COT control does not need any slope compensation. The frequency can change to accommodate any perturbations on a cycle-by-cycle basis, as illustrated in Fig. 10.26c).

COT converters usually restrict the minimum off-time to keep the duty cycle resolution in case of low values of V_{in} when the duty cycle reaches an upper bound. A standard approach is to reduce the switching frequency to typically 10–20%. It can be accomplished by either increasing the trip point voltage V_{trip} or by decreasing the charging current of capacitor C when V_{in} gets close to V_{out}.

The COT converter may even enter DCM and operate in PFM burst mode to further reduce the losses as described in Section 10.9. The current sensing information is lost in DCM since the inductor current is zero. A maximum off-timer can resolve this.

Similar to constant on-time control, also constant off-time control exists. It incorporates peak-current detection to set the on-time while the *constant* off-time is set dependent on the voltage conditions. The main drawback of constant off-time control is that the frequency reduces for positive load steps. This way, the load transient response, and light-load efficiency are typically worse than for constant on-time control. Also, the on-time may get too small at high V_{in}, limiting the duty cycle.

10.8   Frequency Compensation

Frequency compensation is achieved by proper design of the controller transfer function $H(s)$. Zeros and poles must be inserted employing resistors and capacitors such that the overall loop gain function $G \cdot H$ maintains sufficient phase margin PM under all conditions. This goal usually corresponds to achieving a slope of -20dB/decade at the 0 dB-crossing of the loop gain in the Bode plot. It typically requires at least one zero to be placed in the compensator.

Another essential goal when designing the compensator is due to the switching nature of the converter. To avoid any in-band noise at the switching frequency f_{sw} of the DC–DC converter, the 0 dB crossover frequency f_c (also called the transit frequency) of the loop gain should be much lower than f_{sw}, typically at 10–30% of f_{sw}.

Summarizing the two general goals for the compensator design:

(1) Choose

$$f_c = 0.1 \dots 0.3 f_{sw}, \tag{10.73}$$

with f_c defined by

$$|G(f_c)H(f_c)| = 0 \text{ dB}. \tag{10.74}$$

(2) Goal: -20 dB/decade crossing at $|G(f_c)H(f_c)| = 0$ dB

10.8.1   Compensator Types

From the analysis of the control-to-output transfer function $G(s)$ for VMC and CMC, we have seen that we usually need a phase boost at the crossover frequency f_c to provide sufficient phase margin. For this purpose, zeros (and poles) must be inserted. From the control theory's point of view, the compensator corresponds to a PID controller comprising a proportional (P), integral (I), and differential (D) frequency behavior. An integrating component is mandatory for achieving high DC gain, good DC regulation, and suppressing low-frequency noise from V_{in}. Integrating behavior requires a pole at DC. With the poles of $G(s)$, multiple poles will be in the loop. To end up at a slope of -20dB/decade we need to insert zeros, typically one or two.

Depending on the plant and its transfer function $G(s)$, different compensators must be implemented. Three types of error amplifier compensation schemes are used. They are classified as type I, type II, and type III. Figure 10.27 shows their implementation using transconductance amplifiers. Type I is a simple active integrator, providing only a pure integral-type compensation. The complexity increases for type II and even more for type III. Type II is a PI controller, while type III provides full PID compensation. As a rule of thumb, the more poles the plant has ($G(s)$), the higher the required compensation type. If we want to use opamps instead of transconductance stages, the passive networks at the output in Fig. 10.27 need to be connected in feedback configuration from the error amplifier output to the inverting input. The circuit shown for type III is actually an opamp implementation. However, it also works with an operational transconductance amplifier (OTA) under certain conditions, which we will investigate further below.

10.8.2   Type I Compensator

The transfer function for the type I compensator in Fig. 10.27a) is simply

$$H(s) = \frac{V_c}{V_{out}} = -\alpha \frac{g_m}{sC_1}, \tag{10.75}$$

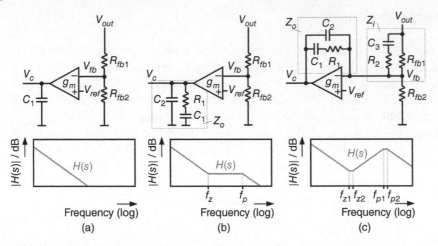

Figure 10.27 Compensator circuits: (a) type I, (b) type II, and (c) type III.

denoting a pole at the origin (DC). Using the transconductance amplifier of Fig. 10.12, the small-signal output impedance $r_{outa} = r_{ds6} \parallel r_{ds7}$ appears in parallel to C1. Hence, at DC, the gain $H_o = H(f = 0)$ will correspond to the DC voltage gain of the transconductance amplifier, $H_o = A_o = -g_m r_{outa}$. The type I compensator does not offer any phase boost. It, therefore, does not reduce the phase shift introduced by the poles of the plant and is usually not applicable in DC–DC converter compensators.

10.8.3 Type II Compensator

Denoting the feedback divider by α (see Eqn. (10.23)), the transfer function can be derived from Fig. 10.27b):

$$H(s) = \frac{V_c}{V_{out}} = -\alpha \, g_m Z_o \tag{10.76}$$

The impedance Z_o of the R–C network is

$$Z_o(s) = \left(\frac{1}{sC_1} + R_1\right) \parallel \frac{1}{sC_2} = \frac{1}{s(C_1 + C_2)} \frac{1 + sR_1 C_1}{1 + sR_1 \frac{C_1 C_2}{C_1 + C_2}}. \tag{10.77}$$

The zero appears at

$$f_z = \frac{1}{2\pi R_1 C_1}, \tag{10.78}$$

and the pole at

$$f_p = \frac{1}{2\pi R_1 \frac{C_1 C_2}{C_1 + C_2}} = \frac{1}{2\pi R_1 C_{12}}. \tag{10.79}$$

The Bode plot in Fig. 10.27b) indicates that the zero f_z must occur at lower frequencies than the pole f_p. For this reason, we require $C_1 \gg C_2$. Under this assumption, Eqn. (10.77) simplifies to

$$Z_o(s) = \frac{1}{sC_1} \frac{1 + sR_1 C_1}{1 + sR_1 C_2}, \tag{10.80}$$

and Eqn. (10.76) becomes

$$H(s) = \frac{V_c}{V_{out}} = -\frac{R_{fb1}}{R_{fb1} + R_{fb2}} \frac{g_m}{sC_1} \frac{1 + sR_1C_1}{1 + sR_1C_2},$$ (10.81)

with the approximated pole-frequency

$$f_p = \frac{1}{2\pi R_1 C_2}.$$ (10.82)

The integrator term g_m/sC_1 describes the DC-pole, similar to Eqn. (10.75). It will be helpful to express $H(s)$ by its zero and pole frequencies:

$$H(f) = \frac{V_c}{V_{out}} = -\alpha \frac{g_m}{j\,2\pi f C_1} \frac{1 + j\frac{f}{f_z}}{1 + j\frac{f}{f_p}}$$ (10.83)

In Eqn. (10.83), we substitute the resistive divider by α as introduced in Eqn. (10.22).

In conclusion, the type II compensator introduces two poles (one at DC) and one zero. Due to the zero, it can achieve a phase boost of up to 90°. Therefore, type II is the preferred choice for compensating plants with 1-pole transfer functions such as a buck converter in CMC (and in VMC when operating in DCM, see Section 10.9).

10.8.4 Type III Compensator

This full PID controller provides two zeros and two poles in addition to the integrating behavior with a pole at DC, Fig. 10.27c). In that sense, the compensator types I and II are subsets of type III. The R-C networks can be mapped to two equivalent impedances, Z_o and Z_i, at the compensator's input and output, as shown in Fig. 10.27c). The configuration forms an inverting gain stage with a transfer function

$$H(s) = \frac{V_c}{V_{out}} = -\frac{Z_o}{Z_i}.$$ (10.84)

Assuming virtual ground at the inverting input of the error amplifier, the feedback resistor R_{fb2} is not part of $H(s)$. The circuit in Fig. 10.27 is a configuration that uses an opamp as the error amplifier because Z_o is connected in a feedback configuration. In a theoretical implementation using an OTA, Z_o would be placed directly at the output V_c toward the ground. However, a type III compensator with an OTA does not offer the same design flexibility as an opamp. Like for the circuit in Fig. 10.27b), $H(s)$ would include the resistive divider of R_{fb1} and R_{fb2}, expressed by scaling factor α (see Eqn. (10.22)) that introduces undesired damping of the signal. In addition, the degree of freedom in placing zeros and poles will be reduced. Fortunately, the circuit of Fig. 10.27c) works also well with a transconductance amplifier (OTA) as long as its transconductance g_m is large and much greater than $1/R_{fb2}$. This is usually fulfilled for large feedback resistors R_{fb2} in the order of $100\,\text{k}\Omega = 1/(10\,\mu\text{A/V})$. Assuming an OTA transconductance g_m, we find the transfer function of Fig. 10.27c) to be

$$H(s) = \frac{V_c}{V_{out}} = \frac{1 - g_m Z_o}{1 + g_m Z_i + \frac{Z_i}{R_{fb2}}},$$ (10.85)

which simplifies to Eqn. (10.84) for large transconductance, i.e., $g_m \gg 1/R_{fb2}$.

Z_o is equal to the expression in Eqn. (10.77). For the R–C network at the input, we find

$$Z_i(s) = R_{fb1} \,\|\, \left(R_2 + \frac{1}{sC_3} \right) = \frac{R_{fb1}(1 + sR_2C_3)}{1 + s(R_{fb1} + R_2)C_3}.$$ (10.86)

Based on the transfer function in Eqn. (10.84), we find a pole f_{p1} and a zero f_{z1} from Z_o (similar to Eqns. (10.78) and (10.79)):

$$f_{p1} = \frac{1}{2\pi R_1 C_{12}} \tag{10.87}$$

$$f_{z1} = \frac{1}{2\pi R_1 C_1} \tag{10.88}$$

Z_i gives another pole-zero pair. Note that Z_i appears in the denominator of Eqn. (10.84). Hence,

$$f_{p2} = \frac{1}{2\pi R_2 C_3} \tag{10.89}$$

$$f_{z2} = \frac{1}{2\pi(R_{fb1} + R_2)C_3} \tag{10.90}$$

The pole at DC is given by

$$f_{po} = \frac{1}{2\pi R_{fb1}(C_1 + C_2)}. \tag{10.91}$$

This is similar to Eqn. (10.75) but with g_m replaced by $1/R_{fb1}$.

The two zeros allow type III compensation to achieve 180° phase boost. This configuration has to be used if the phase shift introduced by the plant can reach −180° like for the buck converter in VMC (and CCM).

Even though we now have derived expressions for the poles and zeros in all three types of compensators, we still need to complete a design guideline on choosing the pole-zero values and how this translates into the resistor and capacitor sizing. Especially when looking at the pole and zero expressions of type III, we notice that the same R and C values are involved multiple times. Due to that interdependence, properly designing the R–C components is not straightforward.

10.8.5 The K-Factor Method

The concept of frequency compensation using the k-factor provides a systematic and easy-to-use methodology for positioning the poles and zeros to ensure a specified phase margin at the targeted crossover frequency f_c. The name stems from the original proposal presented in 1983 by Dean Venable [11]. The k-factor denotes the necessary separation between the zeros and poles introduced by the compensation network. It aims to position the zeros and poles such that the open-loop crossover frequency f_c occurs at the geometric mean in between. Cao [12] proves analytically that this corresponds to the maximum phase boost. The theory is based on the relation between the plant and controller transfer functions $G(s)$ and $H(s)$, respectively, given by Eqn. (10.74) at the crossover frequency f_c,

$$|H(f_c)| = \frac{1}{|G(f_c)|}. \tag{10.92}$$

We will outline the design procedure below for type II and type III. Type I does not provide any phase boost and will not be covered.

10.8.6 The K-Factor for Type II Compensation

Figure 10.28 depicts the Bode plot of $G(s)$ and $H(s)$ as well as of the loop gain $G(s)H(s)$. We note that $H(s)$ has a low-frequency phase shift of −90° due to the integrating behavior corresponding to a pole at the origin. For f_c to be located at the geometric mean between f_z and f_p, the k-factor needs

Figure 10.28 Bode plot of the plant and controller transfer functions $G(s)$ and $H(s)$ as well as the loop gain $G(s)H(s)$ for the k-factor method.

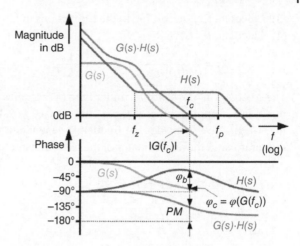

to fulfill these conditions:

$$f_z = \frac{f_c}{k}, \quad f_p = k \cdot f_c \tag{10.93}$$

From the Bode plot or from the expressions for $G(s)$ and $H(s)$, we can extract the given parameters, which are

- the crossover frequency f_c,
- the desired phase margin PM of the loop gain $G \cdot H$ at f_c,
- the magnitude $|G(f_c)|$, and
- the phase shift $\varphi_c = \varphi(G(f_c))$.

The following design procedure allows calculating R_1, C_1, and C_2 in Fig. 10.27b). We successively determine the values of the required phase boost φ_b, the k-factor, and the corresponding locations of the zero f_z and the pole f_p.

(1) Calculate the required phase boost ϕ_b:

$$\varphi_c - 90° + \varphi_b = -180° + PM \rightarrow \varphi_b = PM - \varphi_c - 90° \tag{10.94}$$

(2) The phase boost is defined by

$$\varphi_b = \arg\left(\frac{1 + j\frac{f}{f_z}}{1 + j\frac{f}{f_p}}\right) = \arctan\left(\frac{f}{f_z}\right) - \arctan\left(\frac{f}{f_p}\right). \tag{10.95}$$

Inserting the k-factor from Eqn. (10.93) we get

$$\varphi_b = \arctan(k) - \arctan\left(\frac{1}{k}\right). \tag{10.96}$$

Applying the trigonometric relationship

$$\arctan(x) + \arctan\left(\frac{1}{x}\right) = 90°, \tag{10.97}$$

the phase boost becomes

$$\varphi_b = 2\arctan(k) - 90°. \tag{10.98}$$

Solving for k,

$$k = \tan\left(\frac{\phi_b}{2} + 45°\right). \tag{10.99}$$

(3) Calculate f_z, f_p according to the expressions in (10.93).

(4) Calculate R_1, C_1, C_2:

$$R_1 = \frac{1}{|G(f_c)|\, \alpha\, g_m}, \quad C_1 = \frac{1}{2\pi R_1 f_z}, \quad C_2 = \frac{1}{2\pi R_1 f_p} \tag{10.100}$$

The calculation of resistor R_1 under item (4) assumes $f_z \gg f_c \gg f_p$, which is usually applicable for this k-factor approach because we aim to place f_c in between f_p and f_c (see also the Bode plot in Fig. 10.28). Now, consider the frequency-dependent terms that describe the zero and pole in the numerator and the denominator of the plant transfer function $G(f)$ in Eqn. (10.83). For $f_z \gg f_c \gg f_p$:

$$j\frac{f}{f_z} \gg 1 \rightarrow 1 + j\frac{f}{f_z} \approx j\frac{f}{f_z} = sR_1C_1$$

$$j\frac{f}{f_p} \ll 1 \rightarrow 1 + j\frac{f}{f_p} \approx 1$$

With these approximations, the controller transfer function of Eqn. (10.81) simplifies to a constant value (corresponding to the plateau in between f_p and f_c, see Fig. 10.28):

$$H = -\alpha\, \frac{g_m}{s\cancel{C_1}}\, s\cancel{R_1}\cancel{C_1} = -\alpha\, g_m\, R_1 = \frac{1}{|G(f_c)|} \tag{10.101}$$

According to Eqns. (10.74) and (10.92), H will be equal to the reciprocal value of $|G(f_c)|$ at the crossover frequency. Solving Eqn. (10.101) for R_1 yields the design equation in (10.100). The capacitance values of C_1 and C_2 follow directly from the zero and pole expressions given in Eqns. (10.78) and (10.82), respectively.

Example 10.8 *Design a type II compensator based on the k-factor method for these parameters:* $f_{sw} = 1\,\text{MHz}, G(f_c)|_{dB} = -8\text{dB}, \varphi_c = -85°, g_m = 275\,\mu\text{A/V}, V_{in} = 3.3\,\text{V}, V_{out} = 1.2\,\text{V}, V_{ref} = 800\,\text{mV}.$

(a) *According to the condition in Eqn. (10.73), we chose the crossover frequency at 10% of f_{sw}:*
$f_c = 100\,\text{kHz}$

(b) *As no phase margin is specified, we aim for PM = 60°.*

(c) *We determine the feedback divider ratio α according to Eqn. (10.22):*

$$\alpha = \frac{V_{ref}}{V_{out}} = \frac{800\,\text{mV}}{1.2\,\text{V}} = 0.667 \tag{10.102}$$

(d) *We calculate the linear plant gain:*

$$G_c = 10^{G(f_c)|_{dB}/20dB} = 0.3981 \tag{10.103}$$

(e) *Equation (10.94) gives the required phase boost φ_b:*

$$\varphi_b = PM - \varphi_c - 90° = 60° - (-85°) - 90° = 55° \tag{10.104}$$

(f) *We determine the k-factor using Eqn. (10.99):*

$$k = \tan\left(\frac{\phi_b}{2} + 45°\right) = \tan\left(\frac{55°}{2} + 45°\right) = 3.17 \tag{10.105}$$

(g) *Now, the zero and the pole frequency follow from Eqn. (10.93) with $f_c = 100\,\text{kHz}$ from item a):*

$$f_z = \frac{f_c}{k} = \frac{100\,\text{kHz}}{3.17} = 315\,\text{kHz}, \quad f_p = 3.17 \cdot 100\,\text{kHz} = 31.7\,\text{kHz} \tag{10.106}$$

(h) *Knowing f_z and f_p we can finally calculate the values of the components R_1, C_1, and C_2 using Eqn. (10.100):*

$$R_1 = \frac{1}{|G(f_c)| \, \alpha \, g_m} = \frac{1}{0.3981 \cdot 0.667 \cdot 275 \, \mu A/V} = 13.7 \, k\Omega$$

$$C_1 = \frac{1}{2\pi R_1 f_z} = \frac{1}{2\pi \cdot 13.7 \, k\Omega \cdot 315 \, kHz} = 368 \, pF$$

$$C_2 = \frac{1}{2\pi R_1 f_p} = \frac{1}{2\pi \cdot 13.7 \, k\Omega \cdot 31.7 \, kHz} = 36.6 \, pF \tag{10.107}$$

10.8.7 The K-Factor for Type III Compensation

Consider the gain plot in Fig. 10.27c). There are two zeros and two poles. As with type II, for type III compensation, the k-factor method as proposed in [11] aims to keep the crossover frequency f_c of the loop gain $G \cdot H$ between the pole-zero pairs. Both zeros will be made equal and, likewise, both poles, leading to a double-zero and a double-pole. This in mind, the definitions given in (10.93) for type II change to

$$f_z = \frac{f_c}{\sqrt{k}}, \quad f_p = \sqrt{k} \cdot f_c. \tag{10.108}$$

For $f_z = f_{z1} = f_{z2}$ and $f_p = f_{p1} = f_{p2}$ the controller transfer function becomes

$$H(f) = \frac{V_c}{V_{out}} = -\frac{1}{j\frac{f}{f_{po}}} \frac{\left(1 + j\frac{f}{f_z}\right)^2}{\left(1 + j\frac{f}{f_p}\right)^2}. \tag{10.109}$$

The design procedure is similar to the one outlined for type II. With the double-zero and double-pole, the phase boost becomes

$$\varphi_b = 2\left(\arctan\left(\sqrt{k}\right) - \arctan\left(\frac{1}{\sqrt{k}}\right)\right). \tag{10.110}$$

Applying the trigonometric relationship of Eqn. (10.97) gives

$$\varphi_b = 4 \arctan\left(\sqrt{k}\right) - 180°. \tag{10.111}$$

Rearranging yields the k-factor

$$k = \left(\tan\left(\frac{\phi_b}{4} + 45°\right)\right)^2, \tag{10.112}$$

with the phase boost defined by Eqn. (10.94).

Equations (10.87)–(10.91) are the five equations we need to determine the values of the components in the R–C networks Z_o and Z_i in Fig. 10.27c). As there are six components in total, the design procedure usually starts with choosing a value for R_{fb1}. Assuming that R_{fb1} and R_{fb2} are in the same order of magnitude and taking the considerations next to Eqn. (10.85) into account, a rule-of-thumb is $R_{fb2} \gg 1/g_m$. Now, the five equations for poles and zeros fully determine the values for R_1, R_2, C_1, C_2, and C_3.

We first derive the absolute value of $H(f)$,

$$|H(f)| = \frac{f_{po}}{f} \frac{1 + \left(\frac{f}{f_z}\right)^2}{1 + \left(\frac{f}{f_p}\right)^2}. \tag{10.113}$$

At $f = f_c$ we can set $|H(f_c)|$ equal to the reciprocal of $|G(f_c)|$ (see Eqn. (10.92)) and with the definitions of 10.108, we find

$$|H(f)| = \frac{f_{po}}{f_c} \frac{1 + \left(\sqrt{k}\right)^2}{1 + \left(\frac{1}{\sqrt{k}}\right)^2} = \frac{f_{po}}{f_c} \, k = \frac{1}{|G(f_o)|}. \tag{10.114}$$

Solving Eqn. (10.114) for f_{po} and inserting into Eqn. (10.91) yields

$$C_1 + C_2 = \frac{1}{2\pi R_{fb1} f_{po}} = \frac{k \, |G(f_c)|}{2\pi R_{fb1} f_c}. \tag{10.115}$$

Now we recall the expression for f_{p1} from Eqn. (10.87) and substitute the term $(C_1 + C_2)$ by Eqn. (10.115),

$$f_{p1} = f_p = \frac{1}{2\pi R_1 C_1 C_2} \frac{k \, |G(f_c)|}{2\pi R_{fb1} f_c}. \tag{10.116}$$

The term $2\pi R_1 C_1$ corresponds to $1/f_{z1}$ (see Eqn. (10.88)). Since $f_{z1} = f_z$ is given by the k-factor definition, we can solve Eqn. (10.116) for C_2:

$$C_2 = \frac{k f_z \, |G(f_c)|}{2\pi R_{fb1} f_c f_p} = \frac{k \left(f_c / \sqrt{k}\right) \, |G(f_c)|}{2\pi R_{fb1} f_c \left(\sqrt{k} f_c\right)} = \frac{|G(f_c)|}{2\pi R_{fb1} f_c} \tag{10.117}$$

Examining Eqn. (10.115) and Eqn. (10.117) leads to

$$C_1 + C_2 = k \, C_2. \tag{10.118}$$

Hence,

$$C_1 = C_2 \, (k - 1). \tag{10.119}$$

Referring to Eqns. (10.90) and (10.89) for f_{z2} and f_{p2}, respectively, we find C_3 from the following expression:

$$\frac{1}{f_{z2}} - \frac{1}{f_{p2}} = 2\pi (R_{fb1} + R_2) C_3 - 2\pi R_2 C_3 = 2\pi R_{fb1} C_3 \tag{10.120}$$

For $f_{z2} = f_z$ and $f_{p2} = f_p$,

$$C_3 = \left(\frac{1}{f_z} - \frac{1}{f_p}\right) \frac{1}{2\pi R_{fb1}}. \tag{10.121}$$

Since we have determined C_1 and C_2, we can calculate R_1 from Eqn. (10.87):

$$R_1 = \frac{1}{2\pi f_p C_{12}} = \frac{1}{2\pi f_p \frac{C_1 C_2}{C_1 + C_2}} \tag{10.122}$$

The resistor value R_2 follows from the definition of $f_{p2}(= f_p)$ in Eqn. (10.89) by inserting C_3 from Eqn. (10.121),

$$R_2 = \frac{1}{2\pi f_p C_3} = \frac{R_{fb1}}{(k - 1)}. \tag{10.123}$$

From Eqn. (10.89) we can find an alternative expression for C_3 which is proposed in [11] as part of the original k-factor method:

$$C_3 = \frac{1}{2\pi f_p R_2} = \frac{1}{2\pi \sqrt{k} f_c R_2} \tag{10.124}$$

After choosing the value for R_{fb1}, Eqns. (10.117), (10.119), (10.122), (10.123), and (10.124) define all passive components C_1, C_2, C_3, R_1, and R_2.

Example 10.9 *Design a type III compensator based on the k-factor method for these parameters: $f_{sw} = 1\,\text{MHz}, G(f_c)|_{dB} = -8\,dB, \varphi_c = -130°, V_{in} = 3.3\,\text{V}, V_{out} = 1.2\,\text{V}, V_{ref} = 800\,\text{mV}.$*

(a) *The first steps are identical to items (a–d) of Example 10.8 for a type II compensator. Hence, $f_c = 100\,\text{kHz}$, $PM = 60°$, $\alpha = 0.667$, and $G_c = 0.3981$. In addition, we chose $70\,\text{k}\Omega$ for feedback resistor R_{fb1}. Only R_{fb1} plays a role in the compensator. For completeness, we can use α to get $R_{fb2} = 140\,\text{k}\Omega$.*

(b) *Equation (10.94) gives the required phase boost φ_b,*

$$\varphi_b = PM - \varphi_c - 90° = 60° - (-130°) - 90° = 100°. \tag{10.125}$$

(c) *The k-factor results from Eqn. (10.112):*

$$k = \left(\tan\left(\frac{\phi_b}{4} + 45°\right)\right)^2 = \left(\tan\left(\frac{100°}{4} + 45°\right)\right)^2 = 7.549 \tag{10.126}$$

(d) *We determine the zero and the pole frequency using Eqn. (10.108) with $f_c = 100\,\text{kHz}$ (see item a)):*

$$f_z = \frac{f_c}{\sqrt{k}} = \frac{100\,\text{kHz}}{\sqrt{7.549}} = 36.4\,\text{kHz}$$

$$f_p = \sqrt{k} \cdot f_c = \sqrt{7.549} \cdot 100\,\text{kHz} = 274.8\,\text{kHz} \tag{10.127}$$

(e) *We can now calculate the values of all passive components based on Eqns. (10.117), (10.119), (10.122), (10.123), and (10.124):*

$$C_2 = \frac{|G(f_c)|}{2\pi R_{fb1}f_c} = \frac{0.3981}{2\pi \cdot 70\,\text{k}\Omega \cdot 100\,\text{kHz}} = 9.05\,\text{pF}$$

$$C_1 = C_2\,(k-1) = 9.05\,\text{pF}\,(7.549 - 1) = 59.3\,\text{pF}$$

$$R_1 = \frac{1}{2\pi f_p \frac{C_1 C_2}{C_1+C_2}} = \frac{1}{2\pi \cdot 274.8\,\text{kHz} \cdot \frac{59.3\,\text{pF} \cdot 9.05\,\text{pF}}{59.3\,\text{pF}+9.05\,\text{pF}}} = 73.8\,\text{k}\Omega$$

$$R_2 = \frac{R_{fb1}}{(k-1)} = \frac{70\,\text{k}\Omega}{(7.549 - 1)} = 10.7\,\text{k}\Omega$$

$$C_3 = \frac{1}{2\pi f_p R_2} = \frac{1}{2\pi \cdot 274.8\,\text{kHz} \cdot 10.7\,\text{k}\Omega} = 409\,\text{pF} \tag{10.128}$$

We notice that the value of $C_3 = 409\,\text{pF}$ is rather large. We may be able to integrate it on-chip, but it will take quite some layout area. If we reconsider the choice of the crossover frequency f_c, we can reduce C_3. As per Eqn. (10.73), we are still on the save side if we set $f_c = 0.3f_{sw} = 300\,\text{kHz}$. This way, C_3 reduces to 136 pF, one-third of the original value. Likewise, C_1 and C_2 get smaller while the resistor values will remain the same.

10.8.8 Capacitance Multiplier

The pole and zero frequencies appear in the denominator of the sizing equations for the passive R–C components. According to Eqn. (10.73), these frequencies scale with the overall switching frequency f_{sw}. Hence, the size of the passives shrinks for higher switching frequencies f_{sw}. However, they might still be too large to be integrated on-chip. On the other hand, users of DC–DC converters often want to avoid designing and placing external resistor-capacitor networks.

For this reason, capacitor multiplier circuits are an attractive alternative. The concept is to use an active circuit that acts to multiply a small on-chip capacitor, which effectively becomes typically more than 10× larger. This way, capacitance multipliers (cap multipliers) enable power management ICs with fully integrated compensation.

There are two options, shown in Fig. 10.29. Rincon-More has explored the fundamental concepts for power management circuits in [13] (2000). The voltage-mode multiplier in Fig. 10.29a) is identical to the Miller effect, well-known in analog amplifier theory. Placing a capacitor C_c between the input and output of a gain stage with negative gain $-A$, the effective capacitance at the input of the gain stage appears to be A-times larger than C_c. For power management circuits, current-mode operation, as shown in Fig. 10.29b), is more beneficial because R-C compensator networks can be implemented more conveniently. In that case, a current-controlled current source (CCCS) is parallel to a capacitor C_c. For this configuration, we can define an equivalent capacitance by the I-V relationship

$$\frac{V_{eq}}{I_{eq}} = \frac{1}{j\omega C_{eq}}. \tag{10.129}$$

If the CCCS provides $K \cdot I_c$ with I_c being the current in capacitor C_c, we can express the equivalent current by

$$I_{eq} = I_c + K \cdot I_c = (K+1)I_c \sim KI_c = KV_{eq}j\omega C_c \rightarrow \frac{V_{eq}}{I_{eq}} = \frac{1}{j\omega KC_c}. \tag{10.130}$$

Comparing Eqns. (10.129) and (10.130),

$$C_{eq} = KC_c. \tag{10.131}$$

Figure 10.29c) shows an implementation example of a type II compensator in a DC–DC converter [14]. Referring to Fig. 10.27b), the capacitor multiplier forms C_1 while C_2 and R_1 are components connected to the output of the error amplifiers. A voltage-follower keeps the voltage across R_x equal to R_y. The current in R_x is equal to I_c. Since R_x is K-times larger than R_y, the current in R_y will be KI_c. It fulfills the condition of Eqn. (10.130), and we get $C_{eq} = C_1 = K \cdot C_c$. The challenge of this concept lies in the voltage follower. It will require a class A-B power stage to drive the large current of KI_c. It also needs to have high unity-gain frequency to ensure proper multiplier operation over the frequency range of interest (beyond the control loop crossover frequency f_c).

In conclusion, a capacitor multiplier allows for an area-efficient on-chip implementation of large compensation capacitors. As an additional advantage, the multiplier can be adaptive and adjust

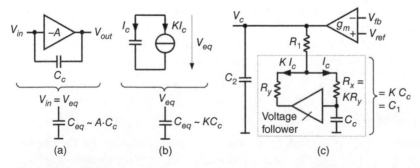

Figure 10.29 Capacitance multiplier circuits: a) voltage mode (Miller effect); b) current mode; c) implementation of a current mode capacitance multiplier as part of the compensator R–C network.

the capacitance value during operation, as proposed in [14]. This way, the phase margin may be reduced during load steps by lowering C_c to get a fast transient response. The converter may get unstable for this short moment, and its compensation will be returned to a steady state with sufficient phase margin shortly after. It can be accomplished by utilizing the on-resistance of transistor switches for R_x and R_y with K-times different W/L-ratio. The multiplier value K can be adjusted by disconnecting segments of the larger transistor (setting its gate voltage to zero).

Example 10.10 *Referring to Example 10.8, $C_1 = 368$ pF turned out to be quite large. The cap-multiplier of Fig. 10.29b) would be suitable to enable on-chip integration of the entire compensator. We chose $R_y = 100\,\Omega$ and $R_x = 10\,R_y = 1\,k\Omega$.*

10.9 Discontinuous Conduction Mode (DCM)

In the above part of this chapter, we always considered the case of continuous inductor current, referred to as continuous conduction mode (CCM). The inductor current $I_L(t)$ ripples around an average value equal to the converter's load current I_{load}. If the output power reduces, the load current decreases to a lower value. For that case, Fig. 10.30a) shows that the inductor current may become negative in converter stages with synchronous rectification (see Section 10.1.10 and Fig. 10.3). However, negative inductor current discharges the output capacitor and adds losses. The zero-crossing needs to be monitored, and the low-side switch needs to be turned off instantaneously to prevent these losses. This technique is called diode emulation. It relates to a converter with asynchronous rectification (see Fig. 10.3) in which the freewheeling diode would block any negative current. If the inductor current reaches zero and cannot go negative, the DC–DC converter enters discontinuous conduction, also called operation in DCM. In general, we can say that a DC–DC converter is usually designed to work in CCM for medium to heavy loads to supply sufficient output power. At the same time, DCM operation is preferred at light loads for high efficiency.

10.9.1 General Constant-Frequency Behavior

We first keep the DC–DC converter operating in constant-frequency PWM control. In the second step, we will consider PFM at varying switching frequencies because of higher power efficiency at light load. Figure 10.30b) shows that the switching period in DCM consists of three sections: (1) the on-time $t_{on} = D\,T$ (high-side switch conducting), (2) the off-time $t_{off} = D_2\,T$ (low-side switch turned on), and (3) the time $(T - (t_{on} + t_{off})) = D_3\,T$ when the inductor current is zero (both switches off, switching node V_{sw} at V_{out}). D is the duty cycle that relates to the same on-time t_{on} as defined for CCM in Eqn. (10.3). In DCM, we set the inductor current ripple ΔI_L equal to the peak current value I_{peak} (see Fig. 10.30b)),

$$D = \frac{t_{on}}{T} = \frac{L\,I_{peak}}{T\,(V_{in} - V_{out})}. \tag{10.132}$$

D_2 denotes the duty cycle of the low-side switch. Referring to Eqn. (10.4) we obtain

$$D_2 = \frac{t_{off}}{T} = \frac{L\,I_{peak}}{T\,V_{out}}. \tag{10.133}$$

Ringing at the Switching Node

As the inductor current reaches zero in DCM, the switching node becomes susceptible to ringing, as indicated in Fig. 10.30b). The inductor forms a resonant tank with the parasitic capacitances of the power transistors. The ringing dies out after a few cycles, and the switching node settles at $V_{sw} = V_{out}$. However, the ringing may slightly change the conversion ratio and cause interference. There are various ways to eliminate oscillations at the switching node. A series damping R–C network (snubber network) can be placed between the switching node and ground at the expense of power loss. Alternatively, some designs place an additional switch in parallel to the inductor. The switch turns on and provides a freewheeling path after the zero crossing of the inductor current has been detected, and both power switches turn off.

Peak Current in DCM

From Fig. 10.30b), we can calculate the average inductor current, which, in steady-state, will be equal to I_{load},

$$\frac{1}{T}\int_0^T I_L(t)\mathrm{d}t = \frac{1}{T}(t_{on} + t_{off})\frac{I_{peak}}{2} = (D + D_2)\frac{I_{peak}}{2} = I_{load}. \tag{10.134}$$

Note that the integral can easily be solved because the area enclosed by $I_L(t)$ is a triangle. We insert D and D_2 from Eqns. (10.132) and (10.133) into Eqn. (10.134) and solve for the peak inductor current:

$$I_{peak} = \sqrt{\frac{2 I_{load}}{L} T \frac{(V_{in} - V_{out})V_{out}}{V_{in}}} \tag{10.135}$$

For a DC–DC converter at a given parameter set of V_{in}, V_{out}, $T = 1/f_{sw}$ and L, Eqn. (10.135) allows us to determine the peak inductor current I_{peak} as a function of I_{load}. Knowing I_{peak}, we can calculate D and D_2 as well as t_{on} and t_{off} from Eqns. (10.132) and (10.133).

Example 10.11 *We want to know I_{peak} and t_{on}, t_{off} in DCM if I_{load} reduces to 50 mA for a DC–DC converter with these parameters: $f_{sw} = 1\,\text{MHz}$, $V_{in} = 3.3\,\text{V}$, $V_{out} = 1.2\,\text{V}$, $L = 2\,\mu\text{H}$.*
We first determine the peak current value from Eqn. (10.135),

$$I_{peak} = \sqrt{\frac{2 I_{load}}{L} T \frac{(V_{in} - V_{out})V_{out}}{V_{in}}}$$

$$= \sqrt{\frac{2 \cdot 50\,\text{mA} \cdot 1\,\mu\text{s}}{2\,\mu\text{H}} \frac{(3.3\,\text{V} - 1.2\,\text{V})1.2\,\text{V}}{3.3\,\text{V}}} = 195\,\text{mA}. \tag{10.136}$$

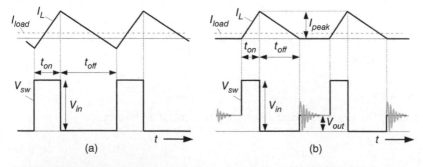

Figure 10.30 a) Waveforms in CCM for small load current: the inductor current gets negative; b) discontinuous conduction mode (DCM): the inductor current remains zero for a fraction of the switching period.

t_{on} and t_{off} *follow from Eqns. (10.3) and (10.4),*

$$t_{on} = \frac{L I_{peak}}{V_{in} - V_{out}} = \frac{2\,\mu H \cdot 195\,mA}{3.3 - 1.2\,V} = 186\,ns, \tag{10.137}$$

$$t_{off} = \frac{L I_{peak}}{V_{out}} = \frac{2\,\mu H \cdot 195\,mA}{1.2\,V} = 326\,ns. \tag{10.138}$$

Finally,

$$D = \frac{t_{on}}{T} = \frac{186\,ns}{1\,\mu s} = 0.19, \tag{10.139}$$

$$D_2 = \frac{t_{off}}{T} = \frac{326\,ns}{1\,\mu s} = 0.33. \tag{10.140}$$

The inductor current will be zero for the remaining time $T - (t_{on} + t_{off}) = 488\,ns$, *which corresponds to almost half the period time* ($D_3 = 0.49$).

Conversion Ratio

In DCM, the conversion ratio will be different from CCM. We introduce M to denote the voltage conversion ratio. In order to ensure the volt-seconds balance, we refine Eqn. (10.7),

$$\left(V_{in} - V_{out}\right) DT = V_{out} D_2 T. \tag{10.141}$$

We rearrange Eqn. (10.134) and substitute I_{peak} from Eqn. (10.133) to find the duty cycle D_2:

$$D_2 = 2\frac{I_{load} L}{V_{out} D_2 T} - D \tag{10.142}$$

Note that D_2 appears in the denominator of the right-hand expression, resulting in a quadratic equation for D_2. Solving for D_2 and inserting it into Eqn. (10.141) yields the conversion ratio of the buck converter in DCM,

$$M = \frac{V_{out}}{V_{in}} = \frac{2}{1 + \sqrt{1 + \frac{8L}{D^2 TR}}}. \tag{10.143}$$

Unlike in CCM (see Eqn. (10.8)), the voltage conversion ratio in DCM depends on the load, expressed by the load resistance $R = V_{out}/I_{load}$.

The average voltage at the switching node depends on V_{out} (see Eqn. (10.141)). Consequently, the AC transfer function becomes a first-order expression (one pole, one zero), which is easier to stabilize (type II instead of type III compensation) [1, 3].

Example 10.12 *Let us verify the voltage conversion ratio of Example 10.11:*

The load resistance in Example 10.11 is $R = V_{out}/I_{load} = 1.2\,V/50\,mA = 24\,\Omega$. *The duty cycle is obtained in Eqn. (10.139) to be* $D = 0.19$. *With all the other parameters from Example 10.11 Eqn. (10.143) yields*

$$M = \frac{V_{out}}{V_{in}} = \frac{2}{1 + \sqrt{1 + \frac{8L}{D^2 TR}}} = \frac{2}{1 + \sqrt{1 + \frac{8 \cdot 2\,\mu H}{(0.19)^2 \cdot 1\,\mu s \cdot 24\,\Omega}}} = 0.36, \tag{10.144}$$

which matches exactly the target voltage conversion ratio $V_{out}/V_{in} = 1.2\,V/3.3\,V = 0.36$.

Equation (10.143) suggests an alternative way to calculate I_{peak} instead of using Eqn. (10.135). We can rearrange Eqn. (10.143) to determine D for the target conversion ratio $M = V_{out}/V_{in}$. Knowing D, we can calculate both t_{on} and I_{peak} from Eqn. (10.132). Inserting I_{peak} into Eqn. (10.133) yields D_2 and t_{off}, respectively. Following this approach, Example 10.11 would give the same results.

10.9.2 Pulse-Frequency Modulation (Burst Mode)

The graphs at the top and in the middle of Fig. 10.31a) illustrate the inductor current in CCM and DCM if the converter keeps operating at a constant frequency. Note that the average of the inductor current equals the load current in a steady state. At some point, the switching losses will dominate the static conduction losses, resulting in a significant efficiency drop at light load. Figure 10.31b) (the curve labeled PWM) illustrates this behavior with the conduction losses dominating at high power. The efficiency roll-off at low power is due to the switching losses. Therefore, when entering DCM, converters are preferably controlled by PFM rather than PWM. This way, the switching activity (frequency) reduces significantly toward the light load. Alternative terms to PFM are pulse-skipping mode or burst mode. Pulse skipping refers to the fact the multiple switching cycles of constant frequency PWM control are skipped. Burst mode describes the generation of one or more pulses that turn on the high-side switch in a PFM fashion if the output voltage has dropped below a minimum value. Figure 10.31a) (bottom) shows the inductor current along with typical durations of the active and passive intervals. In consequence, the light load efficiency significantly increases, as shown in Fig. 10.31b) (PFM + I_{bias}). If the load decreases further, the total bias current I_{bias} of circuits in the control loop and the power stage will impact efficiency. Turning off all unnecessary blocks will further improve the efficiency as illustrated by the third efficiency curve in Fig. 10.31b) (PFM + $\bcancel{I_{bias}}$).

Figure 10.32 shows an example implementation of PFM control for the CMC DC–DC converter of Fig. 10.15. Starting from a DC–DC converter in VMC or CMC, we need to add these functions to support PFM:

(1) Detect discontinuous conduction: Various ways exist.
   (a) Zero crossing detection (ZCD) of the inductor current (shown in Fig. 10.32): The switching node voltage during the dead time after the low-side switch turns off can be detected. If the inductor current is still greater than zero, V_{sw} will be one (body) diode forward drop below ground. Else, V_{sw} will be higher and eventually equal to V_{out}, see Fig. 10.30b). For stable operation, some designs require ZCD to occur multiple times (e.g., five times) before entering PFM. Sensing concepts are discussed in Section 6.7. In particular, the circuit in Fig. 6.22b) will be well-suitable.
   (b) DCM can also be detected if the error signal V_c falls below a minimum level.

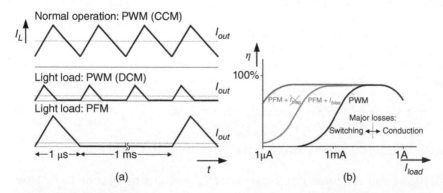

Figure 10.31 a) Inductor current for CCM (top) and DCM (middle) in PWM operation and PFM (bottom); b) impact on conversion efficiency η for PWM as well as PFM with and without disabling bias currents in the control part.

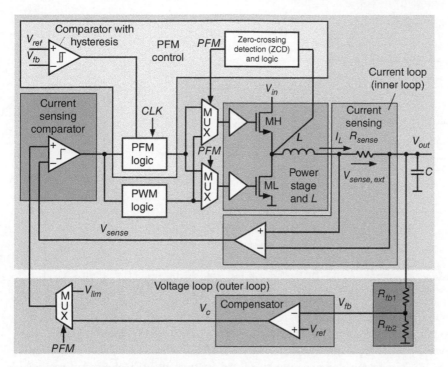

Figure 10.32 By adding PFM control, the current-mode controlled DC–DC converter of Fig. 10.15 can support power-efficient light-load operation.

(2) After entering PFM, the main control loop will be disabled, and a comparator with hysteresis (upper and lower trip point V_{tpu}, V_{tpl}, see also Fig. 10.13) will monitor the output voltage. There are now two phases:

   (a) Passive phase or skipping interval ($V_{fb} > V_{tpu}$): No switching activity. Even though it is very low, the load current will be delivered by the output capacitor. Thereby, V_{out} gradually discharges.

   (b) Active phase ($V_{fb} < V_{tpl}$): One or multiple pulses (bursts) charge up the output capacitor until $V_{fb} = \alpha V_{out}$ exceeds the upper comparator threshold V_{tpu}. There are many different ways to generate the pulse. A single-shot timer with a fixed duration can be used. In constant on-time (COT) converters (see Section 10.7), the pulse generation in DCM can be the same as for CCM, which makes COT control attractive for light-load operation. For converters that support PWM CMC at regular load, the pulse can be generated by reusing the existing peak-current control. However, the current sensing signal V_{sense} is compared to a voltage limit V_{lim} rather than to the error voltage V_c (see Fig. 10.16 for CMC). The converter in Fig. 10.32 incorporates this method (shown at the bottom as part of the outer control loop). In PFM mode, an analog multiplexer connects V_{lim} to the current sensing comparator (which reuses the existing comparator of the pulse-width modulator for PWM mode).

(3) Ultra-light load operation (PFM + I_{bias} in Fig. 10.31b)): During the passive phase, all internal circuits can be turned off. Just the comparator and the voltage reference circuit will remain active to monitor the output voltage and wake up the converter (active phase) once V_{out} has dropped below the threshold defined by V_{tpl}. It results in minimum current consumption (typ. a few microamperes). Instead of V_{fb}, it can be beneficial to monitor V_c (an amplified version of V_{fb}), but at the expense of somewhat larger current consumption.

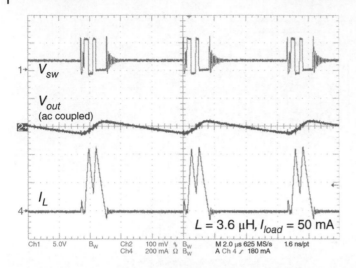

Figure 10.33 Measured waveforms of a DC–DC buck converter in PFM mode (from [15]) for $L = 3.6\,\mu H$, $C = 10\,\mu F$, and $I_{load} = 50\,mA$. The system clock (active phase) is $f_{sw} = 1.6\,MHz$.

Figure 10.33 shows measured waveforms [15] of V_{sw}, V_{out}, and of the inductor current I_L for a converter according to Fig. 10.32 in PFM mode. The output voltage shows the expected ripple between two boundaries given by the comparator hysteresis, which is 55 mV in this case. If the converter is in the passive phase, V_{out} ramps down and eventually falls below the lower threshold, initiating the active phase. The high-side switch turns on and energizes the inductor (I_L ramps up) synchronized to the system clock *CLK*, causing V_{out} to rise. Once the current sensing signal V_{sense} hits V_{lim}, the high-side switch turns off, and the low-side switch turns on (with a small dead time in between). The next rising edge of the clock *CLK* stops the cycle. As long as V_{out} is below the upper threshold, pulses are generated by repeating the sequence of turning on the high- and low-side again. Afterward, the converter enters the passive phase again and stays there while V_{out} remains above the lower threshold. Figure 10.33 also shows the ringing at the switching node voltage V_{sw} once the inductor current has reached zero (passive phase). In this case, 50 mA of load current requires two switching cycles in each active phase. For higher load current, the converter will eventually go into CCM and provide a continuous series of pulses while the inductor current does not go down to zero.

Can we estimate the number of pulses during the active phase? We know that a certain amount of charge Q is delivered onto the output capacitor C per cycle. Q is the difference between the charge delivered from the inductor L and the charge dissipated by the load. We assume the load current I_{load} to be constant over one switching period, hence,

$$Q = \int_0^T I_L(t)\,dt - \int_0^T I_{load}(t)\,dt = \frac{(t_{on} + t_{off})I_{peak}}{2} - I_{load}T. \tag{10.145}$$

The net charge Q translates into an increase in the output voltage by an amount $\Delta V_{out} = Q/C$. Therefore, multiple pulses will result in a voltage swing of $V_{hyst} = n \cdot \Delta V_{out}$ defined by the comparator hysteresis V_{hyst} (referred to V_{out}, not to V_{fb}). Each pulse will take roughly one clock period $T = 1/f_{sw}$. t_{on} and t_{off} are defined by Eqns. (10.3) and (10.4). Replacing the ripple ΔV_{out} by the comparator hysteresis V_{hyst} and T by nT we can determine the number of required pulses from Eqn. (10.145),

$$n = \frac{CV_{hyst}}{\dfrac{LI_{peak}^2}{2}\dfrac{V_{in}}{(V_{in} - V_{out})V_{out}} - I_{load}T}. \tag{10.146}$$

For I_{peak}, we need to insert the upper current limit defined by V_{lim} in Fig. 10.32.

Example 10.13 *Verify the number of pulses for the measurements in Fig. 10.33 using Eqn. (10.146).*
According to [15] the parameters are $V_{in} = 5.5\,V$, $V_{out} = 1.63\,V$, $V_{hyst} = 55\,mV$, $I_{load} = 50\,mA$,
$I_{peak} = 440\,mA$, $L = 3.6\,\mu H$, $C = 10\,\mu F$, $f_{sw} = 1.6\,MHz$, and Eqn. (10.146) yields:

$$
n = \frac{CV_{hyst}}{\dfrac{LI_{peak}^2}{2} \dfrac{V_{in}}{(V_{in} - V_{out})V_{out}} - \dfrac{I_{load}}{f_{sw}}}
$$

$$
= \frac{10\,\mu F \cdot 55\,mV}{\dfrac{3.6\,\mu H \cdot (440\,mA)^2}{2} \dfrac{5.5\,V}{(5.5\,V - 1.63\,V)1.63\,V} - \dfrac{50\,mA}{1.6\,MHz}} = 2.0177 \sim 2.0. \qquad (10.147)
$$

This result matches with Fig. 10.33. Two switching cycles are required to recharge the output capaci-
tor such that V_{out} exceeds the threshold V_{hyst}. If the load current reaches 180 mA, further measurements
in [15] show that the number of cycles increases to four. If, in addition, the inductor is changed to
$L = 0.82\,\mu H$, even eight cycles are needed. The pulses stop when the converter enters its passive phase
after V_{out} has increased more than the comparator's hysteresis V_{hyst}.

10.10 The Boost Converter

Figure 10.34 shows the fundamental boost converter, also called a step-up converter because it
generates an output voltage higher than the input voltage. Like the buck converter (see Fig. 10.1),
the power stage in Fig. 10.34a) consists of two power switches and an inductor. Both switches are
controlled complementary (with a dead time in between as explained in Section 2.8). The output
voltage is buffered by a capacitor C. As the major difference to the buck converter, the inductor is connected to the input voltage V_{in}. Because of the steady inductor current I_L, there is a
non-pulsating input current ripple, which relaxes the requirements on the input buffer capacitor
(omitted in Fig. 10.34).

10.10.1 Inductor Current

The inductor voltage will change in polarity over one switching cycle and cause the inductor current to ramp up and down, as shown in Fig. 10.34b). For $PWM = 1$, switch M1 conducts and

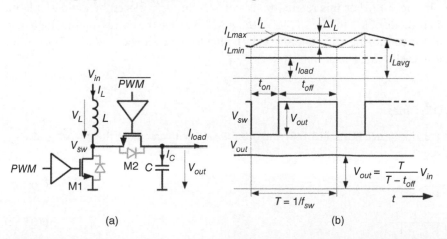

(a)   (b)

Figure 10.34 a) Basic boost converter with b) waveforms of the inductor current I_L and typical node
voltages V_{sw}, V_{out}.

energizes the inductor L. It corresponds to the on-time t_{on} of the boost converter, as indicated in Fig. 10.34b). Neglecting the on-resistance of M1, the switching node is at ground potential while the inductor voltage gets equal to V_{in}. Note that it is essential to keep the polarity of M2, as shown in Fig. 10.34b). Otherwise, M2's body diode would conduct and drain out the output capacitor C during t_{on}. The inductor current increases by

$$\frac{\Delta i_L}{\Delta t} = \frac{V_{in}}{L}. \tag{10.148}$$

During t_{on}, no energy is delivered to the output, and the output capacitor C gets discharged by the load current I_{load}. This behavior differs significantly from the buck converter, where the inductor always delivers current to the output. For this reason, the boost converter belongs to the class of converters with indirect energy transfer.

If the inductor current reaches its maximum value I_{Lmax}, the PWM changes to $PWM = 0$ and activates M2 (after turning off M1). As the inductor current continues to flow, it pulls up the switching node voltage V_{sw}. During the dead time at the transition between M1 and M2, the switching node is clamped to $V_{out} + V_F$ where V_F is the forward voltage of M2's body diode. Once M2 turns on, neglecting its on-resistance, V_{sw} reaches V_{out}. Because of $V_{out} > V_{in}$ the inductor voltage $V_L = V_{in} - V_{out}$ is negative and the inductor current will ramp down,

$$\frac{\Delta i_L}{\Delta t} = \frac{V_{in}}{L} = -\frac{V_{out} - V_{in}}{L}. \tag{10.149}$$

Unlike in the buck converter, the average inductor current is not equal to the load current I_{load} because the inductor delivers current to the output only during the off-time. From the charge balance, we get

$$I_{load} T = \int_0^{t_{off}} I_L(t)\mathrm{d}t = I_{Lavg} \, t_{off} \rightarrow I_{Lavg} = \frac{I_{load}}{D'}, \tag{10.150}$$

where $D' = t_{off}/T$. As derived below, the ratio I_{Lavg}/I_{load} is identical to the conversion ratio V_{out}/V_{in}.

10.10.2 Asynchronous Rectification

The boost converter can be implemented with asynchronous rectification similar to the buck converter, as shown in Fig. 10.3. For this reason, transistor M2 can be replaced by a diode (placed exactly like the body diode of M2 in Fig. 10.34). Without M2, we save the corresponding high-side gate driver and its gate control logic. On the downside, higher losses are associated with the rectifier diode, like for the asynchronous buck converter.

10.10.3 On-/Off-Times

For a given current ripple ΔI_L Eqns. (10.148) and (10.149) allow to calculate the on-/off-times t_{on} and t_{off}, respectively:

$$t_{on} = \Delta I_L \frac{L}{V_{in}}, \tag{10.151}$$

$$t_{off} = \Delta I_L \frac{L}{V_{out} - V_{in}}. \tag{10.152}$$

10.10.4 Current Ripple

The inductor current ramps up and down and leads to a well-defined inductor current ripple $\Delta I_L = I_{Lmax} - I_{Lmin}$. We can determine the current ripple by adding t_{on} and t_{off} from Eqns. (10.151) and (10.152), respectively. The result corresponds to the period time $T = 1/f_{sw}$. Rearranging gives an expression for the current ripple,

$$\Delta I_L = \frac{V_{in}}{Lf_{sw}}\left(1 - \frac{V_{in}}{V_{out}}\right), \tag{10.153}$$

which is similar to the current ripple of the buck converter in CCM as given in Eqn. (10.9).

10.10.5 Inductor Sizing

Equation (10.153) allows deriving the required inductor value for a given current ripple,

$$L = \frac{V_{in}}{\Delta I_L f_{sw}}\left(1 - \frac{V_{in}}{V_{out}}\right). \tag{10.154}$$

As expected, operating at a larger switching frequency reduces the inductor size for given input and output voltages.

10.10.6 Voltage Conversion Ratio

The volt-second balance of Eqn. (10.5) must be fulfilled for any DC–DC converter. It allows us to derive the voltage conversion ratio $M = V_{out}/V_{in}$. Similar to Eqn. (10.7) for the buck converter, we can put together the condition for the boost converter:

$$V_{in}\,t_{on} + \left(V_{in} - V_{out}\right)t_{off} = 0 \tag{10.155}$$

Rearranging yields the voltage conversion ratio M of the boost converter (in CCM),

$$M = \frac{V_{out}}{V_{in}} = \frac{T}{T - t_{on}} = \frac{1}{1 - D} = \frac{1}{D'}. \tag{10.156}$$

Since $D' = t_{off}/T \leq 1$, the conversion ratio is always greater than unity – as expected for the boost step-up converter. The parasitic resistance in the boost converter (e.g., the on-resistance of the switches) limits the conversion ratio to typically 4–5.

Example 10.14 *Calculate the inductance L, the on-time t_{on}, the off-time t_{off} and the voltage conversion ratio M of a boost converter for $V_{in} = 1.2\,\text{V}$, $V_{out} = 3.3\,\text{V}$, $I_{load} = 1\,\text{A}$, $\Delta I_L = 0.4 \cdot I_{load} = 0.4\,\text{A}$, and $f_{sw} = 1\,\text{MHz}$. What is the average inductor current? Note that the parameters are identical to Example 10.1 for the buck converter, except that V_{in} and V_{out} are swapped.*

Equation (10.154) gives the inductance:

$$L = \frac{V_{in}}{\Delta I_L f_{sw}}\left(1 - \frac{V_{in}}{V_{out}}\right) = \frac{1.2\,\text{V}}{0.4\,\text{A} \cdot 1\,\text{MHz}}\left(1 - \frac{1.2\,\text{V}}{3.3\,\text{V}}\right) = 1.909\,\mu\text{H} \tag{10.157}$$

Interestingly, this is the same value obtained for the buck converter in Example 10.1.

The on- and off-times follow from Eqns. (10.151) and (10.152):

$$t_{on} = \Delta I_L \frac{L}{V_{in}} = 0.4\,\text{A}\frac{1.909\,\mu\text{H}}{1.2\,\text{V}} = 636\,\text{ns}, \tag{10.158}$$

$$t_{off} = \Delta I_L \frac{L}{V_{out} - V_{in}} = 0.4\,\text{A}\frac{1.909\,\mu\text{H}}{3.3\,\text{V} - 1.2\,\text{V}} = 364\,\text{ns}. \tag{10.159}$$

Compared to Example 10.1, the on- and off-time have swapped.

According to Eqn. (10.155) the conversion ratio is

$$M = \frac{V_{out}}{V_{in}} = \frac{1}{D'} = \frac{T}{T - t_{on}} = \frac{1\,\mu s}{1\,\mu s - 636\,ns} = 2.75. \tag{10.160}$$

We find the average inductor current using Eqn. (10.150),

$$I_{Lavg} = \frac{I_{load}}{f_{sw}\,t_{off}} = \frac{1\,A}{1\,MHz \cdot 364\,ns} = 2.75\,A. \tag{10.161}$$

10.10.7 Output Voltage Ripple

The input current ripple in a boost converter is low, thanks to the inductor. However, this situation at the output is much different. The energy is stored in the inductor during t_{on} and then suddenly dumped onto the output buffer capacitor C during t_{off}. This current step translates into significant current and output voltage ripple. During t_{off} the capacitor current I_C is the difference between the average inductor current I_{Lavg}, given by Eqn. (10.149), and the load current I_{load}:

$$I_C = I_{Lavg} - I_{load} = I_{load}\left(\frac{1}{D'} - 1\right) \tag{10.162}$$

Following the I–V relation at the output capacitor, we describe the output voltage ripple by

$$\Delta V_{out} = I_C \frac{t_{off}}{C}. \tag{10.163}$$

We can substitute I_C by Eqn. (10.162) and t_{off} by D'/f_{sw} and find a surprisingly simple expression for the output voltage ripple,

$$\Delta V_{out} = \frac{I_{load}\,D}{f_{sw}\,C}. \tag{10.164}$$

10.10.8 Capacitor Sizing

Based on Eqn. (10.164), we can calculate the minimum output capacitance for a target ripple specification of ΔV_{out},

$$C = \frac{I_{load}\,D}{f_{sw}\,\Delta V_{out}}. \tag{10.165}$$

Example 10.15 *Calculate the output capacitance C of the boost converter of Example 10.14 for an output ripple specificantion of $\Delta V_{out} = 10\,mV$?*
  With $I_{load} = 1\,A, f_{sw} = 1\,MHz$, and $D = t_{on}/T = 0.636$ from Example 10.14 Eqn. (10.165) gives

$$C = \frac{I_{load}\,D}{f_{sw}\,\Delta V_{out}} = \frac{1\,A \cdot 0.636}{1\,MHz \cdot 10\,mV} = 6.4\,\mu F. \tag{10.166}$$

We can chose $C = 6.8\,\mu F$ as a suitable output capacitor.

10.10.9 Control Loop

Like the buck, the boost converter can operate in VMC and CMC. In VMC, the control loop is very similar to the buck converter in Fig. 10.7, including PWM based on a sawtooth signal V_s that is compared to the compensated error voltage V_c. The only difference is in the control logic that converts the PWM signal of the comparator into control signals for the power switches M1 and M2. $PWM = 1$ ($V_s < V_c$) turns on M1 while $PWM = 0$ ($V_s \geq V_c$) turns on M2. In CMC, the ramp of the sawtooth generator is replaced by the inductor current signal V_{sense}. Current sensing is more straightforward than in the buck converter because the inductor current is equal to the drain current of the low-side switch M1. Replica sensing can be implemented relatively easily, as described in Section 6.6. If the duty cycle D exceeds 50%, the boost converter in CMC gets prone to subharmonic oscillations as described in Section 10.6 for the buck converter (see Fig. 10.20). A compensation ramp must either be subtracted from V_c or added to V_{sense}. The circuits of Fig. 10.24 can be adapted. Figure 10.24a) can be converted to low-side replica sensing. The design in Fig. 10.24b) will become even simpler if R_s gets replaced by power transistor M1. Then, the V–I converter and the current mirror connected to V_{sense} are not required anymore.

10.10.10 The PWM Switch Model

To derive the boost converter's DC operating point and various AC transfer functions, we generate a PWM switch model as introduced by Vorpérian [16]. The model is widely applicable to analyze various switched-mode converter topologies. The model can also be used with the buck converter. However, the buck converter is an exception, which can be analyzed directly, as done earlier in the chapter. The modeling approach is elaborated in detail in [3]. The key idea is to replace the power stage with an equivalent three-terminal model. It is worth the effort because the model can be applied to any DC–DC converter topology. We start with the relation between the average inductor current I_{Lavg} and the load current I_{load} given by Eqn. (10.150). Assuming averaged steady-state quantities, we replace the average inductor current I_{Lavg} by I_L. Rearranging Eqn. (10.150) describes the load current I_{load} as a function of the average inductor current $I_{Lavg} = I_L$,

$$I_{load} = (1 - D)I_L. \tag{10.167}$$

This behavior can be modeled by an ideal transformer with a turns ratio $1 : D$, as shown in Fig. 10.35a). Additional components are added to represent the entire power stage of the boost converter completely. R_L is the inductor series resistance. R represents the load (V_{out}/I_{load}).

In an ideal transformer, the ratio of the current I_1 on the primary side to the current I_2 on the secondary side is $I_1 = -I_2/D$ (see inset in Fig. 10.35a)). In our model, the current on the secondary side is the (average) inductor current I_L. Hence, the primary side current is $-DI_L$. According to the transformer connection in Fig. 10.35a), both currents are added and source the load current

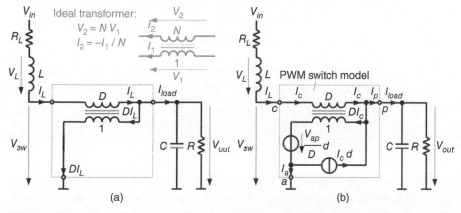

(a)   (b)

Figure 10.35 Boost converter with PWM switch model: a) large-signal model; b) combined large-signal and small-signal model.

of the boost converter. It matches with Eqn. (10.167) and confirms the PWM switch model based on the ideal transformer. We need to check if the model also represents the voltage conversion. The primary-side voltage equals $V_1 = V_{out}$. With a turns ratio $1 : D$, the voltage on the secondary side will be $V_2 = DV_{out}$. Considering the loop equation, the secondary side also needs to fulfill this condition:

$$V_2 = DV_{out} = V_{out} - \overline{V_{sw}} \tag{10.168}$$

By inspection, we find the average switching node voltage $\overline{V_{sw}}$ from Fig. 10.34b),

$$\overline{V_{sw}} = \frac{1}{T} \int_0^T V_{sw}(t)\mathrm{d}t = \frac{t_{off}}{D} V_{out} = (1 - D)V_{out}. \tag{10.169}$$

We can confirm Eqn. (10.168) by inserting $\overline{V_{sw}}$ of Eqn. (10.169) into it. In conclusion, we found a PWM switch model that correctly represents the DC operating point.

As the main benefit of the model, the AC transfer functions can be derived with low effort using small-signal analysis. By *small-signal*, we mean that there is only a slight variation with respect to the DC operating point; for example, the duty cycle D varies by a small amount ΔD. Before we can apply the model, we need to validate it concerning the small-signal behavior. For this purpose, we derive the small-signal representation of Eqn. (10.167) by calculating the partial differential for the two variables D and I_L,

$$\Delta I_{load} = \frac{\partial I_{load}}{\partial D}\Delta D + \frac{\partial I_{load}}{\partial I_L}\Delta I_L. \tag{10.170}$$

For simplification, we introduce lower-case variables d and i_L to denote the Δ-terms of the duty cycle D and the inductor current I_L. This way, Eqn. (10.170) turns into

$$i_{load} = \frac{\partial I_{load}}{\partial D}d + \frac{\partial I_{load}}{\partial I_L}i_L. \tag{10.171}$$

By applying the differentiation on Eqn. (10.167), (10.171) becomes

$$i_{load} = -I_L d + (1 - D)i_L. \tag{10.172}$$

Comparing Eqns. (10.167) and (10.172) reveals that the small-signal expression for i_{load} has one term $-I_L d$ more. We need to add $I_L d$ to the load current Figure 10.35b) contains the three-terminal PWM switch model, enclosed by the dashed box, with its terminals named according to [16]: $a =$ active (switch, the bottom connection of M1), $p =$ passive (switch, the output connection of M2), and $c =$ common (connection between M1 and M2, i.e., the switching node). The additional current source $I_L d = I_c d$ (referring to the current at terminal c of the general switch model) contributes to the load current and creates a correct small-signal representation of I_{load} (i_{load}) given by Eqn (10.172).

As shown in Fig. 10.35b), there is also a voltage source $\frac{V_{ap}}{D}d$ required to represent the transformer's voltage conversion correctly and to match Eqn. (10.168) in the small-signal domain. We obtain this voltage by partially differentiating Eqn. (10.168),

$$v_2 = \frac{\partial V_2}{\partial D}d + \frac{V_2}{\partial V_{out}}v_{out} = V_{out}d + Dv_{out}. \tag{10.173}$$

Again, lower-case symbols represent small-signal parameters (v_2 stands for ΔV_2, etc.). To ensure that the general expression in Eqn. (10.168) matches the small-signal representation in Eqn. (10.173), the transformer's secondary voltage V_2 needs to be reduced by $V_{out}d$. It is equal to subtracting $\frac{V_{ap}}{D}d$ at the primary side as depicted in Fig. 10.35b) with $V_{ap} = -V_{out}$ referring to terminals a and p.

Now, the model is complete. We use the conventions for the terminal names within the PWM model (inside the dashed box). Current I_c equals the (average) inductor current I_L while current I_p corresponds to the load current I_{load}. I_a equals the average current in switch M1 over one switching period. Since M1 only conducts during the on-time, the current at terminal a becomes $I_a = DI_L$.

10.10.11 Steady-State Analysis

The PWM switch model allows for steady-state operating point analysis in an easy way. For steady-state analysis (DC), we set all small-signal quantities to zero, i.e., we can use the large-signal model of Fig. 10.35a). This model allows us to derive the general voltage conversion ratio M alternatively to the approach that results in Eqn. (10.156). As a strength of the model, we can also investigate the influence of parameters like the series resistance R_L of the inductor on the conversion ratio M. To do this, we apply Kirchhoff's voltage law (KVL) on the boost converter model of Fig. 10.35a):

$$-V_{in} + (R_L \cdot I_L) - D \cdot V_{out} + V_{out} = 0. \tag{10.174}$$

In the DC case, the voltage drop across the inductor is zero and can be omitted. We can now express the inductor current I_L in terms of I_{load} using Eqn. (10.167). We further substitute I_{load} by V_{out}/R where R is the equivalent output resistance of the boost converter. Rearranging yields

$$M = \frac{V_{out}}{V_{in}} = \frac{1}{1-D} \frac{1}{\left(1 + \frac{R_L}{R(1-D)^2}\right)}. \tag{10.175}$$

For $R_L = 0$, Eqns. (10.156) and (10.175) are equal. Figure 10.36 plots the voltage conversion ratio M versus duty cycle D according to Eqn. (10.175).

Example 10.16 *We repeat calculating the voltage conversion ratio of Example 10.14 in the presence of an equivalent inductor resistance $R_L = 100\,\text{m}\Omega$. In Example 10.14 the conversion ratio is $M = 2.75$ at a load current of $I_{load} = 1\,\text{A}$.*

With $V_{out} = 3.3\,\text{V}$ we obtain an equivalent load resistance of $R = V_{out}/I_{load} = 3.3\,\Omega$.
With the results from Example 10.14 the duty cycle is

$$D = t_{on} f_{sw} = 636\,\text{ns} \cdot 1\,\text{MHz} = 0.636. \tag{10.176}$$

Figure 10.36 Voltage conversion ratio $M = V_{out}/V_{in}$ of the boost converter as a function of the duty cycle D for different ratios between the resistance R_L of the inductor and the equivalent load resistance $R = V_{out}/I_{load}$. The graph assumes a constant load resistance $R = 10\,\Omega$ and a variable $R_L = 50, 100, 200$, and $500\,\text{m}\Omega$.

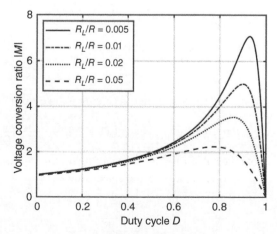

We can now insert R, R_L, and D into Eqn. (10.175),

$$M = \frac{V_{out}}{V_{in}} = \frac{1}{1-D} \frac{1}{\left(1 + \frac{R_L}{R(1-D)^2}\right)}$$

$$= \frac{1}{1-0.636} \frac{1}{\left(1 + \frac{100\,\text{m}\Omega}{3.3\,\Omega\,(1-0.636)^2}\right)} = 2.237. \tag{10.177}$$

Compared to the case for $R_L = 0$ in Example 10.14 (M = 2.75), the conversion ratio has dropped significantly. It is one of the main reasons why the conversion ratio of boost converters is limited. Typical maximum ratios are M = 4-5.

10.10.12 Small-Signal Analysis

Using the model in Fig. 10.35b), we can derive various transfer functions such as the line-to-output, control-to-output behavior, and open-loop input and output impedance. To demonstrate the procedure, we analyze the duty-cycle-to-output transfer function. In this case, the input voltage V_{in} is constant and needs to be set to zero since we perform a small-signal analysis. With this in mind, we define the impedance of the inductor

$$Z_L = R_L + sL, \tag{10.178}$$

and write down the loop equation similar to Eqn. (10.174),

$$Z_L i_L - D v_{out} - V_{out}\, d + v_{out} = 0. \tag{10.179}$$

Solving for the inductor current,

$$i_L = -\frac{v_{out}(1-D) + V_{out}\, d}{Z_L}. \tag{10.180}$$

We will use the expression shortly.

From Fig. 10.35b), we can derive a small-signal expression of the load current,

$$i_{load} = i_L - D i_L + I_L\, d = (1-D)i_L + I_L\, d. \tag{10.181}$$

The average inductor current I_L in steady state can be substituted by Eqn. (10.170). We can further express the load current by the equivalent load resistance, $I_{load} = V_{out}/R = M V_{in}/R$,

$$i_{load} = (1-D)i_L + \frac{M V_{in}}{R(1-D)}\, d. \tag{10.182}$$

Employing the output impedance

$$z_{out} = sC \parallel R = \frac{R}{1+sRC}, \tag{10.183}$$

Equation (10.182) allows to calculate the output voltage,

$$v_{out} = z_{out} i_{load} = z_{out}\left((1-D)i_L + \frac{M V_{in}}{R(1-D)}\, d\right). \tag{10.184}$$

Now we substitute the inductor current i_L in Eqn. (10.184) by Eqn. (10.180) and rearrange to find the duty-cycle-to-output transfer function,

$$\frac{v_{out}}{d} = V_{in}M^2 \frac{\left(1 - \frac{Z_L M^2}{R}\right)}{1 + \frac{Z_L M^2}{z_{out}}}. \tag{10.185}$$

10.10.13 Control-to-Output Transfer Function

It is straightforward to derive the control-to-output transfer function $G(s) = v_{out}/v_c$, knowing the duty-cycle-to-output transfer function, Eqn. (10.185). v_c denotes the compensated (small-signal) error voltage. All we need to do is to multiply Eqn. (10.185) by the modulator's transfer function $d/v_c = 1/V_{s,max}$, which is identical to Eqn. (10.25) as derived in the buck converter section. Please refer to the corresponding considerations regarding $G(s)$ of the buck converter in Section 10.5, see Eqn. (10.39). For the boost case, we use lower-case symbols because we derive all expressions by small-signal analysis. Finally, we obtain

$$G(s) = \frac{v_{out}}{v_c} = \frac{v_{out}}{d} \cdot \frac{d}{v_c} = \frac{v_{out}}{d} \cdot \frac{1}{V_{s,max}} \tag{10.186}$$

with v_{out}/d given in Eqn. (10.185). Inserting the impedances Z_L and Z_{out}, as given by Eqns. (10.178) and (10.183), results in the final expression for the control-to-output transfer function $G(s)$ in VMC,

$$G(s) = \frac{v_{out}}{v_c} = G_o \frac{1 - \frac{s}{\omega_z}}{1 + \frac{s}{\omega_o Q} + \frac{s^2}{\omega_o^2}}, \tag{10.187}$$

where,

$$G_o = \frac{V_{in}}{V_{s,max}(1-D)^2} \frac{1 - \frac{R_L}{R(1-D)^2}}{1 + \frac{R_L}{R(1-D)^2}} \approx \frac{V_{in}}{V_{s,max}(1-D)^2}, \tag{10.188}$$

$$\omega_z = \frac{R(1-D)^2 - R_L}{L} \approx \frac{R}{L}\left(\frac{V_{in}}{V_{out}}\right)^2, \tag{10.189}$$

$$\omega_o = \frac{1}{\sqrt{LC}}\sqrt{\frac{R_L + R(1-D)^2}{R}} \approx \frac{1}{\sqrt{LC}} \cdot \frac{V_{in}}{V_{out}}, \tag{10.190}$$

$$Q = \frac{\omega_o}{\frac{1}{RC} + \frac{R_L}{L}}. \tag{10.191}$$

It is possible to include the equivalent series resistance of the output capacitor in the model of Fig. 10.35b). In that case, a second zero occurs similar to the buck converter as considered in Eqn. (10.41).

$G(s)$ in Eqn. (10.187) is a second-order transfer function. Therefore, a type III compensator will be required as described in Section 10.8.1. Like in the buck converter, in CMC, the inner loop (current loop) eliminates the double-pole formed by L and C. Slope compensation similar to the buck converter may be applied to suppress subharmonic oscillations (see Section 10.6).

10.10.14 Right Half-Plane Zero (RHPZ)

The negative sign in the numerator of the boost converter's VMC control-to-output transfer function $G(s)$ in Eqn. (10.187) reveals that the zero ω_z occurs in the right-half section of the s-plane. The s-plane depicts the locations of the poles and zeros with respect to their real and imaginary coordinates. The same RHPZ occurs in CMC, identical to Eqn. (10.189). A zero in the right half-plane causes a phase lag of $-90°$, similar to a pole. Consequently, an RHPZ harms the dynamic stability of the control loop, in contrast to a regular left half-plane zero, which is usually beneficial because it gives $+90°$ phase shift. An RHPZ is typical for converters with indirect energy transfer (see the comments at the very beginning of this section on boost converters). The interpretation is as follows: The average current that is delivered to the load is proportional to $D' = (1 - D)$ (see Eqn. (10.150)). In other words, energy is delivered to the load only during the off-time t_{off}. Now assume a positive load current step, which causes the output voltage to drop (slightly). Consequently, the compensation voltage V_c at the output of the error amplifier will increase and enlarge the duty cycle D. It is intuitive since D determines the duration of the inductor's energizing cycle (the on-time t_{on}). Larger D will further reduce $(1 - D)$ and, consequently, the current delivered to the output. Hence, V_{out} will drop even further over several switching cycles due to the RHPZ. If the converter maintains stability in the presence of an RHPZ, the current toward the output will eventually go up and correct for the load step. However, this is usually a case that we want to avoid. Since it is impossible to get rid of the RHPZ, the standard approach is to design the crossover frequency f_c of the loop gain to be sufficiently below the RHPZ given by $f_z = \frac{\omega_z}{2\pi}$, i.e.,

$$f_c = 0.1 \ldots 0.3 f_z. \tag{10.192}$$

This approach is similar to Eqn. (10.73). The crossover frequency f_c will be set to the lower value determined from Eqns. (10.73) and (10.192). Unfortunately, as indicated by Eqn. (10.189), $f_z = \frac{\omega_z}{2\pi}$ depends on the load resistance R, i.e., on the operating point, in particular on the load current I_{load}. For low-load currents, the RHPZ will move to higher values, while high load currents negatively influence the stability of the boost converter. Therefore, Compared to the buck converter, the bandwidth of the boost converter is typically much lower.

Example 10.17 *Determine the crossover frequency f_c for the boost converter of Example 10.16. From Eqn. (10.73),*

$$\cdot f_c = 0.1 f_{sw} = 0.1 \cdot 1\,\text{MHz} = 100\,\text{kHz}. \tag{10.193}$$

According to Eqn. (10.189) the RHPZ frequency is

$$\begin{aligned} f_z = \frac{\omega_z}{2\pi} &= \frac{R(1-D)^2 - R_L}{2\pi L} \\ &= \frac{3.3\,\Omega\,(1 - 0.636)^2 - 100\,\text{m}\Omega}{6.28 \cdot 1.9\,\mu\text{F}} = 28.0\,\text{kHz}. \end{aligned} \tag{10.194}$$

Eqn. (10.192) yields

$$f_c = 0.1 f_z = 2.8\,\text{kHz}. \tag{10.195}$$

This crossover frequency is much lower than the result of Eqn. (10.193). We can increase f_c and, with it, the dynamic response of the boost converter by reducing the load current (higher load resistance R). We can also relax the safety margin by setting f_c to $0.3 f_z$. For $I_{load} = 100$ mA (10x higher load resistance R) Eqn. (10.194) gives

$$f_c = 0.3 f_z = 107\,\text{kHz}. \tag{10.196}$$

Now, we can achieve a crossover frequency at around 100 kHz as suggested by Eqn. (10.193).

10.10.15 Discontinuous Conduction Mode (DCM)

In DCM, there will be an idle time after t_{on} and t_{off} in which the inductor current is zero, both switches turn off, and V_{sw} gets pulled to V_{in} by the inductor. Like the buck converter case in Section 10.9, we now associate the off-time with a new duty cycle D_2. The volt-second balance of Eqn. (10.155) holds, and we can express the output voltage as a function of D and D_2

$$V_{out} = V_{in}\frac{D + D_2}{D_2}. \tag{10.197}$$

Repeating the charge balance of Eqn. (10.150) in DCM yields

$$I_{load} = \frac{V_{out}}{R} = \frac{V_{in}DD_2 T}{2L}, \tag{10.198}$$

with resistance R representing the load. Solving Eqns. (10.197) and (10.198) for D_2 and setting the two equations equal allows deriving the voltage conversion ratio of the boost converter in DCM,

$$M = \frac{V_{out}}{V_{in}} = \frac{1}{2}\left(1 + \sqrt{1 + \frac{2}{L}D^2 TR}\right) \approx D\sqrt{\frac{TR}{2L}}. \tag{10.199}$$

Similar to the buck converter, the conversion ratio in DCM depends on the load current expressed by the load resistance $R = V_{out}/I_{load}$. For the same duty cycle D, the conversion ratio in DCM can be larger than in CCM.

The control-to-output transfer function of the boost converter in DCM is given by

$$G(s) = \frac{v_{out}}{v_c} = G_0 \frac{1}{1 + \frac{s}{\omega_p}}, \tag{10.200}$$

where,

$$G_o = 2\frac{V_{out}}{DV_{s,max}}\frac{M-1}{2M-1} \approx \frac{V_{out}}{DV_{s,max}}, \tag{10.201}$$

and

$$\omega_p = \frac{2M-1}{(M-1)RC} \approx \frac{2}{RC}. \tag{10.202}$$

The approximations assume $M \gg 1$. Unlike in the CCM case (see Eqn. (10.187)), the control-to-output transfer function is a first-order expression. Therefore, the boost converter in DCM can be compensated much easier by a type II compensator. See Section 10.8 for details.

10.11 The Buck-Boost Converter

Figure 10.37 shows the inverting buck-boost DC–DC converter. In addition to the buck converter (Fig. 10.1) and the boost converter (Fig. 10.34), it represents the third option of placing the inductor, in this case, between the switching node and ground. The converter does not benefit from the

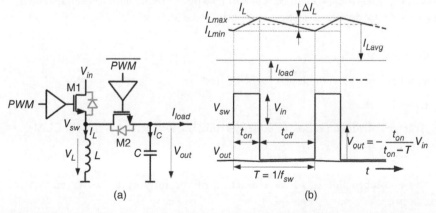

Figure 10.37 a) The inverting buck-boost converter with b) typical waveforms of the inductor and load currents I_L, I_{load}, and voltages V_{sw}, V_{out}.

filtering property of the inductor because L is neither connected to the input nor the output. Therefore, the inverting buck-boost converter usually requires larger input and output buffer capacitors.

During the on-time, transistor M1 connects the switching node to V_{in} as illustrated in Fig. 10.37b). The inductor gets energized, and its current $I_L(t)$ ramps up. Afterwards, M1 turns off, and M2 turns on for the remainder of the switching period $T = 1/f_{sw}$. There needs to be a short dead time at the transition to avoid cross-conduction between input and output of the converter via M1 and M2. During t_{off}, the inductor current I_L pulls the switching node potential below ground, and the converter maintains a negative output voltage V_{out}. In the steady state, the switching node voltage is limited by the negative output voltage stored on the output capacitor C.

Since the output voltage is always *negative* with respect to the input voltage, the topology is called *inverting* buck-boost DC–DC converter. But why *buck-boost*? Because the output voltage can be either between zero and $-V_{in}$ (inverting buck: $|V_{out}| \leq V_{in}$) or lower than $-V_{in}$ (inverting boost: $|V_{out}| \geq V_{in}$). Like for the buck and the boost converter, the level of V_{out} is determined by PWM. Only during t_{off} (M2 turned on) current is delivered to the output of the DC–DC converter. That is why the buck-boost converter belongs to the class of topologies with indirect energy transfer, like the boost converter.

Since the output voltage is always negative, only a few applications need an inverting buck-boost converter. Instead, the four-switch buck-boost converter (covered further below) is more popular because of its positive output voltage. However, the flyback converter (Section 10.12), widely used in state-of-the-art off-line power supplies, comprises the inverting buck-boost at its secondary side.

10.11.1 Inductor Current

By inspection, we find from Fig. 10.37 that the load current is inverted to the inductor current. Applying the charge balance,

$$I_{load} T = \int_0^{t_{off}} -I_L(t)\mathrm{d}t = -I_{Lavg}\, t_{off} \rightarrow I_{Lavg} = -\frac{I_{load}}{D'}. \qquad (10.203)$$

Except for the sign, this result is identical to Eqn. (10.150) of the boost converter. Note that I_{load} will be negative since $V_{out} < 0$.

10.11.2 Asynchronous Rectification

For asynchronous rectification, switch M2 can be replaced by a power diode with the same polarity as M2's body diode (see also the details in Section 10.1 and Fig. 10.3).

10.11.3 Voltage Conversion Ratio

Based on the volt-second balance given by Eqn. (10.5), we find

$$V_{in}\, t_{on} - V_{out}\, t_{off} = 0, \qquad (10.204)$$

which results in the voltage conversion ratio M of the inverting buck-boost converter (in CCM),

$$M = \frac{V_{out}}{V_{in}} = -\frac{D}{1-D}. \qquad (10.205)$$

Figure 10.38 plots the absolute value of M as a function of the duty cycle D. We can solve Eqn. (10.205) for the duty cycle:

$$D = \frac{t_{on}}{T} = \frac{V_{out}}{V_{out} - V_{in}} = \frac{|V_{out}|}{|V_{out}| + V_{in}} \qquad (10.206)$$

Similar to Eqn. (10.175) of the boost converter, the equivalent load resistance and the inductor resistance can be included. In that case, the maximum absolute value of M will be limited and drops if D increases further (similar to Figure 10.36 for the boost converter).

10.11.4 On-/Off Times

For a given current ripple ΔI_L the on-/off-times are easily found,

$$t_{on} = DT = \Delta I_L \frac{L}{V_{in}}, \tag{10.207}$$

$$t_{off} = (1 - D)T = \Delta I_L \frac{L}{-V_{out}}. \tag{10.208}$$

10.11.5 Inductor Sizing

Adding both equations (10.207) and (10.208), we find $T = 1/f_{sw} = t_{on} + t_{off}$, which can be solved for the inductor current ripple ΔI_L and, further on, yield the sizing equation of the inductor,

$$L = \frac{V_{in} V_{out}}{V_{out} - V_{in}} \frac{1}{\Delta I_L f_{sw}} = \frac{V_{in} D}{\Delta I_L f_{sw}}. \tag{10.209}$$

10.11.6 Output Voltage Ripple and Capacitor Sizing

The output current relationship is exactly like for the boost converter (see Eqn. (10.203)). The only exception is the negative sign, which can be ignored since we are only interested in the ripple magnitude. Consequently, the output voltage ripple and the capacitor sizing equations are identical to the boost case as given in Eqns. (10.164) and (10.165), repeated here for convenience,

$$\Delta V_{out} = \frac{I_{load} D}{f_{sw} C}, \tag{10.210}$$

$$C = \frac{I_{load} D}{f_{sw} \Delta V_{out}}. \tag{10.211}$$

10.11.7 Discontinuous Conduction Mode (DCM)

As for the buck and boost converter, the voltage conversion ratio of the buck-boost converter changes when operating in DCM. As defined in Fig. 10.30 for the buck, there will be a third phase in addition to t_{on} and t_{off} in which the inductor current is zero and the switching node goes to ground potential (0 V). We define a duty cycle D_2 corresponding to the off-time t_{off} in DCM. The volt-second balance changes to

$$D V_{in} + D_2 V_{out} = 0. \tag{10.212}$$

We can derive D_2 from the charge balance during the off-time. As $I_L(t)$ has a triangular shape, the integral can be easily determined:

$$\frac{\Delta I_L D_2 T}{2} = I_{load} T \tag{10.213}$$

We can express the load current by the equivalent output resistance $I_{load} = V_{out}/R$ and substitute ΔI_L by Eqn. (10.207). Solving for D_2 gives

$$D_2 = \frac{t_{off}}{T} = \frac{2 V_{out} L}{R D V_{in} T}. \tag{10.214}$$

Inserting D_2 of Eqn. (10.214) into Eqn. (10.212) and rearranging yields

$$D^2 V_{in}^2 = -\frac{2 V_{out}^2 L}{R T}. \tag{10.215}$$

In the final step, we extract the voltage conversion ratio of the inverting buck-boost converter in DCM,

$$M = \frac{V_{out}}{V_{in}} = -D \sqrt{\frac{R T}{2 L}}. \tag{10.216}$$

In comparison to CCM (see Eqn. (10.205)), the conversion ratio in DCM depends on the load current (because of the equivalent load resistance $R = V_{out}/I_{load}$). At a given duty cycle D, the absolute conversion ratio in DCM can be higher than in CCM. Interestingly, M is linearly dependent on the duty cycle D like for the buck converter in CCM (see Eqn. (10.8)).

Example 10.18 *We investigate the voltage conversion ratio in CCM and DCM depending on the duty cycle D. We also consider two load cases: $I_{load} = 10$ mA and 100 mA. Given parameters: $V_{out} = -3$ V, $L = 22\,\mu H, f_{sw} = 1$ MHz.*

In CCM, the voltage conversion ratio M depends only on the duty cycle as given by Eqn. (10.205). We can directly plot it, as shown in Fig. 10.38. For convenience, we show the absolute value $|M|$. The curve is highly nonlinear. At $D = 0.6$, we can read off $M = -1.5$. In order to achieve $V_{out} = -3$ V we require the input voltage to be $V_{in} = 2$ V, see Eqn. (10.205). Note that we do not include the equivalent load and inductor resistors. They would cause $|M|$ to decrease with D after reaching a maximum value.

In DCM, Eqn. (10.216) defines the conversion ratio. We need to determine the equivalent load resistance $R = -V_{out}/I_{load}$ for the load cases, which gives $R = 30$ and $300\,\Omega$. For the two cases, Fig. 10.38 shows $|M|$ versus duty cycle D. Unlike in CCM, the curves are linear. For the same duty cycle, $|M|$ is greater compared to CCM for $I_{load} = 10$ mA and, vice versa, $|M|$ is lower for $I_{load} = 100$ mA. At $D = 0.6$, the conversion ratios are -2.1 and -0.67, respectively.

10.11.8 Control-to-Output Transfer Function

The inverting buck-boost converter can be controlled by applying VMC and CMC similar to the buck converter described in Sections 10.5 and 10.6. Due to the similarity to the boost converter, the control-to-output transfer function has the same form as Eqn. (10.187). The DC gain G_o, the second-order pole ω_o, and the quality factor Q are identical; see Eqns. (10.188) to (10.191).

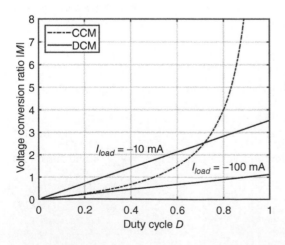

Figure 10.38 Absolute value $|M|$ of the voltage conversion ratio $M = V_{out}/V_{in}$ of the inverting buck-boost converter as a function of the duty cycle D in continuous (CCM) and discontinuous conduction mode (DCM). DCM is shown for two different load currents, $I_{load} = -10$ mA and -100 mA. The remaining parameters are given in Example 10.18.

Only the RHPZ is different. For the inverting buck-boost converter in CCM, it is given by

$$\omega_z = \frac{R(1-D)^2}{LD} \approx \frac{RD}{L}\left(\frac{V_{in}}{V_{out}}\right)^2.\tag{10.217}$$

In comparison to ω_z of the boost converter given in Eqn. (10.189), we multiply by the duty cycle D. Because $D \leq 1$, the RHPZ of the buck-boost converter will occur at even lower frequencies than in the boost case.

In DCM, the control-to-output transfer function is identical to the one-pole expression of the boost convert as given in Eqn. (10.200). The parameters G_o and ω_p are the same, and no RHPZ occurs.

Example 10.19 *Determine the RHPZ-frequency f_z of an inverting buck-boost converter with these parameters: $R = V_{out}/I_{load} = 3.3\,\Omega$, $L = 1.9\,\mu H$, $D = 0.636$. Compare the result with the RHPZ of the boost converter in Example 10.17. We neglect the inductor's series resistance R_L.*

We can directly insert all parameters into Eqn. (10.217) and determine the RHPZ,

$$f_z = \frac{\omega_z}{2\pi} = \frac{R(1-D)^2}{2\pi LD} = \frac{3.3\,\Omega \cdot (1-0.636)^2}{2.28 \cdot 1.9\,\mu H \cdot 0.636} = 9.1\,kHz.\tag{10.218}$$

The RHPZ of the boost converter in Example 10.17 gives $f_z = 28.0\,kHz$ for the same parameters (but including $R_L = 100\,m\Omega$). In conclusion, the RHPZ frequency of the buck-boost converter is approximately three times lower compared to the boost converter. It forces us to choose a lower crossover frequency f_c.

10.11.9 The Non-Inverting Buck-Boost Converter

The inverting buck-boost converter provides a negative output voltage, which makes the topology unsuitable for many applications. The four-switch buck-boost DC–DC converter shown in Fig. 10.39a) forms an alternative converter that generates a positive output voltage. Therefore,

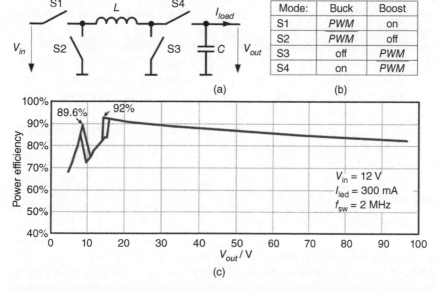

Figure 10.39 The four-switch buck-boost converter: a) topology; b) control signals of switches S1 to S4 in buck and boost operating mode; c) measured efficiency versus V_{out} of the converter. Source: Adapted from [17].

this topology is also called a non-inverting buck-boost converter. It combines a single-stage buck and boost converter utilizing the same inductor in the center. It can either operate in buck mode ($V_{out} \leq V_{in}$) or in boost mode ($V_{out} \geq V_{in}$) depending on the switch control signals, as shown in Fig. 10.39b). Therefore, all the abovementioned characteristics discussed in the buck and the boost converter sections also apply here.

In either mode, there is always one additional switch in series (S4 in buck and S1 in boost mode), contributing additional conduction losses. Hence, the four-switch buck-boost topology typically has lower conversion efficiency than a single-stage buck or boost converter. As an example, Fig. 10.39c) shows a measured efficiency curve of a GaN-based four-switch buck-boost LED driver presented in [17].

The most difficult design challenge is usually to achieve stable operation in the crossover region between buck and boost mode. The crossover point corresponds to the condition $V_{in} = V_{out}$. A common practice is alternating between buck and boost mode after a few switching cycles. It causes additional losses such that the conversion efficiency is worse in the crossover region. However, this usually differs from a typical buck-boost converter operating point. In Fig. 10.39c), the crossover point is at $V_{in} = V_{out} = 12\,\text{V}$. The efficiency drop within the range of 9–15 V is visible. Some power management designs decide on alternative topologies to avoid issues related to the crossover region. Specifically, the SEPIC converter (single-ended primary inductance converter) provides a positive output voltage and shows a continuous transition between buck and boost mode. It comes at the expense of additional passive components in the power path. See, for instance, [3] for details.

10.12 The Flyback Converter

The flyback converter shown in Fig. 10.40 provides an isolated DC–DC conversion from V_{in} to V_{out}. In a typical application, the flyback converter can be connected to a DC-link voltage of $V_{in} = 400\,\text{V}$. Combined with a rectifier stage (see Section 10.13), an AC–DC converter is formed (V_{ac} to V_{out}). For this reason, the flyback converter is a popular off-line power supply topology for power levels in the 50 W range and above, especially in chargers and adapters for laptops and tablets, smartphones, drones, game consoles, etc.

In addition to a transformer, the main components are the power transistor M1 on the primary side (connected to the primary winding of the transformer) and the diode D on the secondary side. The clamping network between the drain of M1 and V_{in} plays an essential role, as we will discuss further below. M1 is controlled by PWM depending on the output voltage V_{out} at the secondary side. All control circuitry is placed on the primary side. Therefore, signal isolation transfers V_{out} from the secondary side back to the primary side. It can be an opto-coupler, a signal transformer, or a capacitive signal isolator. In addition to the details covered below, [3] represents a valuable source on a broad range of topics related to the flyback converter.

10.12.1 The Transformer

The transformer provides galvanic isolation required for safety in high-voltage off-line power converters (V_{ac} typically connected to the 110/230 V grid). Isolation will also be beneficial for noise immunity. The transformer has a dot on one end of each winding; see Fig. 10.40. These dotted ends are mutually equivalent. If the voltage rises at one dotted end, it also goes up at the other dotted connection. The same applies to the non-dotted ends. The transformer is modeled on the primary side by its magnetizing inductance L_m and its leakage inductance L_{lk}. There is also a leakage

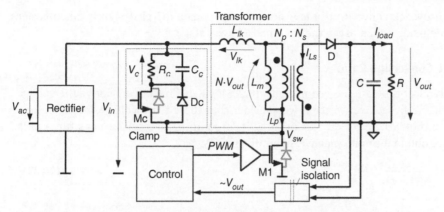

Figure 10.40 The flyback converter: The transformer provides isolated DC–DC conversion based on the concept of the inverting buck-boost. A rectifier in front turns the flyback topology into an AC–DC converter for off-line power adapters.

inductance on the secondary side, which is negligible for understanding the general operation. However, it can be significant and may need to be included in detailed calculations. The primary magnetizing inductance L_m accounts for the fact that some energy is required to magnetize the magnetic material (represented by a current that flows through L_m in parallel to the primary winding of the ideal transformer). The leakage inductance L_{lk} relates to the magnetic flux that is not fully linked to the primary and secondary winding. The transformer's turns ratio N is defined by

$$N = \frac{N_p}{N_s},$$
(10.219)

where N_p and N_s are the number of primary and secondary side turns, respectively. Since $V_{in} > V_{out}$, we usually have $N > 1$. From the fundamental transformer relationships, we also know the ratio between the primary and secondary side magnetizing inductance L_p, L_s,

$$L_s = L_p \left(\frac{N_s}{N_p} \right)^2 = \frac{L_p}{N^2}.$$
(10.220)

10.12.2 Fundamental Operation

The power stage operates like in the inverting buck-boost converter as described in Section 10.11, except that a transformer replaces the inductor. However, the flyback topology of Fig. 10.40 eliminates the inherent voltage inversion of the *inverting* buck-boost topology because the dotted ends are connected oppositely. When M1 conducts during the on-time t_{on}, the non-dotted end of the primary side gets pulled to ground. Consequently, the corresponding end at the secondary side also goes low. The diode D blocks, and the load current $I_{load} = V_{out}/R$ (with the equivalent load resistance R) is delivered exclusively by the buffer capacitor C. During t_{on}, the primary inductance gets energized from V_{in}, and the inductor current reaches its maximum I_{Lmax}. As soon as M1 turns off, the drain voltage of M1 will instantaneously rise because the primary-side inductor current I_{Lp} continues to flow into the high-impedance drain node of M1. It means the non-dotted node at the secondary side will see a steep positive voltage swing. Diode D gets forward-biased. Its current delivers I_{load} and recharges C. In other words, due to the transformer polarity, the inverting behavior of the conventional buck-boost converter is eliminated, and the voltage conversion ratio gets a positive sign. Moreover, compared to the inverting buck-boost converter of Fig. 10.37, we see that M1 in the flyback converter is a low-side switch. Its gate driver design is much easier compared to the high-side switch M1 in the buck-boost power stage.

10.12.3 Voltage Conversion Ratio

Based on the buck-boost expression given in Eqn. (10.205), we can intuitively derive the voltage conversion ratio of the flyback converter if we consider that there is no voltage inversion anymore. Omitting the minus sign in Eqn. (10.205) and incorporating the transformer's turns ratio of Eqn. (10.219), we obtain the voltage conversion ratio in CCM,

$$M = \frac{V_{out}}{V_{in}} = \frac{D}{N(1-D)}. \tag{10.221}$$

In DCM, we find

$$M = \frac{V_{out}}{V_{in}} = D\sqrt{\frac{RT}{2L_p}}. \tag{10.222}$$

where L_p denotes the primary-side magnetizing inductance.

10.12.4 Control-to-Output Transfer Function

Due to the similarity, the transfer function is identical to the inverting buck-boost converter and, in turn, similar to the boost converter in CCM as given by Eqns. (10.187) to (10.191) with the RHPZ according to Eqn. (10.217). In DCM, Eqns. (10.200) to (10.202) of the boost converter also apply to the flyback converter. However, there are two exceptions in both CCM and DCM:

(1) The inductance L corresponds to the secondary side magnetizing inductance L_s of the transformer. Based on Eqn. (10.220), we can substitute L_s by the primary side magnetizing inductance, $L_s = L_p/N^2$.
(2) The DC gain G_o has to be divided by the turns ratio N of the transformer.

10.12.5 Control

The flyback converter is preferably operated in DCM because of easier frequency compensation and better light-load performance (see Section 10.9).

VMC or CMC can control the entire flyback converter. In VMC, for instance, the control block in Fig. 10.40 is identical to $H(s)$ in Fig. 10.5 plus the pulse-width modulator block. A signal proportionally to V_{out}, provided by the signal isolator, feeds into the controller. The PWM signal at the controller output controls the gate of transistor M1.

A clamp circuit is usually implemented as a snubber that limits the voltage spike at the drain of the power switch M1 when it turns off. It is mainly affected by the transformer's leakage inductance L_{lk}. The stray inductance of interconnects and the printed circuit board (PCB) usually occurs in series to L_{lk}, increasing the effective parasitic inductance. There are two kinds of clamps, resulting in a passive clamp flyback (PCF) and an active clamp flyback converter (ACF).

10.12.6 The Passive Clamp Flyback Converter (PCF)

This topology uses a passive clamp circuit on the primary side consisting of a parallel connection of a resistor R_c and a capacitor C_c in series with diode Dc, as shown in Fig. 10.40, also called RCD clamp. The transistor Mc is not used (it is only present in the ACF to replace the diode DC; see Section 10.12.7). Consider the waveforms in Fig. 10.41a) for the typical case of DCM operation.

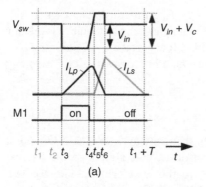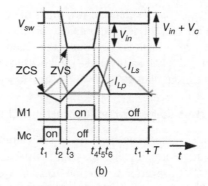

Figure 10.41 Transient waveforms of a) the passive (PCF) and b) the active clamp flyback (ACF) converter. The ACF converter incorporates zero-voltage switching (ZVS) and zero-current switching (ZCS) and recycles some of the transformer's leakage energy between t_1 and t_2.

Refer to Section 10.9 for a description of the general behavior in DCM. Between t_1 and t_3, the currents in both L_p and L_s are zero. Because there is no energy stored in any inductance, the switching node voltage V_{sw} (the drain node of M1) is equal to V_{in}. We can ignore t_2 as it is only relevant for the ACF. At t_3, the power switch M1 turns on and pulls V_{sw} to zero. Switching losses occur, which is one of the drawbacks compared to the ACF topology. The magnetizing inductance L_m gets energized while the current I_{Lp} ramps up. On the secondary side, the diode D is reverse biased such that $I_{Ls} = 0$. The PWM control turns off M1 at t_4 if I_{Lp} has reached its peak value. I_{Lp} continuous to flow and pulls up V_{sw}, which exceeds V_{in}. The passive clamp limits the switching node voltage to $V_{in} + V_c$ (neglecting the forward drop of diode Dc), reaching t_5. During clamping, the diode D forms a freewheeling path through the passive clamp such that I_{Lp} charges the capacitor C_c in the clamp. For the understanding of the clamping function, it is helpful to consider the voltage V_c at capacitor C_c in the clamp as a constant voltage source. It can be adjusted by R_c.

During clamping, the magnetic coupling to the secondary side limits the voltage across the primary winding to $N \cdot V_{out}$. Hence, V_{sw} settles at $V_{in} + V_c = V_{in} + N \cdot V_{out} + V_{clp}$ where N is the turns ratio according to Eqn. (10.219) and V_{clp} defines a clamping voltage level. This maximum switching-node voltage (plus some margin) needs to be handled by M1 without violating its safe operating area (SOA) (see Section 3.4). Lower maximum voltages at the switching node allow higher turns ratios N such that the voltage ratings on the secondary side can be lowered. Alternatively, a transistor type with lower maximum ratings can be used for M1.

Due to the negative voltage across the primary side of the transformer, I_{Lp} ramps down after t_5 and reaches zero at t_6 when V_{sw} falls to V_{in}. In parallel, the secondary-side current I_{Ls} ramps up with its magnitude defined by the turns ratio. At t_6, I_{Ls} starts ramping down again and reaches zero at the end of the switching cycle ($t = t_1 + T$).

We can estimate the losses P_c in the clamp by considering the loop formed by the transformer's primary side and the clamp:

$$V_{lk} + N \cdot V_{out} - V_c = 0 \tag{10.223}$$

The ratio of $P_c = V_c \cdot I_{Lp}$ to the power $P_{lk} = V_{lk} \cdot I_{Lp}$ in the leakage inductance L_{lk} is equal to the ratio V_c/V_{lk}. With V_{lk} from Eqn. (10.223) we obtain

$$P_c = P_{lk} \frac{V_c}{V_{lk}} = P_{lk} \frac{V_c}{V_c - N \cdot V_{out}} = \frac{1}{2} I_{Lp}^2 L_{lk} f_{sw} \frac{V_c}{V_c - N \cdot V_{out}}. \tag{10.224}$$

For $V_c = 2 \cdot N \cdot V_{out}$, P_c corresponds to two times the energy $\frac{1}{2} I_{Lp}^2 L_{lk}$ stored in the leakage inductance. Note that Eqn. (10.224) is an approximation assuming the magnetizing inductance is greater than the total leakage inductance (including the secondary side). It also considers C_c and the output buffer capacitor C a constant voltage source. Together with all other losses, the PCF topology can hardly achieve more than 90% conversion efficiency.

10.12.7 The Active Clamp Flyback Converter (ACF)

The ACF topology replaces the diode D in the clamp with an *active* transistor Mc, and there is no resistor R_c anymore (see Fig. 10.40) [3, 18]. The operation is quite similar to the passive clamp flyback (PCF) converter in Section 10.12.6. We refer to the waveforms in Fig. 10.41b), which includes the gate control signal of Mc. Like in the PCF case, the cycle starts with zero current in both transformer windings. At t_1, Mc turns on. Note that Mc benefits from zero-current switching (ZCS), which reduces its switching losses (see Section 2.9) and, in particular, improves the light-load efficiency [19]. As for the PCF case, we can assume V_c as a constant voltage source. Hence, V_{sw} reaches $V_{in} + V_c$ such that L_m sees a negative voltage, and the current I_{Lp} goes negative. As a major advantage over the PCF converter, a fraction of the leakage energy of the transformer, which is stored on C_c (between t_5 and t_6), is recycled because it gets transferred to V_{out} and V_{in} while Mc is conducting and improves the power efficiency. Mc turns off at t_2 at sufficient negative current to bring the switching node voltage V_{sw} down to zero because I_{Lp} commutates to the main power switch M1 and discharges its parasitic drain-source capacitance. We can achieve zero-voltage switching (ZVS) of M1 when turning it on at t_3 ($V_{sw} = 0$).

Beyond t_3, the behavior is nearly identical to the PCF converter. During clamping, between t_5 and t_6, the body diode of Mc forms the freewheeling path between V_{sw} and the clamping network such that I_{Lp} charges C_c. In other words, C_c stores the leakage energy of L_{lk}, and effectively suppresses ringing of the switching node voltage V_{sw}. The leakage energy gets recycled between t_1 and t_2 when L_m gets energized and a negative current I_{Lp} builds up.

ZVS of M1 at t_3 can be achieved by the sensing circuits presented in Section 6.7. This information can be used to control the turn-off instant of Mc. In steady state, the integral of I_{Lp} between t_1 and t_2 needs to be equal to the charge ($\propto C_{oss}$ of M1) required to bring the switching node from its maximum level at approximately $V_{in} + V_c$ down to zero.

In conclusion, there are two main advantages of the ACF approach over the conventional PCF: (1) higher efficiency due to ZVS (and ZCS), recycling of leakage energy, (2) Reduced ringing improves electromagnetic interference (EMI) (see Section 12.3) and parasitic coupling to sensitive circuits. This way, the ACF topology allows expanding the power range from below 50 W (PCF) toward 150 W. On the other hand, the precise control of M1 and Mc is one of the main challenges in ACF converters. Advanced power management ICs provide precise ZVS control and improvements to recycle the transformer's leakage energy efficiently [20]. High-voltage IC technologies allow the implementation of ACF converters with fully integrated power stages and controls. The design in [19] supports input voltages up to 300 V and achieves power levels of 500 mW at switching frequencies up to 1 MHz.

10.13 Rectifier Circuits

A rectifier has to be inserted as a preceding stage in front of the DC–DC converter to process AC input voltages. See, for instance, the flyback converter in Fig. 10.40 with an initial rectifier stage. The most straightforward solution would be to connect the AC input via a diode. This way, every second half-wave would charge the buffer capacitor at the input of the DC–DC converter. It works

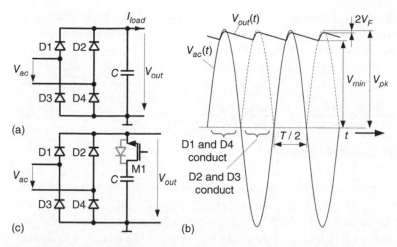

Figure 10.42 Full-wave rectifier: a) schematic; b) waveforms; c) rectifier with active zero-crossing buffer. Source: Adapted from [22].

but requires a large buffer capacitor because the energy is delivered in relatively long time steps. The well-established full-wave rectifier in Fig. 10.42a) transfers energy at every sine half-wave from the AC input voltage V_{ac} to the buffer capacitor C. Four diodes, D1 to D4, are required. They conducted in pairs. Figure 10.42b) illustrates the operation. During the positive half-wave of V_{ac}, D1 and D4 get forward biased and form a closed loop that connects V_{ac} with C. Subsequently, D2 and D3 conduct during the negative half-wave of the AC input voltage V_{ac}. The diode pairs conduct when $|V_{ac}|$ exceeds the minimum output voltage (buffered by C) by more than two times the forward voltage V_F of the diodes. The diode conduction and recharging of C stops at $V_{ac} = V_{pk}$. Due to the diode forward voltage drop, the peak voltage of V_{out} is reduced by $2V_F$ with respect to V_{pk}, as indicated in Fig. 10.42b). All four diodes must block the AC input peak voltage V_{pk} and be rated accordingly. On-chip integration is challenging because parasitic bipolar effects need to be suppressed (see Section 3.3.1). Silicon-on-insulator (SOI) technologies provide a clear advantage for the on-chip integration of high-voltage interfaces [21] (see Section 3.2.3).

10.13.1 The Buffer Capacitor

While delivering the load current, the capacitor discharges from $V_{pk} - 2V_F$ to V_{min}. The latter is the minimum input voltage of the subsequent DC–DC converter. We can neglect the diode forward voltage for high-voltage applications such as off-line converters and assume the maximum capacitor voltage to be equal to V_{pk}. The energy ΔE delivered by the capacitor when discharged from V_{pk} to V_{min} is

$$\Delta E = \frac{C}{2} \left(V_{pk}^2 - V_{min}^2 \right). \tag{10.225}$$

The maximum storing time corresponds to half of the sine period (the time between two successive peaks of the rectified sine wave), see Fig. 10.42b). For a constant output power P_{out}, we can substitute the energy in Eqn. (10.225) by $\Delta E = P_{out} \frac{T}{2}$. Solving for C gives its sizing equation,

$$C = \frac{P_{out} \, T}{V_{pk}^2 - V_{min}^2}. \tag{10.226}$$

Note that this is a worst-case calculation. As V_{min} reduces and approaches zero, the storing time gets $T/4$, and C can become half the size.

Example 10.20 *What is the required value of C if the full-wave rectifier is used in an off-line converter at the 230 V/50 Hz AC grid? The minimum output voltage is $V_{min} = 100$ V. We consider two cases, $P_{out} = 10$ W and a relatively low power of $P_{out} = 50$ mW.*

The peak voltage of the 230 V grid is $V_{pk} = \sqrt{2} \cdot 230$ V $= 325$ V. Further, $T = 1/50$ Hz $= 20$ ms. For $P_{out} = 10$ W Eqn. (10.226) gives

$$C = \frac{P_{out}\, T}{V_{pk}^2 - V_{min}^2} = \frac{10\,\text{W} \cdot 20\,\text{ms}}{(325\,\text{V})^2 - (100\,\text{V})^2} = \frac{10\,\text{VA} \cdot 20 \times 10^{-3}\,\text{s}}{105.625\,\text{V}^2 - 10.000\,\text{V}^2} = 2.09\,\mu\text{F}. \tag{10.227}$$

For $P_{out} = 50$ mW the same calculation gives a much lower value of $C = 10.5$ nF.

There are two ways to reduce the size of the buffer capacitor further. We can lower the minimum input voltage and also reduce the storing time. In [22], V_{min} of the DC–DC converter that is connected to the rectifier is reduced to 12.5V. A significantly smaller storing time is achieved by introducing an active zero crossing buffer, as shown in Fig. 10.42c). The capacitor C charges during the rising transition of V_{out} via the body diode of M1 until V_{out} reaches its maximum. Afterward, the body diode blocks, and M1 remains off while V_{out} drops. M1 turns on when V_{out} reaches $V_{min} = 12.5$V. Now, the energy stored on C is released to buffer the zero crossing of the AC input voltage. Consequently, V_{out} experiences a short spike with a magnitude of approximately V_{pk}. M1 turns off at $V_{ac} = 150$V. This concept reduces the buffering time to ∼0.25 ms. If we repeat Example 10.20 for this low storing time in combination with $V_{min} = 12.5$V we obtain $C = 0.65$ nF at $P_{out} = 50$ mW. In [22], this capacitor is fully integrated on-chip.

10.13.2 Low-Voltage Rectifier Circuits

For low-voltage AC sources, the forward voltage drop of the rectifier diode in Fig. 10.42 significantly influences the functionality and conversion efficiency. It applies, in particular, to energy harvesting front-ends for AC sources from mechanical/vibration and RF inputs. The underlying idea is to replace all diodes with transistors, which result in much lower voltage drops if activated.

Figure 10.43 shows rectifier circuits suitable for low-voltage operation. The CMOS full-wave rectifier is a transistor-based version of Fig. 10.42a). Based on the early work in [24], it has been used in [25] and [26] for vibration energy harvesting. Large W/L-ratios of the transistors minimize the voltage drop and power consumption. For example, [25] uses $W/L \approx 750$. The threshold voltages of the transistors determine the start-up voltage, i.e., the minimum amplitude of the AC input voltage.

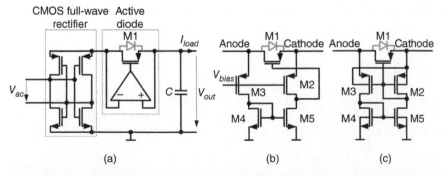

Figure 10.43 Low-voltage rectifier circuits: a) CMOS full-wave rectifier followed by an active diode; b) zero-bias active diode (M1) with low-power amplifier; c) ultra-low power version of the active diode [23].

The full-wave rectifier in Fig. 10.42a) cannot block reverse current (from V_{out} back to the source). Therefore, an active diode is implemented as a second stage. It operates like the body diode of M1, but the PMOS transistor M1 makes it an *active* diode with minimum voltage drop and power dissipation. The gate of M1 is controlled by a differential stage, which acts as a comparator. If the AC input is larger than V_{out}, the gate voltage of M1 is low, and the transistor turns on. Hence, it shorts the body diode, and the AC source can charge capacitor C with the minimum forward voltage drop across M1. If the AC voltage gets lower than V_{out}, the differential stage pulls the gate of M1 to high potential. Consequently, M1 turns off, and its diode voltage blocks.

There are various ways to implement the differential stage. Especially in energy harvesting, it needs to be supplied from the only available energy sources, the AC input and the output buffer capacitor C (that is, in turn, also charged from the AC source). For this reason, common-gate gain stages are widely used. Figure 10.42b) shows a conceptual circuit. Maurath et al. [27] provides an implementation example, including start-up and self-biasing. An ultra-low power version of the active diode is shown in Fig. 10.42c) [23]. It is fully self-biased from both the anode and the cathode. The control current I_C is a fraction of the diode current I_D. Depending on the sizing of M1 to M3, I_C can be as low as 0.5% of I_D. Note that the bulk of M3 is biased from the cathode. For low diode currents I_D, M1 to M3 operate in weak inversion. Forward voltages as low as 20 mV can be achieved. As a primary benefit of this circuit, the control current is zero during the blocking phase.

10.14 Multi-phase Converters

The multiphase (MP) interleaved DC–DC converter topology overcomes the limitations of conventional single-phase converters regarding power efficiency and switching frequencies. Figure 10.44 shows a typical MP buck converter in VMC. Multiple power stages (1 to n) connect to one single output V_{out}. Each phase comprises high-side and low-side power switches, including

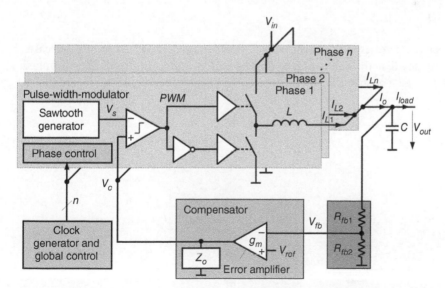

Figure 10.44 A multiphase buck converter in voltage-mode control: n power stages connect to a common output voltage V_{out}. They are controlled in a time-interleaved manner using one central compensator. Each stage gets its separate clock and control signals. Per phase, the PWM signal is generated by comparing the global compensation voltage V_c with a sawtooth voltage V_s.

gate control and one power inductor. The average phase current is I_{load}/n. MP converters operate in a time-interleaved way and combine high output power with a fast transient response. Because of these advantages, MP converters have been widely used in (integrated) voltage regulators, particularly for compute power delivery (see Section 12.8).

At higher frequencies, some MP converters utilize a delay-locked loop to synchronize the phases automatically, as demonstrated in [28] at $f_{sw} = 25 - 70$ MHz. Current-mode PWM control (CMC) would require a high-bandwidth current sensor (discussed, for instance, in [29]). Therefore, many MP converters implemented hysteretic control [30–34] and VMC similar to Fig. 10.44 [29, 35, 36]. Nevertheless, at moderate switching frequencies, few designs implement CMC like [37] (operating at 2.25 MHz).

10.14.1 Ripple Cancellation

The main advantage of the MP topology is the current ripple cancellation between the different phase currents. Figure 10.45 shows the inductor currents in each phase and the resulting output current for a four-phase interleaved buck converter at two different duty cycle ratios. The phases are shifted by $360°/n$. In the case of Fig. 10.45, the stages are operated at $0, 90°, 180°$, and $270°$.

The inductor current ripple in each phase is defined by Eqn. (10.9). With the duty cycle $D = V_{out}/V_{in}$ from Eqn. (10.8) we can express the current ripple as a function of D,

$$\Delta I_L = \frac{V_{out}}{Lf_{sw}}(1-D) = \frac{DV_{in}}{Lf_{sw}}(1-D) = \frac{V_{in}}{Lf_{sw}}\left(D - D^2\right). \tag{10.228}$$

By inspection, we find that zero ripple corresponds to $D = 0$ and $D = 1$. Actual designs will operate in between these duty cycle boundaries. Differentiating Eqn. (10.228) and setting the resulting expression equal to zero reveals that the maximum ripple occurs at $D = 0.5$.

The current I_o that flows into the output is the sum of all phase currents,

$$I_o(t) = \sum_{i=1}^{n} I_{Li}(t). \tag{10.229}$$

The interleaved operation greatly reduces the current ripple at I_o as illustrated for the two cases in Fig. 10.45. We also see that the fundamental ripple frequency of I_o is effectively multiplied by the number of phases. The behavior is identical to a single-phase converter operating at

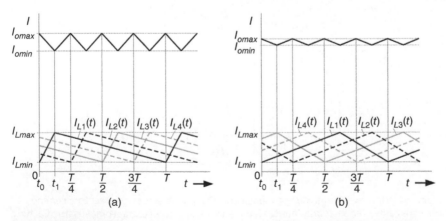

Figure 10.45 Phase currents $I_{Lx}(t)$ and output current I_o for a four-phase buck converter for different duty cycles: a) $D = 0.125$; b) $D = 0.625$.

$n \cdot f_{sw}$. The input root-mean-square (RMS) currents reduce similarly. Consequently, MP converters allow smaller input and output buffer capacitors but require multiple inductors (see design choices below). Lower ripple also means lower RMS current and lower power loss in the equivalent series resistances in each buffer capacitor.

We now derive the output current ripple $\Delta I_o = I_{omax} - I_{omin}$ [38]. Because of the piecewise-linear shape of the current waveforms, I_{omax} and I_{omin} correspond to the time when $I_o(t)$ reaches an inflection point. Therefore, the output ripple can be referred to as the interval between t_o and t_1,

$$\Delta I_o = I_{omax} - I_{omin} = I_o(t_1) - I_o(t_0). \tag{10.230}$$

During the time interval $(t_o - t_1)$, $I_{L1}(t)$ in Fig. 10.45a) rises but all other phase currents $I_{L2}(t)$ to $I_{L4}(t)$ undergo a falling transition. In Fig. 10.45b) $I_{L2}(t)$ ramps down while, during the same interval $(t_o - t_1)$, the other three phase currents go up. We can, therefore, expect that the current ripple ΔI_o in an MP converter will be a function of the duty cycle exactly like for a single-phase converter (see in Eqn. (10.228)). For n phases and a given duty cycle D, there will be p phases in the buck converter with $\Delta I_{Li} \geq 0$ and $(n - p)$ phases with $\Delta I_{Li} < 0$. If we define m to be the integer of $n \cdot D$ then,

$$p = m + 1. \tag{10.231}$$

For the four-phase example of Fig. 10.45a), we get $n \cdot D = 0.5$ and $m = 0$. Hence, there is $p = 1$ phase with a positive slope and a total of $(n - p) = 3$ phases with $\Delta I_{Li} < 0$. Also, Fig. 10.45b) can be confirmed by this calculation.

Inserting Eqn. (10.229) into Eqn. (10.230):

$$\Delta I_o = \sum_{i=1}^{n} I_{Li}(t_1) - \sum_{i=1}^{n} I_{Li}(t_o) = \sum_{i=1}^{n} \left(\Delta I_{Li} \right) \tag{10.232}$$

The current ripple ΔI_{Li} within the interval $(t_o - t_1)$ can be determined from a linear expression,

$$\Delta I_{Li} = \begin{cases} \dfrac{V_{out}(1 - D)}{DL} \left(t_1 - t_o \right) & \text{if } \Delta I_{Li} \geq 0 \\[2mm] \dfrac{-V_{out}}{L} \left(t_1 - t_o \right) & \text{if } \Delta I_{Li} < 0. \end{cases} \tag{10.233}$$

We can now combine Eqns. (10.232) and (10.233) while observing Eqn. (10.231):

$$\Delta I_o = p \frac{V_{out}(1 - D)}{DL} \left(t_1 - t_o \right) + (n - p) \frac{-V_{out}}{L} \left(t_1 - t_o \right) \tag{10.234}$$

As the last step, we need to know the time interval $(t_1 - t_o)$ defined as the duration from I_{L1} ramping up from zero to the first inflection point of one of the other inductor currents:

$$\left(t_1 - t_o \right) = \frac{1}{f_{sw}} \frac{n \cdot D - m}{n} \tag{10.235}$$

Substituting $(t_o - t_1)$ of Eqn. (10.235) into Eqn. (10.234) and rearranging yields

$$\Delta I_o = \frac{V_{out}}{f_{sw}L} (1 + m - n \cdot D) \left(1 - \frac{m}{n \cdot D} \right). \tag{10.236}$$

We can use DV_{in} instead of V_{out} for the buck converter in CCM. This way, we can directly compare the output ripple in MP operation to the single-phase ripple given by Eqn. (10.2). For $D \leq 0.5$, m becomes zero (see Eqn. (10.231)) and Eqn. (10.236) simplifies to

$$\Delta I_o = \frac{V_{out}}{f_{sw}L} (1 - n \cdot D). \tag{10.237}$$

Similarly, [38] derives the current ripple in DCM.

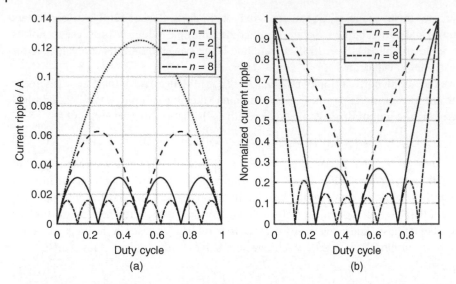

Figure 10.46 Current ripple as a function of the DC–DC converter's duty cycle for $n = 1$ to $n = 8$ phases: a) current ripple for $V_{in} = 5\,V$, $L = 10\,\mu H$, and $f_{sw} = 1\,MHz$; b) current ripple normalized to the ripple in single-phase operation.

Figure 10.46 plots the current ripple based on Eqn. (10.236). It confirms that multiphase topologies will significantly reduce the output current ripple. Even in the worst case, the ripple is significantly lower. The more phases, the smaller the ripple. Coming back to the four-phase operation of Fig. 10.45 with $D = 0.125$ and $D = 0.625$, respectively, we can read off from Fig. 10.46b) a ripple reduction to 0.58 and 0.27 in comparison to single-phase operation.

10.14.2 Design Choices

Understanding the effect of ripple cancellation and that one phase carries an average $(1/n)$-fraction of the load current, there are three design options for multiphase DC–DC converters. We discuss them with respect to a single-phase converter with a baseline sizing of the power switches R_{DSon}, the inductor L (with series resistance R_L), and the output capacitor C (with equivalent series resistance R_{ESR}).

(1) *Low volume (high power density)*: The switch resistance per phase can be increased to $n \cdot R_{DSon}$ (smaller W/L-ratio). We keep the same L per phase, but its series resistance R_L can be n-times higher. It means the coil wire can have a smaller diameter, and the volume of each inductor reduces. As a rule of thumb, the total volume of all inductors remains similar to that of the original inductor L. So far, there are no savings in size. However, the output capacitance gets significantly lower due to the ripple cancellation effect. We can use the original single-phase design for each phase and deliver n-times the output current to increase the power density further.

(2) *High conversion efficiency*: We follow case 1 but use the original inductor (L, R_L). This way, the conduction losses in the inductor are minimized. Also, the ripple-related losses will be at a minimum. The ripple-related losses in the output capacitor's R_{ESR} can be further reduced by placing multiple capacitors with smaller capacitance in parallel.

(3) *Fast transient response*: This is often the preferred design choice in MP DC–DC converters. Fast response to load transients means high di/dt per phase. The output current ripple remains

identical to the single-phase converter. We can reduce the inductor per phase significantly, perhaps even to L/n. Even though the output current ripple will be identical to the single-phase design, we may reduce the buffer capacitor because the converter benefits from faster transient response. Like in all other cases, we can reduce the switch resistance to R_{DSon}/n because the average phase current remains I_{load}/n. The fast transient response comes at the expense of higher losses because the more significant ripple contributes conduction losses in the power switches and the series resistances of L and C.

10.14.3 Enhanced Transient Response

In addition to the design option in case (3) above, some designs turn on all phases in parallel without time-shift immediately if a load step happens or, vice versa, the phase controller turns off all stages during a load release. In the case of a positive load step, this reduces the effective inductance by a factor of n. This way, an additional charge can be delivered to the output more quickly, reducing the droop at V_{out}.

10.14.4 Power Efficiency

Single-phase converters deliver all the output power through a single inductor and a pair of power transistors. Multiphase topologies distribute the power loss evenly across all phases. With the output power $P_{out} = V_{out}I_{load}$ the conversion efficiency is

$$\eta = \frac{P_{out}}{P_{loss} + P_{out}}. \tag{10.238}$$

The losses P_{loss} are described in Section 2.7. To understand how the power efficiency scales over load current, we consider only the conduction losses according to Eqn. (2.50) and the switching losses as given by Eqn. (2.51),

$$P_{loss} = P_{cond} + P_{sw}. \tag{10.239}$$

We have seen that the current ripple in each phase plays an important role in an MP converter. According to Eqn. (2.50), the ripple will contribute to the conduction losses. In fact, at light load, the conduction losses will be dominated by the ripple. Note that the ripple, as defined by Eqn. (10.228), does not scale with the load current,

$$P_{cond}(n) = nR_{DSon}\left(\left(\frac{I_{load}}{n}\right)^2 + \frac{\Delta I_L^2}{12} \right). \tag{10.240}$$

At a given switching frequency, the switching losses increase proportionally to the average current per switch, which is equal to the total load current I_{load} divided by the number of phases n,

$$P_{sw} = 2V_{in}t_{tr}f_{sw}\frac{I_{load}}{n}. \tag{10.241}$$

Figure 10.47 plots the efficiency according to Eqn. (10.238) with the losses from Eqns. (10.239), (10.240), and (10.241). The parameters are $V_{in} = 3.3\,\text{V}$, $V_{out} = 1.2\,\text{V}$, $L = 80\,\text{nH}$, $f_{sw} = 5\,\text{MHz}$, $R_{DSon} = 50\,\text{m}\Omega$, and $t_{rf} = 1\,\text{ns}$. Operating all eight phases ensures high efficiency at medium and high loads. At lower currents, the ripple-related conduction loss and the switching losses dominate over the pure conduction loss. Hence, higher efficiency is achieved if fewer phases are used, down to single-phase operation at a very light load. The impact will be even higher if gate charge losses are also considered. We also see that single-phase operation results in much lower efficiency over

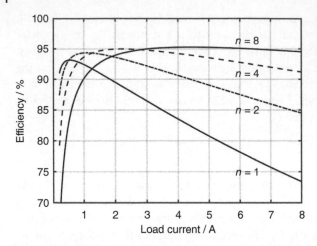

Figure 10.47 Conversion efficiency of a multiphase DC–DC converter with n phases enabled (phase shedding). At full load, maximum efficiency is achieved if all phases are active. At lower loads, the efficiency benefits from operating fewer phases, down to a single-phase configuration.

almost the entire load range. For the parameters used in Fig. 10.47, the current ripple per phase is $\Delta I_L = 1.9$ A. For $I_{load} < \Delta I_L/2 = 0.85$ A, the converter would enter DCM, not considered in the example. Note that a synchronous buck allows staying in CCM even if the inductor current gets negative. It is the opposite of diode emulation described for DCM in Section 10.9. It can implement high-side ZVS as described in Section 2.9.

10.14.5 Phase Shedding

The efficiency curves in Eqn. (10.238) suggest that phases should be added depending on the load current starting from single-phase operation at a very light load. The optimum point to turn on/off a stage corresponds to the intersection of two efficiency curves. This mechanism is called phase shedding. It ensures maximum efficiency over the entire load range. On the design side, this requires measuring the load current I_{load} at the output of the MP DC–DC converter, which can be accomplished using a sense resistor. For phase balancing (see Section 10.14.6), the current in each phase needs to be measured. In that case, phase shedding can be controlled by the phase current signals. The phase current can be measured using DCR-sensing. Details are described in Section 6.6 along with several other current sensing concepts.

10.14.6 Phase Balancing

A major challenge in multiphase converters is phase management. Besides phase shedding, this requires balancing the current evenly among all active phases to avoid thermal stress and excessive conduction losses in any one phase. The root cause will be slight differences in the duty cycle of each phase. This effect can happen due to offset in gain stages and comparators in the control loop and mismatch between passive components, including power inductors. Inter-phase mismatch of the sawtooth generator (VMC) or current-ramp sensing (CMC) also causes a current mismatch. Each phase represents a finite output resistance R that causes the real output voltage V_{outr} to be slightly lower than the ideal output voltage $V_{out} = D \cdot V_{in}$. We consider V_{outr} and V_{out} the average voltages because there will be some ripple. Consider a dual-phase converter with an identical output resistance R per phase for simplicity. The average phase current becomes

$$\overline{I_{L1,2}} = \frac{1}{R}\left(D_{1,2} \cdot V_{in} - V_{outr}\right). \tag{10.242}$$

Figure 10.48 Influence of phase duty cycle mismatch ΔD on the average phase currents $\overline{I_{L1,2}}$ and total conduction loss P_{cond} in a dual-phase DC–DC converter. The parameters are given in Example 10.21. A small mismatch causes a large difference between phase currents and significantly increases the conduction loss.

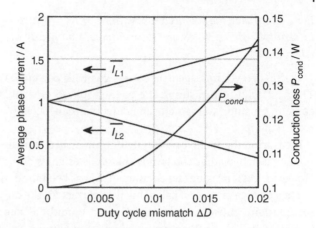

with the individual duty cycles D_1 and D_2 that are affected by duty cycle mismatch ΔD,

$$D_1 = D + \frac{1}{2}\Delta D, \tag{10.243}$$

$$D_2 = D - \frac{1}{2}\Delta D. \tag{10.244}$$

Figure 10.48 shows that a small duty cycle mismatch will cause a significant current offset between the phases. The parameters are taken from Example 10.21. As the main drawback of phase mismatch, also shown in Fig. 10.48, the conduction loss will significantly increase:

$$P_{cond} = \left(\overline{I_{L1}}^2 + \overline{I_{L2}}^2\right) R_{DSon} \tag{10.245}$$

Example 10.21 *We want to find out the current mismatch as well as the increase in conduction loss for a duty cycle mismatch of $\Delta D = 0.02$ for $I_{load} = 2\,\text{A}$. The other parameters are $V_{in} = 3.3\,\text{V}$, $V_{out} = 1.2\,\text{V}$, $V_{outr} = 1.15\,\text{V}$, and $R_{DSon} = 50\,\text{m}\Omega$.*

In the ideal case without duty cycle mismatch, each phase conducts an average current of $\frac{I_{load}}{2} = 1\,\text{A}$. From Eqn. (10.242), we can determine the output resistance:

$$R = \frac{1}{I_{load}/2}\left(V_{out} - V_{outr}\right) = \frac{1}{1\,\text{A}}\left(1.2\,\text{V} - 1.15\,\text{V}\right) = 50\,\text{m}\Omega \tag{10.246}$$

According to Eqns. (10.243) and (10.244), we get

$$D_1 = \frac{V_{out}}{V_{in}} + \frac{1}{2}\Delta D = 0.364 + 0.01 = 0.374, \tag{10.247}$$

$$D_2 = \frac{V_{out}}{V_{in}} - \frac{1}{2}\Delta D = 0.364 - 0.01 = 0.354. \tag{10.248}$$

With these values, we can determine the average phase currents using Eqn. (10.242):

$$\overline{I_{L1}} = \frac{1}{R}\left(D_1 \cdot V_{in} - V_{outr}\right) = 1.66\,\text{A}, \tag{10.249}$$

$$\overline{I_{L2}} = \frac{1}{R}\left(D_2 \cdot V_{in} - V_{outr}\right) = 0.34\,\text{A}. \tag{10.250}$$

For $\Delta D = 0.02$, $\overline{I_{L1}}$ is almost five times greater than $\overline{I_{L2}}$.

The conduction loss follows from Eqn. (10.245):

$$P_{cond} = \left(\overline{I_{L1}}^2 + \overline{I_{L2}}^2\right) R_{DSon} = \left(\overline{1.66\,\text{A}}^2 + \overline{0.34\,\text{A}}^2\right) 50\,\text{m}\Omega = 143.6\,\text{mW} \tag{10.251}$$

For comparison, the balanced case with $\overline{I_{L1}} = \overline{I_{L1}} = 1$ A *causes* $P_{cond} = 100$ mW. *Hence, as also shown in example of Fig. 10.48, the losses increase by nearly 50% at a duty cycle mismatch as small as* $\Delta D = 0.02$.

Various power management IC designs with current balancing functions have been published. They all require measuring the average phase current. Unlike cycle-by-cycle high-bandwidth current sensing, average current detection can be implemented at much more relaxed speed requirements.

Measuring the average phase current can be accomplished by replica sensing. Full-wave current sensing is achieved by combining the replica signals of the high-side and the low-side switch [29]. However, DCR sensing (see Section 6.6.4) has been used more extensively [31–33, 37].

The key idea of current balancing is to correct for the duty cycle mismatch by adjusting the intersection point between the sawtooth ramp V_s and the compensator voltage V_c as shown for VMC in Fig. 10.8. This can be accomplished by modifying the slope of V_s [36, 39] or by adjusting the compensator voltage per phase [29, 32–34]. The latter approach is illustrated in Fig. 10.49. The concept is similar to [29] but has been simplified to three phases, and instead of replica sensing, DCR sensing is applied. DCR sensing can be implemented as described in Section 6.6 with the sensing circuit of Fig. 6.17d). The average current can be detected if a filter capacitor is placed in parallel to the resistor R_2 at the output of the sensing circuit as proposed in [37]. The first phase is defined as the master. Its PWM signal is directly generated by comparing the global compensation voltage V_c with the local phase-1 sawtooth signal. However, for phases 2 and 3, the phase current errors $(V_{L1} - V_{L2})$ and $(V_{L1} - V_{L3})$ with respect to the master phase V_{L1} are detected by amplifiers

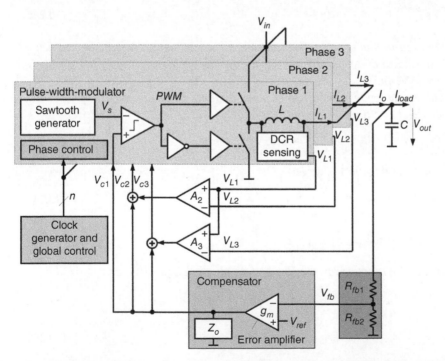

Figure 10.49 A multi-phase DC–DC converter with phase balancing: DCR sensing determines the average phase current signals V_{L1} to V_{L3}. Phase 1 forms the master. The difference between its current signal V_{L1} and V_{L2}, V_{L3}, respectively gets added to the master compensation voltage V_{c1} to generate the corresponding signals V_{c2} and V_{c3}.

A2 and A3. These signals get added to the global compensation voltage V_c, resulting in dedicated compensation signals V_{c2} and V_{c3} for phases 2 and 3. This way, negative feedback forces phase current balancing. Assume that the average current in phase 2 increases. Because V_{L2} exceeds V_{L1}, the compensation voltage V_{c2} get smaller than V_c. This way, in phase 2, the intersection between its local sawtooth ramp and V_{c2} will happen earlier and reduce the average current in phase 2. Adding the average phase current errors and V_c can be done in the current domain using V–I-converters similar to the concepts for slope compensation in Fig. 10.24. Instead of the difference to V_{L1}, the average phase current comparison can also be made in reference to the average V_L of all (three) phases. Then, also V_{c1} refers to the sum of the global V_c signal and the difference between $(V_L - V_{L1})$ [33, 34].

10.14.7 Coupled Inductors

Multiphase converters may benefit from magnetic coupling since there are at least two inductors. Wong et al. [40] investigates how integrated inductors (see also Section 4.2) can be utilized in multiphase converters. In particular, inverse (negative) coupling can improve both the steady-state and transient performances of multiphase converters. The design in [40] reduces the current ripple ΔI_L to lower than 60% compared to the non-coupled case and demonstrates up to 10% efficiency improvement. Future advances in integrated inductor technology will provide even more benefits, especially in fully integrated voltage regulators for point-of-load (PoL) processor core supplies (see Sections 1.7 and 12.8).

10.15 Single-Inductor Multiple-Output Converters (SIMO)

Almost every state-of-the-art electronic system needs multiple supply voltages. See, for instance, the application overview in Chapter 1 and the block diagrams in Fig. 1.7. The straightforward way to generate these voltages would be to implement multiple single-output DC–DC converters, sometimes called single-inductor single-output converters (SISO). However, this is usually not a cost-effective solution because every SISO converter needs its passive components, traditionally dominated by large inductors, typically placed off-chip. Besides cost, this solution does not achieve a small power module volume. Moreover, each off-chip inductor requires dedicated pins. More pins add more challenges due to electrostatic discharge (ESD) and EMI (see Section 12.3).

Since the late 1990s, single-inductor multiple-output (SIMO) switching converters have been explored as an attractive way to provide multiple output rails while requiring only one single inductor. Thanks to its cost and size benefit, the SIMO approach has become popular in System-on-Chip (SoC) designs for applications like microcontrollers, wireless transceivers, and smartphones.

10.15.1 Introduction

As the basic idea, SIMO converters employ time-multiplexing control. The fundamental SIMO buck and boost topologies are shown in Fig. 10.50a,b). Switches S3x (buck) and S2x (boost) have been added to the known SISO topologies to implement time-multiplexing. Other topologies, such as buck-boost, can extend to a SIMO converter. Often, SIMO designs have two outputs, referred to as single-inductor dual-output (SIDO) DC–DC converter. Early SIMO examples include a 4-output SIMO buck [41], a dual-output boost [42, 43], and a five-output boost [44].

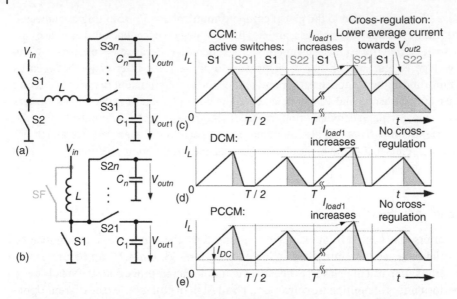

Figure 10.50 Single-Inductor Multiple Output (SIMO) converters: a) buck; b) boost, and inductor current in c) CCM; d) DCM, and e) PCCM (pseudo-continuous conduction mode). The second period illustrates the effect of cross-regulation for each mode, assuming a positive load step at output V_{out1}.

SIMO topologies can be operated in CCM and DCM, like the buck and the boost converter in Fig. 10.50a,b). In DCM, a free-wheeling switch SF is placed parallel to the inductor (activated at zero current as described for DCM in Section 10.9, also needed in PCCM, see Section 10.15.1). Typical waveforms of the inductor current for the case of a boost converter with two outputs are shown in Fig. 10.50c,d). Each switching period (duration T) is subdivided into two phases, one for each output. Consider the case of continuous conduction in Fig. 10.50c). Switch S1 is activated at the beginning, and inductor current builds up following the slope given by $\Delta I_L/\Delta t = V_{in}/L$. If I_L reaches the peak-current limit (the compensator voltage V_{c1} from the V_{out1} control loop), S1 turns off and S21 connects the coil to V_{out1}. The current ramps down by $\Delta I_L/\Delta t = (V_{out1} - V_{in})/L$. At $t = T/2$, S21 disconnects, and S1 turns on again to re-energize the inductor to control the second output V_{out2}. After reaching the peak-current value of the control loop for V_{out2}, S1 turns off, and S22 ties the coil to V_{out2}. In steady-state, the inductor current at the beginning and the end of one switching period must be equal. This way, the volt-second balance will be fulfilled, causing zero average voltage across the inductor (see also Eqn. (10.7)).

V_{out1} and V_{out2} can have different values, set by corresponding feedback dividers like in the SISO case. Like in this example, the PWM values and the peak-current values (V_{cx}) do not need to be identical. In other words, all rails can be defined independently. The same requirements for stable operation and fast transient response as for the SISO counterparts apply. However, a new parameter, called cross-regulation, occurs that specifically describes SIMO DC–DC converters.

10.15.2 Cross-Regulation

This effect is defined as a change in the output voltage of one sub-converter by a load change of another sub-converter. Applied to a dual-output converter, the cross-regulation seen at V_{out2} for a load change ΔI_{out1} at V_{out1} is defined as

$$Z_{21} = \frac{\Delta V_{out2}}{\Delta I_{load1}}. \tag{10.252}$$

We can explore the cross-regulation effect considering the inductor current for the three cases shown in Fig. 10.50c–e). We always assume that the current at V_{out1} increases after the first switching period ($t = T$). Subsequently, the compensated error voltage V_{c1}, which sets the peak current for V_{out1}, increases as indicated. It happens exactly like in a SISO converter in CMC (see Section 10.6 and Fig. 10.16). Due to the limited response time of the outer voltage loop, V_{c1} will take a few switching cycles (not shown in Fig. 10.50c–e) for the sake of clarity) to settle at its higher value. The energizing cycle starts by turning on S1. As the input voltage did not change, the slope of the current ramp remains $\Delta I_L / \Delta t = V_{in}/L$ as before, but it reaches a higher peak value. When S21 gets turned on, its slope is also unchanged $\Delta I_L / \Delta t = (V_{out1} - V_{in})/L$. Hence, the inductor current ends up at a somewhat higher value. Due to the higher current demand, this is part of the control behavior at V_{out1}. However, the higher current value is the starting level of the energizing phase for V_{out2}. Its rising and falling slopes, as well as its peak-current value, do not change. But, since the initial current level has increased, the average inductor current reduces. Starting and end values are no longer equal, violating the volt-second balance discussed above. Hence, cross-regulation happens at V_{out2}. We can conclude that SIMO converters in CCM may exhibit severe cross-regulation.

We now analyze the case of DCM shown in Fig. 10.50d). The sequence of activating the switches remains identical to CCM, as shown in Fig. 10.50c). Typical for DCM, the current reaches zero at the end of each sub-period dedicated to each output voltage. For this reason, in DCM, a load change does not influence cross-regulation. The second part of Fig. 10.50d) illustrates that higher I_{load1} shortens the zero current phase at the end of each sub-cycle, but the inductor current remains at zero. We can conclude that the DCM does not show any cross-regulation. Why not operate all SIMO converters in DCM? Section 10.9 on DCM operation provides the answer. The main drawback is the larger output voltage ripple. The converter requires a large current ripple and peak inductor current to ensure DCM over the entire load range. It harms voltage ripple and switching noise performance. Enlarging the output buffer capacitor helps but leads to a slower dynamic response. Hence, the output power is limited in DCM.

An interesting alternative is an operation mode called pseudo-continuous conduction mode (PCCM) proposed in [43]. This approach combines the advantages of CCM (higher power) and DCM (no cross-regulation). The waveform in Fig. 10.50e) indicates that the zero DC current in a DCM converter is now replaced by a constant DC current I_{DC}. In other words, the floor of the inductor current is raised by a DC level that can be adjusted depending on the load and current ripple specification. The dynamic behavior is still identical to a DCM converter because I_{DC} is only a DC component. However, I_{DC} increases the average inductor current. Therefore, as an advantage of PCCM, a more minor current ripple can be achieved. To maintain the DC current in the inductor, a free-wheeling switch SF has to be added in parallel to the inductor, as shown in Fig. 10.50b). Switch SF gets activated at the end of each energizing cycle when the inductor current reaches I_{DC}. Besides the need for an additional power switch, this reveals the main drawback of PCCM operation. It unnecessarily dissipates power in the resistance of the inductor and free-wheeling switch.

10.15.3 Single Energizing Cycle per Switching Period

The above SIMO concepts implement multiple energizing cycles per switching period, one per output. Alternatively, starting each switching period with one common energizing cycle is possible. The goal is to build up sufficient energy to supply all outputs afterward in a time-multiplexed manner, like for the approach with multi-energizing cycles per period.

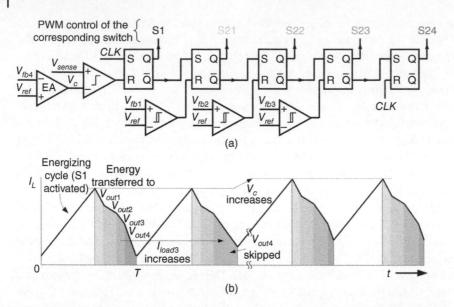

(a)

(b)

Figure 10.51 SIMO converter with a single energizing cycle per switching period: a) Control circuit with hysteretic control for V_{out1} to V_{out3} and peak-current CMC for V_{out4}; b) inductor current over multiple switching cycles. After the first switching period, the control operation is demonstrated for the case that the load I_{load3} at V_{out3} increases.

Figure 10.51a) shows a suitable control circuit for four outputs [45]. This general control circuit would be suitable for the SIMO with *ordered power-distributed control* presented in [44]. The converter in [44] comprises five boost converter outputs, four outputs with hysteretic control (a comparator with hysteresis, connected to V_{fb1} to V_{fb3}, see Fig. 10.51a)) and the last stage with linear control using an error amplifier EA and followed by a comparator (shown on the left of Fig. 10.51a)). Implementing CMC type II compensation will be applicable. Instead of a resistive feedback divider, the output voltages can be directly connected to the comparators. In that case, different reference voltage levels need to be generated.

While the hysteretic approach keeps the control effort for the first four outputs low, the last stage compensates for the errors transferred and accumulated from the comparator-controlled stages V_{out1} to V_{out3}. Only the last output V_{out4} requires a compensation network. The first switching period in Fig. 10.51b) shows the energizing cycle in which the inductor current is charged to a peak value. During this time, switch S1 turns on until the R–S flip-flop goes into reset when the current sensing signal V_{sense} of the inductor current exceeds the compensation voltage V_c of V_{out4}. It is exactly like the SISO converter operating in CMC with the waveforms shown in Fig. 10.16. The second flip-flop turns on S21 to deliver energy to V_{out1}. The inductor current starts to ramp down. S21 remains active until V_{out1} has sufficiently increased such that V_{fb1} reaches V_{ref}. Now S21 turns off, and S22 turns on to transfer energy from the inductor toward V_{out2}. The procedure continues for S23 and V_{out3}. Finally, S24 turns on and recharges V_{out4}. It stops with the rising edge of the switching clock *CLK* that marks the beginning of the next switching period. Now, S1 turns on again and initiates the next energizing cycle.

The role of linear control by the error amplifier at the fourth output is illustrated in the subsequent switching cycles in Fig. 10.51b). We assume that the load current I_{load3} at output V_{out3} increases. The time-multiplexed energy transfer toward V_{out1} and V_{out2} is not affected. However, V_{out3} will take much longer to reach its target value. Due to the hysteretic control, S23 will remain active for the remainder of the switching period, while S24 will not turn on at all. Since the energy transfer to V_{out4} is skipped at all (marked at the end of the second period in Fig. 10.51b)), V_{out4}

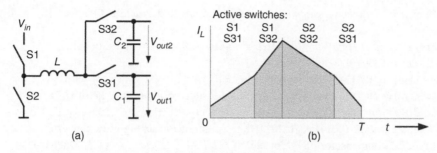

Figure 10.52 A single-inductor multiple output (SIMO) converter operated in energy-conservation mode (ECM): a) Dual-output buck converter; b) ECM sequencing and inductor current.

and, subsequently, V_{fb4} will drop. It clearly indicates cross-regulation (like in CCM for multiple energizing cycles per period, see Fig. 10.50c)). The error amplifier causes V_c to rise. It will not happen instantaneously, but after a few cycles, higher V_c will cause the inductor to reach a higher peak current value while S1 is turned on. This behavior is indicated in the third switching cycle in Fig. 10.51b). With the higher peak current, I_{out3} can be fully supported such that V_{out3} can ramp up within the period. There is now even time left to transfer energy to V_{out4} at the end of the switching cycle. We have reached steady-state, and the fourth cycle in Fig. 10.51b) continues like the period before. If I_{out3} drops, the behavior would be vice versa; S24 would turn on for a longer time, causing too much energy into V_{out4}. V_c goes down after some delay, properly adjusting the turn-on time of S24. Le et al. [44] suggests variants of *ordered power-distributed control* in which only some outputs are recharged per cycle. They can also be dynamically reordered depending on the load distribution.

With a single energizing cycle, the same observations for operation in CCM, DCM, and PCCM as for multiple energizing cycles apply. However, while each output receives all the energy stored in the inductor in the multi-energizing approach, the outputs share the same single energy. In other words, each output in the single-energizing approach receives less energy per cycle but more often, which means there is less time for each output to droop, causing a lower ripple in general. In the end, this depends on the operation parameters.

General SIMO converter designs aim to combine and optimize the pros and cons in CCM, DCM, and PCCM depending on their particular specifications of V_{in}, V_{out}, and load range at every output. Lee et al. [46] proposes a SIMO converter in Energy-Conservation Mode (ECM) that achieves low output voltage ripple and high power conversion efficiency while minimizing cross-regulation. The design comprises a buck converter according to Fig. 10.50a) with two outputs, as shown in Fig. 10.52a). Each of the two outputs is controlled by a dedicated error amplifier in peak-current CMC, including slope compensation (see Section 10.6). Figure 10.52b) shows the sequence of the energy delivery paths and the corresponding inductor current. The two error signals are converted into current by voltage-to-current converters and summed up to set the common peak-current level, i.e., the maximum inductor current in Fig. 10.52b). Therefore, the average inductor current in ECM is lower than, for instance, in PCCM, which results in a better trade-off between cross-regulation and energy efficiency.

References

1 Erickson, R.W. and Maksimovic, D. (2001) *Fundamentals of Power Electronics*, Springer.
2 Mohan, N., Undeland, T.M., and Robbins, W.P. (2003) *Power Electronics. Converters, Applications and Design*, John Wiley & Sons, Inc., 3rd edn.

3 Basso, C.P. (2008) *Switch-Mode Power Supplies: Spice Simulations and Practical Designs*, McGraw-Hill.

4 Wittmann, J. (2019) *Integrated High-Vin Multi-MHz Converters*, Springer International Publishing, Cham, doi: 10.1007/978-3-030-25257-1.

5 Wittmann, J. and Wicht, B. (2015) A configurable sawtooth based PWM generator with 2 ns on-time for >50 MHz DCDC converters, in *2015 11th Conference on Ph.D. Research in Microelectronics and Electronics (PRIME)*, pp. 41–44, doi: 10.1109/PRIME.2015.7251089.

6 Cheng, L., Liu, Y., and Ki, W.H. (2014) A 10/30 MHz fast reference-tracking buck converter with DDA-based type-III compensator. *IEEE Journal of Solid-State Circuits*, 49 (12), 2788–2799, doi: 10.1109/JSSC.2014.2346770.

7 Lee, C.F. and Mok, P. (2004) A monolithic current-mode CMOS DC-DC converter with on-chip current-sensing technique. *IEEE Journal of Solid-State Circuits*, 39 (1), 3–14, doi: 10.1109/JSSC.2003.820870.

8 Herzer, S., Kulkarni, S., Jankowski, M., Neidhardt, J., and Wicht, B. (2009) Capacitive-coupled current sensing and auto-ranging slope compensation for current mode SMPS with wide supply and frequency range, in *2009 Proceedings of ESSCIRC*, pp. 140–143, doi: 10.1109/ESS-CIRC.2009.5326034.

9 Sakurai, H. and Sugimoto, Y. (2005) Design of a current-mode, MOS, DC-DC buck converter with a quadratic slope compensation scheme, in *48th Midwest Symposium on Circuits and Systems, 2005*, pp. 671–674, Vol. 1, doi: 10.1109/MWSCAS.2005.1594190.

10 Umimura, K., Sakurai, H., and Sugimoto, Y. (2007) A CMOS current-mode DC-DC converter with input and output voltage-independent stability and frequency characteristics utilizing a quadratic slope compensation scheme, in *ESSCIRC 2007 - 33rd European Solid-State Circuits Conference*, pp. 178–181, doi: 10.1109/ESSCIRC.2007.4430274.

11 Venable, D. (1983) The K-factor: a new mathematical tool for stability analysis and synthesis, in *Proceedings of Powercon 10*, pp. 1–12.

12 Cao, L. (2011) Type III compensator design for power converters. *Power Electronics Technology Magazine*, Vol. 37.

13 Rincon-Mora, G. (2000) Active capacitor multiplier in miller-compensated circuits. *IEEE Journal of Solid-State Circuits*, 35 (1), 26–32, doi: 10.1109/4.818917.

14 Chen, K.H., Chang, C.J., and Liu, T.H. (2008) Bidirectional current-mode capacitor multipliers for on-chip compensation. *IEEE Transactions on Power Electronics*, 23 (1), 180–188, doi: 10.1109/TPEL.2007.911776.

15 Angkititrakul, S. and Hu, H. (2008) Design and analysis of buck converter with pulse-skipping modulation, in *2008 IEEE Power Electronics Specialists Conference*, pp. 1151–1156, doi: 10.1109/PESC.2008.4592085.

16 Vorperian, V. (1990) Simplified analysis of PWM converters using model of PWM switch. CCM and II. DCM. *IEEE Transactions on Aerospace and Electronic Systems*, 26 (3), 490–496 and 497–505, doi: 10.1109/7.106126, 10.1109/7.106127.

17 Ke, X., Liu, W.C., Song, M.K., Xue, J., Zheng, C., Liu, K., Leng, Y., and Chen, M. (2021) 33.3 An automotive-use 2MHz 100VOUT flicker-free frequency-modulated GaN-based buck-boost LED driver achieving bootstrap charge balancing and 16.8dbuV radiated EMI noise reduction, in *2021 IEEE International Solid- State Circuits Conference (ISSCC)*, Vol. 64, pp. 464–466, doi: 10.1109/ISSCC42613.2021.9365739.

18 Zaman, A. and Radic, A. (2020) How to design and implement an adapter power supply with active clamp flyback: an all silicon design methodology. *IEEE Power Electronics Magazine*, 7 (4), 36–43, doi: 10.1109/MPEL.2020.3033608.

19 Rindfleisch, C., Otten, J., and Wicht, B. (2022) A highly-integrated 20-300V 0.5W active-clamp flyback DCDC converter with 76.7% peak efficiency, in *2022 IEEE Custom Integrated Circuits Conference (CICC)*, pp. 1–2, doi: 10.1109/CICC53496.2022.9772834.

20 Kuo, C.C., Lee, J.J., He, Y.H., Wu, J.Y., Chen, K.H., Lin, Y.H., Lin, S.R., and Tsai, T.Y. (2021) A dynamic resonant period control technique for fast and zero voltage switching in GAN-based active clamp flyback converters. *IEEE Transactions on Power Electronics*, 36 (3), 3323–3334, doi: 10.1109/TPEL.2020.3016324.

21 Rindfleisch, C. and Wicht, B. (2021) A resonant one-step 325 V to 3.3-10 V DC-DC converter with integrated power stage benefiting from high-voltage loss-reduction techniques. *IEEE Journal of Solid-State Circuits*, 56 (11), 3511–3520, doi: 10.1109/JSSC.2021.3098751.

22 Rindfleisch, C. and Wicht, B. (2022) A 110/230 V AC and 15-400 V DC 0.3 W power-supply IC with integrated active zero-crossing buffer. *IEEE Journal of Solid-State Circuits*, 57 (12), 3816–3824, doi: 10.1109/JSSC.2022.3199653.

23 van Liempd, C., Stanzione, S., Allasasmeh, Y., and van Hoof, C. (2013) A 1μW-to-1mW energy-aware interface IC for piezoelectric harvesting with 40nA quiescent current and zero-bias active rectifiers, in *2013 IEEE International Solid-State Circuits Conference Digest of Technical Papers*, pp. 76–77, doi: 10.1109/ISSCC.2013.6487644.

24 Ghovanloo, M. and Najafi, K. (2004) Fully integrated wideband high-current rectifiers for inductively powered devices. *IEEE Journal of Solid-State Circuits*, 39 (11), 1976–1984, doi: 10.1109/JSSC.2004.835822.

25 Rao, Y. and Arnold, D.P. (2011) An input-powered vibrational energy harvesting interface circuit with zero standby power. *IEEE Transactions on Power Electronics*, 26 (12), 3524–3533, doi: 10.1109/TPEL.2011.2162530.

26 Leicht, J. and Manoli, Y. (2017) A 2.6 μ-1.2 mW autonomous electromagnetic vibration energy harvester interface IC with conduction-angle-controlled MPPT and up to 95% efficiency, *Journal of Solid-State Circuits*, 52 (9), 2448–2462, doi: 10.1109/JSSC.2017.2702667.

27 Maurath, D., Becker, P.F., Spreemann, D., and Manoli, Y. (2012) Efficient energy harvesting with electromagnetic energy transducers using active low-voltage rectification and maximum power point tracking. *IEEE Journal of Solid-State Circuits*, 47 (6), 1369–1380, doi: 10.1109/JSSC.2012.2188562.

28 Li, P., Xue, L., Hazucha, P., Karnik, T., and Bashirullah, R. (2009) A delay-locked loop synchronization scheme for high-frequency multiphase hysteretic DC-DC converters. *IEEE Journal of Solid-State Circuits*, 44 (11), 3131–3145, doi: 10.1109/JSSC.2009.2033508.

29 Huang, C. and Mok, P.K.T. (2013) A 100 MHz 82.4% efficiency package-bondwire based four-phase fully-integrated buck converter with flying capacitor for area reduction. *IEEE Journal of Solid-State Circuits*, 48 (12), 2977–2988, doi: 10.1109/JSSC.2013.2286545.

30 Abu-Qahouq, J., Mao, H., and Batarseh, I. (2004) Multiphase voltage-mode hysteretic controlled DC-DC converter with novel current sharing. *IEEE Transactions on Power Electronics*, 19 (6), 1397–1407, doi: 10.1109/TPEL.2004.836639.

31 Hazucha, P., Schrom, G., Hahn, J., Bloechel, B., Hack, P., Dermer, G., Narendra, S., Gardner, D., Karnik, T., De, V., and Borkar, S. (2005) A 233-MHz 80%-87% efficient four-phase DC-DC converter utilizing air-core inductors on package. *IEEE Journal of Solid-State Circuits*, 40 (4), 838–845, doi: 10.1109/JSSC.2004.842837.

32 Song, M.K., Sankman, J., Lee, J., and Ma, D. (2016) A 200-MHz 4-phase fully integrated voltage regulator with local ground sensing dual loop ZDS hysteretic control using 6.5nH package bondwire inductors on 65nm bulk CMOS, in *2016 21st Asia and South Pacific Design Automation Conference (ASP-DAC)*, pp. 9–10, doi: 10.1109/ASPDAC.2016.7427976.

33 Lee, B., Song, M.K., Maity, A., and Ma, D.B. (2017) 10.7 A 25MHz 4-phase SAW hysteretic DC-DC converter with 1-cycle APC achieving 190 ns tsettle to 4A load transient and above 80% efficiency in 96.7% of the power range, in *2017 IEEE International Solid-State Circuits Conference (ISSCC)*, pp. 190–191, doi: 10.1109/ISSCC.2017.7870325.

34 Lee, B., Song, M.K., and Ma, D.B. (2017) On-chip inductor DCR self-calibration technique for high frequency integrated multiphase switching converters, in *2017 IEEE Applied Power Electronics Conference and Exposition (APEC)*, pp. 2449–2452, doi: 10.1109/APEC.2017.7931042.

35 Burton, E.A., Schrom, G., Paillet, F., Douglas, J., Lambert, W.J., Radhakrishnan, K., and Hill, M.J. (2014) FIVR - Fully integrated voltage regulators on 4th generation Intel(R) Core(TM) SoCs, in *2014 IEEE Applied Power Electronics Conference and Exposition - APEC 2014*, pp. 432–439, doi: 10.1109/APEC.2014.6803344.

36 Su, Y.P., Chen, W.C., Huang, Y.P., Lee, Y.H., Chen, K.H., and Luo, H.Y. (2014) Pseudo-ramp current balance (PRCB) technique with offset cancellation control (OCC) in dual-phase DC-DC buck converter. *IEEE Transactions on Very Large Scale Integration (VLSI) Systems*, 22 (10), 2192–2205, doi: 10.1109/TVLSI.2013.2283606.

37 Roh, Y.S., Moon, Y.J., Park, J., Jeong, M.G., and Yoo, C. (2015) A multiphase synchronous buck converter with a fully integrated current balancing scheme. *IEEE Transactions on Power Electronics*, 30 (9), 5159–5169, doi: 10.1109/TPEL.2014.2368130.

38 Yang, X., Zong, S., and Fan, G. (2017) Analysis and validation of the output current ripple in interleaved buck converter, in *IECON 2017 - 43rd Annual Conference of the IEEE Industrial Electronics Society*, pp. 846–851, doi: 10.1109/IECON.2017.8216146.

39 Huang, Y.P., Su, Y.P., Lee, Y.H., Chu, K.Y., Shih, C.J., Chen, K.H., Du, M.J., and Cheng, S.H. (2011) Single controller current balance (SCCB) technique for voltage-mode multiphase buck converter, in *2011 IEEE International Symposium of Circuits and Systems (ISCAS)*, pp. 761–764, doi: 10.1109/ISCAS.2011.5937677.

40 Wong, P.L., Xu, P., Yang, P., and Lee, F. (2001) Performance improvements of interleaving VRMs with coupling inductors. *IEEE Transactions on Power Electronics*, 16 (4), 499–507, doi: 10.1109/63.931059.

41 Belloni, M., Bonizzoni, E., Kiseliovas, E., Malcovati, P., Maloberti, F., Peltola, T., and Teppo, T. (2008) A 4-output single-inductor DC-DC buck converter with self-boosted switch drivers and 1.2A total output current, in *2008 IEEE International Solid-State Circuits Conference - Digest of Technical Papers*, pp. 444–626, doi: 10.1109/ISSCC.2008.4523248.

42 Ma, D., Ki, W.H., Tsui, C.Y., and Mok, P. (2003) Single-inductor multiple-output switching converters with time-multiplexing control in discontinuous conduction mode. *IEEE Journal of Solid-State Circuits*, 38 (1), 89–100, doi: 10.1109/JSSC.2002.806279.

43 Ma, D., Ki, W.H., and Tsui, C.Y. (2003) A pseudo-CCM/DCM SIMO switching converter with freewheel switching. *IEEE Journal of Solid-State Circuits*, 38 (6), 1007–1014, doi: 10.1109/JSSC.2003.811976.

44 Le, H.P., Chae, C.S., Lee, K.C., Wang, S.W., Cho, G.H., and Cho, G.H. (2007) A single-inductor switching DC-DC converter with five outputs and ordered power-distributive control. *IEEE Journal of Solid-State Circuits*, 42 (12), 2706–2714, doi: 10.1109/JSSC.2007.908767.

45 Huang, M.H. and Chen, K.H. (2009) Single-inductor multi-output (SIMO) DC-DC converters with high light-load efficiency and minimized cross-regulation for portable devices. *IEEE Journal of Solid-State Circuits*, 44 (4), 1099–1111, doi: 10.1109/JSSC.2009.2014726.

46 Lee, Y.H., Yang, Y.Y., Wang, S.J., Chen, K.H., Lin, Y.H., Chen, Y.K., and Huang, C.C. (2011) Interleaving energy-conservation mode (IECM) control in single-inductor dual-output (SIDO) step-down converters with 91% peak efficiency. *IEEE Journal of Solid-State Circuits*, 46 (4), 904–915, doi: 10.1109/JSSC.2011.2108850.

11

Hybrid DC–DC Converters

Hybrid DC–DC converters combine capacitive and inductive voltage conversion concepts to mutually eliminate each other's drawbacks. All kinds of combined architectures with parallel or cascaded capacitive and inductive conversion stages are possible. Nevertheless, hybrid usually refers to elaborated topologies with merged L-C arrangements. Chapters 9 and 10 demonstrate that capacitor-based and inductor-based concepts can provide much higher conversion efficiencies than the linear regulator in Chapter 7. Hybrid converters, as the youngest class of DC–DC converters, go one step further and enable a drastic reduction of the inductor size while minimizing switching losses and improving the overall conversion efficiency.

Table 11.1 compares capacitive and inductive DC–DC converters with the hybrid approach. The table illustrates how hybrid converters can outperform other conversion concepts. They support a wide operating range of V_{in}, V_{out}, and I_{load} and provide high efficiency from low to high load.

Hybrid converters allow for operation at switching frequencies in the multi-megahertz range (>10 MHz) while keeping the dynamic losses low. They can operate at the resonance of the combined L-C components and in continuous conduction mode (CCM). Various other operating options and even a combination (called hybrid mode) are possible. At resonance, the inductor current $I_L(t)$ becomes sinusoidal as illustrated in Fig. 11.1. The hybrid converter enters CCM at higher frequencies, and $I_L(t)$ approaches a triangular current ripple pattern, similar to a pulse-width modulation (PWM) controlled inductive converter. We will look into the details in the upcoming parts of this chapter.

The need for an additional inductor may be considered a disadvantage. However, excellent performance can be achieved for small inductors in the single-digit nanohenry range with quality factors of ~4, opening up opportunities for miniaturization. In particular, hybrid converters are well-suitable to leverage the potential of integrated inductors (see Chapter 4) and, consequently, enable fully integrated DC–DC converters, including all passive components either on-chip or by co-integration in the same package. In contrast, conventional capacitive and inductive converters operate in the lower MHz range with passive components in the order of µH and µF.

Design of Power Management Integrated Circuits, First Edition. Bernhard Wicht.
© 2024 John Wiley & Sons Ltd. Published 2024 by John Wiley & Sons Ltd.

Table 11.1 Hybrid converters combine the advantages and solve the drawbacks of capacitive and inductive DC–DC converters.

	Capacitive DC–DC conversion	Inductive DC–DC conversion	Hybrid DC–DC conversion
Pros	Efficient at low power	Efficient at low power	✓
	can be fully integrated	simple regulation	keeps the
	high conversion ratios	wide range of V_{in}, V_{out}	advantages
Cons	Charge sharing losses	Large L or very high f_{sw}	× eliminates the drawbacks

11.1 Hybridization of Capacitive and Inductive Concepts

Figure 11.2 illustrates how a hybrid converter can be derived in two ways. Interestingly, either approach results in the same fundamental hybrid DC–DC converter of Fig. 11.2:

(1) We take the fundamental 2:1 switched-capacitor (SC) converter of Fig. 9.1, repeated in Fig. 11.2a). Adding an inductor between the flying and the output capacitor results in Fig. 11.2b). Thus, the inductor replaces the equivalent output resistance R_{out}, which is responsible for most power losses in the SC converter. This way, the energy is stored in L instead of converted to thermal energy in R_{out}. This hybrid converter is often named a resonant switched-capacitor (ReSC) converter because a resonant tank is formed. The losses will ideally be zero if operated at the resonance frequency. The real hybrid converter will still show a finite equivalent output resistance, but it will be much lower compared to the SC counterpart. We will investigate this in detail below.

(2) We start from the inductor-based buck DC–DC converter of Fig. 10.1 and add a capacitor by modifying the power stage as shown in Fig. 11.2. The power stage now comprises two more switches, S2 and S4. Even though the resulting topology is identical to option (1), the switches are controlled differently (in four phases, whereas (1) uses a two-phase control). The ideal capacitor voltage in steady-state will be $V_{in}/2$. Hence, the voltage across the inductor will be reduced by half compared to the original inductive converter. Consequently, it allows us to reduce the inductor size by a factor of two, or we can reduce the switching frequency and benefit from lower switching losses.

To understand the benefits and potential of hybrid conversion in more detail, we follow item (1) and analyze the charge redistribution in Section 11.2.

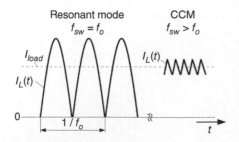

Figure 11.1 Inductor current $I_L(t)$ at different switching frequency f_{sw}: At the resonance point, $f_{sw} = f_o$, the inductor current represents a full-wave rectified sine wave. For higher frequencies, $f_{sw} > f_o$, the inductor current approaches a triangular shape like in a PWM-modulated inductive converter. The average inductor current always equals the load current I_{load} in steady-state.

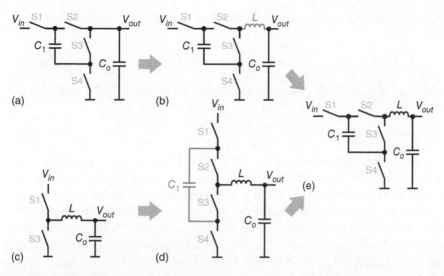

Figure 11.2 "Hybridization" can be applied to a) an SC converter by b) adding an inductor L or to c) an inductive buck converter by d) adding two more switches and inserting a capacitor C_1. e) Both approaches result in the same fundamental hybrid step-down converter.

11.2 The Benefit of Soft-Charging

It is instructive to consider charging a capacitor with and without a series inductor, also called hard-charing and soft-charging.

11.2.1 Hard-Charging

Consider the two capacitors in Fig. 11.3a) connected by a switch at $t = 0$. In the case of an SC converter, C_1 could be a flying capacitor, while C_2 represents the output buffer capacitor. For simplicity, we assume both capacitances to have the same value of $C = C_1 = C_2 = 1$ F. Initially, C_1 is pre charged to $V_1 = 2$ V while C_2 is fully discharged, i.e., $V_2 = 0$. At $t = 0$, charge transfers from C_1 to C_2 until $V = V_1 = V_2 = 1$ V. From electrical engineering theory, we recall that half the energy will be lost when charging a capacitor. We can easily verify this for the example of Fig. 11.3a). At

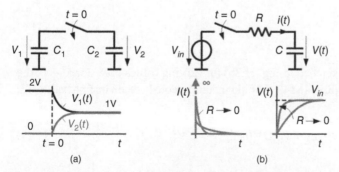

Figure 11.3 a) Charge redistribution between two capacitors; b) hard-charging of a capacitor from a voltage source V_{in} via an equivalent resistance R.

$t \le 0$, the total energy in the setup is equal to the energy stored in C_1 (because zero energy is stored in C_2 due to $V_2 = 0$):

$$E_1 = \frac{C_1}{2} V_1^2 = \frac{1\,\text{F}}{2} (2\,\text{V})^2 = 2\,\text{J} \qquad (11.1)$$

When the switch is closed at $t = 0$, both capacitors are connected in parallel. The total energy is now

$$E_2 = \frac{C_1}{2} V_1^2 + \frac{C_2}{2} V_2^2 = CV^2 = 1\,\text{F} \cdot (1\,\text{V})^2 = 1\,\text{J}. \qquad (11.2)$$

Comparing the results in Eqns. (11.1) and (11.2) confirms that half the energy is lost. Why is this? The whole procedure is called the hard-charging of a capacitor in contrast to soft-charging, covered further below. The losses due to hard charging are called charge redistribution or capacitive charging losses. They are due to the fact that the voltage across the capacitors cannot change instantaneously. In the example above, the voltage difference of $V_1 - V_2 = 2\,\text{V}$ will drop across the switch (and any interconnect resistance) at $t = 0$. The switch and the connecting wires will always have finite resistance in an actual circuit. Even if an extremely low-resistive switch is used, there will be a voltage drop for a short time at $t = +0$. The value of the resistance does not influence the capacitive charging losses.

We can investigate this behavior using the equivalent circuit in Fig. 11.3b). A voltage source V_{in} charges a capacitor C via an equivalent resistance R. In an SC converter, V_{in} would correspond to the ideal output voltage V_{int} as shown in the model of Fig. 9.2. Starting from the I-V relationship of the capacitor,

$$I(t) = C \frac{dV(t)}{dt}, \quad V(t) = \frac{1}{C} \int I(t) dt, \qquad (11.3)$$

we can put together the loop equation,

$$V_{in} = I(t) \cdot R + \frac{1}{C} \int I(t) dt. \qquad (11.4)$$

Differentiating both sides of Eqn. (11.4) and rearranging yields a homogeneous first-order differential equation,

$$0 = \frac{dI(t)}{dt} + \frac{1}{RC} I(t). \qquad (11.5)$$

For the boundary condition $v(t = 0) = 0$ and applying Eqn. (11.3), we find the capacitor current and voltage as a solution of Eqn. (11.4):

$$I(t) = I_0 \cdot e^{-\frac{t}{\tau}} = \frac{V_{in}}{R} \cdot e^{-\frac{t}{RC}}$$
$$V(t) = \frac{1}{C} \int_0^\infty I(t) dt = V_{in} \cdot \left(1 - e^{-\frac{t}{RC}} \right) \qquad (11.6)$$

The expressions for $I(t)$ and $V(t)$ are sketched in Fig. 11.3b) for varying values of R. For $R \to 0$, the initial current I_0 would be an infinite spike, and $V(t)$ would instantaneously reach its final value V_{in}.

The energy emitted by the source is

$$E_{in} = \int_0^\infty V_{in} I(t) \, dt = V_{in} \int_0^\infty I(t) \, dt = V_{in} C V(t) \,|_{t=0}^{t \to \infty} = C V_{in}^2, \qquad (11.7)$$

while the energy stored in C is

$$E_C = \int_0^\infty V(t) I(t) dt = \int_0^\infty V(t) C \frac{dV(t)}{dt} dt = \frac{1}{2} C V(t)^2 \,|_{t=0}^{t \to \infty} = \frac{C}{2} V_{in}^2. \qquad (11.8)$$

Hence, the energy loss is

$$E_{loss} = E_{in} - E_C = \frac{C}{2} V_{in}^2, \tag{11.9}$$

which is, as expected, half the energy delivered by the voltage source V_{in}. E_{loss} is dissipated as heat in the equivalent resistance R:

$$E_R = \int R I(t)^2 dt = E_{loss} = \frac{C}{2} V_{in}^2 \tag{11.10}$$

Interestingly, E_{loss} does not depend on R. We can explain this result from the $I(t)$-curves in Fig. 11.3b). The peak value I_o at $t = +0$ will vary with R, but the integral will always be the same as given by Eqn. (11.10). Only C and V_{in} determine the losses.

11.2.2 Soft-Charging

If we insert an inductor L, the loss energy E_{loss} can be preserved by storing it in L. Consider the setup in Fig. 11.4a). It corresponds to Fig. 11.3a) with an inductor in series to the switch. We start from the same initial conditions, $V_1 = 2\,\text{V}$ and $V_2 = 0$. The initial voltage difference drops across the inductor if the switch closes at $t = 0$. In this example, $v_L(0) = V_1 - V_2 = 2\,\text{V}$. Now, L and C form a resonant circuit in which the energy oscillates between the energy-storing components. The inductor L maintains the charging current into C_2 even when $V_1 = V_2 = 1\,\text{V}$. Consequently, V_2 increases further and reaches a peak voltage of $V_1(0) = 2\,\text{V}$ while C_1 fully discharges to $V_1 = 0$. If we turn off the switch exactly at $V_2 = 2\,\text{V}$, the energy from C_1 is completely transferred to C_2 without any losses. This mechanism is the key idea behind the soft charging of a capacitor, sometimes called inductive charging.

A sinusoidal oscillation can be observed if the switch is turned on. A damped oscillation will occur since there will always be some series resistance in an actual circuit. If the resistance is reasonably low, we can assume that the peak value of V_2 reaches the initial voltage $V_1(0)$ of C_1. If R cannot be neglected, the losses may still be much lower than in the hard-switching case. However, if we keep the switch closed and let the oscillation run, V_2 will settle at $V_1(0)/2 (= 1\,\text{V}$ for the example in Fig. 11.4a)) and half the energy will be lost. This mechanism is the same as in the hard-charging case shown in Fig. 11.3.

We can analyze the soft-charging behavior more in detail based on the equivalent circuit in Fig. 11.4b). We recall the I-V expression of an inductor,

$$V(t) = L \frac{dI(t)}{dt}, \quad I(t) = \frac{1}{L} \int V(t) dt. \tag{11.11}$$

Figure 11.4 a) Charge transfer between two capacitors via an inductor; b) soft-charging of a capacitor from a voltage source V_{in} via an inductor L; c) energy $E_{in}(t)$ delivered by V_{in} and energy $E_C(t)$ stored on C for the setup in b). Lossless energy transfer occurs when E_{in} and E_C reach their peak value.

Similar to Eqns. (11.4) and (11.5), we can put together the loop equation and derive a differential equation,

$$V_{in} = L\frac{dI(t)}{dt} + I(t) \cdot R + \frac{1}{C}\int I(t)dt, \tag{11.12}$$

$$0 = \frac{d^2I(t)}{dt^2} + \frac{R}{L}\frac{dI(t)}{dt} + \frac{1}{LC}I(t). \tag{11.13}$$

Equation (11.13) is a homogeneous second-order differential equation with the generic solution

$$I(t) = I_o\, e^{-\delta t}\, \sin(\omega_d t), \tag{11.14}$$

where $\delta = R/2L$ is the damping coefficient and ω_d the natural frequency. In a suitable real design, we can neglect the damping in the first half-wave, i.e., $\delta = 0$. We can further assume the resonant frequency

$$\omega_o = \frac{1}{\sqrt{LC}}, \tag{11.15}$$

to be equal to ω_d. Hence, we can simplify Eqn. (11.14) and obtain

$$I(t) = I_o\, \sin(\omega_o t). \tag{11.16}$$

The amplitude value I_o can be determined from the initial conditions $v(t = 0) = 0$ and $i(t = 0) = 0$. First, we differentiate Eqn. (11.16),

$$\frac{dI(t)}{dt} = I_o\, \omega_o\, \cos(\omega_o t). \tag{11.17}$$

Applying the initial conditions to Eqn. (11.12) and inserting Eqn. (11.17) yields

$$V_{in} = L\frac{dI(t)}{dt} = LI_o\, \omega_o\, \cos(\omega_o t) \rightarrow V_{in} = LI_o\, \omega_o. \tag{11.18}$$

Rearranging the result of Eqn. (11.18) and applying the definition of ω_o from Eqn. (11.15) gives the amplitude of the current $I(t)$:

$$I_o = \frac{V_{in}}{L\omega_o} = V_{in}\sqrt{\frac{C}{L}} \tag{11.19}$$

The capacitor charging current $I(t)$ as given in Eqn. (11.16) is now fully defined,

$$I(t) = I_o\, \sin(\omega_o t) = V_{in}\sqrt{\frac{C}{L}}\, \sin(\omega_o t). \tag{11.20}$$

The capacitor voltage $V(t)$ follows from Eqn. (11.3) by integrating the current $I(t)$,

$$V(t) = \frac{1}{C}\int_0^\theta I(t)dt = \frac{I_o}{\omega_o C}\left[-\cos(\omega_o t)\right]_0^\theta = \frac{I_o}{\omega_o C}\left[-\cos(\omega_o t) + 1\right]. \tag{11.21}$$

The integration boundary θ can be replaced by the time t. It is also convenient to substitute I_o by the expression of Eqn. (11.19). This gives the final solution for the capacitor voltage $V(t)$ according the Fig. 11.4b):

$$V(t) = V_{in}\left(1 - \cos(\omega_o t)\right) \tag{11.22}$$

Both $I(t)$ and $V(t)$ are shown in Fig. 11.4b). At half the period, the voltage reaches its maximum, equal to $2V_{in}$. At this point, the current $I(t)$ is zero, which enables zero-current switching with minimum switching losses (see Section 2.9).

The energy delivered by the voltage source V_{in} is

$$E_{in} = \int V_{in} I(t)\,dt = \int V_{in} I_o \sin(\omega_o t)\,dt = E_o \left(1 - \cos(\omega_o t)\right), \tag{11.23}$$

where E_o corresponds to the energy amplitude

$$E_o = V_{in}^2 C. \tag{11.24}$$

Like in the hard-switching case, the energy E_C stored in capacitor C is given by Eqn. (11.8). We can insert the capacitor voltage of Eqn. (11.22) and obtain

$$E_C = \frac{E_o}{2} \left(1 - \cos(\omega_o t)\right)^2. \tag{11.25}$$

The energy loss E_{loss} corresponds to the difference between E_{in} and E_C. Unlike for the hard-switching case, E_{loss} is now a function of time because the energy commutes between the inductor and the capacitor. Figure 11.4c) shows that zero losses can be achieved if the switch turns off exactly at half the period (π/ω_o). We conclude that proper switching enables lossless energy transfer to the capacitor. Hybrid converters extensively utilize this effect to obtain higher conversion efficiency than purely capacitive or inductive DC–DC converters.

11.3 Basic Resonant SC Converter Stages

From the Section 11.2, we know that loss-less charging of a capacitor can be achieved if the charging path has an inductor in series. It is fulfilled in the fundamental hybrid converter stage in Fig. 11.2e) that we derived in Section 11.1. We pick this up and apply a two-phase gate control φ_1 and φ_2 to the switches S1 to S4 as shown in Fig. 11.5. The non-overlapping signals φ_1 and φ_2

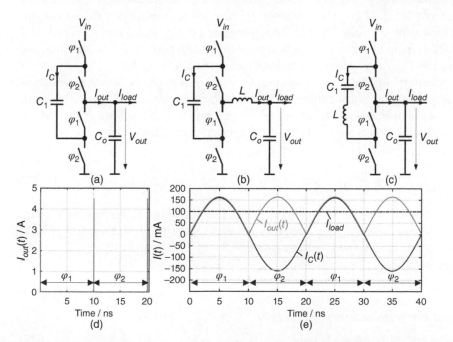

Figure 11.5 Fundamental 2:1 converter stages: a) capacitive (SC); b) direct resonant SC; c) indirect resonant SC; d) simulated output current $I_{out}(t)$ for a); e) simulated output current $I_{out}(t)$ and capacitor current $I_C(t)$ for both b) and c). Parameters: $I_{load} = 100\,\text{mA}, f_{sw} = 20\,\text{MHz}, C_1 = 1\,\text{nF}, C_o = 9\,\text{nF}$.

are the same as defined in Fig. 8.1c) for charge pump circuits and SC converters. For comparison, the basic 2:1 SC converter is depicted in Fig. 11.5a).

Two kinds of basic resonant SC converters can be obtained by adding an inductor as shown in Fig. 11.5b,c). Both topologies provide the same ideal voltage conversion ratio 2:1, like the original SC stage. If operated in resonant mode, they are named direct and indirect ReSC converters as introduced in [1]. The control signals φ_1 and φ_2 operate the converter at the resonance frequency f_o. The output buffer capacitor C_o is usually much larger than the flying capacitor C_1. Since C_o appears in series to C_1, its influence on the oscillation can be neglected. Hence, the resonance frequency is given by

$$f_o = \frac{\omega_o}{2\pi} = \frac{1}{2\pi\sqrt{LC_1}}. \tag{11.26}$$

The direct resonant SC converter in Fig. 11.5b) provides *direct* conversion because the inductor is always connected to the output capacitor C_o. The inductor current is unidirectional and always flows to the output. As an alternative, the inductor can also be placed in series to the flying capacitor C_1, which results in the indirect resonant SC converter shown in Fig. 11.5c). This topology operates with a bidirectional inductor current.

Figure 11.5d) shows the simulated current I_{out} that flows from the flying capacitor toward V_{out} in the SC converter of Fig. 11.5a). As expected, due to hard-charging, there are current spikes of high magnitude. Charge redistribution losses occur, which can be expressed by the equivalent output resistance R_{out} (see Section 9.5.4). Moreover, these current spikes are usually a root cause of electromagnetic interference (EMI).

In contrast, both resonant converters show sinusoidal current waveforms as depicted in Fig. 11.5e). In both types, the current I_C in the flying branch is fully sinusoidal while the current I_{out} into the output node is a full-wave rectified sine wave. Note the difference in the inductor current $I_L(t)$. In the direct topology of Fig. 11.5b), $I_{out}(t)$ is equal to the inductor current as shown initially in the left part of Fig. 11.1. The major power transfer in the inductor is at dc, resulting in a lower impact of inductor AC losses [1]. The indirect topology of Fig. 11.5c) sees a bidirectional sinusoidal inductor current. Since its average is zero, power components occur at harmonics of the resonance frequency f_o, causing higher AC losses than in the direct topology.

The simulated waveform in Fig. 11.5e) shows that I_{out} reaches zero precisely at the transitions between φ_1 and φ_2. Therefore, as an additional advantage over conventional SC conversion, both fundamental resonant converters provide zero-current switching (ZCS). As outlined in Section 2.9, lower switching losses due to ZCS further improve the conversion efficiency.

The mean value of I_{out} is equal to the load current I_{load},

$$I_{load} = \frac{1}{T}\int_0^T I_{out}(t)\mathrm{d}t = \frac{2I_o}{T}\int_0^{T/2}\sin(\omega_o t)\mathrm{d}t = \frac{2I_o}{\pi} \;\rightarrow\; I_o = \frac{\pi}{2}I_{load}. \tag{11.27}$$

Example 11.1 *The hybrid resonant SC converters in Fig. 11.5 operate at the resonance frequency $f_o = 50\,\text{MHz}$ and at a load current of $I_{load} = 100\,\text{mA}$. The value of the flying capacitor is $C_1 = 1\,\text{nF}$. Calculate the inductance L and the peak inductor current I_o.*

Rearranging Eqn. (11.26):

$$L = \frac{1}{C_1(2\pi f_o)^2} = \frac{100}{1\,\text{nF}(2\pi \cdot 50\,\text{MHz})^2} = 10.13\,\text{nH} \sim 10\,\text{nH} \tag{11.28}$$

The inductor peak current follows from Eqn. (11.27),

$$I_o = \frac{\pi}{2}I_{load} = 1.57 \cdot 100\,\text{mA} = 157\,\text{mA}. \tag{11.29}$$

This value matches the simulation in Fig. 11.5e).

The maximum voltage drop across the switches is either equal to V_{out} or $V_{in} - V_{out}$. With the ideal conversion ratio 2:1, the voltage stress becomes $V_{in}/2$. However, as described in [2], the direct topology of Fig. 11.5b) will see higher voltages across the switches. The maximum switch voltage increases by the voltage V_C across the flying capacitor C_1. V_C depends on L, C_1, the parasitic resistance (Q of the resonant tank), and the load current I_{load}. Nevertheless, the switch ratings may need to be chosen equal to V_{in} to ensure a safe start-up.

11.4 Frequency Generation and Tuning

Hybrid converters can achieve high conversion efficiencies with low inductance values, which requires high-frequency operation in the multi-megahertz range. For on-chip frequency generation, various oscillator types can be used. Figure 11.6a) shows a relaxation oscillator used in [3] at frequencies up to 50 MHz. The two NAND gates in the lower center form an RS flipflop. It toggles every time one of the capacitor voltages V_1 or V_2 exceed the reference level V_{ref} (typically a bandgap voltage, see Section 6.4). Due to the symmetrical structure, V_1 ramps up while V_2 is pulled to zero or vice versa. The capacitors charge at a constant rate $\Delta V/\Delta t = I_{osc}/C$ with $\Delta V = V_{ref}$. Two subsequent V_1 and V_2 charging cycles define one square wave period $T = 1/f_{sw}$ of the oscillator output voltage V_{out}. Hence, we find

$$f_{sw} = \frac{I_{osc}}{2\,C\,V_{ref}}. \tag{11.30}$$

Example 11.2 *Determine the capacitance C for the relaxation oscillator in Fig. 11.6a) operating at $f_{sw} = 50$ MHz. The current and reference voltage are $I_{osc} = 1\,\mu A$ and $V_{ref} = 0.5$ V, respectively. What area is required using a MOM capacitor (see Section 4.1) with a density of 0.3 fF/μm?*
  From Eqn. (11.30),

$$C = \frac{I_{osc}}{2f_{sw}\,V_{ref}} = \frac{1\,\mu A}{2 \cdot 50\,\text{MHz} \cdot 0.5\,\text{V}} = 20\,\text{fF}. \tag{11.31}$$

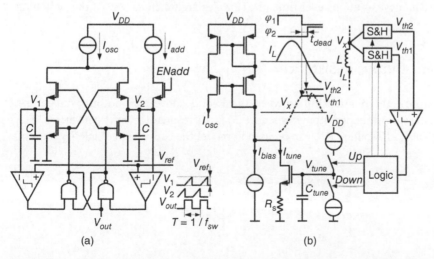

(a)   (b)

Figure 11.6 Frequency generation: a) relaxation ocscillator. Source: Adapted from [3]; b) frequency tuning circuit. Source: Adapted from [4].

This capacitance value requires an area of

$$A = \frac{C}{0.3\,\text{fF}/\mu\text{m}} = 67\,\mu\text{m}^2, \tag{11.32}$$

which corresponds to a square of about 8 μm *by* 8 μm.

As the frequency depends on the absolute value of C, it can be trimmed by adjusting I_{osc}. If I_{osc} is generated from a bandgap reference voltage, variations of V_{ref} and I_{osc} cancel out. The relaxation oscillator in Fig. 11.6a) shows an additional current I_{add} that feeds only into the right-hand branch. Enabled by a signal *ENadd*, it accelerates the charging speed of V_2 and provides the necessary duty cycle asymmetry required for some topologies. See, for instance, the 3:1 and 3:2 series parallel topologies discussed in Section 11.7 (Eqns. (11.61) and (11.62)).

Minimum R_{out} and ZCS can only be achieved if the ReSC converter operates exactly at the resonance frequency f_o according to Eqn. (11.26). Because L and C are often subject to large variation, tuning circuits are required. Figure 11.6b) shows an autotuning concept based on [4]. It incorporates a predictive control that controls the falling instant of clock φ_1, exactly to happen when the inductor current reaches zero. The procedure is similar to predictive dead-time control, as Section 2.8 explains. Consider the timing shown in the upper center of Fig. 11.6b). A sample-and-hold stage (S&H) stores the voltage at one terminal of the inductor (V_x) at the falling edge of φ_1, named V_{th1} in Fig. 11.6b). In [4] V_x is the voltage across a small ground-referred auxiliary winding wound on the inner radius of the actual on-chip resonator based on the indirect topology of Fig. 11.5c). In the direct stage of Fig. 11.5b) V_x corresponds to the switching node to the left of the inductor. During the dead-time interval between φ_1 and φ_2 the inductor voltage will rise if the inductor current is still positive (this case is shown in Fig. 11.6b)). Therefore, the second sampling voltage V_{th2}, taken shortly after the first sample, will be greater than V_{th1}. The opposite case, $I_L < 0$, leads to $V_{th1} < V_{th2}$. Along with some logic, a clocked comparator translates the difference between the sampled voltages into an up or down-pulse that controls a charge pump. Its output voltage V_{tune} connects to a transistor that operates as a simple voltage follower. It determines the current $I_{tune} \propto V_{tune}/R_s$ that is combined with a constant bias current I_{bias}. A cascode current mirror provides the tuned current I_{osc} that can directly be connected to the relaxation oscillator of Fig. 11.6a). In [4], the on-times of φ_1 and φ_2 are tuned individually. Due to the low switching frequency variations over time, the samples are taken at 1/32 of the switching frequency.

11.5 Equivalent Output Resistance

Figure 11.7a) shows how the relationship between the hybrid converter's output voltage and load current defines the equivalent output resistance R_{out}. In resonant hybrid converters, R_{out} changes depending on f_{sw}. Therefore, R_{out} can be used to model the conversion efficiency and the

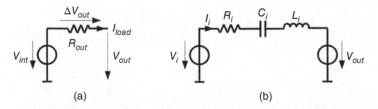

(a)　　　　　(b)

Figure 11.7 a) Equivalent output resistance R_{out}; b) hybrid converter equivalent circuit in any phase *i*.

regulation characteristics. We first analyze R_{out} for the case that the hybrid converter operates exactly at its resonance frequency. As investigated in Section 11.2.2, this will result in minimum charge redistribution losses and enable ZCS. In the second step, we expand the analysis and derive the equivalent output resistance when operating off the resonance point. This way, we can generate a plot of R_{out} over frequency and conclude how to control the hybrid resonant converter.

Each switching phase of the hybrid converter can be described by an equivalent R-L-C circuit according to Fig. 11.7b). The circuit is similar to Fig. 11.4b). Based on Fig. 11.7b), R_{out} can be derived. The approach is similar to the R_{out} analysis of SC converters in Section 9.5.4. We follow the full derivation of R_{out} presented at the end of Section 9.5.4. The equivalent circuits in Figs. 9.9 and 11.7 are equal except for the inductor. While this will result in different current equations for $I_i(t)$, the overall calculation procedure is identical. Therefore, Eqn. (9.55) applies to hybrid converters as well, repeated here for convenience,

$$R_{out} = \frac{\sum_i E_i}{f_{sw} \cdot q_{out}^2 \cdot \left(\sum a_{out}^i\right)^2}. \tag{11.33}$$

11.5.1 Equivalent Output Resistance at Resonance

From Section 11.2.2, we know that the current $I_i(t)$ in the R-L-C loop of Fig. 11.7b) can be modeled by Eqn. (11.14). As resonant operation requires a reasonable Q factor, the damping coefficient δ will be very low. Assuming $\delta = 0$, the current in the charge transfer loop becomes

$$I_i(t) = A_i \sin(\omega_i t), \tag{11.34}$$

where ω_i represents the angular resonance frequency $1/\sqrt{L_i C_i}$ of the equivalent R-L-C tank in phase i (similar to Eqn. (11.26)). Note the similarity to Eqn. (11.16). Depending on the equivalent configuration, ω_i will differ in each phase (see Section 11.7.1). The only unknown parameter in Eqn. (11.34) is A_i. It can be determined from the following consideration (see also Eqns. (9.57) and (9.58) in Section 9.5.4). The integral of $I_i(t)$ over the duration t_i of phase i is identical to the charge q_{out}^i delivered to the converter's output in phase i. For a two-phase converter, each phase takes $t_i = \frac{1}{2f_{sw}} = \frac{\pi}{\omega_o}$ and, thus,

$$a_{out}^i \cdot q_{out} = \int_0^{t_i} I_i(t)dt = \frac{A_i}{\omega_o} \left[-\cos(\omega_o t)\right]_0^{\pi/\omega_o} = \frac{2A_i}{\omega_o}. \tag{11.35}$$

The charge flow coefficients a_{out}^i are the same as for the corresponding SC converter, i.e., the SC circuit that remains if L is replaced by a short. Refer to Chapter 9, Section 9.5.1 for details. Rearranging Eqn. (11.35) and setting $\omega_o = \omega_i$ gives

$$A_i = \frac{1}{2} a_{out}^i q_{out} \omega_i. \tag{11.36}$$

With respect to Eqn. (11.33), we finally need to determine the energy E_i that is dissipated during phase i (similar to Eqn. (9.59) in the SC case),

$$E_i = \int_0^{t_i} R_i \cdot I_i^2(t)dt = R_i \cdot A_i^2 \left[\frac{t}{2} - \frac{\sin(2\omega_i t)}{4\omega_i}\right]_0^{\pi/\omega_i} = \frac{R_i A_i^2 \pi}{2\omega_i}. \tag{11.37}$$

Now we insert E_i from Eqn. (11.37) into Eqn. (11.33) and substitute A_i by Eqn. (11.36) to obtain

$$R_{out} = \frac{\sum_i R_i \omega_i \pi (a_{out}^i)^2}{8 f_{sw} \left(\sum a_{out}^i\right)^2}. \tag{11.38}$$

Since we consider the resonant case where $f_{sw} = \omega_i / 2\pi$ the R_{out} expression simplifies to

$$R_{out} = \frac{\sum_i R_i \pi^2 (a_{out}^i)^2}{4 \left(\sum a_{out}^i \right)^2}. \tag{11.39}$$

References [5] and [6] take the damping coefficient δ into account and derive R_{out} from the current, according to Eqn. (11.14), resulting in

$$R_{out} = \frac{\sum_i \frac{(a_{out}^i)^2}{C_i} \tanh \frac{R_i \pi}{4L_i \omega_i}}{2 f_{sw} \left(\sum a_{out}^i \right)^2} = \frac{\sum_i \frac{(a_{out}^i)^2}{C_i} \tanh \frac{\pi}{4Q_i}}{2 f_{sw} \left(\sum a_{out}^i \right)^2}. \tag{11.40}$$

Q_i is the quality factor of the resonant tank in phase i,

$$Q_i = \frac{1}{R_i} \sqrt{\frac{L_i}{C_i}}. \tag{11.41}$$

If Q_i is reasonably large, the tanh function can be approximated by its argument, which results in the same expression for R_{out} as given in Eqn. (11.39). As outlined in [6], the approximation of Eqn. (11.40) by Eqn. (11.39) gives only 2% error for $Q = 3$, and 20% error for $Q = 1$.

11.5.2 R_{out} of the Fundamental 2:1 Hybrid Converter

We can apply Eqn. (11.39) to the two types of fundamental hybrid resonant converters in Fig. 11.5. The direct and indirect stages have an identical R-L-C equivalent circuit according to Fig. 11.7b) in each phase.

The charge flow vectors of the 2:1 stage can be taken from Eqn. (9.17) in Section 9.5.1:

$$\begin{aligned}
\boldsymbol{a^1} &= \begin{bmatrix} a_{out}^1 & a_{c,1}^1 & a_{in}^1 \end{bmatrix} = \begin{bmatrix} \frac{1}{2} & \frac{1}{2} & \frac{1}{2} \end{bmatrix} \\
\boldsymbol{a^2} &= \begin{bmatrix} a_{out}^2 & a_{c,1}^2 & a_{in}^2 \end{bmatrix} = \begin{bmatrix} \frac{1}{2} & -\frac{1}{2} & 0 \end{bmatrix}
\end{aligned} \tag{11.42}$$

Consequently,

$$\begin{aligned}
\sum_i a_{out}^i &= a_{out}^1 + a_{out}^2 = \frac{1}{2} + \frac{1}{2} = 1 \\
\sum_i (a_{out}^i)^2 &= (a_{out}^1)^2 + (a_{out}^2)^2 = \frac{1}{4} + \frac{1}{4} = \frac{1}{2}.
\end{aligned} \tag{11.43}$$

Applied to Eqn. (11.39), we find

$$R_{out} = \frac{R_i \pi^2}{8} \approx R_i. \tag{11.44}$$

R_i corresponds to the sum of all resistances in the R-L-C loop per phase i. If neglecting other resistive components like interconnect resistance and the capacitor's equivalent series resistance, R_i is determined by the on-resistance R of the two active switches in each phase. Hence, $R_i = 2R$ if all switches have equal resistance R. In that case, R_{out} is close to the minimum achievable output resistance of the equivalent SC converter in fast-switching limit (FSL) operation given in Eqn. (9.46).

The equivalent output resistance of the SC converter in SSL (slow-switching limit) is given in Section 9.5.4, Eqn. (9.34). For the 2:1 stage at resonance with $f_{sw} = f_o$ Eqn. (11.26) yields

$$R_{out} = \frac{1}{4 C_1 f_{sw}} = \frac{\pi}{2} \sqrt{\frac{L}{C_1}}. \tag{11.45}$$

It is interesting to compare R_{out} of the SC and resonant SC 2:1 converter. With $R_{out,ReSC}$ from Eqn. (11.44) and $R_{out,SC}$ from Eqn. (11.45) we obtain

$$\frac{R_{out,SC}}{R_{out,ReSC}} = \frac{4}{\pi R_i} \sqrt{\frac{L}{C_1}} = \frac{4}{\pi} Q \approx Q, \tag{11.46}$$

where Q is the quality factor of the equivalent R-L-C loop given by Eqn. (11.41).

This result implies two options for hybrid resonant converters (at resonance frequency):

(1) It achieves Q times lower R_{out} at the same frequency.
(2) It can operate at Q times lower frequency with the same R_{out}.

As a benefit of (2), switching losses scale down by a factor of Q. It also relaxes the design requirements of critical circuit blocks regarding speed and bandwidth. Even low-quality factors $Q < 10$ make a big difference. We also note that for $Q = 1$, both converter types approach the same equivalent output resistance.

Example 11.3 *Calculate R_{out} at resonance and the quality factor Q of the 2:1 ReSC converter with equally sized switches of 250 mΩ each ($R_i = 500$ mΩ). We use the parameters $C_1 = C_i = 1$ nF and $L = 10.13$ nH of Example 11.1 which result in the resonance frequency $f_o = 50$ MHz. Compare R_{out} to a 2:1 SC converter at the same frequency.*

We can determine R_{out} from Eqn. (11.44),

$$R_{out} = \frac{R_i \pi^2}{8} = \frac{500 \text{ m}\Omega \cdot 9.87}{8} = 617 \text{ m}\Omega. \tag{11.47}$$

Based on Eqn. (11.41),

$$Q = \frac{1}{R_i} \sqrt{\frac{L}{C_1}} = \frac{1}{500 \text{ m}\Omega} \sqrt{\frac{10.13 \text{ nH}}{1 \text{ nF}}} = 6.4. \tag{11.48}$$

For the SC converter Eqn. (11.45) gives

$$R_{out} = \frac{1}{4 C_1 f_{sw}} = 3.5 \text{ }\Omega. \tag{11.49}$$

In accordance with Eqn. (11.46) this value is approximately Q times larger compared to $R_{out} = 617$ mΩ of the ReSC converter.

11.5.3 Equivalent Output Resistance Over Frequency

The expression for R_{out} in Eqn. (11.38) can be expanded beyond operation at the resonance point. For frequencies f_{sw} below the resonance frequency f_o, there are mainly two operating modes: discontinuous conduction mode (DCM) or subharmonic operation. A widely used operating mode in DCM is dynamic off-time modulation (DOTM). Above f_o, the inductor current will not reach zero anymore, resulting in CCM operation.

Dynamic Off-Time Modulation (DOTM)

For $f_{sw} < f_o$, a dead time is inserted between two resonant half-cycles where the inductor current has reached zero. Figure 11.8 shows how this reduces the effective switching frequency f_{sw} and, concurrently, the average inductor current. Regulation of the ReSC converter is achieved by varying the dead time (off-time), called dynamic off-time modulation (DOTM). As the off-time can be regulated continuously over a wide range, DOTM enables high-resolution voltage regulation.

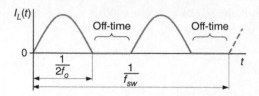

Figure 11.8 Inductor current $I_L(t)$ during dynamic off-time modulation (DOTM).

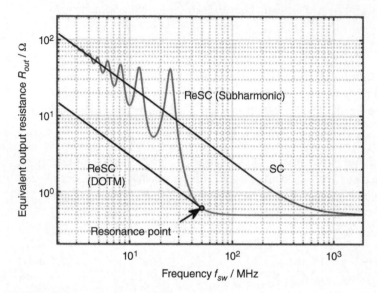

Figure 11.9 Equivalent output resistance R_{out} versus switching frequency f_{sw} of the hybrid resonant SC 2:1 converter in subharmonic mode and dynamic off-time modulation (DOTM) in comparison to a 2:1 SC converter.

Figure 11.9 depicts the equivalent output resistance of the basic 2:1 resonant converter over the switching frequency for various operating modes, including DOTM.

It is interesting to consider the load regulation at a given voltage conversion ratio, i.e., the required change of R_{out} in response to a varying load current I_{load}. The load current at $f_{sw} = f_o$ is described according to Eqn. (11.27):

$$I_{load,o} = 2f_o \int_0^{1/(2f_o)} I_o \sin(\omega_o t)\mathrm{d}t \rightarrow \int_0^{1/(2f_o)} I_o \sin(\omega_o t)\mathrm{d}t = \frac{I_{load,o}}{2f_o} \tag{11.50}$$

In a similar way and based on Fig. 11.8, we can determine the load current I_{load} in DOTM mode,

$$I_{load} = 2f_{sw} \int_0^{1/(2f_o)} I_o \sin(\omega_o t)\mathrm{d}t. \tag{11.51}$$

The integral term in Eqn. (11.51) can be substituted by Eqn. (11.50),

$$I_{load} = \frac{f_{sw}}{f_o} \cdot I_{load,o}. \tag{11.52}$$

We find the equation for R_{out} in DOTM by inserting I_{load} into the expression of the equivalent output resistance as defined in Fig. 11.7a),

$$R_{out} = \frac{\Delta V_{out}}{I_{load}} = \frac{\Delta V_{out}}{I_{load,o}} \frac{f_o}{f_{sw}} = R_{out,o} \frac{f_o}{f_{sw}}, \tag{11.53}$$

where $R_{out,o}$ represents the output resistance at resonance according to Eqn. (11.38) or, in case of the basic 2:1 stage, Eqn. (11.44). Interestingly, in DOTM, the ReSC converter has the same $1/f_{sw}$ dependency of R_{out} as the SC converter operating in SSL. However, R_{out} is offset by Q, as shown in Eqn. (11.46) and in Fig. 11.9.

As one of the advantages of DOTM, lowering the switching frequency toward lower power reduces the power loss at light load. The operation is similar to the pulse-frequency modulation (PFM) mode in purely inductive DC–DC converters (see Section 10.9.2).

Subharmonic Operation

Instead of DOTM, the R-L-C circuit can continue oscillating over the full period of $1/f_{sw}$. Because $f_{sw} < f_o$ multiple periods $1/f_o$ of the resonant half-sine fit into $1/f_{sw}$. It represents subharmonic oscillation with respect to f_{sw}. We can obtain R_{out} from the general expression in Eqn. (11.33). The required current equation can be found by solving the corresponding differential equation [3]. However, this leads to very complex expressions and tools like MATLAB® can be used to solve the system of equations and the final equation for R_{out}. This approach enables calculation R_{out} for different duty cycles and conversion ratios [3].

For the 2:1 converter, Pasternak et al. [6] present a general solution for R_{out}:

$$R_{out} = \frac{1}{4\,Cf_{sw}} \left(\frac{\sinh\left(\frac{R_i}{4\,Lf_{sw}}\right) + \frac{R_i}{4\,\pi Lf_o} \sin\left(\frac{\pi f_o}{f_{sw}}\right)}{\cosh\left(\frac{R_i}{4\,Lf_{sw}}\right) - \cos\left(\frac{\pi f_o}{f_{sw}}\right)} \right) \tag{11.54}$$

This equation is not only valid in subharmonic mode. It also applies at the resonance point as a specific form of Eqn. (11.40), and it holds in CCM ($f_{sw} > f_o$). However, the expression in Eqn. (11.54) does not cover DOTM due to the inserted off-time.

As an interesting observation, we can compare DOTM and subharmonic operation to DCM and forced continuous conduction mode in a purely inductive DC–DC converter (Chapter 10).

Continuous Conduction Mode (CCM)

For $f_{sw} > f_o$, the inductor current will be continuous. R_{out} reaches its minimum similar to R_{FSL} in an SC converter (see Section 9.5.4). If f_{sw} is slightly above f_o, it will still show a sine-wave-shaped ripple. For $f_{sw} \gg f_o$, the current will presume a triangular shape exactly like in an inductor-based DC–DC converter (see Chapter 10). In that case, the converter can be controlled by varying the duty cycle (PWM). R_{out} is determined by the total effective resistance in the resonant loop, mainly determined by the on-resistance of the active switches. Moreover, R_{out} is identical to the output resistance of the SC converter operating in FSL as given in Eqn. (9.44), repeated here for convenience:

$$R_{out,min} = 2 \sum_{i=1}^{n_{ph}} \sum_{j=1}^{n_{sw}} R_j \left(a_{rj}\right)^2 \tag{11.55}$$

The charge flow vector elements a_{rj} are defined in Section 9.5. For the 2:1 converter with equally sized switches of resistance R, we find $R_{out,min} = 2R$.

Due to the continuous current, there will be no benefit from ZCS anymore. However, soft-charging can often be achieved. In addition, the presence of one or more flying capacitors reduces the voltage across the inductor, enabling smaller inductance values L.

Typical hybrid converters with CCM operation are the 3L-buck, the FCML, and the double step-down (DSD) converters covered in Sections 11.10–11.12.

11.5.4   Equivalent Output Resistance Diagram

Figure 11.9 shows the relationship between the equivalent output resistance and switching frequency for the basic 2:1 resonant converter in Fig. 11.5b) in subharmonic and DOTM mode. There will be a similar plot for any hybrid ReSC converter. Like for the SC case in Chapter 9 (see Fig. 11.10, for instance), it is convenient to use a double-log scale. For comparison, the curve for a pure SC converter is included. At the resonance frequency $f_o = 50$ MHz, the output resistance reaches $R_{out,o} = 617$ mΩ, which is in line with Eqn. (11.40). This value is $Q = 6.4$ times lower than the SC converter. Likewise, to obtain the same effective resistance, the SC converter requires Q times higher switching frequencies in the order of $f_{sw} = 500$ MHz. Below the resonance point R_{out} gets larger as f_{sw} decreases. In the DOTM case, it follows the $1/f_{sw}$ dependency described by Eqn. (11.53). Subharmonic operation shows the typical behavior according to Eqn. (11.54). For $f_{sw} > f_o$, the output resistance approaches its minimum of $R_{out} = 2R$, which is identical to R_{FSL} of the SC converter. Note that the series resistance of the inductor may lead to a somewhat larger output resistance compared to an SC converter.

In the indirect topology, the output resistance does not stay flat beyond f_o and increases again [7] because the inductor current waveform is different compared to the direct converter (see Fig. 11.5e)). A negative inductor current means current backflow, which increases the effective output resistance for $f_{sw} > f_o$. Thus, beyond f_o, the graph in Fig. 11.9 only applies to the direct hybrid SC converter.

11.6   Control of Hybrid Converters

The conversion ratio of the hybrid ReSC, such as 2:1 for the basic topology of Fig. 11.2, provides a coarse control of the output voltage. At a given ideal conversion ratio N_i, fine control is achieved by modifying the equivalent output resistance R_{out} depending on the input voltage V_{in}, the load current I_{load}, and the target output voltage V_{out}. Equation (11.56) of the SC converter holds and can be rearranged to express R_{out} as a function of N_i, V_{in}, V_{out}, and I_{load},

$$R_{out} = \frac{N_i V_{in} - V_{out}}{I_{load}}. \tag{11.56}$$

The different mechanisms that determine R_{out}, described in Section 11.5, offer several control options for the converter. The most popular methods are DOTM, PWM, and switch conductance regulation (SwCR). At light load, hybrid converters may operate in SC mode controlled by frequency modulation [8, 9]. Each control method introduces losses associated with R_{out}.

11.6.1   Dynamic Off-Time Modulation (DOTM)

Dynamic off-time modulation (see Section 11.5.3) allows reducing the effective switching frequency starting from the resonance frequency f_o. Since the components in the resonant tank remain unchanged and result in sine wave oscillation at a fixed frequency f_o, a dead time is introduced after a half-cycle of the inductor current, as shown in Fig. 11.8.

We investigate how varying the switching frequency modulates R_{out} and the waveform of $I_{out}(t)$. In contrast to Eqns. (11.50) to (11.53) in Section 11.5.3, we keep the load current I_{load} constant. The charge redistribution losses in R_{out} are proportional to the root-mean-square (RMS) output current. Therefore, it is interesting to study how the RMS value of the output current $I_{out(t)}$ in DOTM changes

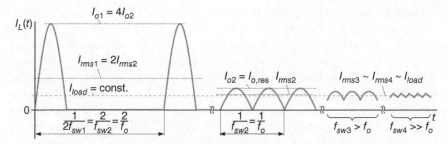

Figure 11.10 Multimode operation at constant load current I_{load} including DOTM ($f_{sw1} = f_o/4$), operation at the resonance point ($f_{sw2} = f_o$), and at frequencies f_{sw3} and f_{sw4} slightly and significantly beyond f_o. In DOTM, the peak current I_o has to increase to deliver the same average current I_{load}, leading to a much higher RMS current than at the resonance point. At $f_{sw} > f_o$, the RMS value is nearly equal to the DC load current.

over frequency (at constant load),

$$I_{out,rms} = \sqrt{2f_{sw} \int_0^{1/(2f_o)} I_o^2 \sin^2 \omega_o t \, dt} = \sqrt{\frac{f_{sw}}{f_o}} \frac{I_o}{\sqrt{2}}, \tag{11.57}$$

where I_o denotes the sine amplitude at the particular switching frequency $f_{sw} \leq f_o$. At f_o, Eqn. (11.57) gives the well-known RMS value of a sinusoidal signal. I_o can be calculated from Eqn. (11.52) with I_{load} from Eqn. (11.27):

$$I_o = I_{load} \frac{\pi}{2} \frac{f_o}{f_{sw}} = I_{o,res} \frac{f_o}{f_{sw}} \tag{11.58}$$

$I_{o,res}$ is the peak output current at the resonance point. Equation (11.58) shows that for a given constant load current I_{load} the amplitude I_o increases if f_{sw} is reduced in DOTM mode. Consequently, the RMS value of I_{out} rises, resulting in higher losses. Figure 11.10 shows the output current at a constant load for various operating modes. In DOTM at $f_{sw1} = f_o/4$ the resonant half-wave takes $1/(2f_o)$. It is followed by an off-time of $3/(2f_o)$. In phase φ_2, this sequence repeats such that the overall period time becomes $1/f_{sw1} = 4/f_o$. According to Eqn. (11.58) the peak value I_{o1} is four times higher than the amplitude $I_{o2} = I_{o,res}$ in resonance shown in the second part of Fig. 11.10 at $f_{sw2} = f_o$.

Figure 11.11 shows an exemplary DOTM control implementation based on hysteretic discrete-time control [3, 8]. The control loop modulates the off-time between the resonant pulses. The design incorporates a clocked comparator, a T-flipflop (toggle flipflop), and two

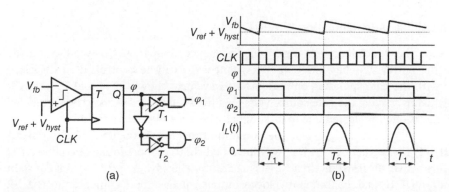

Figure 11.11 a) Implementation of DOTM control; b) timing diagram.

pulse-generators for each phase φ_1 and φ_2. There is also a system clock CLK from a global oscillator. V_{fb} is a measure of the converter's output voltage V_{out} generated by a resistive divider in the same way as in a linear regulator (R_1, R_2 in Fig. 7.2) or in an inductive DC–DC converter (R_{fb1} and R_{fb2} in Fig. 10.5). If V_{fb} drops below the reference value plus a hysteresis, i.e., ($V_{ref} + V_{hyst}$), the next rising CLK edge triggers the T-flipflop. Every change of its input logic level causes the T-flipflop to toggle its output level. The timing diagram in Fig. 11.11b) shows that the decrease in V_{fb} sets the general phase signal φ to 1. The generation of the switch control signals φ_1 and φ_2 is implemented by an AND gate and a delay chain, which can be tuned to exactly match the duration of the resonant half-sine. As investigated in Section 11.7.1, this duration may be different in each phase depending on the topology. Therefore, different pulse lengths T_1, T_2 can be adjusted. In [8], the delay chain consists of a current-starved inverter with a tunable delay, followed by a capacitor that connects to a Schmitt trigger and a final inverter. After the φ_1 pulse, V_{fb} exceeds ($V_{ref} + V_{hyst}$) again, and the inductor current remains zero during the off-time. The load current is supplied out of C_o such that V_{out} and V_{fb} gradually decrease. The rate of discharge and the duration of the off-time depend on the load current. Once the comparator detects $V_{fb} < (V_{ref} + V_{hyst})$ the next rising clock edge with toggle the output signal φ of the T-flipflop again to initiate the T_2 pulse and φ_2 as shown in Fig. 11.11b). The entire sequence of hysteretic control repeats alternating between φ_1 and φ_2.

11.6.2 Pulse-Width Modulation (PWM)

As the frequency increases beyond f_o, such as f_{sw3} and f_{sw4} in Fig. 11.10, the root-mean-square (RMS) current in the inductor decreases continuously and approaches DC (I_{load}). At $f_{sw4} \gg f_o$, the inductor current shows a triangular current ripple with a mean value equal to I_{load}. Hence, the circuit operates like inductor-based DC–DC converters covered in Chapter 10. Consequently, pulse-width modulation (PWM) can be applied to control the power conversion. For PWM control, more phases need to be introduced to connect the coil to ground between transitions as explained in Sections 11.10 and 11.11. Since there is no resonant operation, the hybrid converter no longer benefits from ZCS. However, soft-charging of the capacitors is still present because the capacitors get charged through the inductor (if the topology supports soft-charging in general, see Section 11.7.2). One of the significant benefits of PWM control is the higher power capability because the RMS inductor current and the related conduction losses get smaller relative to the output power (i.e., I_{load}, see Fig. 11.10). However, switching losses will become more dominant because f_{sw} increases (see the 3-level and FCML converters in Sections 11.10 and 11.11). At frequencies slightly above f_o, the converter may be operated in quasi-resonant mode with duty-cycle control but partially sinusoidal inductor current [2].

11.6.3 Switch Conductance Regulation (SwCR)

This method keeps the converter operating at the resonance frequency f_o. By adjusting the resistance R_i of the power switches, the output resistance R_{out} can be controlled as suggested by Eqn. (11.40) with the corresponding output current waveform shown in Fig. 11.12a). SwCR can be implemented by varying the gate-source voltage of the switch transistors or by segmented switches. The latter is often preferred due to its simplicity because the switch is built as a binary weighted array of parallel switch segments. The minimum resistance is obtained if all segments are turned on. If only the segment of the least-significant bit (LSB) or a predefined baseline offset resistance R_{offset} are active, the resistance reaches its maximum. The advantage of SwCR is constant frequency operation (EMI) and, consequently, lower output voltage ripple compared to DOTM, which operates at a lower frequency. If output voltage ripple is of concern, DOTM requires larger

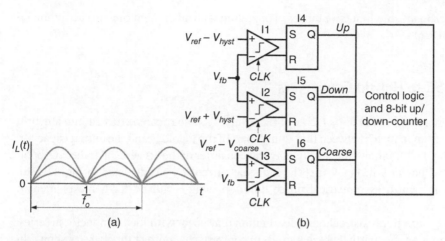

Figure 11.12 Switch conductance regulation (SwCR): a) inductor current $I_L(t)$ for varying switch on-resistance and b) implementation of the control loop.

output buffer capacitance [3]. On the downside, operating at resonance, switching losses are more dominant in comparison to DOTM.

Figure 11.12b) shows how SwCR can be implemented using hysteretic control (based on [3, 8]). The control loop aims to keep V_{fb} within a hysteresis window $V_{ref} \pm V_{hyst}$. For this reason, I1 and I2 form a window comparator. Implemented as clocked comparators, I1 and I2 achieve minimum propagation delay to support hybrid converter designs with switching frequencies in the range of 50 MHz. The output signals *up* and *down* of I1 and I2 are connected to a control logic block containing an 8-bit synchronous binary up/down counter. Its counter value determines the number of activated parallel switch segments. The higher the count, the lower the on-resistance. If V_{out} experiences a significant drop and falls below a threshold $V_{ref} - V_{coarse} < V_{ref} - V_{hyst}$ the control logic enters a coarse mode in which only the upper five bit of the counter are controlled. Renz and Wicht [3] investigate the transient behavior for various bit resolutions in the coarse mode and demonstrate that five-bit results in the best load step response, much better than using the full 8-bit counter.

The converter can go into SC mode to enhance light-load efficiency and apply frequency modulation [8, 9]. The transition happens if the load current has reached such a low value that all switch segments are disabled and only the baseline switch resistance R_{offset} is active. Setting this condition equal to the overdamped case of the equivalent resonant tank allows us to determine R_{offset}. The overdamped case corresponds to $Q = \frac{1}{2}$, hence,

$$Q = \frac{1}{R}\sqrt{\frac{L}{C}} = \frac{1}{2} \rightarrow R_{offset} = 2\sqrt{\frac{L}{C}}. \tag{11.59}$$

Example 11.4 *For $L = 10\,\text{nH}$ and $C = 1\,\text{nF}$ we calculate*

$$R_{offset} = 2\sqrt{\frac{L}{C}} = 2\sqrt{\frac{10\,\text{nVs/A}}{1\,\text{nAs/V}}} = 6.32\,\Omega. \tag{11.60}$$

11.6.4 Multi-Mode Operation

A wide load range can be covered by combining the above control methods as illustrated in Fig. 11.10. Multi-mode control [2] leads to a single converter that can transition between resonant, quasi-resonant, and inductive modes of operation, and all of them can also be combined with the DOTM (DCM).

11.7 From SC to Hybrid Converters

Based on the approach shown in Fig. 11.2, any SC converter topology can be turned into a hybrid converter by adding an inductor between the "flying" part of the topology and the output capacitor. As an example, Fig. 11.13a) shows the resulting hybrid converter of the 3:1 series-parallel SC topology introduced in Chapter 9 in Fig. 9.4a). The inductor current $I_L(t)$ is depicted in Fig. 11.13b). We immediately recognize asymmetrical phase duration, which will be further explained in Section 11.7.1.

If multiple flying capacitors exist, charge redistribution may occur without the inductor in series. Soft-charging cannot or only partially be achieved. In contrast, the indirect topology, as shown in Fig. 11.5c), always supports soft-charging. However, it requires placing an inductor in series to every flying capacitor. It is usually not an option, especially in topologies with higher-order conversion ratios incorporating many flying capacitors. Fortunately, some standard *direct* conversion stages, such as the series-parallel and Fibonacci topologies, support full soft-charging as analyzed below in Section 11.7.2. Other direct topologies, Dickson in particular, can be modified to become soft-charging capable. This topic will be explored in Section 11.7.3.

11.7.1 Asymmetrical Phases

The inductor current of the 3:1 converter in Fig. 11.13b) exhibits asymmetrical phases because, unlike the basic 2:1 hybrid converter, the L-C-R component values in the resonant loop are different in each switching phase. The inductor will always be the same, but the resistance and the capacitance may change. Hence, during phase φ_i the resonant loop of the converter oscillates at

$$f_{o,i} = \frac{1}{2\pi\sqrt{LC_{eff,i}}}, \tag{11.61}$$

where $C_{eff,i}$ is the effective capacitance. The frequency $f_{o,i}$ corresponds to a period time equal to $1/f_{oi}$ of phase φ_i. The total period time is the sum of the period times of all phases.

Consider the 3:1 hybrid series-parallel converter of Fig. 11.13a). Its equivalent circuits in both phases can be taken from the SC case in Fig. 9.7a). During φ_1 the flying capacitors C_1 and C_2 appear in series resulting in $C_{eff1} = C/2$ for equal capacitors $C_1 = C_2 = C$. In phase φ_2, both capacitors are

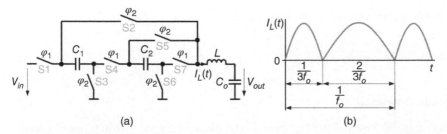

(a)   (b)

Figure 11.13 a) A hybrid 3:1 converter based on the series-parallel SC topology of Fig. 9.4a); b) inductor current $I_L(t)$ at resonance f_o.

connected in parallel. Hence, $C_{eff2} = 2C$. Inserting into Eqn. (11.61) gives

$$f_{o1} = \frac{1}{2\pi\sqrt{L}}\sqrt{\frac{2}{C}} \quad \text{and} \quad f_{o2} = \frac{1}{2\pi\sqrt{L}}\frac{1}{\sqrt{2C}} = \frac{1}{2\pi\sqrt{L}}\frac{1}{2}\sqrt{\frac{2}{C}}, \tag{11.62}$$

which indicates that f_{o2} is only half the value of f_{o1} and its period time is twice as large. Figure 11.13b) illustrates the resulting current waveforms in the 3:1 converter. The effect of unequal period times is not restricted to the series-parallel configuration and applies to any topology. The oscillator needs to provide a time base for different period times (see Section 11.4). For this reason, the relaxation oscillator in Fig. 11.6a) has a separate current source I_{add} that shortens the duration of φ_2.

Example 11.5 *What are the resonance frequencies f_{o1} and f_{o2} and the corresponding period times if $L = 10\,nH$ and $C = 1\,nF$?*

We can directly insert the given values into Eqn. (11.62) and obtain

$$f_{o1} = \frac{1}{2\pi\sqrt{L}}\sqrt{\frac{2}{C}} = \frac{1}{2\pi\sqrt{10\,nH}}\sqrt{\frac{2}{1\,nF}} = 71.2\,MHz, \tag{11.63}$$

$$f_{o2} = \frac{1}{2\pi\sqrt{L}}\frac{1}{\sqrt{2C}} = \frac{1}{2\pi\sqrt{10\,nH}}\frac{1}{\sqrt{2\cdot 1\,nF}} = 35.6\,MHz. \tag{11.64}$$

These results correspond to period times of

$$T_{o1} = \frac{1}{f_{o1}} = \frac{1}{71.2\,MHz} = 14.05\,ns \quad \text{and} \quad T_{o2} = \frac{1}{f_{o2}} = \frac{1}{35.6\,MHz} = 28.1\,ns. \tag{11.65}$$

11.7.2 Analysis of the Soft-Charging Capability

We can analyze the soft-charging capability using a method similar to the charge-flow analysis introduced in Section 9.5 for SC converters. The method determines the voltage changes ΔV across all flying capacitors for each phase due to the charge being delivered to the output. Knowing ΔV and the charge flow for the particular flying capacitor, the capacitance $C = \Delta V/q$ can be calculated. Soft-charging cannot be achieved if the analysis results in infinite or negative values of C.

We will see that the relative normalized capacitance will be sufficient to assess whether the topology is soft-charging compatible. Likewise, the absolute value ΔV is not required. ΔV is instead an indicator which flying capacitor sees which voltage change. Before we investigate common SC topologies based on Fig. 11.14, we need to define some assumptions:

(1) The capacitor voltage change must sum to zero over all switching phases. For the typical case of two phases: $\Delta V_{c,i}^1 + \Delta V_{c,i}^2 = 0$, where the exponents 1 and 2 indicate the phases φ_1 and φ_2, respectively.
(2) For each loop of the topology, Kirchhoff's voltage law (KVL) has to be fulfilled in all phases.
(3) ΔV is usually small such that the rules of a small-signal analysis can be applied. Consequently, the constant input voltage source V_{in} is replaced by a short as depicted in Fig. 11.14 in phase φ_1.
(4) The inductor together with V_{out} is replaced by a constant current source in between the switching node V_{sw} and ground as shown in Fig. 11.14. The current source carries I_{load} because the average inductor current equals the load current I_{load}. The small-signal analysis would remove the constant current source, leaving V_{sw} floating. While this would represent the circuit behavior well, we keep the current source to indicate the inductor current toward the load, independent of V_{sw} in each phase.

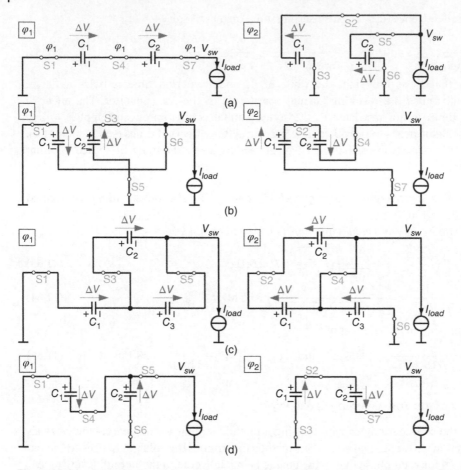

Figure 11.14 Analyzing the voltage changes across each flying capacitor of common 3:1 topologies: a) series-parallel; b) Dickson; c) ladder; d) Fibonacci. See also Figs. 9.4 and 9.7 in Chapter 9.

Item (4) is the main benefit of a hybrid converter over a standard SC topology because the voltage across the inductor can vary. It adds a degree of freedom in KVL, and soft-charging may be achieved.

Figure 11.14 shows all common SC topologies covered in Chapter 9, see Figs. 9.4 and 9.7. Consider the equivalent circuits of the series-parallel converter in phases φ_1 and φ_2 in Fig. 11.14a). With the load current supplied toward the output, the voltage change ΔV across C_1 will be positive from the converter's input to the output as indicated. Due to assumption (1) (see above), we can immediately draw the voltage drop in phase φ_2 in opposite polarity compared to φ_1. C_1 and C_2 are in parallel during φ_2, which defines the same voltage drop ΔV to occur across C_2 (fulfilling KVL in line with the assumption (2)). According to assumption (1), we can go back to phase φ_1 and add ΔV at C_2 in opposite polarity (with respect to φ_2).

After the voltage changes across each flying capacitor C_j have been determined, we can summarize them in a vector similar to the charge flow vector introduced in Section 9.5.1. Since the voltage change is expressed in multiples of ΔV and we are not interested in the actual value of ΔV, we normalize the voltage change to ΔV,

$$V = \begin{bmatrix} \Delta V_{c,1} & \Delta V_{c,2} & \cdots & \Delta V_{c,n} \end{bmatrix} / \Delta V. \tag{11.66}$$

The vector V denotes the normalized voltage changes across the flying capacitors. Because the voltage change differs only by its sign in each phase, it is sufficient to consider one phase only. We go for the first phase.

For the series-parallel converter of Fig. 11.14a) we find

$$V = \left[\Delta V_{c,1} \;\; \Delta V_{c,2}\right]/\Delta V = \left[+\Delta V \;\; +\Delta V\right]/\Delta V = \begin{bmatrix}1 & 1\end{bmatrix}. \tag{11.67}$$

We find the capacitances from $C_j = q_{c,j}/\Delta V_{c,j}$ or, in normalized form,

$$C_j = \frac{q_{c,j}}{q_x} \cdot \frac{\Delta V}{\Delta V_{c,j}}, \tag{11.68}$$

where $q_{c,j}$ is the corresponding element of the charge flow vector (see Eqn. (9.12)) and q_x the corresponding charge quantity for this topology (see Eqn. (9.16) and Fig. 9.7 in Section 9.5.1).

For the 3:1 topologies, the charge flow vectors are listed in Table 9.3 and repeated in Table 11.2. For the series-parallel converter in phase φ_1, we obtain

$$\boldsymbol{q}_c = \left[q_{c,1} \;\; q_{c,2}\right]/q_x = \begin{bmatrix}1 & 1\end{bmatrix}, \tag{11.69}$$

where $\boldsymbol{q}_c$ refers to the normalized capacitor charge flow vector (i.e., the inner part of the general charge flow vector without the elements referring to V_{in} and V_{out}).

Inserting Eqns. (11.69) and (11.67) into Eqn. (11.68) gives

$$C_1 = \frac{1}{1} = 1, \quad C_2 = \frac{1}{1} = 1, \tag{11.70}$$

or in vector form

$$\boldsymbol{C} = \begin{bmatrix}C_1 & C_2\end{bmatrix} = \begin{bmatrix}1 & 1\end{bmatrix}. \tag{11.71}$$

This result proves that the hybrid 3:1 series-parallel converter is fully soft-charging capable if both flying capacitors have the same capacitance. This constraint is in line with the optimum capacitor sizing outlined in Section 9.8.1 (see Example 9.4). For other topologies and configurations, the soft-charging constraint may not align with the requirements for the sizing according to the charge flow. However, this is usually acceptable, given the benefit of achieving soft charging. Further analysis proves that higher-order series-parallel converters are fully soft-charging capable as long as all flying capacitors have identical sizes [10].

We can analyze the remaining topologies following the same procedure. Table 11.2 lists the capacitor charge flow vectors $\boldsymbol{q}_c$, the voltage change vectors V, and the capacitance vectors, all in normalized form. We immediately recognize that, like Series-Parallel, the 3:1 Fibonacci and Dickson topologies achieve soft-charging if all capacitors are equal. The Ladder converter is unsuitable for soft-charging because it would require a negative capacitance of C_2, i.e., a capacitor, which increases its voltage whenever it is discharged. Such a capacitor does not exist.

Table 11.2 Voltage-change analysis of common 3:1 hybrid SC topologies (referred to Fig. 11.14).

Series-Parallel	$\boldsymbol{q}_c = \begin{bmatrix}1 & 1\end{bmatrix}, V = \begin{bmatrix}1 & 1\end{bmatrix}, C = \begin{bmatrix}1 & 1\end{bmatrix}$
Dickson	$\boldsymbol{q}_c = \begin{bmatrix}1 & -1\end{bmatrix}, V = \begin{bmatrix}1 & -1\end{bmatrix}, C = \begin{bmatrix}1 & 1\end{bmatrix}$
Ladder	$\boldsymbol{q}_c = \begin{bmatrix}1 & -1 & 2\end{bmatrix}, V = \begin{bmatrix}1 & 1 & 1\end{bmatrix}, C = \begin{bmatrix}1 & -1 & 2\end{bmatrix}$
Fibonacci	$\boldsymbol{q}_c = \begin{bmatrix}1 & -1\end{bmatrix}, V = \begin{bmatrix}1 & -1\end{bmatrix}, C = \begin{bmatrix}1 & 1\end{bmatrix}$

As an alternative to assessing the soft-charging capabilities based on the voltage change presented above, it can also be mathematically analyzed as suggested in [11]. The procedure starts with assembling a linear system of equations based on the KVL loop matrix in each switching phase. After bringing the matrix in each phase into its reduced echelon form, the basis vectors for the null space can be found, and subsequently, the voltage changes across each capacitor. The capacitances are calculated using Eqn. (11.68). While the mathematical approach of [11] will be faster once a script in MATLAB® or Python® is available, the ΔV-method from [10] provides intuition about the topology. Both methods lead to the same results.

Note that the results in Table 11.2 apply to the standard stages with a voltage conversion ratio of 3:1. For higher conversion ratios and any other topology, the soft-charging capabilities need to be evaluated based on one of the described methods. It can be shown that all series-parallel and Fibonacci topologies are naturally soft-charging compatible if the flying capacitors have the same size. In contrast, the ladder topology never achieves soft charging. Therefore, this topology will have minimal benefit from adding an inductor at the output node. Soft-charging can be achieved by implementing the ladder topology as an indirect conversion stage with the inductor added in series to every flying capacitor [12, 13].

It is interesting to study higher-order Dickson conversion stages. Whether this topology can achieve soft charging depends on the conversion ratio. While the 3:1 converter is fully soft-charging capable if the flying capacitors are equally sized, the 4:1 implementation in Fig. 11.15 is not. Let us do the analysis starting with the capacitor charge flow vector in phase φ_1 (see Fig. 11.15b)),

$$\boldsymbol{q}_c = \begin{bmatrix} 1 & -1 & 1 \end{bmatrix}. \tag{11.72}$$

During phase φ_2, the vector will be identical except that all elements must be multiplied by -1. To determine the voltage change, we start with the voltage drop of $+\Delta V$ across C_3, see Fig. 11.15b). During φ_2, the polarity of the voltage at C_3 will change as indicated in Fig. 11.15c). Note that the absolute value of the voltage change does not vary. To determine the voltage changes at C_1 and C_2, we put together the KVL loop equations in each phase:

$$\begin{aligned} \varphi_1 &: \ \Delta V_{c,3}^1 = \Delta V_{c,1}^1 - \Delta V_{c,2}^1 \\ \varphi_2 &: \ \Delta V_{c,3}^2 = \Delta V_{c,1}^2 + \Delta V_{c,2}^2 \end{aligned} \tag{11.73}$$

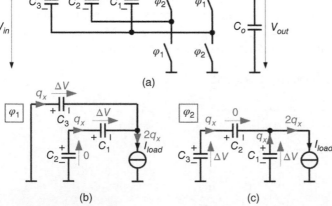

Figure 11.15 Soft-charging analysis of the hybrid 4:1 Dickson converter: a) Topology and b,c) equivalent circuits during phases φ_1, φ_2.

From assumption (1) above, we require $\Delta V^1_{c,i} + \Delta V^2_{c,i} = 0$. Hence, adding up both equations in (11.73) yields $\Delta V_{c,2} = 0$ as indicated in the equivalent circuits in Fig. 11.15. This result also determines $\Delta V_{c,1}$ in both phases, shown in Fig. 11.15. The voltage change vector in phase φ_1 becomes

$$V = \begin{bmatrix} 1 & -0 & 1 \end{bmatrix}. \tag{11.74}$$

The signed zero indicates that the required capacitance to ensure soft-charging is not negative [10], which can be the case for other topologies such as ladder. As a result of Eqn. (11.74), the 4:1 Dickson topology would require an infinitely large capacitor $C_2 = q_{c,2}/\Delta V_{c,2}$. Consequently, the topology cannot achieve soft-charging. However, the Dickson converter approaches soft-charging as C_2 increases to about five times higher than the capacitance of C_1 and C_3. This effect is investigated in [10] by comparing the ratio between the RMS current and the average current in C_3. Full soft charging means equal RMS and average current.

Toward higher conversion ratios, it can be found that the Dickson topology can achieve soft-charing for any odd conversion ratios (5:1, 7:1) but not for any even ratio (4:1, 6:1). Dickson topologies with even ratios may be modified to achieve full soft-charging by adding additional switching phases as described in Section 11.7.3.

11.7.3 Split-Phase Control

The analysis of the hybrid 4:1 Dickson converter shows it is not inherently soft-charging. By applying the technique of split-phase control as proposed in [14], any Dickson converter can turn into a soft-charging topology. However, it is not required for Dickson converters with odd conversion ratios because they are inherently fully soft-charging compatible (see Section 11.7.2). The key idea is to introduce two secondary switching phases in addition to the initial phases. We will explore this technique for the 4:1 Dickson converter. Figure 11.16 shows the equivalent circuits for the four phases called φ_{1a}, φ_{2a} (identical to the original phases φ_1, φ_2 in Fig. 11.15), and φ_{1b}, φ_{2b} (secondary phases). To understand the operation of split-phase control, it appears easier to consider the large-signal behavior, which includes the input voltage source V_{in}. Note that V_{in} is only connected during phase φ_{1a}. Operating the converter with conventional two-phase control KVL results in two constraints for complete soft-charging:

$$\begin{aligned} \varphi_{1a} &: V_{in} - V_{c,3} = V_{c,2} - V_{c,1} \\ \varphi_{2a} &: V_{c,3} - V_{c,2} = V_{c,1} \end{aligned} \tag{11.75}$$

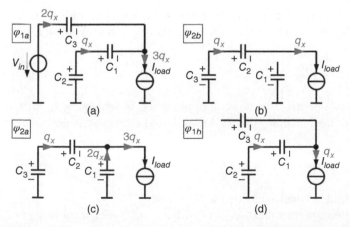

(a)

(b)

(c)

(d)

Figure 11.16 The hybrid 3:1 Dickson converter with split-phase control: Adding two secondary phases φ_{1b} and φ_{2b} to the exiting phases φ_{1a} and φ_{2a} (see Fig. 11.15) enables soft-charging.

This relationship is similar to the conditions in Eqn. (11.73) found by small-signal analysis. Both constraints can be satisfied at zero load current, for instance, for $V_{in} = 4$ V, $V_{c,1} = 1$ V, $V_{c,2} = 2$ V, and $V_{c,3} = 3$ V. However, if a load current is present, a charge has to be transferred through the capacitors during the switching phases, leading to a voltage drop across the capacitors. With this additional voltage drop, both constraints cannot be satisfied simultaneously [10, 11]. At the transition from φ_{2a} to φ_{1a} the voltages $V_{in} - V_{c,3}$ and $V_{c,2} - V_{c,1}$ are different. Likewise, the second condition is not met when entering φ_{2a}.

With split-phase control, the secondary phases occur between the transitions and ensure that the conditions of Eqn. (11.75) are fulfilled. The switching sequence is $\varphi_{1a} \rightarrow \varphi_{2b} \rightarrow \varphi_{2a} \rightarrow \varphi_{1b}$ in cyclic order. When entering φ_{2b} from φ_{1a} the capacitors are rearranged such that C_2 and C_3 are already in series as required in the subsequent phase φ_{2a}. C_1 gets disconnected such that its voltage $V_{c,1}$ remains the same as in phase φ_{1a}. During φ_{2b}, the load current I_{load} charges C_2 and discharges C_3 until $V_{c,3} - V_{c,2}$ reaches exactly $V_{c,1}$ such as to fulfill the second condition in Eqn. (11.75). Full soft-charging is achieved as C_2 and C_3 charge via a current source, i.e., the inductor in a real hybrid converter. Now, the converter can transition into phase φ_{2a} without any charge redistribution. In general, the "b"-phases always precede the actual "a"-phases. So does also φ_{1b} with respect to φ_{1a}. In phase φ_{1b}, C_3 is disconnected while C_1 and C_2 get charged and discharged, respectively. The soft-charging applies here also because C_1 and C_2 are in series to the current source (the inductor). Once the first condition of Eqn. (11.75) is met, phase φ_{1a} can be entered.

To ensure that the conditions of Eqn. (11.75) are met at the end of both "b"-phases, their duration must be controlled precisely and different from the "a"-phases. This result can be derived from the charge flow analysis over all phases of Fig. 11.16. See Section 9.5 for details on charge flow analysis. The investigation is more challenging as we have four phases instead of two conventional switching phases. However, this approach leads to the results indicated by the charge flow quantity q_x in Fig. 11.16:

(1) The charge flow associated with C_1 and C_3 is zero in phases φ_{1b} and φ_{2b}, respectively, because of the disconnected capacitor.
(2) The series connection of C_1 and C_2 in phases φ_{1a} and φ_{1b} sees the same charge flow q_x. Likewise, the chargeflow in C_2 and C_3 is q_x during φ_{2a} and φ_{2b}.
(3) The charge flow in each capacitor over all phases needs to sum up to zero (similar to Eqn. (9.15) in Section 9.5). Hence, in phase φ_{1a} capacitor C_3 requires a charge flow of $2q_x$. We also find for C_1 during φ_{2a} a charge flow of $-2q_x$.

If we sum up the total charge delivered to the output over one switching cycle with all four phases, we obtain $q_{out} = 8q_x$. Similar to Eqn. (9.13) from the general charge flow analysis, for a constant load current, we can define

$$I_{load} = \frac{q_{out}}{T} = \frac{q_{1a}}{t_{1a}} = \frac{q_{2a}}{t_{2a}} = \frac{q_{1b}}{t_{1b}} = \frac{q_{2b}}{t_{2b}}, \tag{11.76}$$

where $T = 1/f_{sw}$ is the duration of one period. The duration of phase φ_{1a} is denoted by t_{1a} with an output charge q_{1a} and so forth for the remaining phases. From Eqn. (11.76), we can derive the following duty ratios:

$$D_{1a} = \frac{t_{1a}}{T} = \frac{q_{1a}}{q_{out}} = \frac{3}{8}, \quad D_{2a} = \frac{3}{8}, \quad D_{1b} = \frac{1}{8}, \quad D_{2b} = \frac{1}{8} \tag{11.77}$$

It means both "b"-phases are one-third in length compared to "a"-phases.

Besides the different phase durations, no modification of the original 4:1 Dickson converter is necessary. The switching sequence only delays the control of the left- and right-most switches in the upper branch of Fig. 11.15a) (between V_{in} and the inductor) [14]. In particular, the left-most switch keeps off during φ_{1b} and turns on in phase φ_{1a}, the right-most switch remains off in phase φ_{2b} and turns on when entering φ_{2a}. A 4:1 hybrid Dickson converter with implemented split-phase control is presented [15]. Operating in CCM with PWM control, the design achieves conversion ratios between 4:1 and 15:1 from a Li–Ion battery input. External passives of $4 \times 22\,\mu F$ and $180\,nH$ support up to 1.5 A of load current. Benefiting from soft-charging, a peak conversion efficiency of 94.2% is achieved.

11.7.4 Boost Converters

As with the SC case, step-up conversion (boost) can be achieved in hybrid converters by swapping the input and output of the converter. The charge pump designs of Chapter 8 can be converted into a hybrid boost converter by adding a series inductor between the *input* voltage and the SC network.

The fundamental indirect topology of Fig. 11.2c) can also be operated in CCM (quasi-resonant, $f_{sw} > f_o$) as a boost converter [16]. The series combination of L and C follows a similar switching sequence as in a purely inductive boost converter (Section 10.10).

11.8 Multi-phase Converters

By operating multiple parallel hybrid conversion stages time-interleaved, the same benefits as for multiphase SC converters can be achieved (see Section 9.12). The main advantages are higher output power and lower output voltage ripple at reduced output buffer capacitance. Several hybrid converter designs implement two indirect stages according to Fig. 11.2c) [7, 16, 17].

Figure 11.17 shows the design published in [7]. It operates at resonance and also supports DOTM with the two power stage branches controlled 180° out of phase at supply voltages of 3.6 to 6 V. All capacitors are implemented on-chip using MIM capacitors (see Section 4.1.3) with a density of 6.6 fF/μm^2. The flying capacitors C_{X1}, C_{X2} are 9 nF. Due to the time-interleaving operation, the bypass capacitors can be minimized to $C_{bp} = 11$ nF, placed at V_{out} and at V_{in}. Two external surface-mounted device (SMD) inductors are mounted on top of the IC. The 2:1 converter achieves efficiencies of 85% for $L_x = 5.5$ nH and 82.5% for $L_x = 1.9$ nH.

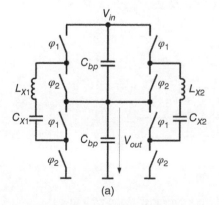
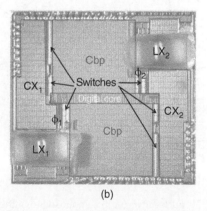

(a)   (b)

Figure 11.17 a) Two-phase hybrid converter and b) corresponding die photo with mounted SMD inductors. Source: Kapil Kesarwani et al. [7]/from IEEE.

The multiphase converter in [16] implements three indirect stages in parallel in quasi-resonant operation. It demonstrates that hybrid operation can be achieved from 1.1 nH printed circuit board (PCB) trace inductors. On the downside, multiphase hybrid converters require a significant number of inductors. In contrast to multiphase SC converters (see Section 9.12) it is usually not desirable to divide their inductance by the number of phases as it increases the resonant frequency per phase.

11.9 Multi-Ratio Converters

The multi-ratio SC converter of Fig. 9.18 is inherently a series-parallel topology. According to the analysis of Section 11.7.2, the topology is soft-charging compatible. It, therefore, can be combined with an inductor toward the output to form a hybrid multi-ratio DC–DC converter that offers three conversion ratios (3:1, 2:1, 3:2) as implemented in [9]. See Fig. 1.9a) for a die photo. As a general approach, in the same way, any multi-ratio SC converter can be converted into a hybrid topology. The resulting hybrid converter offers multiple conversion ratios and high conversion efficiency over a broader input and output voltage range. This approach combines the advantages of both the multi-ratio and the hybrid conversion.

11.10 The Three-Level Buck Converter

The three-level buck converter (3L-buck converter) in Fig. 11.18a) is a popular hybrid converter that operates at frequencies far beyond the resonance frequency f_o. Therefore, it can be controlled by pulse-width modulation and maintains a continuous inductor current I_L, i.e., the converter operates in CCM similar to a purely inductive converter as described in Chapter 10. The 3L-buck

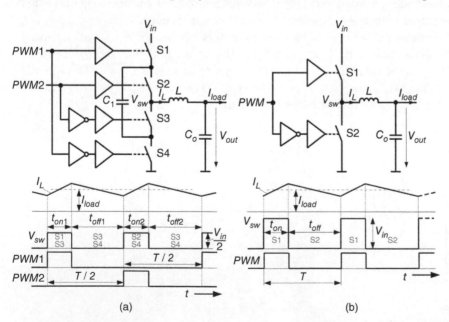

(a)   (b)

Figure 11.18 a) The 3L-buck converter in comparison to b) the 2L-buck converter along with their waveforms of inductor current I_L, switching node voltage V_{sw} and PWM control signals (shown for the 3L-buck converter for the case of $V_{out} < V_{in}/2$).

converter is an excellent example of the hybridization of inductive DC–DC converters, as discussed in Section 11.1, see also Fig. 11.2. In that sense, the fundamental inductive converter of Fig. 10.1 represents a two-level converter, shown in Fig. 11.18b) for comparison.

An excellent introduction to the three-level operation can be found in [18] and [19]. The 3L-buck converter evolves from the 2L-buck by doubling the low-side and high-side switches and adding a flying capacitor C_1. C_1 can be considered as a voltage source that is balanced at $V_{in}/2$ in an ideal steady state. This way, the switching node voltage V_{sw} alternates not only between V_{in} and ground like for the 2L-buck. It can also assume $V_{in}/2$ as a third level, which explains the name *three-level buck converter*.

The three-level operation brings several advantages over the conventional 2L-buck converter: (1) Because the voltage across the switches is reduced by half, transistors with lower voltage ratings can be used. It leads to (2) lower switching losses due to lower gate charge and because the voltage swing at the switching nodes of the power stage is reduced by half. (3) Lower voltage across the inductor decreases the inductor current ripple, which translates into a lower output voltage ripple and allows for smaller passive components (L, C_o).

11.10.1 Pulse-Width Modulation

While the general structure of the 3L-buck in Fig. 11.18a) is identical to the basic 2:1 hybrid converter in Fig. 11.5b), its control is different. There are two pulse-width modulated signals, of which $PWM1$ and $PWM2$ control the outer and inner pair of switches, respectively. Each signal is identical to PWM in the 2L-buck in Fig. 11.18b). $PWM1$ and $PWM2$ have the same duty cycle but are shifted apart by 180°. Ideally, the on- and off-times are identical. Using the same duty cycle definition $D = V_{out}/V_{in}$ as introduced for the inductive buck converter in Eqn. (10.8), the on-times are defined by:

$$t_{on} = t_{on1} = t_{on2} = \begin{cases} DT & D \le 0.5 \\ (D - 0.5)\,T & D > 0.5 \end{cases} \tag{11.78}$$

The off-times are defined as the remaining time until half period time is reached, i.e., $t_{off} = t_{off1} = t_{off2} = (T/2 - t_{on})$. Equation (11.78) indicates that there are two cases in the 3L-buck converter depending on whether V_{out} is lower or greater than $V_{in}/2$.

11.10.2 Switching Sequences

Not only the on- and off-times are different in each case. Also, the switching sequences change. Fig. 11.18a) shows the control signals for the case $V_{out} \le V_{in}/2$ with the activated switches annotated in the V_{sw} waveform. There are three states:

State (1) During t_{on1} S1 and S3 connect C_1 between V_{in} and the inductor L. Assuming $V(C_1) = V_{in}/2$ the switching node gets pulled to $V_{in}/2$. Due to the positive voltage across the inductor, its current I_L will ramp up.

State (2) When both PWM signals are in off-time, S3 and S4 turn on and provide a free-wheeling path for the inductor by connecting the switching node V_{sw} to ground. Like in the basic 2L-buck converter, the inductor current will decrease because of the inverted polarity of the inductor voltage. The upper terminal of C_1 is not connected, so its voltage is preserved.

State (3) As soon as t_{on2} starts switches S2 and S4 get activated. V_{sw} goes to $V_{in}/2$ causing I_L to ramp up before the converter enters State (2) once more with S3 and S4 turned on.

In summary, the switching sequence is State (1) → State (2) → State (3) → State (2). We can conclude that V_{sw} alternates between $V_{in}/2$ and 0 (ground) if $D \leq 0.5$.

During the other case, $D > 0.5$, the switching node alternates between V_{in} and $V_{in}/2$. A new state occurs, which can be considered the inverted case of State (2).

State (4) Both *PWM1* and *PWM2* are at logical "1" level. Switches S1 and S2 conduct and connect the left-hand terminal of the inductor to V_{in}. At the same time, the bottom plate of C_1 keeps floating.

For $D > 0.5$ ($V_{out} > V_{in}/2$), the switching sequence starts at the rising edge of *PWM1* with State (4) following the sequence State (4) → State (1) → State (4) → State (3).

11.10.3 Voltage Conversion Ratio

The volt-second balance introduced in Section 10.1 allows deriving the voltage conversion ratio of the 3L-buck converter. Based on Eqn. (10.5) and reading off the voltage across the inductor L from Fig. 11.18a) we find for $D \leq 0.5$,

$$t_{on1}\left(\frac{V_{in}}{2} - V_{out}\right) + t_{off1}(-V_{out}) + t_{on2}\left(\frac{V_{in}}{2} - V_{out}\right) + t_{off2}(-V_{out}) = 0. \tag{11.79}$$

In the ideal case, all on- and off-times will be equal. Also, $t_{on} + t_{off} = \frac{T}{2}$. Rearranging yields the voltage conversion ratio,

$$\frac{V_{out}}{V_{in}} = \frac{t_{on}}{T} = D, \tag{11.80}$$

where $D = D_1 = D_2$ denotes the duty cycles of *PWM1* and *PWM2* which are ideally equal. The same voltage conversion ratio can be obtained in the case of $D > 0.5$. Hence, the 3L-buck converter has the same conversion ratio as the 2L-buck converter; see Eqn. (10.8) for comparison.

11.10.4 Inductor Current and Output Voltage Ripple

Following the same approach outlined in the context of Eqn. (10.9), we can determine the current ripple ΔI_L in the 3L-buck converter. Among the various ways of calculating the ripple, we use the first interval related to t_{on1} (see Fig. 11.18a)),

$$\Delta I_L = t_{on1}\frac{V_L}{L}. \tag{11.81}$$

For duty cycles below and above $D = 0.5$ the on-time t_{on1} and the voltage across the inductor V_L will be different:

$$t_{on1} = \begin{cases} DT = \frac{V_{out}}{V_{in}f_{sw}} & D \leq 0.5 \\ DT - \frac{T}{2} = \left(D - \frac{1}{2}\right)\frac{1}{f_{sw}} & D > 0.5 \end{cases} \tag{11.82}$$

$$V_L = \begin{cases} \frac{V_{in}}{2} - V_{out} & D \leq 0.5 \\ V_{in} - V_{out} & D > 0.5 \end{cases} \tag{11.83}$$

Inserting t_{on1} and V_L into Eqn. (11.81) yields

$$\Delta I_L = \begin{cases} \frac{V_{in}D}{Lf_{sw}}(0.5 - D) & D \leq 0.5 \\ \frac{V_{in}(1-D)}{Lf_{sw}}(D - 0.5) & D > 0.5 \end{cases}. \tag{11.84}$$

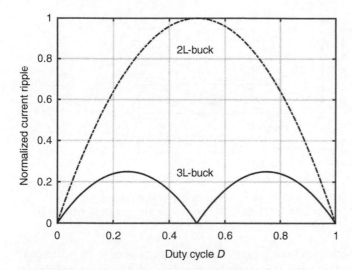

Figure 11.19 Current ripple of the fundamental 2L-buck converter and the 3L-buck converter normalized to the maximum ripple of the 2L-buck converter.

Figure 11.19 shows ΔI_L compared to the current ripple of the 2-level buck converter according to Eqn. (10.9). The diagram indicates that the maximum current ripple of the 3L-buck converter is four times lower.

Consequently, the output voltage ripple will be lower. When calculating the charge Q delivered to recharge the output capacitor according to Eqn. (10.15), we recognize that it is reduced by half compared to the 2L-buck converter (because the charging time is only $T/4$ instead of $T/2$, but repeats two times per period). We find the voltage ripple similar to Eqn. (10.16),

$$\Delta V_{out} = \frac{Q}{C} = \frac{\Delta I\, T}{16\, C}. \tag{11.85}$$

Equation (11.85) confirms the benefit of the 3L converter in terms of low output voltage ripple. Compared to Eqn. (10.16) ΔV_{out} is reduced by $1/2$ while the current ripple ΔI_L according to Eqn. (11.84) is $1/4$ compared to the fundamental 2L-buck converter. Hence, the output voltage ripple reduces to $1/8$ (12.5%). As the 3L converter needs a flying capacitor, this comparison may not be fair. Nevertheless, [18] still confirms a ΔV_{out} reduction to 25% if the same total capacitance is allocated.

11.10.5 Control of the 3L-Buck Converter

From Eqn. (11.80), we know that the voltage conversion ratio of the 3L-buck converter is identical to the fundamental inductive buck converter. Therefore, we can apply similar control techniques outlined for inductive converters in Chapter 10. Figure 11.20a) shows the 3L-buck converter with voltage-mode control (VMC) [18, 20]. Compared to the 2L-buck converter in VMC described in Section 10.5, two saw-tooth generators and two clock signals are needed operating with a 180° phase shift ($= T/2$). The resistive divider and the error amplifier (including frequency compensation) are identical to the inductive buck converter.

The waveforms in Fig. 11.20b) show how the rising edge of $CLK1$ sets the output of corresponding RS flipflop to $PWM1 = 1$. $PWM1$ gets reset once the ramp signal V_{s1} exceeds the compensated error voltage V_c. The same sequence happens for $PWM2$, just shifted by 180°. See also Fig. 10.8 for a

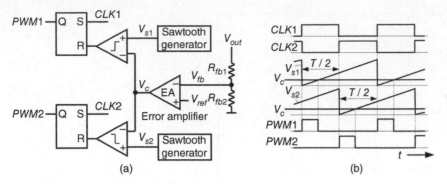

(a) (b)

Figure 11.20 a) Voltage-mode PWM control of the 3L-buck converter and b) signal waveforms. Two saw-tooth signals and clocks with a 180° phase shift ($T/2$) are used to generate the control signals *PWM1* and *PWM2*.

comparison to the fundamental inductive buck converter. The same circuits from Section 10.5 can be used for the saw-tooth generator (Fig. 10.9), the error amplifier (Fig. 10.12), and comparator (Fig. 10.13).

As for the 2L-buck, various other control methods, such as hysteretic and constant on-/off-time control in both voltage and current mode operation, can be applied. The design in [21] changes between adaptive on- and off-time control (AOOC) to expand the duty cycle resolution and achieve a wide voltage conversion range at high switching frequencies up to 11.8 MHz. The converter enters DCM at light-load to enhance the conversion efficiency as outlined for inductive converters in Chapter 10 (see Section 10.9).

11.10.6 Flying Capacitor Balancing

There must be zero net charge accumulation at the flying capacitor with the voltage at C_1 balanced exactly at $V_{in}/2$. It requires the on-times t_{on1} and t_{on2} of States (1) and (2), respectively, to be equal, which can never be achieved in real converter designs. One of the reasons is asymmetrical charge injection from the low-side and high-side switches and gate drivers. Moreover, the voltage $V(C_1)$ on C_1 is unconstrained because the volt-seconds balance in Eqn. (11.79) holds even if $V(C_1)$ is not equal to $V_{in}/2$ [22]. Therefore, active control of the capacitor voltage is required to balance $V(C_1)$. It can be accomplished by adding a dedicated control loop for flying capacitor balancing to the main control loop. Figure 11.21 shows an implementation based on the circuits from Fig. 11.20. It keeps the error amplifier (EA_M, note the reversed polarity), the comparators, and the saw-tooth generators. The balancing loop consists of a second error amplifier EA_B implemented as a fully differential circuit with common-mode feedback CMFB. A sense amplifier detects the flying capacitor voltage $V(C_1)$ and feeds it to the inverting input of EA_B. A reference voltage V_{mid} provides the set value for $V(C_1)$ at the non-inverting input of EA_B. Its target value is $V_{mid} = V_{in}/2$ generated by a simple 2:1 SC stage or by a linear regulator (see Fig. 11.22). The sense amplifier measures the capacitor voltage either at a fixed point in time (per period like in [21]) or continuously [23]. The shunt current sensing circuits of Section 6.6 can be utilized because they detect the current through a voltage drop. The implementation in [23] uses the sensing circuit of Fig. 6.17b). Due to the differential output of EA_B, the pulse generation for PWM1 and PWM2 sees different comparator trip points V_{bal1} and V_{bal2} instead of V_c in Fig. 11.20.

The operation of the flying capacitor balancing loop is as follows. Assume that $V(C_1) > V_{mid} = V_{in}/2$. It will cause V_{bal1} to go down while V_{bal2} increases, as shown by the curves in the middle-left

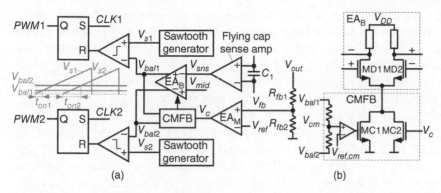

Figure 11.21 a) Flying capacitor balancing added to the 3L-buck converter with VMC (see Fig. 11.20); b) conceptual implementation of error amplifier EA_C including common-mode feedback CMFB.

of Fig. 11.21a). Consequently, t_{on1} will be shortened and t_{on2} extended. As t_{on1} determines the charging time of C_1, the voltage across the flying capacitor will decrease and ideally settle at $V_{in}/2$. The regular voltage-mode control loop is incorporated into the common-mode feedback CMFB shown in Fig. 11.21b). CMFB is a local control loop that maintains the common-mode voltage V_{cm} at the output of EA_B at $V_{ref,cm}$. If V_{cm} starts drifting to higher values, transistor MC1 increases its drain current and pulls both output voltages back down until $V_{cm} = V_{ref,cm}$. Transistor MC2 performs a similar task as part of the main voltage-mode control loop. Assume that V_{out} rises. Then, V_{fb} and the compensated error voltage V_c increase. As V_c controls MC2 its drain current increases and pulls down both output voltages V_{bal1} and V_{bal2} of EA_B. Consequently, both on-times t_{on1} and t_{on2} get reduced. According to Eqn. (11.80), this reduces V_{out} (at a given input voltage V_{in}). This way, Fig. 11.21 provides a solution for a 3L-buck converter with VMC and flying capacitor balancing.

11.10.7 An Implementation Example with V_{mid}-Generation

Many 3L-buck designs use p-type (PMOS) transistors for the high-side switches as shown in Fig. 11.22 [20, 21, 24]. A design with n-type (NMOS) high-side switches is presented in [25]. The major advantage of an implementation with PMOS devices lies in the easier design of the gate driver and gate supply. The drawback of the larger gate charge (dynamic losses) of the p-type

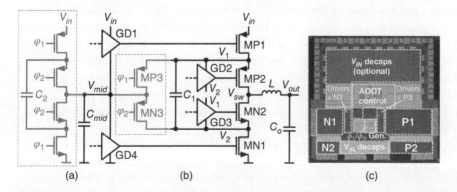

Figure 11.22 Implementation example of a 3L-buck converter with two ways of generating $V_{mid} = V_{in}/2$ using a) a switched-capacitor 2:1 step-down stage or b) two switches MP3, MN3 connected to C_{mid}. Source: Adapted from [21]. c) die photo of [21] in 0.25 μm CMOS with an area of 2.4 mm by 2.4 mm.

switches compared to an all-NMOS power stage is tolerated because thin-oxide devices rated at $V_{in}/2$ can be used. These devices have very low parasitic capacitance, resulting in low switching losses (see Sections 2.3 and 2.7). The gate driver design is even easier if an intermediate supply rail V_{mid} is available. As shown in Fig. 11.22, V_{mid} will be the ground potential of the upper gate driver GD1 and the positive gate supply for the lower low-side driver GD4. The inner gate drivers GD2 and GD3 are fully supplied from the flying capacitor C_1. There are various ways to generate V_{mid}. A basic 2:1 SC stage (see Section 9.1) as shown in Fig. 11.22a) provides $V_{mid} = V_{in}/2$ with high power efficiency but at the expense of an additional flying capacitor C_2. Figure 11.22b) shows the concept proposed in [21] that implements the 2:1 stage using the existing flying capacitor C_1 and two additional switches MP3, MN3. Alternatively, a linear regulator can be used that operates from V_{in} (not shown in Fig. 11.22). However, it suffers lower power efficiency than the charge pump approach (see Chapter 7). Fig. 11.22c) shows the die photo of the 3L-buck converter of [21] designed in a 0.25µm CMOS technology. P1, P2, P3, and N1, N2, P3 correspond to the switches MP1-MP3, MN1-MN3. AOOT denotes the adaptive on-/off-time block, as discussed in the control section above. The design uses external passives $C_1 = 1\,\mu F$, $C_{mid} = 220\,nF$, and $L = 220\,nH$ and operates at $f_{sw} = 4.4 - 11.8\,MHz$. The maximum values of input voltage and load current are 5.0 V and 800 mA, respectively. The output voltage range is $0.34 - 4.5$ V.

11.10.8 Start-up

One of the challenging issues of three-level converters is their start-up. As the switches can ideally be rated to $V_{in}/3$, it needs to be ensured that the drain-source voltage does not exceed $V_{DS,max}$ when ramping up the input voltage rail V_{in} at the start-up of the converter. One approach can be a current source that precharges C_1 to $V_{in}/2$ [26]. A comparator monitors its voltage $V(C_1)$ and turns the current source off if $V(C_1)$ exceeds $V_{in}/2$. This way, the power transistors never experience a drain-source voltage higher than $V_{in}/2$.

11.11 The Flying-Capacitor Multi-Level Converter (FCML)

Flying-capacitor multi-level (FCML) DC–DC converters are a class of hybrid converters that expand the idea of the three-level converter by adding more flying capacitors. The concept of multi-level converter architectures has existed since the 1990s [27]. They recently gained attention as an interesting approach for compact and efficient DC–DC converters that can regulate over a wide conversion range of V_{in} to V_{out} [22]. Both step-up and step-down topologies are possible. We will concentrate on the down-conversion to explore the fundamental concepts of FCML converters. Figure 11.23 shows the four-level buck converter (4L-buck converter) as an example of a FCML topology. It can be derived from the 3L-buck shown in Fig. 11.18 by adding a second flying capacitor C_2 and one more high-side and low-side switch. In the same way, the topology can be expanded to achieve 5-level and higher-order FCML converters.

The advantages are similar to the 3L-buck converter: (1) The inherent SC stage down-converts V_{in} by a fixed ratio such that the inductor and the capacitors see a much lower voltage across. Concurrently, the number of levels at the switching node is increased (multi-level). The higher the number of levels, the lower the current ripple in the inductor and the output voltage ripple. This way, the ripple frequency in the inductor is effectively increased (without increasing the switching frequency). Consequently, the inductor and the capacitors can be reduced in size. Alternatively, the converter can be operated at a lower switching frequency. (2) The maximum voltage across

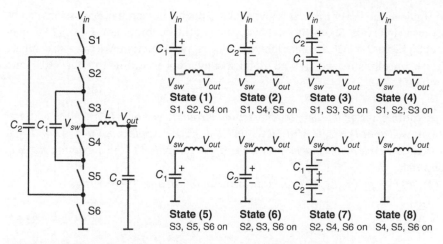

Figure 11.23 The 4L-buck converter and its eight switching states.

the switches is reduced in steady-state. Low-voltage transistors that require less area may be used, which leads to lower switching loss and fast transient response.

Figure 11.23 shows the possible states of the 4L-buck converter. Depending on the required duty cycle, they will be activated in a dedicated order. Generally, in an N-level FCML converter, $(N - 1)$ duty-cycle intervals can be distinguished. Table 11.3 shows the three intervals of the 4L-buck converter and the corresponding switching sequence of the different states according to Fig. 11.23. One switch pair changes at a time.

Consider the lower interval $0 < D \leq \frac{1}{3}$, for example. The sequence starts with State (2), which charges C_2 and energizes the inductor L. In steady-state, $V(C_2)$ will ideally reach $\frac{2}{3}V_{in}$ while $V(C_1) = \frac{1}{3}V_{in}$. After a free-wheeling phase in State (8), the converter enters State (7), which puts C_1 and C_2 in series. Due to the different polarities of both capacitors, the switching node voltage V_{sw} will not exceed $V_{in}/3$. After another free-wheeling cycle (State (8)), C_1 connects between V_{sw} ground during State (5). The sequence will repeat, starting with State (2).

By inspection we find that V_{sw} alternates between 0 (ground) and $V_{in}/3$. In the same way, the middle interval $\frac{1}{3} < D \leq \frac{2}{3}$ confines V_{sw} between $\frac{1}{3}V_{in}$ and $\frac{2}{3}V_{in}$. During the upper interval of $\frac{2}{3} < D \leq 1$ we identify $V_{sw} \in (\frac{2}{3}V_{in}, V_{in})$. In conclusion, the maximum swing of V_{sw} is $V_{in}/3$ or, in its general form, $V_{in}/(N - 1)$. Consequently, the current and voltage ripple reduces to

$$\Delta I_L = \frac{V_{in} D}{L f_{sw}} \left(\frac{1}{N - 1} - D \right) \quad \text{for} \quad 0 < D \leq \frac{1}{N - 1},$$

$$\Delta V_{out} = \frac{1}{N - 1} \frac{\Delta I_L}{8 L f_{sw}}. \tag{11.86}$$

Table 11.3 Switching sequences of the 4L-buck converter.

Duty cycle range	Switching sequence (State (1) to (8))
$0 < D \leq \frac{1}{3}$	$(2) \rightarrow (8) \rightarrow (7) \rightarrow (8) \rightarrow (5) \rightarrow (8)$
$\frac{1}{3} < D \leq \frac{2}{3}$	$(1) \rightarrow (2) \rightarrow (3) \rightarrow (5) \rightarrow (6) \rightarrow (7)$
$\frac{2}{3} < D \leq 1$	$(1) \rightarrow (4) \rightarrow (3) \rightarrow (4) \rightarrow (6) \rightarrow (4)$

Only the current ripple in the lower interval is given for simplicity, but similar expressions can be found for all duty cycle intervals. Since the level N appears in the denominator, Eqn. (11.86) confirms the benefit of higher-order FCML converters regarding ripple reduction. We can also compare the ripple with the expressions for the 2L- and 3L-buck converters according to Eqns. (10.9) and (11.84) (plotted in Fig. 11.19).

Example 11.6 *According to Eqn. (11.86), the maximum current ripple of a 4-level buck converter occurs at $D = 1/6$ (as well as at $D = 3/6, 5/6$). How much lower is the ripple compared to the 2L and 3L converters for that duty cycle? How much lower is the ripple compared to the maximum ripple of the 2L and 3L converters?*

Dividing Eqn. (11.86) by Eqn. (11.84) gives the ratio between the 4L and 3L converter,

$$\frac{\Delta I_{L,4L}}{\Delta I_{L,3L}} = \frac{\frac{1}{3} - \frac{1}{6}}{\frac{1}{2} - \frac{1}{6}} = \frac{1}{2}. \tag{11.87}$$

Similarly, the ratio between the 4L and 2L converter is

$$\frac{\Delta I_{L,4L}}{\Delta I_{L,2L}} = \frac{\frac{1}{3} - \frac{1}{6}}{1 - \frac{1}{6}} = \frac{1}{5}. \tag{11.88}$$

The maximum ripple of the 2L and 3L converters occurs at $D = 1/2$ and $D = 1/4$, respectively. With the maximum ripple of the 4L converter at $D = 1/6$, we get

$$\frac{\Delta I_{L,4\,Lmax}}{\Delta I_{L,3\,Lmax}} = \frac{\frac{1}{6}\left(\frac{1}{3} - \frac{1}{6}\right)}{\frac{1}{4}\left(\frac{1}{2} - \frac{1}{4}\right)} = \frac{16}{36} = 0.44. \tag{11.89}$$

and

$$\frac{\Delta I_{L,4\,Lmax}}{\Delta I_{L,2\,Lmax}} = \frac{\frac{1}{6}\left(\frac{1}{3} - \frac{1}{6}\right)}{\frac{1}{2}\left(1 - \frac{1}{2}\right)} = \frac{4}{36} = 0.11. \tag{11.90}$$

In conclusion, the maximum current ripple of the 4L-buck converter is approximately 45% or 10% compared to a 3L and 2L converter, respectively (see also Fig. 11.19).

Regarding flying capacitor balancing and maximum-ratings protection of the switches at start-up, the same requirements as for the 3L-buck converter apply.

The major drawback of FCML converters is the large number of switches and flying capacitors. Therefore, these converter types are only preferred if a wide voltage conversion range needs to be supported. If the ratio between V_{out} and V_{in} is relatively fixed, other hybrid topologies may benefit more. An alternative topology is the double step-down converter described in Section 11.12.

11.12 The Double Step-Down (DSD) Converter

This hybrid converter, also called a series capacitor buck converter or extended duty ratio converter, is well-suitable for point-of-load voltage regulators with large voltage conversion ratios. Figure 11.24a) shows its topology, which is very similar to the interleaved, two-phase buck converter in Fig. 11.24b) (see also Section 10.14) but with an added energy transfer capacitor C_1. This

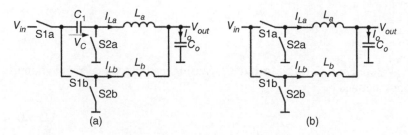

Figure 11.24 a) The double step-down converter; b) the two-phase inductive buck converter for comparison.

capacitor ensures soft-charging and causes a lower inductor current ripple similar to a 3L-buck converter (see Section 11.10) by dividing the input voltage by half before applying it to the inductors. In steady-state, the voltage at C_1 is constant and does not need voltage balancing.

After being introduced in 2005 as a discrete implementation by [28], monolithically integrated DSD converters came up ten years later [29, 30]. Recent designs include 48 V-to-1 V converters using GaN power switches [31].

11.12.1 General Operation

The two pairs of power switches, S1a/S2a and S1b/S2b are operated with 180° phase shift. There are three switching phases indicated by States (1) to (3) in Fig. 11.25. Figure 11.25d) shows the idealized waveforms of the inductor currents along with the output buffer capacitor current. The duration of State (1) is equal to the on-time t_{on} of S1a, which results in an inductor current ripple

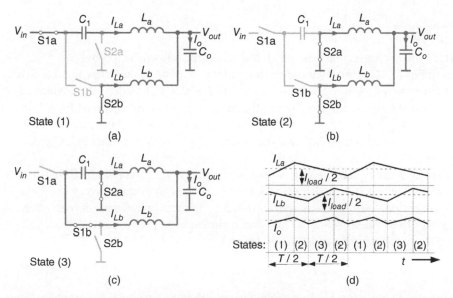

Figure 11.25 The three switching states of the double step-down converter: a) State (1); b) State (2); c) State 3; and d) the waveforms of the inductor currents I_{La}, I_{Lb}, and the current I_C into the output bypass capacitor C_o with an indication of the switching states.

in L_a of

$$\Delta I_{La} = \frac{(V_{in} - V_C - V_{out})t_{on}}{L_a}. \tag{11.91}$$

During State (2), the energy stored in L_a is transferred to C_o, causing the inductor current to ramp down with the current ripple defined by

$$\Delta I_{La} = \frac{V_{out}t_{off}}{L_a}. \tag{11.92}$$

During State (3) energy is transferred from C_1 to L_b causing the inductor current I_{Lb} to rise,

$$\Delta I_{Lb} = \frac{(V_C - V_{out})t_{on}}{L_b}. \tag{11.93}$$

Note that States (1) and (3) have equal lengths defined by t_{on}, which relates to the duty cycles $D = t_{on}/T$ of switches S1a and S1b, respectively. During States (1) and (2), the inductor L_b is connected to ground, causing its current to ramp down:

$$\Delta I_{Lb} = \frac{(V_{out})t_{off}}{L_b} \tag{11.94}$$

In steady-state, the current ripple ΔI_{La} in Eqns. (11.91) and (11.92) needs to be equal, which yields the volt-second balance of the DSD converter,

$$(V_{in} - V_C - V_{out})t_{on} = V_{out}t_{off}. \tag{11.95}$$

Likewise, ΔI_{Lb} given by Eqns. (11.91) and (11.92) has to be the same,

$$(V_C - V_{out})t_{on} = (V_{out})t_{off}. \tag{11.96}$$

For Eqns. (11.95) and (11.96) to be equal, the capacitor voltage needs to be $V_C = V_{in}/2$ in steady-state. Inserting this condition into Eqn. (11.95) allows us to derive the voltage conversion ratio,

$$\frac{V_{out}}{V_{in}} = \frac{1}{2}\frac{t_{on}}{t_{on} + t_{off}} = \frac{D}{2}. \tag{11.97}$$

The DSD converter's voltage conversion ratio is half that of a conventional buck converter.

The waveforms in Fig. 11.25d) show that the on-times of the main power switches S1a and S1b are not allowed to overlap, which restricts the duty cycle to $D \leq 0.5$. In addition, C_1 provides an inherent 2:1 step-down voltage conversion. Hence, the minimum input voltage must be at least four times greater than V_{out}, which restricts the application space of the DSD converter to large down-conversion ratios from high input voltages such as the 12 V-to-1.2 V converter in [30] or the 48 V-to-1 V design in [31].

The output bypass capacitor C_o sees the sum of both currents I_{La} and I_{Lb}. As a result, the current ripple cancels similar to an inductive multiphase converter like the two-phase topology shown in Fig. 11.24b). The current ripple ΔI_o in the output capacitor C_o reduces because the currents from both inductors add in a time-interleaved way. This effect is similar to an inductive multiphase converter like the two-phase topology shown in Fig. 11.24b),

$$\Delta I_o = \frac{V_{out}}{f_{sw}L}(1 - 4D). \tag{11.98}$$

Comparing this result to the ripple of an inductive multiphase buck converter given by Eqn. (10.237) (for $D \leq 0.5$) reveals that the ripple behavior of the DSD converter is the same as that of a four-phase inductive converter.

11.12.2   Inductor Current Balancing

Like in multiphase inductive converters discussed in Section 10.14, phase current balancing has to be ensured. As a major benefit over a purely inductive multiphase implementation, the DSD converter provides inherent balancing of the inductor currents because an inherent feedback loop exists between the inductor currents I_{La}, I_{Lb} and the capacitor voltage V_C [28, 32]. According to Fig. 11.25, inductor L_a charges C_1 during State (1) and inductor L_b discharges C_1 in State (3). Assuming, for example, that the average inductor current in L_a is larger than that in L_b, the capacitor voltage V_C would gradually increase. Hence, during State (1), the voltage across L_a will decrease, causing the inductor current to decrease. L_b, in contrast, will see a larger voltage during State (3), causing an increase in the average L_b current. Hence, the DSD converter automatically reaches a steady state where both average inductor currents are equal, and the capacitor voltage V_C is constant.

11.12.3   Start-Up

Like the 3L-buck converter in Section 11.10, the DSD converter requires careful start-up. [30] describes a 10 mA current source that smoothly charges capacitor C_1 to $V_{in}/2$ from a pre-regulator similar to [26].

11.13   Inductor-First Topologies

Moving the inductor in the fundamental inductive buck converter toward the input forms an interesting converter configuration with several advantages. Above all, the inductor will carry a lower average current. Consider the conventional inductive buck converter as introduced in Chapter 10 (see Section 10.1), shown again in Fig. 11.26a). Due to the switching nature, large current pulses occur between zero and I_{load} as indicated at both the V_{in} and ground connections. Especially when using miniaturized inductors, as discussed in Section 4.2, with relatively large series resistance R_{DC} this will cause large conduction loss in the coil:

$$P_{cond} = I_{load}^2 R_{DC} \tag{11.99}$$

For simplicity, we neglect the current ripple and assume $I_{load} = I_{L,RMS}$ (see Section 2.7). The inductor-first approach can significantly reduce the losses.

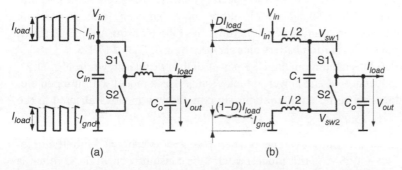

Figure 11.26   a) Conventional inductive buck converter and b) modification to an inductor-first topology.

The conventional inductive converter can be transformed into an inductor-first topology by swapping the locations of the power switches and the inductor as proposed in [33]. For this reason, the inductor splits into two parts of half the inductance, as shown in Fig. 11.26b). Now the input bypass capacitor C_{in} becomes a flying capacitor C_1. The upper inductor voltage changes between $(V_{in} - V_{out})$ (during $t_{on} = DT$) and $(V_{in} - V_C - V_{out})$ (during $t_{off} = (1 - D)T$). Likewise, the bottom inductor switches between $(V_{out} - V_C)$ and V_{out}. To ensure that the average voltage across the inductors remains zero, the voltage at capacitor C_1 must equal V_{in}. The volt-seconds balance for both inductors yields an identical voltage conversion ratio

$$\frac{V_{out}}{V_{in}} = D, \tag{11.100}$$

which is the same as for the inductive buck converter in Fig. 11.26a).

Depending on the duty cycle D, the average current inductor current will be much lower, as shown in Fig. 11.26b). Similar to Eqn. (11.99), we can calculate the losses,

$$
\begin{aligned}
P_{cond} &= I_{in}^2 R_{DC} + I_{gnd}^2 R_{DC} \\
&= (DI_{load})^2 R_{DC} + (1 - D)^2 I_{load}^2 R_{DC} \\
&= n I_{load}^2 R_{DC} \quad \text{where} \quad n = (1 - 2D + 2D^2).
\end{aligned} \tag{11.101}
$$

Comparing the conduction losses of the conventional buck converter in Eqn. (11.99) with the losses in Eqn. (11.101) reveals the losses scale by a factor n, which is between 0.5 (at $D = 0.5$) and 1 (at $D = 0$ and $D = 1$). Depending on the design constraints, like the total inductor volume, R_{DC} will also scale and offset the advantage given by n. However, significant power loss savings can be achieved, as demonstrated in [33]. The inductor-first topology achieves the same losses as a conventional two-phase converter at $D = 0.5$ without needing phase balancing when using identical inductors.

As an additional benefit of placing the inductor directly at the input and ground, the continuous current substantially reduces the noise and EMI. However, the noise at the output increases because the inductor currents add in phase.

Utilizing the Cable Inductance

A promising application of the inductor-first approach is to utilize the parasitic inductance of the cable that connects V_{in} and ground. Such a design has been presented in discrete form [34] and later as an IC-level design in [35]. The USB cable, for instance, has a significant inductance in the range between hundreds of nanohenries and several microhenries [34]. By using a 1 m USB cable, [34] demonstrates a 15W inductorless Li-ion battery charger operating at $f_{sw} = 2$ MHz.

Based on this idea, [36] develops a bidirectional USB power delivery cable shown in Fig. 11.27. It uses the parasitic cable inductance as the power inductor (shown as a model with lumped elements). Two hybrid step-down converter configurations support the two modes indicated at the bottom of Fig. 11.27:

- *Forward mode*: The mains adapter output voltage of $V_A = 16 - 25$ V is converted into the notebook supply voltage $V_{NB} = 8 - 12.6$ V. In this mode, switch S1 is constantly on while S3 remains off. The inductor (i.e., the cable), the remaining switches, and C_p form an inductor-first topology.
- *Reverse mode*: This configuration supports device-to-device operation using the notebook to charge a smartphone, for instance, from an output voltage $V_A = 5 - 9$ V. S4 keeps permanently

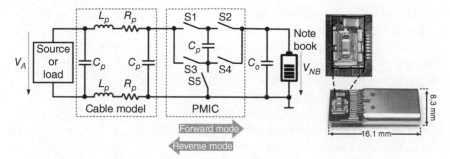

Figure 11.27 A bidirectional power delivery, voltage-regulating USB cable utilizing the cable as the power inductor [36]. The forward-mode implements a hybrid inductor-first converter. The reverse mode operates as a three-level buck converter. Right: The power management IC and all passives integrated into the USB Type-C connector. Source: Adapted from [36].

turned off, and a 3L-buck converter is formed with the cable acting as the power inductor. Note that this is not an inductor-first topology, and the 3L-buck converter operates as described in Section 11.10. The design includes flying-capacitor balancing and a technique called cable emulation to enable closed-loop control without the need to feed back the load side to the controller (keeping in mind that there might be a ground shift between the connectors on each side).

Operating at $f_{sw} = 5\,\text{MHz}$, the converter delivers up to $42\,\text{W}$ (forward) and $18\,\text{W}$ (reverse) of power for fast charging of the device's Li-ion battery. As a significant advantage, the cable also serves as a heat sink in both operating modes (directions). The photos on the right of Fig. 11.27 show the assembly of the converter IC in a Type-C USB connector together with all passives [36].

References

1 Kesarwani, K. and Stauth, J.T. (2015) The direct-conversion resonant switched capacitor architecture with merged multiphase interleaving: cost and performance comparison, in *2015 IEEE Applied Power Electronics Conference and Exposition (APEC)*, pp. 952–959, doi: 10.1109/APEC.2015.7104464.

2 Schaef, C., Rentmeister, J., and Stauth, J.T. (2018) Multimode operation of resonant and hybrid switched-capacitor topologies. *IEEE Transactions on Power Electronics*, 33 (12), 10512–10523, doi: 10.1109/TPEL.2018.2806927.

3 Renz, P. and Wicht, B. (2021) *Integrated Hybrid Resonant DCDC Converters*, Springer International Publishing, doi: 10.1007/978-3-030-63944-0.

4 McLaughlin, P.H., Datta, K., and Stauth, J.T. (2022) A monolithic 3:1 resonant Dickson converter with variable regulation and magnetic-based zero-current detection and autotuning, in *2022 IEEE International Solid-State Circuits Conference (ISSCC)*, vol. 65, pp. 304–306, doi: 10.1109/ISSCC42614.2022.9731601.

5 Evzelman, M. and Ben-Yaakov, S. (2013) Average-current-based conduction losses model of switched capacitor converters. *IEEE Transactions on Power Electronics*, 28 (7), 3341–3352, doi: 10.1109/TPEL.2012.2226060.

6 Pasternak, S., Schaef, C., and Stauth, J. (2016) Equivalent resistance approach to optimization, analysis and comparison of hybrid/resonant switched-capacitor converters, in *2016*

IEEE 17th Workshop on Control and Modeling for Power Electronics (COMPEL), pp. 1–8, doi: 10.1109/COMPEL.2016.7556737.

7 Kesarwani, K., Sangwan, R., and Stauth, J.T. (2014) 4.5 A 2-phase resonant switched-capacitor converter delivering 4.3W at $0.6W/mm^2$ with 85% efficiency, in *2014 IEEE International Solid-State Circuits Conference Digest of Technical Papers (ISSCC)*, pp. 86–87, doi: 10.1109/ISSCC.2014.6757349.

8 Renz, P., Lueders, M., and Wicht, B. (2020) A 47 MHz hybrid resonant SC converter with digital switch conductance regulation and multi-mode control for Li-ion battery applications, in *2020 IEEE Applied Power Electronics Conference and Exposition (APEC)*, IEEE, New Orleans, LA, USA, pp. 15–18, doi: 10.1109/APEC39645.2020.9124238.

9 Renz, P., Kaufmann, M., Lueders, M., and Wicht, B. (2019) 8.6 A fully integrated 85%-peak-efficiency hybrid multi ratio resonant DC-DC converter with 3.0-to-4.5 V input and 500 μA-to-120 mA load range, in *2019 IEEE International Solid-State Circuits Conference (ISSCC)*, IEEE, San Francisco, CA, USA, pp. 156–158, doi: 10.1109/ISSCC.2019.8662491.

10 Henriksen, M.M., Otten, J., Gehl, A., and Wicht, B. (2023) The ΔV-method: an intuitive method for analyzing soft-charging capabilities of hybrid switched-capacitor DC-DC converters, in *2023 IEEE 24th Workshop on Control and Modeling for Power Electronics (COMPEL)*.

11 Lei, Y. and Pilawa-Podgurski, R.C.N. (2015) A general method for analyzing resonant and soft-charging operation of switched-capacitor converters. *IEEE Transactions on Power Electronics*, 30 (10), 5650–5664, doi: 10.1109/TPEL.2014.2377738.

12 Stauth, J.T., Seeman, M.D., and Kesarwani, K. (2012) A resonant switched-capacitor IC and embedded system for sub-module photovoltaic power management. *IEEE Journal of Solid-State Circuits*, 47 (12), 3043–3054, doi: 10.1109/JSSC.2012.2225731.

13 Schaef, C., Din, E., and Stauth, J.T. (2017) 10.2 A digitally controlled 94.8%-peak-efficiency hybrid switched-capacitor converter for bidirectional balancing and impedance-based diagnostics of lithium-ion battery arrays, in *2017 IEEE International Solid-State Circuits Conference (ISSCC)*, pp. 180–181, doi: 10.1109/ISSCC.2017.7870320.

14 Lei, Y., May, R., and Pilawa-Podgurski, R. (2016) Split-phase control: achieving complete soft-charging operation of a Dickson switched-capacitor converter. *IEEE Transactions on Power Electronics*, 31 (1), 770–782, doi: 10.1109/TPEL.2015.2403715.

15 Liu, W.C., Assem, P., Lei, Y., Hanumolu, P.K., and Pilawa-Podgurski, R. (2017) 10.3 A 94.2%-peak-efficiency 1.53 A direct-battery-hook-up hybrid Dickson switched-capacitor DC-DC converter with wide continuous conversion ratio in 65nm CMOS, in *2017 IEEE International Solid-State Circuits Conference (ISSCC)*, pp. 182–183, doi: 10.1109/ISSCC.2017.7870321.

16 Schaef, Ċ., Kesarwani, K., and Stauth, J.T. (2015) 20.2 A variable-conversion-ratio 3-phase resonant switched capacitor converter with 85% efficiency at $0.91W/mm^2$ using 1.1nH PCB-trace inductors, in *2015 IEEE International Solid-State Circuits Conference - (ISSCC) Digest of Technical Papers*, pp. 1–3, doi: 10.1109/ISSCC.2015.7063075.

17 McLaughlin, P.H., Xia, Z., and Stauth, J.T. (2020) 11.2 A fully integrated resonant switched-capacitor converter with 85.5% efficiency at 0.47W using on-chip dual-phase merged-LC resonator, in *2020 IEEE International Solid-State Circuits Conference - (ISSCC)*, pp. 192–194, doi: 10.1109/ISSCC19947.2020.9062901.

18 Liu, X., Mok, P.K.T., Jiang, J., and Ki, W.H. (2016) Analysis and design considerations of integrated 3-level buck converters. *IEEE Transactions on Circuits and Systems I: Regular Papers*, 63 (5), 671–682, doi: 10.1109/TCSI.2016.2556098.

19 Piqué, G.V. and Alarcón, E. (2011) *CMOS Integrated Switching Power Converters - A Structured Design Approach*, Springer, New York, Dordrecht, Heidelberg, London.

20 Liu, X., Huang, C., and Mok, P.K.T. (2016) A 50 MHz 5 V 3W 90% efficiency 3-level buck converter with real-time calibration and wide output range for fast-DVS in 65nm CMOS, in *2016 IEEE Symposium on VLSI Circuits (VLSI-Circuits)*, pp. 1–2, doi: 10.1109/VLSIC. 2016.7573475.

21 Karasawa, Y., Fukuoka, T., and Miyaji, K. (2018) A 92.8% efficiency adaptive-on/off-time control 3-level buck converter for wide conversion ratio with shared charge pump intermediate voltage regulator, in *2018 IEEE Symposium on VLSI Circuits*, pp. 227–228, doi: 10.1109/VLSIC. 2018.8502403.

22 Kesarwani, K. and Stauth, J.T. (2015) Resonant and multi-mode operation of flying capacitor multi-level DC-DC converters, in *2015 IEEE 16th Workshop on Control and Modeling for Power Electronics (COMPEL)*, pp. 1–8, doi: 10.1109/COMPEL.2015.7236511.

23 Hardy, C., Pham, H., Jatlaoui, M.M., Voiron, F., Xie, T., Chen, P.H., Jha, S., Mercier, P., and Le, H.P. (2023) 11.1 A scalable heterogeneous integrated two-stage vertical power-delivery architecture for high-performance computing, in *2023 IEEE International Solid-State Circuits Conference (ISSCC)*, pp. 182–184, doi: 10.1109/ISSCC42615.2023.10067315.

24 Chu, L.C., Yang, W.H., Zhang, X.Q., Lai, Y.J., Chen, K.H., Wey, C.L., Lin, Y.H., Lin, S.R., and Tsai, T.Y. (2017) 10.5 A three-level single-inductor triple-output converter with an adjustable flying-capacitor technique for low output ripple and fast transient response, in *2017 IEEE International Solid-State Circuits Conference (ISSCC)*, pp. 186–187, doi: 10.1109/ISSCC.2017.7870323.

25 Lee, B. and Ma, D.B. (2021) A 20 MHz on-chip all-NMOS 3-level DC-DC converter with interception coupling dead-time control and 3-switch bootstrap gate driver. *IEEE Transactions on Industrial Electronics*, 68 (7), 6339–6347, doi: 10.1109/TIE.2020.2996148.

26 Reusch, D., Lee, F.C., and Xu, M. (2009) Three level buck converter with control and soft startup, in *2009 IEEE Energy Conversion Congress and Exposition*, pp. 31–35, doi: 10.1109/ECCE. 2009.5316265.

27 Meynard, T. and Foch, H. (1992) Multi-level conversion: high voltage choppers and voltage-source inverters, in *PESC 1992 Record. 23rd Annual IEEE Power Electronics Specialists Conference*, pp. 397–403, Vol. 1, doi: 10.1109/PESC.1992.254717.

28 Nishijima, K., Harada, K., Nakano, T., Nabeshima, T., and Sato, T. (2005) Analysis of double step-down two-phase buck converter for VRM, in *INTELEC 05 - 27th International Telecommunications Conference*, pp. 497–502, doi: 10.1109/INTLEC.2005.335149.

29 Shenoy, P.S., Amaro, M., Freeman, D., and Morroni, J. (2015) Comparison of a 12 V, 10 A, 3 MHz buck converter and a series capacitor buck converter, in *2015 IEEE Applied Power Electronics Conference and Exposition (APEC)*, pp. 461–468, doi: 10.1109/APEC.2015.7104391.

30 Shenoy, P.S., Lazaro, O., Ramani, R., Amaro, M., Wiktor, W., Khayat, J., and Lynch, B. (2016) A 5 MHz, 12 V, 10 A, monolithically integrated two-phase series capacitor buck converter, in *2016 IEEE Applied Power Electronics Conference and Exposition (APEC)*, pp. 66–72, doi: 10.1109/ APEC.2016.7467853.

31 Yan, D., Ke, X., and Ma, D.B. (2020) Direct 48-/1-V GaN-based DC-DC power converter with double step-down architecture and master-slave AO2T control. *IEEE Journal of Solid-State Circuits*, 55 (4), 988–998, doi: 10.1109/JSSC.2019.2957237.

32 Shenoy, P.S., Lazaro, O., Amaro, M., Ramani, R., Wiktor, W., Lynch, B., and Khayat, J. (2015) Automatic current sharing mechanism in the series capacitor buck converter, in *2015 IEEE Energy Conversion Congress and Exposition (ECCE)*, pp. 2003–2009, doi: 10.1109/ECCE. 2015.7309943.

33 Abdulslam, A. and Mercier, P.P. (2019) A passive-stacked third-order buck converter with inherent input filtering achieving 0.7 – W/mm^2 power density and 94% peak efficiency. *IEEE Solid-State Circuits Letters*, 2 (11), 240–243, doi: 10.1109/LSSC.2019.2935563.

34 Seo, G.S. and Le, H.P. (2017) An inductor-less hybrid step-down DC-DC converter architecture for future smart power cable, in *2017 IEEE Applied Power Electronics Conference and Exposition (APEC)*, pp. 247–253, doi: 10.1109/APEC.2017.7930701.

35 Hardy, C. and Le, H.P. (2019) 8.3 A 10.9W 93.4%-efficient (27W 97%-efficient) flying-inductor hybrid DC-DC converter suitable for 1-cell (2-cell) battery charging applications, in *2019 IEEE International Solid-State Circuits Conference - (ISSCC)*, pp. 150–152, doi: 10.1109/ISSCC.2019.8662432.

36 Tong, Z., Huang, J., Lu, Y., and Martins, R.P. (2023) A 42W reconfigurable bidirectional power delivery voltage-regulating cable, in *2023 IEEE International Solid-State Circuits Conference (ISSCC)*, pp. 192–194, doi: 10.1109/ISSCC42615.2023.10067491.

12

Physical Implementation

This chapter covers the system integration aspects of power management ICs. We will discuss critical topics like top-level layout floor planning, finding the correct pinout, and IC-level layout.

Understanding different package types and how the die is assembled within the IC is crucial since it impacts the electrical and thermal behavior of the IC.

General noise coupling mechanisms related to interconnections will also be examined, and we will establish general rules to minimize noise coupling due to fast voltage and current transients. We will also derive guidelines for PCB layout. Properly designed interconnections from the die through the package to the PCB layout are essential.

The topic of power delivery is explored, and how vertical power delivery and integrated voltage regulators (IVRs) help reduce the impedance of the power delivery network (PDN).

Finally, we discuss thermal design, which helps optimize heat transfer from the IC to the environment and identify hot spots on the die where sensors for thermal protection should be placed.

12.1 Layout Floor Planning

A typical power management IC consists of three functional domains: High-voltage/power and the low-voltage analog and digital portions of the IC. Figure 12.1 provides an example. As an initial rule of top-level layout floor planning, these domains should be separated as much as possible. In particular, the high-voltage blocks should be kept close and not intermixed with the low-voltage domains. Because high-voltage isolation is always achieved by spacing, multiple high-voltage domains require more chip area, resulting in higher costs.

The power stage not only sees high DC voltages. It also experiences high dv/dt and di/dt transients, which may couple to sensitive blocks in the low-voltage analog domain of the IC. Likewise, the digital part is usually a noise emitter. Hence, spacing and isolation between the analog and the power/digital parts are essential. See Section 3.3 for parasitic coupling effects and isolation techniques. Each power transistor with a drain node connected to an IC pin must be surrounded by a guard ring to avoid parasitic bipolar transistor effects and latch-up.

Analog circuits may also be affected by mechanical stress. Hence, critical blocks such as the bandgap reference should not be placed at the corners and edges of the die because the bending will be higher than in the center of the die. This is considered in the example floor plan in Fig. 12.1. Mechanical stress occurs during die attach. Even slight strain may shift reference voltages by about 10 mV compared to the measurement results from wafer probing.

Design of Power Management Integrated Circuits, First Edition. Bernhard Wicht.
© 2024 John Wiley & Sons Ltd. Published 2024 by John Wiley & Sons Ltd.

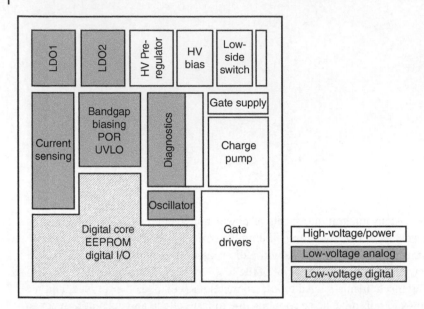

Figure 12.1 Floor plan of a power management IC with the three main areas for high-voltage/power, low-voltage analog, and digital.

The power stages may be rearranged if a thermal analysis (see Section 12.9) indicates a hot spot with excessive overheating. The particular power transistor may also be increased in size to distribute the dissipated power across a larger area.

12.2 Packaging

After manufacturing, the chips diced from a wafer are packaged. The package is usually highly critical for the electrical and thermal performance of power management ICs.

The IC package serves three primary functions:

(1) Protection and mechanical stability
(2) It interconnects the IC electrically to the PCB, redistributing the connections to a manageable fine pitch.
(3) Heat dissipation by directing the heat away from the die

12.2.1 Package Types

Figure 12.2 shows some package examples. Popular package families are Small Outline Integrated Circuit (SOIC), Thin Small Outline Package (TSOP), and Quad Flat Pack (QFP) in various types such as the (Thin) Shrink Small Outline Package (TSSOP, SSOP), low profile QFP (LQFP), or thin QFP (TQFP, 1 mm body thickness). Quad flat no-lead (QFN) packages offer a leadless, low profile, and small package size.

Within the package, the die is attached to the center pad of a substrate. Bond wires connect the metal pads of the IC to a lead frame, which, at its other end, forms the package pins that get soldered to the PCB. Figure 12.3 shows a lead frame with the die attached in the center and Fig. 12.4 a package cross-section. The lead frame has to accommodate handling during the assembly, high thermal conductivity to dissipate heat generated by the IC during operation and good electrical properties with low parasitics. In the final step, the IC is encapsulated by a mold compound.

5mm

(a)   (b)   (c)   (d)

Figure 12.2 IC packages: a) SOIC, b) SSOP, c) QFP, d) QFN (shown approximately to scale).

Figure 12.3 A lead frame with an IC attached in the center.

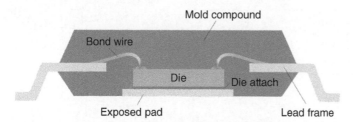

Figure 12.4 Cross-section of an IC package with an exposed pad.

Thermally enhanced packages contain an exposed pad, as shown in Fig. 12.4. Further details on the thermal behavior are covered in Sections 12.2.4 and 12.9.

For large pin counts, advanced package types like Ball Grid Array (BGA) or Wafer-level Chip-Scale Package (WCSP) replace wire bonding with a copper redistribution layer (RDL) that connects to an area array of solder balls (bumps). These technologies, in turn, enable flip-chip assembly to achieve smaller sizes at lower costs. The die is placed face down, with the bumps connecting directly to the carrier. The solder dots are typically 100 μm in diameter. Flip-chip assembly achieves very low parasitic resistance and inductance between the die and the package pin.

12.2.2   Bond Wires

Wire bonding uses gold, copper, and aluminum with typical wire diameters of 0.8, 1.0, and 1.3 mil (1 mil = 1/1000 µm). The former is the standard wire for any low-current signal and supply pin. 1.3 mil-wires are typically used for larger currents, such as the output pins of regulators and DC–DC converters. Table 12.1 gives an overview of the current handling capabilities of gold (Au) and copper (Cu) bond wires along with their parasitic resistance (at 1 GHz) [1, 2]. These values approximate typical bond wire lengths of 1–3 mm. The current-carrying capability varies significantly with the wire length [2]. The work in [3] provides more theoretical background and experimental data on the fusing current of bond wires.

Copper has a lower resistivity and offers a significant cost advantage over gold (more than 60%) [1]. The parasitic inductance and capacitance (to ground) are nearly identical for gold and copper wires. From the discussion related to integrated inductors in Section 4.2, we know that the wire inductance is approximately 1 nH/mm, see Eqn. (4.10). With values in the order of 80 fF per mm length, the capacitance is usually negligible compared to the lead frame capacitance (see Section 12.2.3) [4]. The bond pad size typically has a side length of 85–100 µm (depending on the bond wire diameter) [1].

12.2.3   Electrical Model

The lead frame, together with the bond wire, can be modeled by a lumped R-L-C equivalent circuit, as shown in Fig. 12.5a) [4]. The (self-)resistance and the (self-)inductance of the lead frame and the bond wire are lumped into one resistor/inductor element R and L. The (self-)capacitance of the bond wire is usually negligible such that only the lead frame's capacitance C is considered.

Table 12.1   Bond wire maximum current I_{max} and resistance R [1, 2].

Diameter		I_{max}/A		R/mΩ/mm	
		Au	Cu	Au	Cu
0.8 mil	20 µm	0.4	0.45	173	145
1.0 mil	25 µm	0.55	0.63	144	116
1.3 mil	33 µm	0.8	0.9	111	89

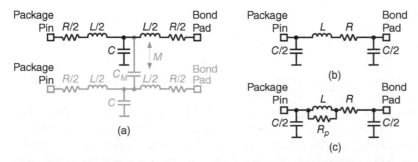

Figure 12.5   Electrical model of an IC package: a) lumped T-equivalent circuit; b) lumped π-equivalent circuit; c) adding $R_p \approx 100\,R$ to obtain a more realistic and lower Q-factor. Typical values of the R, L, and C components are given in Table 12.2.

Table 12.2   Parameters of standard IC packages with an exposed pad [4–9].

Type	Size body (Pad)/mm	R/mΩ	L/nH	C/pF	$R_{th,ja}$/K/W
TQFP-48	7×7 (5×5)	12–13.8	0.96 1.1	0.20–0.23	25
TSSOP-28	4.4×9.7 (3.0×5.5)	7–16	0.7–2.1	0.18–0.37	35
SOIC-8	4.9×3.8 (2.3×2.3)	5.1–8.2	0.7–1.3	0.22–0.26	55

If cross-talk between neighboring pins is a concern, a more accurate model may include mutual inductance and capacitance elements M, C_M, as shaded symbols indicate. Their influence is usually neglected in a first-order approximation.

The simple T-model of Fig. 12.5a) allows approximating the influence of the package interconnect parasitics despite its limited accuracy. Alternatively, the π-model of Fig. 12.5b) may be used. In that case, the capacitance of the bond pad and any structure connected at the IC pin can be added to the capacitances at the IC pin and the bond pad. Denoting these capacitances by C_{pin} and C_{pad}, rule-of-thumb-values of the equivalent parameters are $R = 100$ mΩ, $L = 1 \ldots 10$ nH, and $C_{pin} \leq 10$ pF, $C_{pad} = 0.1 \ldots 1$ pF. Advanced packages like BGA achieve much lower parasitics.

Table 12.2 gives an overview of parasitic elements for typical packages (not including the bond wire). The parameters are taken from various data sheets and application notes as specified. The mutual elements M and C_M are not listed. M is typically 30–50% of L. C_M can be three times larger than C (more emphasized for center pins and low-pin count packages where the spacing is smaller).

Example 12.1   *Estimate the R-L-C components of the electrical model for a 28-pin TSSOP package with a copper bond wire of 2 mm length. Use worse-case parameters for the package.*

We can take the parameters of the TSSOP-28 package from Table 12.2. The resistance of the bond wire is listed in Table 12.1. For 2 mm we obtain a resistance of $2 \cdot 89$ m$\Omega = 178$ mΩ. With 1 nH/mm the bond wire inductance is 2 nH. We assume that we can neglect the capacitance of the bond wire. Adding the equivalent components of the package and the bond wire yields $R = 178 + 16 = 194$ mΩ, $L = 2 + 4.1 = 6.1$ nH, and $C = 0.37$ pF.

Adding the lumped R-L-C equivalent circuit at each pin connection is often mandatory when running top-level simulations of power management ICs. The resistance causes a voltage drop in the package. All parasitics together form local resonant tanks that cause oscillations, possibly leading to malfunctions like coupling, debiasing, false switching, toggling, etc. However, the models in Fig. 12.5a,b) neglect the AC resistance determined by the skin effect (see Fig. 4.4c)). They overestimate the Q-factor (see Eqn. (4.7)), leading to excessive ringing such that circuit simulations may suffer from convergence issues. As a hands-on solution, a resistor R_p can be placed parallel to L, as shown in Fig. 12.5c). R_p bypasses ωL at high frequencies and limits the nummerator in Eqn. (4.7). R_p can be estimated from Eqn. (11.41) for $Q > 5$. As a rule-of-thumb, $R_p = 100\ R$, i.e. typically $R_p = 10\ \Omega$.

12.2.4   Thermal Behavior

There are usually considerable losses in power management ICs that cause the die to heat up. Thermally enhanced packages are used to keep the die temperature (specifically, the junction temperature T_j) at a reasonable level. Figure 12.4 shows the package cross-section with the exposed

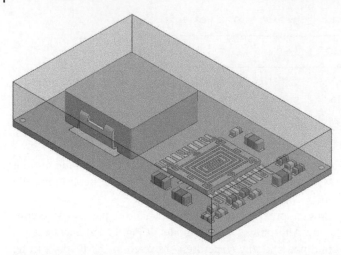

Figure 12.6 System-in-package with a power management IC, active devices, a power inductor, and capacitors mounted on a substrate; packaged in plastic using injection molding. Source: Laili Wang et al. [10]/from IEEE.

pad. Such packages are similar to standard packages but have an exposed pad providing a heat sink. The pad, also called thermal or power pad, is a metal block attached to the die on one side and exposed to the environment on the other. Soldered directly to the PCB, the exposed pad can significantly increase the heat dissipation over standard packages. As a rule-of-thumb, the thermal resistance reduces by 50% [8]. The thermal pad can be connected to ground, thereby reducing each pin's parasitic loop inductance. The thermally enhanced package aims to keep the temperature difference between the environment and the junction of devices integrated on the die as low as possible. This behavior is characterized by the thermal resistance defined in Section 12.9. The right column in Table 12.2 provides values of the junction-to-ambient thermal resistance $R_{th,ja}$ of standard IC packages.

12.2.5 System-in-Package

System-in-package (SiP) modules address the demand for higher system integration by implementing a complete power management solution in a single package using heterogeneous integration [11, 12]. As shown in Fig. 12.6, one or multiple DC–DC converter ICs, power transistors, input and output capacitors, control-loop components, and the power inductor are combined in one module that fits into a surface-mount package such as QFN or BGA [10, 13]. This approach is called 2.5-D packaging, with all parts soldered or wire-bonded on one substrate or connected to a lead frame through wire bonding. 3-D packaging technologies enable stacking multiple ICs and components [12]. The whole module is packaged using injection molding. Due to their compactness and low footprint, SiP modules save board space and achieve high power density. In addition, using SiP modules reduces the manufacturing effort because of fewer assembly steps. The increased integration level also improves the system's reliability. See also Section 1.8 in Chapter 1 where Fig. 1.19 provides an example of a DC–DC converter with a package-integrated inductor.

12.3 Electromagnetic Interference (EMI)

Physical implementation and, in particular, the upcoming topics on interconnections, IC pinout, and PCB layout are closely related to electromagnetic interference (EMI). This section briefly introduces the fundamental EMI mechanisms. For in-depth coverage of this complex topic, refer to the excellent electromagnetic compatibility (EMC) and EMI books, such as [14], which provide an invaluable resource.

In switched-mode power conversion, noise can be injected in the supply and ground lines, referred to as conducted EMI. There is also radiated EMI (noise) coupled through the air. A pulse-width modulation (PWM)-controlled voltage can be approximated by a trapezoidal signal as defined in Fig. 12.7a) by the rise time t_r, the fall time t_r, and the pulse-width time t_p (sometimes called the on-time t_{on}). For simplicity, we assume the rise and fall time to be equal, expressed by the rise-fall time $t_{rf} = t_r = t_f$. The PWM signal runs at a specific switching frequency f_{sw}, corresponding to a period time $T = 1/f_{sw}$. The related electromagnetic emissions can be predicted by the frequency spectrum of that signal [14]:

$$V_n = 20\,\mathrm{dB}\mu\mathrm{V}\left[\log\left(2\frac{\hat{V}}{1\,\mu\mathrm{V}}\right) + \log\left|\frac{\sin(\pi t_p f)}{\pi t_p f}\right| + \log\left|\frac{\sin(\pi t_{rf} f)}{\pi t_{rf} f}\right|\right] \tag{12.1}$$

This expression is plotted in Fig. 12.7b). It is the Fourier transform of the time-domain PWM waveform in Fig. 12.7a) and represents the envelope of the harmonics included in the trapezoidal signal. Note that the unit is dBμV because the voltage is referred to 1μV, which is commonly used when measuring conducted emissions.

The noise spectrum consists of three parts, which show quite intuitive behavior. The first part is a constant emission level $V_{n,o}$ defined by the magnitude $\hat{V}$ and the period time T of the PWM voltage. It confirms that the EMI emissions increase with the magnitude $\hat{V}$ and the switching frequency $f_{sw} = 1/T$. The second and third terms represent asymptotes with a slope of -20 and -40 dB per decade. The first corner frequency is defined by the PWM pulse width t_p. The shorter the pulses, the higher the corner frequency. A similar relationship applies to the second corner frequency, related

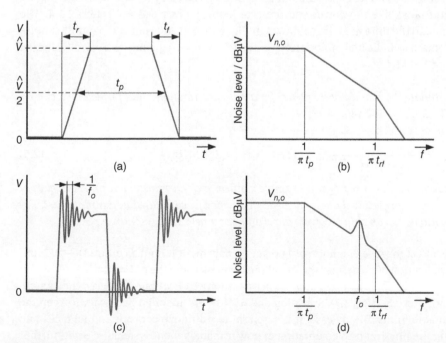

Figure 12.7 Electromagnetic interference (EMI): a) PWM waveform in the time-domain and b) corresponding noise spectrum; c) PWM signal with parasitic ringing at the resonance frequency f_o, which d) adds a peak in the spectrum at f_o.

to the rise-fall time t_{rf}. Steep switching transitions move the corner frequency to higher values, making the EMI noise worse.

Example 12.2 *We want to estimate the magnitude $V_{n,o}$ and the corner frequencies according to Eqn. (12.1) for a DC–DC converter operating at $f_{sw} = 1\,MHz$ ($T = 1\,\mu s$) from a battery input voltage $V_{bat} = 50\,V$. The converter's duty cycle is 20%, which corresponds to a pulse width of $t_p = 0.2\,\mu s$. The transition slope is 50 V/ns (see also Table 5.3).*

The input voltage V_{bat} corresponds to the high-level $\hat{V}$ of the PWM signal in Fig. 12.7a). Hence, the magnitude becomes

$$
\begin{aligned}
V_{n,o} &= 20\ \text{dB}\,\mu\text{V}\log\left(2\frac{V_{bat}}{1\,\mu\text{V}}\right) = 20\ \text{dB}\,\mu\text{V}\log\left(2\frac{50\,\text{V}}{1\,\mu\text{V}}\right) \\
&= 20\ \text{dB}\,\mu\text{V}\log\left(10^8\right) = 160\ \text{dB}\,\mu\text{V}.
\end{aligned}
\tag{12.2}
$$

The first corner frequency depends on the pulse-width t_p,

$$
f_{c1} = \frac{1}{\pi t_p} = \frac{1}{3.14 \cdot 0.2\,\mu\text{s}} = 1.59\,\text{MHz}.
\tag{12.3}
$$

The rise-fall time t_{rf} is required for the second corner frequency. We derive it from the transition slope, which gives $t_{rf} = 1$ ns. Hence,

$$
f_{c2} = \frac{1}{\pi t_{rf}} = \frac{1}{3.14 \cdot 1\,\text{ns}} = 318\,\text{MHz}.
\tag{12.4}
$$

The EMI noise can occur in both the voltage and the current domain. Voltage-domain noise is due to fast voltage transitions that impose currents in the parasitic capacitance of the power devices and the interconnections. A broad-band emission spectrum like in Fig. 12.7b) occurs. Fast changes in current excite parasitic inductances and cause voltage spikes expressed by $V_L = L\,di/dt$ and ringing in parasitic resonant tanks like the power and the gate loop, as described in Section 2.5.4. The ringing will be visible in the time domain, as shown in Fig. 12.7c). It adds narrow-band EMI noise in the frequency domain as indicated in Fig. 12.7d) at the resonance frequency f_o. Typical values of f_o are in the range of 100 MHz.

Example 12.3 *Estimate the resonance frequency for a parasitic resonant tank formed by a combination of $L = 10\,nH$ and $C = 100$ pF.*

The resonant tank oscillates at

$$
f_o = \frac{1}{2\pi\sqrt{LC}} = \frac{1}{2\pi\sqrt{10\,\text{nH} \cdot 100\,\text{pF}}} = 159.15\,\text{MHz} \approx 159\,\text{MHz}.
\tag{12.5}
$$

The typical values of L and C used in this example confirm that current-domain noise occurs at about 100 MHz. f_o occurs just below the second corner frequency f_{c2} calculated in Example 12.2. For slower voltage transitions, f_{c2} can shift to lower values and occur below the resonance peak f_o.

EMI can severely affect the save operation of a power management IC and an entire DC–DC converter. While the application often defines the switching frequency and the pulse width, the rise-fall times t_{rf} should only be as short as necessary. Larger t_{rf} improves EMI but increases the switching losses (I–V overlap, see Section 2.7). Shielding and EMI filters are often necessary to keep the noise within acceptable limits. In any case, it is essential to minimize parasitic inductances and capacitances during the physical implementation of power management systems. We will further consider this aspect.

12.4 Interconnections

12.4.1 Classification

To achieve a physical connection with minimum parasitic effect, we distinguish three categories of interconnections:

- high-di/dt loops, usually called *hot loops*, with high continuous and pulsating current
- high-dv/dt nodes with fast voltage transients (power and digital)
- sensitive signal nodes, analog signals

These classifications are fundamental for pinout assignment, on-chip wiring, and PCB layout, which are covered in Sections 12.5–12.8 of this chapter.

Sensitive nodes include the control loop, feedback divider, error amp, current sensing, and reference and biasing circuits (e.g., the bandgap reference in Section 6.4). These parts of power management ICs are the victims of any noise that couples from high-di/dt loops and high-dv/dt nodes. We, therefore, study the underlying mechanisms of noise coupling due to current and voltage transients and derive design guidelines. We use the inductive buck converter in Fig. 12.8a) as a representative circuit containing all three interconnection types.

12.4.2 Continuous Current

The finite interconnection resistance leads to unwanted voltage drops in case of a large continuous current. A critical scenario in Fig. 12.8a) includes the switching phases when M1 or M2 are

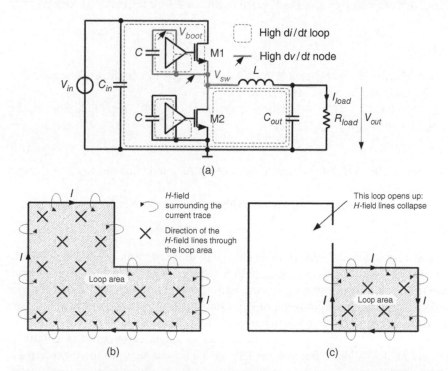

Figure 12.8 a) High-di/dt loops and high-dv/dt nodes in a buck converter; b) high-di/dt loop with corresponding $\vec{H}$-field passing through the loop area during the on-time (M1 turned on) and c) during the off-time (M2 turned on).

turned on and conduct the inductor current, corresponding on average to the load current. All interconnections in that loop should be sufficiently wide to reduce the parasitic resistance.

12.4.3 High-di/dt Loops

The two switches M1 and M2 and the input bypass capacitor C_{in} form a hot loop with high di/dt transients. It corresponds to the power loop discussed in Section 2.5.4, see also Fig. 2.10. Hot loops radiate wide-band magnetic fields $\vec{H}$ proportional to the loop area and the current. According to the law of Biot–Savart, any current flow I is associated with a magnetic field

$$\vec{H} = \frac{I}{4\pi} \int_C \frac{\mathrm{d}\vec{x} \times \vec{r}}{r^3} \propto (\vec{x} \times \vec{r}), \tag{12.6}$$

where $\vec{x}$ corresponds to the direction of the current flow in an infinitesimal length vector d$\vec{x}$ of a wire and $\vec{r}$ denotes the distance to the center of the wire (the radius).

Based on Eqn. (12.6), we can apply the right-hand rule to identify the relation between the current flow and the direction of the magnetic field. If the thumb points toward the current flow $\vec{x}$, the index finger in the direction of the $\vec{r}$, the vector of the magnetic field $\vec{H}$ corresponds to the direction of the angled middle finger. In a more simplified way, keeping the thumb parallel to the wire in the direction of the current flow ($\vec{x}$), the remaining fingers wrap around the wire with the fingertips pointing in the direction of the $\vec{H}$ field. The magnetic field around the loop traces in both switching phases is indicated in Fig. 12.8b,c) [15].

We can use Ampere's law to calculate the $\vec{H}$ field. Refer to Section 4.2.1 (integrated inductors) for details. Equation (4.2) provides Ampere's law in its general form. In case of a concentric circle of radius $\vec{r}$ that encloses the wire (see also Fig. 4.4a) in Section 4.2.1 for illustration) Ampere's law simplifies to

$$\oint_C \vec{H}\,\mathrm{d}\vec{l} = H \cdot 2\pi r = I, \tag{12.7}$$

where the term $2\pi r$ describes the perimeter of the circle. This equation confirms the linear relationship between the magnetic field and the current.

Referring again to Section 4.2.1, Eqn. (4.3) defines the magnetic flux Φ. The $\vec{H}$ field lines cut the loop surface at a right angle in the loop. Hence,

$$\Phi = B \cdot A = \mu \cdot H \cdot A, \tag{12.8}$$

where H is the absolute value of $\vec{H}$, B the absolute value of the magnetic flux density $\vec{B}$, A the loop area, and μ the permeability.

Changing the flux induces a voltage v_i anywhere in the network, expressed by Faraday's law:

$$v_i = -\frac{\mathrm{d}\Phi}{\mathrm{d}t} = -\frac{\mathrm{d}(B \cdot A)}{\mathrm{d}t}, \tag{12.9}$$

This voltage v_i is the noise source. Minimizing the change of magnetic flux (and minimizing the flux at all) reduces EMI proportionally. According to Eqn. (12.9), this can be achieved by lowering the rate of change of the current transition (proportionally changing the field $\vec{H}$) and of the loop area. We see both effects in the buck converter of Fig. 12.8. During the on-time, when transistor M1 conducts, the total inductor current flows in the loop. This current drops to zero once the converter enters the off-time phase where M1 stops conducting and the inductor current commutates to transistor M2 (and its body diode). But also the loop area changes between the two phases as illustrated in Fig. 12.8b,c).

The loop consisting of both switches M1, M2, and capacitor C_{in} experiences high-di/dt pulsating currents leading to high dΦ/dt. It forms the most critical loop, usually considered *the* hot loop [16, 17]. This loop requires the most attention during PCB design (see Section 12.7).

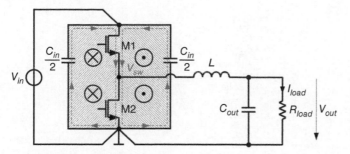

Figure 12.9 Placing two input capacitors on opposite sides between V_{in} and ground cancels the magnetic fields $\vec{H}$. Based on the right-hand rule, the $\vec{H}$ field orientation is into the plane $\otimes$ on the left and out of plane $\odot$ on the right.

Magnetic flux causes (parasitic) inductance $L = \Phi \cdot I$ and magnetic coupling. Due to the parasitic inductance, fast current transients cause ringing and voltage spikes in the attached interconnections (die, package, PCB), resulting in EMI, see Section 12.3.

Referring to the considerations of Section 2.5.4, the gate loops of the low-side and high-side switch see high di/dt transients as indicated in Fig. 12.8a). Due to longer interconnections, the impact worsens for discrete power transistors connected to a gate driver in a separate IC.

An effective technique to reduce the noise emission of the main hot loop comprising M1, M2, and C_{in} is shown in Fig. 12.9. The input capacitor C_{in} is split into two parts, each placed symmetrically at both sides of the power switches [17, 18]. Assuming the peak currents in both loops decrease by half compared to the original current, each capacitor can also be reduced by half. According to Eqn. (12.7), half the current reduces the H-field already by a factor of 2. As another great advantage of this technique, the field lines cancel because the circulating currents flow in opposite directions, indicated by $\otimes$ and $\odot$ in Fig. 12.9. The opposing $\vec{H}$-fields from the loops cancel each other out, resulting in an ideally zero net magnetic field.

12.4.4 High-dv/dt Nodes

Voltage changes on the switching node couple directly across parasitic capacitances in every direction and can disturb sensitive circuit nodes. The switching node voltage toggles between ground and supply at a high dv/dt rate. In the case of a bootstrap supply (see Section 5.11), the V_{boot} node will see the same steep dv/dt transition as the switching node. Like a simple plate capacitor, reducing the layout area of the high-dv/dt nodes and increasing the separation from other circuit parts is key for minimizing the noise coupling [16, 17].

12.5 Pinout

Due to the mechanisms related to the different kinds of interconnections described in Section 12.4, finding an optimum pinout of a power management IC can be challenging. Figure 12.10 shows typical pinouts of DC–DC converters with external power transistors. The converter schematic explains the names used for the package pins. See Chapter 10 for details on the fundamental inductive buck converter. The floor plan of the IC usually defines the first approach of a pin assignment. Based on the floor plan considerations in Section 12.1, the analog, digital, and power domains should be separated, which naturally also groups the pins of the IC related to these domains. As a general rule, use separate supply pins for analog, digital, and power domains if the pin count allows. Ways for interconnections between supply domains are presented in Section 12.6.2.

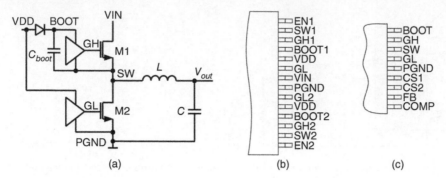

(a)   (b)   (c)

Figure 12.10 IC pinout examples for buck converters with external power switches in TSSOP packages: a) Schematic indicating the pin names; b) power pins of a dual converter; c) a single step-down converter with separated power and analog pins (top to bottom).

High-di/dt loops should be kept as small as possible. For this reason, all supply pins should be placed next to their ground return. This way, the bypass capacitor can be connected between both pins as close as possible, achieving minimum parasitic loop inductance. The ideal location for supply and any high-di/dt pins is in the center of one side of the package because this gives the shortest leads and bond wires that connect to the die (minimum parasitic inductance and resistance). In Fig. 12.10b), the input voltage and power ground pins, VIN and PGND, are next to each other and in the middle. Because of various constraints, this cannot always be achieved, and a trade-off is usually required.

Pins with high continuous current may require double or triple bond wire connections; see Fig. 12.13a) and the die photo in Fig. 7.2c) (Chapter 7). Table 12.1 provides the current capabilities of different bond wires.

High-dv/dt pins, such as the switching node and the bootstrap voltage, cause strong (capacitive) coupling to adjacent pins and should be separated from sensitive nodes. A good practice is to place them at corners of the package and next to ground or supply voltages that are not too sensitive to disturbances. In Fig. 12.10, these pins are kept at the upper and lower end.

Sensitive and quiet signals can be assigned to corner pins. In the case of a dual-side package like the TSSOP, all non-power pins can be placed on the opposite side. Furthermore, sensitive pins can be separated by ground and stable supply pins from noisy interconnections. In Fig. 12.10c), the input pins CS1, CS2 of the current sensing circuit (shunt or DCR-sensing, see Section 6.6) are separated from the low-side gate signal GL by power ground PGND. Other sensitive signals, such as the feedback voltage FB and the compensator output voltage COMP, follow below CS2.

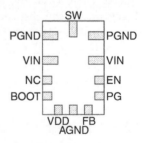

Figure 12.11 An ultra-low-noise DC–DC converter IC with integrated power stage in a QFN package. It supports the low-noise technique of Fig. 12.9 by a symmetrical set of VIN and PGND pins (left and right).

Some packages support the concept of Fig. 12.9 for DC–DC converter designs with extremely low noise and ultra-low-EMI emissions. Figure 12.11 shows an example DC–DC converter IC in a QFN package with a symmetrical set of VIN and PGND pins to the left and right of the package. The switching node pin SW is placed at the center top of the package. Such designs often have a fully integrated power stage, reducing noise emissions. Examples are LT8614 (Silent Switcher architecture, Analog Devices) and LMR36015 (Texas Instruments).

12.6 IC-Level Wiring

On the top level, the interconnections should be made in groups reflecting the three categories identified in Section 12.4: (1) Supply lines (high current and high-di/dt transients), (2) noisy lines (digital, PWM), and (3) sensitive lines (analog).

12.6.1 Shielding

High dv/dt lines should not be routed parallel to a sensitive analog signal line over a long distance. A significant parasitic capacitance forms between the metal traces, causing displacement currents to flow depending on the direction and magnitude of the dv/dt transition, as shown in Fig. 12.12a). These currents induce unwanted noise and ripple on the analog side. If we cannot avoid the two lines to get close, a ground line in between can provide shielding, shown in Fig. 12.12b). Likewise, a shield plate can provide proper isolation in the vertical direction, as shown in Fig. 12.12c). For instance, a grounded shield plate in MET2 can be inserted at the crossing of the bandgap reference voltage in MET1 (sensitive analog) and a digital clock line routed in MET3.

12.6.2 Grounding and Supply Connections

The essential rule is to avoid combining noisy, digital, and high-power supply domains with sensitive analog supplies. Digital, power, and analog ground lines should only be connected at the final ground plane on the PCB. Electrically, this forms a star connection with minimum interaction between the ground domains as depicted in Fig. 12.13a). In contrast, Fig. 12.13b) illustrates how the finite wiring resistance causes a ground shift and noise coupling between the domains. There may also be a separation within each domain like AGND in Fig. 12.13a), which has two bond pads. Double or triple bonding supports higher currents, used for the PGND connections on the right of Fig. 12.13a).

Especially if the IC is pin-limited, some domains need to be merged. Fig. 12.14 shows three different options for supply interconnections [19]. The case with fully separated supplies is depicted in Fig. 12.14a). The second best option in Fig. 12.14b) keeps separate bond pads but connects both supplies to a common package pin. Interestingly, the equivalent circuit of this configuration forms an L-C low-pass filter that prevents noise coupling. If the die is pad-limited, the option in Fig. 12.14c) combines the supplies on the chip level, resulting in the lowest isolation between the supply domains.

Example 12.4 *Estimate the filter cut-off frequency for the supply options in Fig. 12.14b) and c). We connect the digital and analog block each to the bond pad with an aluminum line that is 5 µm wide and 500 µm long. The sheet resistance is 70 mΩ/□. DVDD and AVDD are each connected with 3 mm bond wires. The bond pad capacitance is 1 pF.*

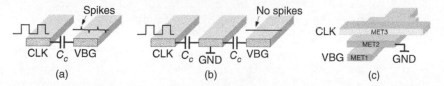

Figure 12.12 a) Capacitive coupling between a noisy line clock line (CLK) and a sensitive bandgap reference voltage (VBG); b) inserting a ground line (GND) in between reduces the noise coupling; c) same scenario for a vertical crossing with a grounded shield plate.

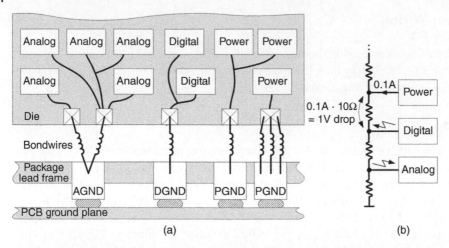

Figure 12.13 Grounding: a) Example with separated analog, digital, and power ground domains (AGND, DGND, PGND) that form a star connection from the IC via leadframe and package to the final ground plane on PCB-level. Sub-domains within AGND, DGND, and PGND on chip-level and package level can help to ensure a clean and quiet ground potential. b) A poor ground connection where the power ground current of 100 mA causes a ground shift of 1 V due to the finite routing resistance of 10 Ω. In addition, excessive ground noise may affect the analog circuits.

We first determine the equivalent R-L-C values. C is given as the bond pad capacitance. Assuming 1 nH/mm, the total bond wire length of 6 mm corresponds to an inductance of L = 6 nH. The resistance can be calculated from the sheet resistance. The wire geometry corresponds to 500/5 = 100 squares, which results in R = 100 · 70 mΩ = 7 Ω.

For Fig. 12.14b), the transfer function from the digital to the analog domain is

$$\frac{V(AVDD)}{V(DVDD)} = \frac{1}{s^2 + s\frac{R}{L} + \frac{1}{LC}}. \tag{12.10}$$

We can now estimate the filter frequency:

$$f_c = \frac{1}{2\pi\sqrt{LC}} = \frac{1}{2\pi\sqrt{6\,\text{nH} \cdot 1\,\text{pF}}} = 2.05\,\text{GHz} \tag{12.11}$$

The cut-off frequency of the R–C low-pass in Fig. 12.14c) is

$$f_c = \frac{1}{2\pi RC} = \frac{1}{2\pi \cdot 7\,\Omega \cdot 1\,\text{pF}} = 22.7\,\text{GHz}. \tag{12.12}$$

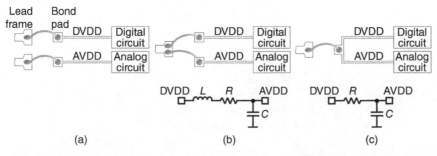

Figure 12.14 Different options of supply connections: a) The digital supply (DVDD) and analog supply (AVDD) are fully separated; b) separate pads and double bond to a shared supply pin; c) DVDD and AVDD share a single bond pad.

The supply option of Fig. 12.14b) is ten times more effective compared to Fig. 12.14c). However, the cut-off frequencies are in the gigahertz range in both cases. PWM control and other signals run in the (lower) megahertz range. Therefore, frequency components will still couple into the analog domain. In conclusion, applications should target separating the supply nets, as shown in Fig. 12.14a).

To further reduce noise coupling and stabilize the supply line against load transients, as much on-chip decoupling capacitance as possible is usually added at each supply. A widely used practice is to fill any empty IC top-level space with capacitance (MOS, MIM, or MOM, depending on voltage ratings and availability, see Section 4.1). Even though the smallest die size is critical in commercial ICs, capacitor values of up to 1 nF can often be accomplished.

Example 12.5 *We again consider the scenario of Example 12.4c). How does the filter capability improve the scenario if the analog block gets a large decoupling capacitor of $C_{in} = 300$ pF placed at the interconnection of the AVDD line to the analog circuit?*

The noise from the digital block propagates through the DVDD and AVDD line, which gives $R_{line} = 14\,\Omega$. It results in this cut-off frequency of the low-pass filter:

$$f_c = \frac{1}{2\pi R_{line} C_{in}} = \frac{1}{2\pi \cdot 14\,\Omega} \cdot 300\,\text{pF} = 37.9\,\text{MHz} \tag{12.13}$$

Compared to the result in Eqn. (12.12), the filtering capability has significantly improved. However, this is only possible if the IC top-level leaves considerable space for placing enough decoupling capacitance.

12.6.3 Thick Copper Metallization

For high-current traces, most bipolar-CMOS-DMOS (BCD) technologies offer one or multiple copper layers. Handling currents in the range of a few hundred milliamperes and above is usually not possible using standard aluminum metal lines. With a typical thickness of 0.5 µm aluminum metalization has a sheet resistance in the order of 50–80 mΩ/□. With lower resistivity and a typical thickness in the range of 3–10 µm, copper metallization achieves sheet resistance values of 1–2 mΩ/□, which is about 50 times lower compared to aluminum.

Thick-copper metallization supports *bond over active circuitry* (BOAC), meaning bond wires can directly connect to the thick copper trace. While BOAC is also possible on aluminum and with any bond wire material (see Section 12.2.2), direct copper-to-copper bonding brings many advantages [20]. Quasi-direct copper-to-copper bonding is realized using an intermediate angstrom-level platinum layer to suppress the deterioration of bonding reliability caused by impurities at the bonding interface. BOAC achieves low-resistive connections directly to the terminals of integrated power transistors. One example can be seen in the lower-right corner of the die photo in Fig. 7.2c) (Chapter 7) where a large power transistor is connected using triple bond wires. Also, electrostatic discharge (ESD) protection cells and any active circuits can be underneath the bond pad using thick copper metallization, which reduces the die size. The drawback is reliability concerns due to higher stress on the pads during bonding, which may cause deformation and cracking of the underlying films [20]. Careful layout design of the pad structure can significantly mitigate these issues.

12.7 PCB Layout Design

The standard printed circuit board (PCB) material is FR-4, a composite material of woven fiberglass cloth with an epoxy resin binder, which is Flame Retardant (leading to the term FR). The current capability of a copper trace correlates to the copper thickness measured in ounces (oz). 1 oz. means that this amount of copper is pressed flat and spread over an area of one square foot (1 oz./1 ft^2.). The resulting copper thickness is approximately 35 μm. Depending on the allowed rise in temperature ΔT (typically 10°C), the IPC-2221 (MIL-STD-275) standard relates the trace width to the current capability. The current levels are approximate worst-case values, usually providing sufficient design margin. For $\Delta T = 10$°C, a trace that is 25 μm wide can carry currents of 1.0 A (1 oz.) and 1.4 A (2 oz.), respectively. The maximum current doubles if the trace width increases by a factor of three.

For very high currents, copper bus bars can be soldered onto the PCB, for instance, in automotive and high-power inverters with currents of more than 100 A. With much higher thickness than the PCB traces, they can carry high currents without any heating issues.

PCB design incorporates all the rules derived in Sections 12.3–12.6 of this chapter. We must properly handle high-di/dt loops, high-dv/dt nodes (power and digital), and sensitive analog signals. Like in Section 12.4, we refer to the DC–DC buck converter in Fig. 12.8 as the representative circuit for deriving the guidelines for PCB layout. One good practice in preparation of a PCB design is to highlight the high-di/dt traces, the high-current traces, the noisy high-dv/dt nodes, and the sensitive traces, as done in Fig. 12.8 (except for the analog signals). Many excellent application notes provide a valuable resource on PCB layout [15–17, 21].

DC–DC converters are usually implemented on a multi-layer PCB with a two-sided assembly such that the power and control-loop components can be placed on opposite sides of the board. A suitable layer assignment of a 4-layer PCB is as follows:

- Layer 1 – Power components
- Layer 2 – Ground plane
- Layer 3 – Analog/control
- Layer 4 – Analog/control

All power components should be placed on the same side of the PCB's top layer (Layer 1). It is similar to the IC floor plan discussed in Section 12.1, where the power domain should not be intermixed with the other domains. The second layer is a solid ground plan to minimize cross-coupling of the switching noise from the top layer into the analog circuits, which are placed on Layers 3 and 4. The ground plane in Layer 2 provides a low-impedance system reference. It should not be segmented.

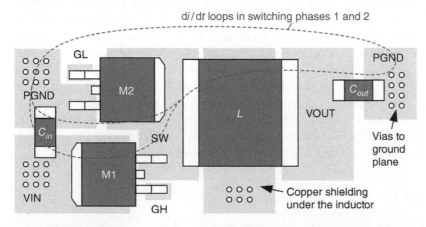

Figure 12.15 Top-layer layout of a 4-layer PCB that implements the DC–DC converter of Fig. 12.10a) with discrete power transistors.

Figure 12.15 shows an example of the power stage layout of a buck converter according to Fig. 12.10a) (with the names of the nodes indicated on the layout). We start the layout with the most critical hot loop consisting of switches M1, M2, and the input capacitor C_{in}. In the case of a power management IC with integrated switches, this rule applies to the corresponding IC pins. M1 and M2 are power transistors in a D2PAK package (see Fig. 3.1a)). C_{in} gets placed close to the two power switches with a low-impedance connection to minimize the loop area. PGND on the top layer is not directly connected to any other ground on that layer. Instead, it is connected through vias to the ground plane in Layer 2 underneath. The control IC should be located at a distance from the noisy switching area, preferably next to the VOUT node (in the case of a buck converter because its output (VOUT) does not see high di/dt).

One root cause of EMI noise due to the change in magnetic flux is the change of the loop area, as studied in Section 12.4 (see Fig. 12.8). Figure 12.15 indicates the two di/dt loops during switching phases 1 and 2. As suggested in [15], we can obtain almost equal loop area in both switching phases utilizing the low-resistive return path via the ground plane.

If implementing the low-noise method of Fig. 12.9 (perhaps using the IC pinout shown in Fig. 12.11), the two hot loops are arranged symmetrically at opposite sides of the IC.

As one of the critical high-dv/dt connections, the switching node SW needs to be as small as possible in area and separated from any sensitive nodes. The high-side gate signal trace GH should be routed together with SW to reduce the inductance and associated dv/dt noise (not shown in Fig. 12.15). The switching node SW connects to the coil. Grounded copper is placed directly under the coil on the top layer. This way, the magnetic field lines from the coil cannot close around the PCB such that all PCB layers below the top layer remain clean.

The output voltage node should be as close to the load as possible to achieve minimum interconnection resistance and voltage drop across the PCB traces, resulting in optimum regulation performance and conversion efficiency.

Thermal vias connect the exposed pad of the power management IC and the power transistors to the ground plane. The heat transfers vertically through the vias into the ground plane. Its massive copper helps significantly to get the heat away from the power stage. A thermal resistance close to the one specified for the package will be reached only with properly placed thermal vias (see Section 12.9).

12.8 Power Delivery

In the introduction part of this book, Fig. 1.14a) in Section 1.7 shows the block diagram of a processor power delivery architecture. The mainboard voltage regulator module (VRM) converts the intermediate voltage of 12 or 48 V down to the microprocessor supply level of about 1 V. Moving the VRM as close as possible to the load reduces the impedance of the power delivery network (PDN, also called the power distribution network). This way, more current can be supplied with lower losses. But also, the transient response can be improved.

The power delivery design has to ensure that the supply voltage seen by the processor core is always within a specified tolerance range of typically 10% of the nominal voltage. A voltage droop below this limit can cause timing issues, resulting in brown-out failures. Likewise, a processor supply voltage that is too high can lead to excessive power dissipation ($\propto V_{DD}^2$, see Eqn. (1.21)) and degrade the device reliability. In this section, we explore the concepts of lateral and vertical power delivery. We will also analyze why inserting an IVR improves performance significantly regarding efficiency and dynamic response.

12.8.1 Lateral and Vertical Power Delivery

Many of today's voltage regulator designs use a lateral power delivery system, where the load (e.g., a microprocessor) and the regulator(s) are placed next to each other on the top side of the PCB [12]. This approach of lateral power delivery for a processor is depicted in Fig. 12.16a). There are several space constraints on the PCB, especially if the VRM does not come in a chip-scale package, and passive L and C components with a rather large footprint are used. Hence, the VRM (the power management IC) and the processor die are typically 1–2 cm apart, limiting the interconnect resistance to more than 0.5 mΩ. At a processor current of 100 A, this corresponds to at least $I^2R = 5\,\mathrm{W}$ power distribution loss and more than $IR = 50\,\mathrm{mV}$ droop.

The concept of vertical power delivery, shown in Fig. 12.16b), facilitates a much shorter power delivery path by placing the voltage converter directly under the processor die, reducing the PDN resistance and the related losses by a factor of 10 [22].

12.8.2 Integrated Voltage Regulators (IVR)

The PDN can be improved by inserting an IVR in between the VRM and the processor with the IVR implemented together on the same die with the processor, as shown in Fig. 1.14b) and Fig. 12.16c). Operating at switching frequencies of more than 100 MHz, the IVR converts the intermediate voltage of around 1.8 V down to <1 V at high efficiency and low output noise, supporting extremely dynamic response. Implementing multiple IVRs on the processor die addresses the demand for fine-grain power management. Typical IVR architectures are multphase inductive and hybrid DC–DC converters, as described in Sections 10.14 and 11.8, respectively.

The IVR's power stage and control loop circuits are part of the processor IC, depicted in Fig. 12.16c) [23]. The output bypass capacitor may combine on-chip MIM and ceramic package capacitors. The power inductor is typically implemented as an air-core device using package or PCB traces. The high switching frequency in the IVR enables the use of inductors with an inductance as low as 1–2 nH. An example implementation of Intel's Fully Integrated Voltage Regulator (FIVR) is shown in Fig. 4.7 [23].

Advanced on-chip inductor technologies with a magnetic core have been developed that show an impressive performance at these high switching frequencies and enable even higher integration density. Details are described in Chapter 4, see Section 4.2.

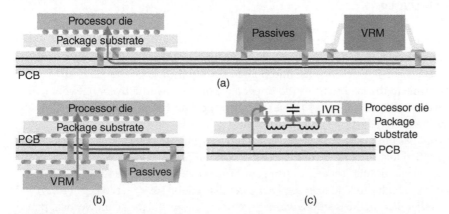

Figure 12.16 Power delivery: a) Lateral power delivery; b) vertical power delivery; c) with integrated voltage regulator (IVR).

As the chip and package area is more expensive than the PCB area, designing an IVR requires assessing the trade-off between the cost adder and the performance gain.

12.8.3 The Power Delivery Network (PDN)

We apply the approach introduced in Section 12.2.3 that leads to the package model in Fig. 12.5 to derive the equivalent circuit of the entire PDN. Figure 12.17 shows the electrical model for both lateral and vertical power delivery according to Fig. 12.16a) and b) with example component values for illustration purposes [12, 24, 25]. For vertical power delivery, the parasitic component values are lower.

C_o is the output buffer capacitor of the voltage regulator VRM, which is often a ceramic capacitor with typical values in the order of 1 μF. Including the equivalent series resistance and inductance of C_o and all other capacitances is essential. C_b is the bypass capacitor next to the processor supply pin with a typical value in the order of 100 nF. An on-die MIM capacitor C_s provides a buffer for the core supply voltage V_{out} next to the load. It has typical values of a few hundred picofarads. The package capacitances shown in Fig. 12.5b) are neglected because they see the larger capacitances C_b or C_s in parallel.

PDN Impedance

The VRM maintains the output voltage for transient events at all frequencies from DC to a few kilohertz. At frequencies above this range, there is a time delay before the voltage regulator can respond to adjust the amount of load current. During this time delay, a voltage droop $\Delta V_{out} = Z_{PDN} \cdot \Delta I_{load}$ occurs, which is determined by the impedance Z_{PDN} of the power distribution network and the magnitude ΔI_{load} of the transient current drawn by the load.

The PDN impedance Z_{PDN} needs to be as low as possible to keep V_{out} equal to the input voltage V_{in} generated by the VRM [12, 26]. This goal can be achieved by reducing the parasitic series inductance (short interconnections) and by maximizing the decoupling capacitance (space limited). The larger these capacitances, the lower the corresponding droop because the impedance is proportional to $1/j\omega C$. The capacitor values in Fig. 12.17 decrease from the VRM to the processor because the available space shrinks significantly. While C_o fits nicely on the PCB, the on-chip capacitance C_s is limited.

Due to the distributed R-L-C configuration, the high-frequency part of sudden steps of the load current I_{load} has to be blocked by C_s because all parasitic series inductances cause an open circuit $\sim j\omega L$. Mid-frequencies are blocked by C_b and low-frequencies by C_o. At the

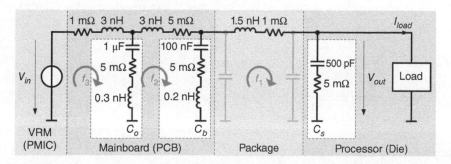

Figure 12.17 Equivalent circuit of a typical power distribution network (PDN) for a one-step conversion using a voltage regulator module (VRM).

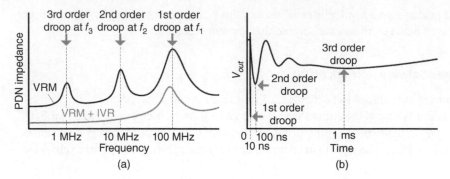

Figure 12.18 a) The impedance of the power delivery network (PDN) over frequency and b) time domain voltage droops in a PDN. Source: Adapted from [12].

lower end of the frequency range, including DC, the VRM handles any variations of the load itself. In the time domain, this behavior is defined as first-order, second-order, and third-order droop.

Figure 12.18a) shows the impedance of the power distribution network over frequency. The three droops correspond to impedance peaks. They are related to frequencies f_1 to f_3 that occur typically in between 100 kHz and 100 MHz. In first order, f_1 to f_3 are the resonance frequencies of the three local resonant tanks indicated in Fig. 12.17. To understand this approximation, we consider the mid-frequency range. We can assume the parasitic inductance to the right of C_b to be a short such that C_s appears parallel to C_b. Because $C_b \gg C_s$, we can neglect C_s. Also, the leg to the left of the large buffer capacitor C_o can be neglected because the VRM does not react in the mid-frequency range. Hence, the inner loop determines the PDN impedance in the mid-frequency range, labeled f_2 in Fig. 12.18a). The resonant tank is formed by C_b and C_o, including their parasitic equivalent series resistance (ESR) and inductance (ESL) components and the printed circuit board (PCB) parasitics between their top-plate connections. Summing up all series inductances and capacitances and denoting them L_2, C_2, the resonance frequency can be calculated by

$$f_2 = \frac{1}{2\pi\sqrt{L_2 C_2}}. \tag{12.14}$$

Since, $C_o \gg C_b$, we can assume $C_2 \approx C_b$. Moreover, the capacitor's ESL is usually negligible such that L_2 is approximately equal to the parasitic inductance of the PCB interconnection. In the same way, the local loops labeled f_1 and f_3 in Fig. 12.17 determine the frequencies of the first and third-order droop. Using the effective L and C values, f_1 and f_3 can be estimated similar to Eqn. (12.14).

Example 12.6 *Estimate the first, second, and third-order droop frequencies for the component values in Fig. 12.17.*

We start with the first-order droop, denoted by f_1. The total loop inductance is

$$L_1 = 1.5\,\text{nH} + 0.2\,\text{nH} = 1.7\,\text{nH}. \tag{12.15}$$

The effective capacitance is the series connection of C_b and C_s:

$$C_1 = \frac{C_b \cdot C_s}{C_b + C_s} = 497.5\,\text{pF} \approx C_s \tag{12.16}$$

Similar to Eqn. (12.14) we find

$$f_1 = \frac{1}{2\pi\sqrt{L_1 C_1}} = \frac{1}{2\pi\sqrt{1.7\,\text{nH} \cdot 497.5\,\text{pF}}} = 173.1\,\text{MHz}. \tag{12.17}$$

For the second-order droop, we determine the total loop inductance and capacitance from Fig. 12.17:
$L_2 = 3.5\,\text{nH}$, $C_2 = 90.9\,\text{nF}$.
Inserting into Eqn. (12.14) yields

$$f_2 = \frac{1}{2\pi\sqrt{L_2 C_2}} = \frac{1}{2\pi\sqrt{3.5\,\text{nH} \cdot 90.9\,\text{nF}}} = 8.9\,\text{MHz}. \tag{12.18}$$

Finally, the third-order droop is determined by the loop on the left of Fig. 12.17: $L_3 = 3.3\,\text{nH}$, $C_3 = C_o = 1\,\mu\text{F}$.
The droop frequency is

$$f_3 = \frac{1}{2\pi\sqrt{L_3 C_3}} = \frac{1}{2\pi\sqrt{3.3\,\text{nH} \cdot 1\,\mu\text{F}}} = 2.8\,\text{MHz}. \tag{12.19}$$

All three frequency values fit the typical values indicated in Fig. 12.18a).

Figure 12.18b) shows the time domain response of the PDN for a load step, indicating the different droops. The first negative spike happens within a few nanoseconds after the load step corresponding to the frequency f_1. As the amount of on-chip capacitance C_s is limited, it is difficult to manage the first droop. The second and third-order droop correspond to the time constants defined by f_2 and f_3. They can usually be reduced to a reasonable level by enlarging C_b and C_o such that their amplitude diminishes compared to the first-order droop.

PDN Impedance with an IVR

The effect on the droop can be significantly improved by using an IVR. Figure 12.19 shows the equivalent circuit for a PDN with an IVR. The VRM and the mainboard (PCB) remain like in Fig. 12.17. The VRM output voltage of typically 1.8 V is the input voltage of the IVR. The IVR is responsible for the last stage of power conversion. It converts the voltage from around 1.8 V down to the core voltage of about 1 V. Some IVRs use higher input voltages of 3 V [12] and 4 V [27] to reduce the input current and, thereby, the PDN routing losses.

Bypass capacitors C_{in1} (a package-integrated ceramic capacitor) and C_{in2} (MIM) at the IVR input reduce the input voltage noise. Seen from the load, the IVR provides a low-impedance termination toward the VRM. A package-integrated ceramic capacitor C_{s2} parallel to the on-chip MIM capacitor

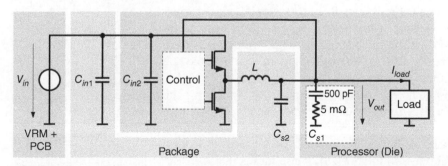

Figure 12.19 Equivalent circuit of a typical power distribution network (PDN) for the combination of a voltage regulator module (VRM) and an integrated voltage regulator (IVR).

C_{s1} results in a large total supply buffer capacitance C_s reducing the first-order droop. Operating at switching frequencies of more than 100 MHz, the IVR decouples the load from the VRM. Compared to the VRM-only architecture in Fig. 12.17, the PDN with an IVR does not show a noticeable second and third-order droop. The IVR's fast dynamic response allows tracking the core voltage in dynamic voltage and frequency scaling (DVFS), as explained in Section 1.7 in the introduction chapter.

12.9 Thermal Design

Most power management ICs have significant power dissipation, which causes the die to heat up. The temperature increase is referred to the junction of any integrated device. For safe operation and well-controlled impact on the IC's reliability over a lifetime, most applications usually limit the junction temperature T_j to 150°C. Some advanced technologies, often qualified for automotive products, support up to 160–170°C.

12.9.1 Thermal Resistance

A temperature difference from a hot to a cold junction causes a heat flow, which describes the amount of thermal energy transferred per time. The heat transfer is, therefore, proportional to the dissipated power (both the heat transfer and the power are measured in Joule). The scenario is conceptionally similar to an electrical network, as shown in Fig. 12.20 [28, 29]. The temperature difference ΔT corresponds to a voltage drop and the power dissipation P_{diss} to the electrical current. The ambient temperature can be considered a constant voltage source (assuming that the IC causes only marginal heating of the environment). Like an electrical resistance, we can define the thermal resistance R_{th} as the ratio between the temperature difference and the power dissipation,

$$R_{th} = \frac{\Delta T}{P_{diss}}. \tag{12.20}$$

The thermal resistance can be seen as the opposition offered by the packaging and assembly materials to the flow of heat energy. The lower the thermal resistance, the smaller the temperature difference. Hence, a small thermal resistance is desired because it leads to a lower junction temperature.

Thermal resistance can be defined for various components involved in heat transfer. The thermal performance of IC packages is typically measured using the junction-to-case resistance $R_{th,jc}$ and the junction-to-ambient resistance $R_{th,ja}$. $R_{th,jc}$ was introduced when metal packages were the standard type. $R_{th,ja}$ is much more useful for state-of-the-art plastic and ceramic packages with non-uniform case temperature distribution. The right-most column in Table 12.2 lists the typical thermal resistance $R_{th,ja}$ of IC packages.

The junction-to-ambient thermal resistance $R_{th,ja}$ measures the ability of the package to dissipate heat from the chip junction to the ambient environment. However, $R_{th,ja}$ values are given for a standardized setup with the IC mounted on a 4-layer PCB with fine-width metal traces in still air [28]. It also means that $R_{th,ja}$ can differ significantly in an application environment. Nevertheless, the standardized $R_{th,ja}$ values allow comparing the thermal performance

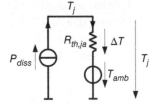

Figure 12.20 Electrical equivalent circuit involving the junction-to-ambient thermal resistance $R_{th,ja}$. Temperature corresponds to voltage. The power dissipation P_{diss} equals the heat flow modeled by a current source. It causes an increase in temperature by ΔT resulting in $T_j = T_{amb} + R_{th,ja} \cdot P_{diss}$.

of various packages. It can also be used for simple hand calculations based on Eqn. (12.20) to estimate the junction temperature for a given power dissipation. However, it requires the power dissipation to be well-known during the design phase. Therefore, the estimation is usually challenging because there is more than one use case, especially in complex ICs with multiple voltage regulators.

Example 12.7 *What is the maximum power dissipation of an IC assembled in a TQFP-48 package if the junction temperature must not exceed $T_{j,max} = 150°C$? The specified ambient temperature range is $T_{amb} = -40 - 125°C$. How does the maximum power dissipation change if a SOIC-8 package is used?*

According to Table 12.2 the package has a thermal resistance of $R_{th,ja} = 25 \, K/W$. The worst case temperature difference is $\Delta T = (T_{j,max} - T_{amb,max}) = 150°C - 125°C = 25 \, K$. Solving Eqn. (12.20) for the power dissipation gives

$$P_{diss} = \frac{\Delta T}{R_{th,ja}} = \frac{25 \, K}{25 \, K/W} = 1 \, W. \qquad (12.21)$$

The SOIC package has a much higher thermal resistance of $R_{th,ja} = 55 \, K/W$. The maximum power reduces by more than half and reaches 455 mW.

12.9.2 Thermal Simulation

Hand calculations, as described in the section above, allow for a first-order estimation of the die temperature. The result can be seen as an average junction temperature across the entire die. Often, the lateral temperature distribution across the IC needs to be known. It will be required to understand where local hot spots occur, and it provides beneficial information during layout floor planning. Excessive overheating can also be addressed by enlarging the area of the related power transistor by increasing its width (which, at the same time, reduces its on-resistance, see Section 3.2). Choosing another package type with lower thermal resistance may also be necessary. Once a suitable floor plan is found, the temperature distribution indicates where sensors for over-temperature protection (see Section 6.3) need to be placed.

Various thermal simulation tools are available that calculate heat transfer, such as Siemens Flotherm and Ansys Icepeak. Matlab can be used as well. Heat transfer takes place through conduction, convection, and radiation. Related to IC assembly, conduction usually dominates, while convection and radiation occur at the boundaries. Thermal simulation tools solve the heat transport equation of a geometry in two or three dimensions, incorporating the die, the packaging, and the PCB assembly. In addition to steady-state simulations, a thermal transient simulation can provide insight into the thermal response if a power stage turns on and off.

In the first step, the simulation region is partitioned into multiple domains that are described by thermal conductivity as well as surface area and material thickness:

1. Define the die topology based on the floor plan
2. Define the size and location of blocks with major power dissipation (power transistors)
3. Define the power dissipation per block
4. Define the package, mount (heat sinking), and PCB parameters

Figure 12.21 provides a thermal simulation example for the IC with the floor plan from Fig. 12.1. The ambient temperature is 125 °C. Six blocks with significant power dissipation are considered. Their geometry and power dissipation are listed in the table, while the floor plan defines their

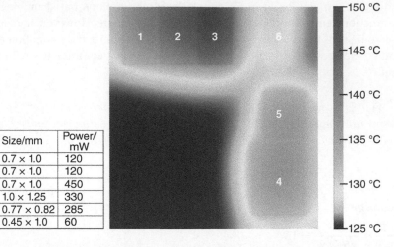

Block	Size/mm	Power/mW
1	0.7 × 1.0	120
2	0.7 × 1.0	120
3	0.7 × 1.0	450
4	1.0 × 1.25	330
5	0.77 × 0.82	285
6	0.45 × 1.0	60

Figure 12.21 Temperature distribution from a thermal simulation of a power management IC for the floor plan defined in Fig. 12.1 at $T_{amb} = 125\,°C$. The table provides the geometry and power dissipation for each relevant block.

location. The simulation can be repeated for various application scenarios with different power dissipation levels. The resulting temperature distribution (heat map) indicates local hot spots. In the example of Fig. 12.1, Block 3 sees the highest temperature with up to $150\,°C$. The temperature sensor for the thermal shutdown should be placed nearby (e.g., a satellite BJT, see Fig. 6.6 in Section 6.3).

References

1 Amkor Technology (2022) *Copper Wirebonding*. TS105H-EN.

2 Shah, J. (2012) Estimating bond wire current-carrying capacity. *Power Systems Design*, pp. 22–25.

3 Flauta, R., Zhou, Z., Fan, H., and Chen, H. (2021) Comparison of experimental and estimated fusing current of gold (Au) and copper (Cu) bonding wires in semiconductor IC packages, in *2021 27th International Workshop on Thermal Investigations of ICs and Systems (THERMINIC)*, pp. 1–6, doi: 10.1109/THERMINIC52472.2021.9626534.

4 Texas Instruments (2004) *AN-1205 Electrical Performance of Packages*. Application Report SNOA405A.

5 Amkor Technology (2019) *LQFP*. DS232G-EN.

6 Amkor Technology (2019) *ExposedPad LQFP/TQFP*. DS231K-EN.

7 Amkor Technology (2021) *TSSOP/MSOP*. DS350T-EN.

8 Amkor Technology (2021) *ExposedPad TSSOP/MSOP/SOIC/SSOP*. DS571N-EN.

9 Amkor Technology (2021) *SOIC*. DS370V-EN.

10 Wang, L., Malcolm, D., and Liu, Y.F. (2016) An innovative power module with power-system-in-inductor structure, in *2016 IEEE Applied Power Electronics Conference and Exposition (APEC)*, pp. 2087–2094, doi: 10.1109/APEC.2016.7468155.

11 Wang, K., Qi, Z., Li, F., Wang, L., and Yang, X. (2017) Review of state-of-the-art integration technologies in power electronic systems. *CPSS Transactions on Power Electronics and Applications*, 2 (4), 292–305, doi: 10.24295/CPSSTPEA.2017.00027.

12 Radhakrishnan, K., Swaminathan, M., and Bhattacharyya, B.K. (2021) Power delivery for high-performance microprocessors - challenges, solutions, and future trends. *IEEE Transactions on Components, Packaging and Manufacturing Technology*, 11 (4), 655–671, doi: 10.1109/TCPMT.2021.3065690.

13 Wang, L., Liu, W., Malcom, D., and Liu, Y.F. (2018) An integrated power module based on the power-system-in-inductor structure. *IEEE Transactions on Power Electronics*, 33 (9), 7904–7915, doi: 10.1109/TPEL.2017.2769681.

14 Paul, C.R. (2006) *Introduction to Electromagnetic Compatibility*, John Wiley & Sons, Inc., Hoboken, NJ, 2nd edn.

15 Barrow, J. (2007) *Reducing Ground Bounce in DC-to-DC Converters - Some Grounding Essentials*, Analog Devices. Analog Dialog 41.

16 Zhang, H.J. (2012) *PCB Layout Considerations for Non-Isolated Switching Power Supplies*, Linear Technology (Analog Devices). Application Note 136.

17 Hedrich, J. (2023) *PCB Design for Low-EMI DC/DC Converters*, Monolithic Power Systems. Article 0030, Rev. 1.0.

18 Armstrong, T. (2019) *Silent Switcher Devices Are Quiet and Simple*, Analog Devices. Analog Dialog 53.

19 Sansen, W. (2006) *Analog Design Essentials*, The Springer International Series in Engineering and Computer Science, Springer, New York, doi: 10.1007/b135984.

20 Hunter, S., Martinez, J., Salas, C., Salas, M., Schofield, J., Sheffield, S., and Wilkins, K. (2011) Bond over active circuitry design for reliability, in *Proceedings of the 44th International Symposium on Microelectronics*, pp. 249–257.

21 Glaser, C. (2015) *Five Steps to a Great PCB Layout for a Step-Down Converter*, Texas Instruments. Analog Applications Journal.

22 Hayes, C. (2018) *Package Prepares for AI Computing*, Electronic Specifier. Vicor.

23 Burton, E.A., Schrom, G., Paillet, F., Douglas, J., Lambert, W.J., Radhakrishnan, K., and Hill, M.J. (2014) FIVR - Fully integrated voltage regulators on 4th generation Intel(R) Core(TM) SoCs, in *2014 IEEE Applied Power Electronics Conference and Exposition - APEC 2014*, pp. 432–439, doi: 10.1109/APEC.2014.6803344.

24 Smith, L.D. and Bogatin, E. (2017) *Principles of Power Integrity for PDN Design*, Prentice-Hall.

25 King, D. (2018) *66AK2G1x: EVMK2GX General Purpose EVM Power Distribution Network Analysis*, Texas Instruments. Application Report SPRACE6.

26 Sturcken, N., Petracca, M., Warren, S., Mantovani, P., Carloni, L.P., Peterchev, A.V., and Shepard, K.L. (2012) A switched-inductor integrated voltage regulator with nonlinear feedback and network-on-chip load in 45 nm SOI. *IEEE Journal of Solid-State Circuits*, 47 (8), 1935–1945, doi: 10.1109/JSSC.2012.2196316.

27 Hardy, C., Pham, H., Jatlaoui, M.M., Voiron, F., Xie, T., Chen, P.H., Jha, S., Mercier, P., and Le, H.P. (2023) 11.1 A scalable heterogeneous integrated two-stage vertical power-delivery architecture for high-performance computing, in *2023 IEEE International Solid- State Circuits Conference (ISSCC)*, pp. 182–184, doi: 10.1109/ISSCC42615.2023.10067315.

28 Edwards, D. and Nguyen, H. (2016) *Semiconductor and IC Package Thermal Metrics*, Texas Instruments. Application Report.

29 Renesas (2019) *Thermal Considerations in Package Design and Selection*. Application Note AN-842, Rev. A.

Index